T0345254

Nonparametric Statistical Methods Using R

Praise for the first edition:

"This book would be especially good for the shelf of anyone who already knows nonparametrics, but wants a reference for how to apply those techniques in R."

-The American Statistician

This thoroughly updated and expanded second edition of **Nonparametric Statistical Methods Using R** covers traditional nonparametric methods and rank-based analyses. Two new chapters covering multivariate analyses and big data have been added. Core classical nonparametrics chapters on one- and two-sample problems have been expanded to include discussions on ties as well as power and sample size determination. Common machine learning topics --- including k-nearest neighbors and trees --- have also been included in this new edition.

Key Features:

- Covers a wide range of models including location, linear regression, ANOVA-type, mixed models for cluster correlated data, nonlinear, and GEE-type.
- Includes robust methods for linear model analyses, big data, time-to-event analyses, timeseries, and multivariate.
- Numerous examples illustrate the methods and their computation.
- R packages are available for computation and datasets.
- Contains two completely new chapters on big data and multivariate analysis.

The book is suitable for advanced undergraduate and graduate students in statistics and data science, and students of other majors with a solid background in statistical methods including regression and ANOVA. It will also be of use to researchers working with nonparametric and rank-based methods in practice.

John D. Kloke is a bit of a jack-of-all-trades as he has worked as a clinical trial statistician supporting industry as well as academic studies and he also served as a teacher-scholar at several academic institutions. He has held faculty positions at the University of California - Santa Barbara, University of Wisconsin - Madison, University of Pittsburgh, Bucknell University, and Pomona College. An early adopter of R, he is an author and maintainer of numerous R packages, including Rfit and npsm. He has published papers on nonparametric rank-based estimation, including analysis of cluster correlated data.

Joseph W. McKean is a professor emeritus of statistics at Western Michigan University. He has published many papers on nonparametric and robust statistical procedures and has co-authored several books, including Robust Nonparametric Statistical Methods and Introduction to Mathematical Statistics. He co-edited the book Robust Rank-Based and Nonparametric Methods. He served as an associate editor of several statistics journals and is a fellow of the American Statistical Association.

CHAPMAN & HALL/CRC

Texts in Statistical Science Series

Recently Published Titles

Stochastic Processes with R
An Introduction
Olga Korosteleva

Design and Analysis of Experiments and Observational Studies using R
Nathan Taback

Time Series for Data Science: Analysis and Forecasting
Wayne A. Woodward, Bivin Philip Sadler and Stephen Robertson

Statistical Theory
A Concise Introduction, Second Edition
Felix Abramovich and Ya'acov Ritov

Applied Linear Regression for Longitudinal Data
With an Emphasis on Missing Observations
Frans E.S. Tan and Shahab Jolani

Fundamentals of Mathematical Statistics
Steffen Lauritzen

Modelling Survival Data in Medical Research, Fourth Edition
David Collett

Applied Categorical and Count Data Analysis, Second Edition
Wan Tang, Hua He and Xin M. Tu

Geographic Data Science with Python
Sergio Rey, Dani Arribas-Bel and Levi John Wolf

Models for Multi-State Survival Data
Rates, Risks, and Pseudo-Values
Per Kragh Andersen and Henrik Ravn

Spatio–Temporal Methods in Environmental Epidemiology with R, Second Edition
Gavin Shaddick, James V. Zidek, and Alex Schmidt

A Course in the Large Sample Theory of Statistical Inference
W. Jackson Hall and David Oakes

Statistical Inference, Second Edition
George Casella and Roger Berger

Nonparametric Statistical Methods Using R, Second Edition
John Kloke and Joesph McKean

For more information about this series, please visit: https://www.routledge.com/Chapman--HallCRC-Texts-in-Statistical-Science/book-series/CHTEXSTASCI

Nonparametric Statistical Methods Using R

Second Edition

John Kloke
Joseph McKean

CRC Press is an imprint of the
Taylor & Francis Group, an **informa** business
A CHAPMAN & HALL BOOK

Designed cover image: © John Kloke and Joseph McKean

First edition published 2015
Second edition published 2024
by CRC Press
2385 NW Executive Center Drive, Suite 320, Boca Raton FL 33431

and by CRC Press
4 Park Square, Milton Park, Abingdon, Oxon, OX14 4RN

CRC Press is an imprint of Taylor & Francis Group, LLC

ISBN: 978-0-367-65135-0 (hbk)
ISBN: 978-1-032-75197-9 (pbk)
ISBN: 978-1-003-03961-7 (ebk)

DOI: 10.1201/9781003039617

Typeset in CMR10
by KnowledgeWorks Global Ltd.

Publisher's note: This book has been prepared from camera-ready copy provided by the authors.

The first author dedicates this edition to his parents.
The second author dedicates this book to Marge.

Contents

Preface

In this second edition, we have substantially expanded the material relative to the first edition. The first chapter has been completely revamped to provide a broader overview. Core classical nonparametrics chapters on one- and two-sample problems have been expanded to include discussions on ties as well as power and sample size determination. We have added common machine learning topics, including k-nearest neighbors and trees. We have placed methods for categorical data in their own chapter and have added a brief introduction to logistic regression. Two new chapters covering multivariate analyses and big data were also added; the latter being a current research topic.

We have reorganized the material somewhat so that the first six chapters tend to require a lower level of statistical preparation, while the remaining five chapters tend to be at a higher level. That said, a few of the sections in the earlier chapters may be considered at a higher level, and so we have flagged these with an asterisk (*).

The first six chapters include an introduction followed by one- and two-sample nonparametric methods for testing and estimation (including confidence intervals), rank-based analyses of regression and ANOVA-type designs, and categorical analysis. These chapters constitute a basis for a one-semester course in applied nonparametrics at the undergraduate and/or graduate level. Such a course naturally serves statistics and data science majors, but also other majors, as well, who have a solid background in statistical methods. Most readers would benefit from having a background in first-semester calculus based statistical methods.

Topics from the later chapters may be selected based on the preferences of the students and the instructor. Prerequisites for the final five chapters include matrix algebra based regression and ANOVA courses. Students completing the entire book should gain a solid background in methods of robust nonparametric analyses appropriate for many designs and datasets that are encountered in statistical practice. These methods provide robust alternatives to traditional least squares methods. We have provided numerous examples of real and simulated datasets that illustrate the methods and their computation in R. Hence, we feel that this book also serves as an informative handbook for the researcher wishing to implement nonparametric and rank-based methods in practice.

Given that R has continued to grow in popularity, we think that many readers will have already taken a course in R or an applied course that used R extensively. For the vast majority of the book, we have used core R (or base

R) rather than add-on packages such as tidyverse. Some will likely consider this omission a mistake, as many users and institutions have fully adopted the paradigm. The two authors, however, have found that the graphics and syntax of base R are suitable for many of the problems found in research and practice that we have encountered.

Throughout the text, we compare methods (e.g., classical least squares or normal theory methods with nonparametric alternatives) under various distributions, often through simulations. As the reader will find, the best choice of method depends on the underlying (typically unknown in practice) distribution from which the data are sampled. Given that the underlying distribution is unknown, often a Wilcoxon method will provide a choice that yields a high level of efficiency over a wide range of distributions. However, we also include some examples of adaptive methods that can be used.

As with the first edition, we have written R packages to aid the reader with computational aspects. The main package is Rfit that was written as a robust, rank-based alternative to lm (lsfit) in core R. The package npsm , along with Rfit, covers most of the code that we developed for the first six chapters. For the software that we developed for later chapters, we ask the reader to refer to the book's web page at https://github.com/kloke/book.

We appreciate the efforts of John Kimmel and Lara Spieker of Chapman & Hall and, in general, the staff of Chapman & Hall for their help in the preparation of this book for publication.

John Kloke
Joe McKean

Preface from the First Edition

Nonparametric statistical methods for simple one- and two-sample problems have been used for many years; see, for instance, Wilcoxon (1945). In addition to being robust, when first developed, these methods were quick to compute by hand compared to traditional procedures. It came as a pleasant surprise in the early 1960s, that these methods were also highly efficient relative to the traditional t-tests; see Hodges and Lehmann (1963).

Beginning in the 1970s, a complete inference for general linear models developed, which generalizes these simple nonparametric methods. Hence, this linear model inference is referred to collectively as rank-based methods. This inference includes the fitting of general linear models, diagnostics to check the quality of the fits, estimation of regression parameters and standard errors, and tests of general linear hypotheses. Details of this robust inference can be found in Chapters 3–5 of Hettmansperger and McKean (2011) and Chapter 9 of Hollander and Wolfe (1999). Traditional methods for linear models are based on least squares fits; that is, the fit which minimizes the Euclidean distance between the vector of responses and the full model space as set forth by the design. To obtain the robust rank-based inference, another norm is substituted for the Euclidean norm. Hence, the geometry and interpretation remain essentially the same as in the least squares case. Further, these robust procedures inherit the high efficiency of simple Wilcoxon tests. These procedures are robust to outliers in the response space, and a simple weighting scheme yields robust inference to outliers in design space. Based on the knowledge of the underlying distribution of the random errors, the robust analysis can be optimized. It attains full efficiency if the form of the error distribution is known.

This book can be used as a primary text or a supplement for several levels of statistics courses. The topics discussed in Chapters 1 through 5 or 6 can serve as a textbook for an applied course in nonparametrics at the undergraduate or graduate level. Chapters 7 and 8 contain more advanced material and may supplement a course based on interests of the class. For continuity, we have included some advanced material in Chapters 1–6, and these sections are flagged with a star (*). The entire book could serve as a supplemental book for a course in robust nonparametric procedures. One of the authors has used parts of this book in an applied nonparametrics course as well as a graduate course in robust statistics for the last several years. This book also serves as a handbook for the researcher wishing to implement nonparametric and rank-based methods in practice.

This book covers rank-based estimation and inference for models ranging from simple location models to general linear and nonlinear models for uncorrelated and correlated responses. Computation using the statistical software system R (R Core Team 2023) is covered. Our discussion of methods is amply illustrated with real and simulated data using R. To compute the rank-based inference for general linear models, we use the R package `Rfit` of Kloke and McKean (2012). For technical details of rank-based methods, we refer the reader to Hettmansperger and McKean (2011); our book emphasizes applications and statistical computation of rank-based methods.

A brief outline of the book follows. The initial chapter is a brief overview of the R language. In Chapter 2, we present some basic statistical nonparametric methods, such as the one-sample sign and signed-rank Wilcoxon procedures, a brief discussion of the bootstrap, and χ^2 contingency table methods. In Chapter 3, we discuss nonparametric methods for the two-sample problem. This is a simple statistical setting in which we briefly present the topics of robustness, efficiency, and optimization. Most of our discussion involves Wilcoxon procedures, but procedures based on general scores (including normal and Winsorized Wilcoxon scores) are introduced. Hogg's adaptive rank-based analysis is also discussed. The chapter ends with discussion of the two-sample scale problem as well as a rank-based solution to the Behrens–Fisher problem. In Chapter 4, we discuss the rank-based procedures for regression models. We begin with simple linear regression and proceed to multiple regression. Besides fitting and diagnostic procedures to check the quality of fit, standard errors and tests of general linear hypotheses are discussed. Bootstrap methods and nonparametric regression models are also touched upon. This chapter closes with a presentation of Kendall's and Spearman's nonparametric correlation procedures. Many examples illustrate the computation of these procedures using R.

In Chapter 5, rank-based analysis and its computation for general fixed effects models are covered. Models discussed include one-way, two- and k-way designs, and analysis of covariance type designs, i.e., robust ANOVA and ANCOVA. The hypotheses tested by these functions are of Type III; that is, the tested effect is adjusted for all other effects. Multiple comparison procedures are an option for the one-way function. Besides rank-based analyses, we also cover the traditional Kruskal–Wallis one-way test and the ordered alternative problem including Jonckheere's test. The generalization of the Fligner–Killeen procedure to the k-sample scale problem is also covered.

Time-to-event analyses form the topic of Chapter 6. The chapter begins with a discussion of the Kaplan–Meier estimate and then proceeds to Cox's proportional hazards model and accelerated failure time models. The robust fitting methods for regression discussed in Chapter 4 are highly efficient procedures, but they are sensitive to outliers in design space. In Chapter 7, high breakdown fits are presented for general regression models. These fits can attain up to 50% breakdown. Further, we discuss diagnostics which measure the difference between the highly efficient fits and the high breakdown fits of general linear models. We then consider these fits for nonlinear and time series models.

Rank-based inference for cluster correlated data is the topic of Chapter 8. The traditional Friedman's test is presented. Computational algorithms using R are presented for estimating the fixed effects and the variance components for these mixed effects models. Besides the rank-based fits discussed in Chapters 3–5, other types of R estimates are discussed. These include, for quite general covariance structure, GEERB estimates which are obtained by a robust iterated re-weighted least squares type of fit.

Besides `Rfit`, we have written the R package `npsm` which includes additional functions and datasets for methods presented in the first six chapters. Installing `npsm` and loading it at the start of each R session should allow the reader to reproduce all of these analyses. Topics in Chapters 7 and 8 require additional packages, and details are provided in the text. The book itself was developed using Sweave (Leisch 2002), so the analyses have a high probability of being reproducible.

The first author would like to thank SDAC in general with particular thanks to Marian Fisher for her support of this effort, Tom Cook for thoughtful discussions, and Scott Diegel for general as well as technical assistance. In addition, he thanks KB Boomer, Mike Frey, and Jo Hardin for discussions on topics of statistics. The second author thanks Tom Hettmansperger and Simon Sheather for enlightening discussions on statistics throughout the years. For insightful discussions on rank-based procedures, he is indebted to many colleagues including Ash Abebe, Yusuf Bilgic, Magdalena Niewiadomska-Bugaj, Kim Crimin, John Kapenga, Josh Naranjo, Jerry Sievers, Jeff Terpstra, and Tom Vidmar. We appreciate the efforts of John Kimmel of Chapman & Hall and, in general, the staff of Chapman & Hall for their help in the preparation of this book for publication. We are indebted to all who have helped make R a relatively easy to use but also very powerful computational language for statistics. We are grateful for our students' comments and suggestions when we developed parts of this material for lectures in robust nonparametric statistical procedures.

John Kloke
Joe McKean

1

Introduction

The popularity of R continues to grow, and we think it is likely many readers have experience working with R. In particular, the following are assumed:

- Readers are familiar with reading external data and package management as well as the Comprehensive R Archive Network (CRAN)[1].
- Readers have knowledge of common object types including lists, matrices, data frames, along with their helper functions, e.g., `length`, `dim`, `[]`.
- Readers have experience with R statistical functionality such as `mean`, `sd` as well as important base R functions such as `c`, `rep`, `seq`, `sample`.

However, we do present a brief overview of some of these topics. Experience with the **dpqr functions** in R would also be helpful.[2]

This first chapter, where we provide a short tour of R, may serve as a gauge of the reader's base R knowledge base. Those interested in a more thorough introduction are referred to a monograph on R (e.g., Chambers 2008). Also, there are a number of free manuals (in pdf) available at the Comprehensive R Archive Network (CRAN). An excellent overview, written by developers of R, is Venables and Ripley (2002).

The built-in documentation system in R can be a helpful first stop for questions.[3] Of course, one may use their favorite search engine often resulting in a `stackoverflow.com` thread.

This chapter covers a range of key prerequisites for the text and presents a general introduction to the types of statistical, graphical, and syntactical approaches commonly used in the text.

The book relies heavily on the R package `npsm`, and most sections assume it is loaded at the start of the R session; we do so in the following code segment.

```
> library(npsm)
```

[1] cran.r-project.org

[2] dpqr refers to density, distribution function, quantile function, and random number generation for a particular distribution. Base R has many such quadruples and more are available via contributed packages. For the normal distribution these functions are: `dnorm`, `pnorm`, `qnorm`, and `rnorm`.

[3] See `help(help)` or `?help` or `help(help.search)`.

1.1 Data and Notation

In this section we present some of the basic data concepts in base R along with some of the handy tools available for working with data and performing other statistical tasks in R. This section may serve as an introduction to R for programmers, R users accustomed to using `tidyverse`[4], or as a review for other R users. Advanced R users may want to skim or skip the section entirely.

1.1.1 Data Types in R

Without going into a lot of detail, R has the capability of handling character (strings), logical (`TRUE` or `FALSE`), and of course numeric data types. The values `TRUE` and `FALSE` are reserved words and represent logical constants. The global variables `T` and `F` are initially defined as `TRUE` and `FALSE`, respectively.[5]

Basic data types in R are vector, matrix, and data frame. For example, the character vector `LETTERS` is the 26 uppercase Roman letters. The reader is encouraged to pull up the help page for `LETTERS` to learn what other built-in constants R has. The following code segment creates a numeric vector of length 3 and assigns[6] it to the variable `x`.

```
> x <- c(161,502,49007)
```

R has a number of functions for generating vectors of numeric values. For example, the colon operator (`:`) is used to generate sequences of integers separated by 1 as illustrated in the following code segment.

```
> 0:9

 [1] 0 1 2 3 4 5 6 7 8 9

> 4.2:0

[1] 4.2 3.2 2.2 1.2 0.2

> -2:1

[1] -2 -1  0  1
```

The `rep` and `seq` commands are also useful for `rep`licating values and generating general `seq`uences. In the following commented[7] code, we provide some illustrative examples.

[4] www.tidyverse.org

[5] When writing production code, one should use the reserved words.

[6] Assignment in R is usually carried out using either the `<-` operator or the `=` operator.

[7] A **comment** in R is the portion of a line to the right of a `#` character.

```
> rep(1,5)              # vector of 5 1's

[1] 1 1 1 1 1

> rep(c(3,2),2)         # the vector [3,2] is repeated 2 times

[1] 3 2 3 2

> seq(1,5)              # create the sequence 1 to 5

[1] 1 2 3 4 5

> seq(1,5,by=2)         # the sequence 1 to 5 in increments of 2

[1] 1 3 5
```

The `sample` function is another handy function we like to use. As an illustrative example, the following code segment generates the result of three tosses of a fair coin where heads is represented by 'H' and tails by 'T'.

```
> sample(c('H','T'),3,replace=TRUE)

[1] "T" "H" "H"
```

To take a random sample (without replacement) of five items from a set of 111 labeled 1001 to 1111, we may combine the `sample` and `seq` commands as illustrated in the following code segment.

```
> sample(seq(1001,1111),5)

[1] 1075 1005 1036 1004 1026
```

1.1.2 Vector and Matrix Notation

We utilize **boldface** font to represent vectors and matrices, with vectors represented by lower-case letters and matrices upper-case letters. For example, a column vector of n observations may be represented as $\boldsymbol{x} = [x_1, \ldots, x_n]^T$ where T denotes transpose. An $n \times p$ matrix may be represented as

$$\boldsymbol{X} = \begin{bmatrix} x_{11} & x_{12} & \cdots & x_{1p} \\ x_{21} & x_{22} & \cdots & x_{2p} \\ \vdots & \vdots & \ddots & \vdots \\ x_{n1} & x_{n2} & \cdots & x_{np} \end{bmatrix}.$$

In R, a matrix may be created using `rbind` or `cbind` for row and column combine, respectively, as illustrated in the next code segment.

```
> X <- rbind(c(11,12),c(21,22))
> X
```

```
     [,1] [,2]
[1,]   11   12
[2,]   21   22

> Y <- cbind(c(11,21),c(12,22))
> Y

     [,1] [,2]
[1,]   11   12
[2,]   21   22
```

At times the `matrix` command is handy as illustrated in the following code.

```
> R <- matrix(seq(1,9),nrow=3)
> R

     [,1] [,2] [,3]
[1,]    1    4    7
[2,]    2    5    8
[3,]    3    6    9
```

Notice the matrix is filled in by columns by default. To override that default and fill the matrix by rows, one may specify the argument `byrow=TRUE` as is demonstrated in the following.

```
> C <- matrix(seq(1,9),nrow=3,byrow=TRUE)
> C

     [,1] [,2] [,3]
[1,]    1    2    3
[2,]    4    5    6
[3,]    7    8    9
```

The **outer product** of two vectors, $\boldsymbol{x}$ and $\boldsymbol{y}$, is the matrix of the product of the elements of the two vectors. Let $\boldsymbol{x} = [x_1, \ldots, x_m]^T$ and $\boldsymbol{y} = [y_1, \ldots, y_n]^T$ then the outer product of $\boldsymbol{x}$ and $\boldsymbol{y}$ is given by

$$\boldsymbol{x}\boldsymbol{y}^T = \begin{bmatrix} x_{11}y_{11} & x_{11}y_{12} & \cdots & x_{11}y_{1n} \\ \vdots & \vdots & \ddots & \vdots \\ x_{m1}y_{11} & x_{m1}y_{12} & \cdots & x_{m1}y_{1n} \end{bmatrix}$$

```
> x <- seq(10,20,by=10)
> y <- seq(.1,.3,by=.1)
> xoy <- outer(x,y)
> xoy

     [,1] [,2] [,3]
[1,]    1    2    3
[2,]    2    4    6
```

A convenient use of the : operator is to specify a range of indices to extract a submatrix as illustrated in the following code segment.

```
> xoy[,1:2]      # extract all rows and columns 1 thru 2

     [,1] [,2]
[1,]    1    2
[2,]    2    4
```

Note, in R, `outer` is a general function where the multiplication operator (*) is the default, but may be overridden by another scalar operator (e.g., +). The **inner product** of two vectors of the same length is the sum of the product of their elements: $\sum_{i=1}^{n} x_i y_i$. In R the `crossprod` command may be used to calculate the inner product.

1.1.3 Data Frames

Data frames are a standard data object in R and are used to combine several variables of the same length, but not necessarily the same type, into a single unit.

In the following code segment, a .csv file is read from the internet and stored in a data frame named `egData`. Here, the commands `dim` and `head` are used to verify the data were read in correctly.

```
> url <-
+ 'https://raw.githubusercontent.com/kloke/book/master/eg2.csv'
> egData <- read.csv(url)
> dim(egData)          # number or rows and columns

[1] 47  4

> head(egData)         # first several records (6 by default)

  cr i0    x1      y
1  B  1  0.42  1.139
2  B  1  0.53  1.458
3  A  1  0.68 -0.057
4  A  1 -0.31 -1.168
5  C  1 -0.17  0.987
6  C  1 -0.25  0.655
```

To access one of the vectors, the `$` operator may be used. For example, to calculate the mean of `x1`, the following code may be executed.

```
> mean(egData$x1)

[1] 0.010638
```

One may also use the column number or column name `D[,3]` or `D[,'x1']` respectively. Omitting the first subscript means to use all rows. The `with` command as follows is another convenient alternative.

```
> with(egData,mean(x1))

[1] 0.010638
```

The $ operator may also be used in an assignment to add a new column to the data frame as illustrated in the following.

```
> egData$x1_2 <- egData$x1^2     # add a column of x1 squared
> egData[1:3,c('x1','x1_2')]    # examine first 3

    x1   x1_2
1 0.42 0.1764
2 0.53 0.2809
3 0.68 0.4624
```

1.1.4 Ranks

Many of the methods in this text are *rank-based*, meaning that the ranks of the observations are used in the computation of the test statistic or estimation procedure. For example, if we have the observations $x_1, x_2, \ldots, x_n$ we denote the **ranks** as $R(x_1), R(x_2), \ldots, R(x_n)$ where $R(x_i)$ is the number of observations less than or equal to x_i; i.e., $R(x_i) = \sum_{j=1}^{n} I(x_j \leq x_i)$ where $I(\cdot)$ is the indicator function.

The base R command `rank` may be used to compute the ranks as is illustrated in the following code segment.

```
> x <- c(7, 17, 19, 18, 10)
> rank(x)

[1] 1 3 5 4 2
```

In the case of tied observations (ties) in the data, one usually assigns the average of the ranks that is allotted to these tied observations.[8] For example, see the following code segment where we use the base R function `rank`, which, by default, uses the average rank.

```
> z<-c(12,18,11,5,11,5,11)
> rank(z)

[1] 6.0 7.0 4.0 1.5 4.0 1.5 4.0
```

[8] Handling of ties is discussed further in Sections 2.5 and 3.4 for the cases of one-sample and two-sample problems, respectively.

1.2 Graphics

We assume that most readers of the text have some statistics or data science background on graphical display of data. Graphics we use frequently in the text are **boxplots** — including **comparisons boxplots** — and **normal** q–q **plots (a.k.a. normal probability plots)**, so we present a brief review of these two types. Boxplots may be created using the base R function `boxplot`, while the base R function `qqnorm` may be used to create normal probability plots. Boxplots[9] allow one to get a general sense of the distribution of the data as well as a tool to compare two or more groups. Normal $q-q$ plots allow one to compare the distribution of the sample to that of a normal reference distribution.

We utilize the `seinfeld` dataset in `npsm` to illustrate the two types of graphics.[10] The `seinfeld` dataset contains the number of viewers (in millions) for each episode of the NBC television show *Seinfeld*. The show ran for 9 seasons from July 1989 to May 1998. In the following code segment, we find the number of episodes per season.

```
> table(seinfeld$season)

 1  2  3  4  5  6  7  8  9
 5 12 23 24 22 24 24 22 24
```

We use comparisons boxplots to examine the distribution of viewers for each of the seasons in Figure 1.1. From the comparisons boxplots we can see viewership increased sometime after season 4. Also apparent are several outliers: there is a low-valued outlier in season 5 and two high-valued outliers in season 9. The two high-valued outliers in season 9 represent the number of viewers for the last two episodes (actually these are each two-part episodes, so there are 4 total observations) of the show which aired the same night (i.e., the night of the series finale). While these outliers are numerically accurate, they are not representative of the general popularity of the individual episodes. The low-valued episode in season 5 was the week of the Winter Olympics that aired on another network, which could be the reason for the outlier.

With $n = 5$ for season 1, there is not much to interpret in the individual boxplot for that season other than noting that the number of viewers for season 1 is not dissimilar to the number for season 2. In practice, it is sometimes helpful to add `n=` to the x axis.[11]

---

[9] Recall that the box contains the middle 50% of the data, the bar in the box is at the median, the whiskers go out to the adjacent points (points nearest to the inner fences — 1.5 interquartile ranges from the quartiles, but within the fences), and the o's denote potential outliers.

[10] See `help(seinfeld)` for code to create the graphics.

[11] In particular, this would help explain the boxplot for season 1.

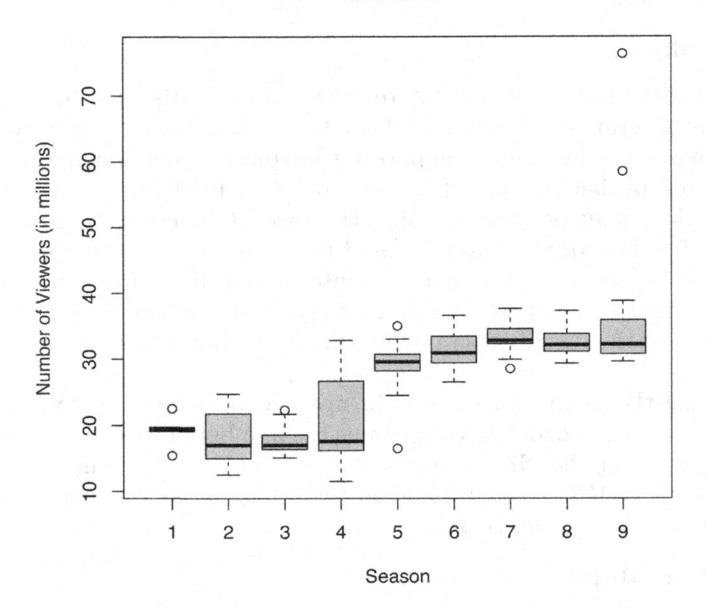

FIGURE 1.1
Comparison boxplots of the number of viewers versus season number.

Next, consider the normal $q-q$ plots in Figure 1.2 for seasons 4, 5, 6, and 9. In addition to plotting the sample quantiles, versus the normal distribution quantiles, we have added a reference line with intercept equal to the median and slope equal to the **MAD (median absolute deviation)**.[12] If the data are drawn at random from a normal distribution they would fall roughly along this line, with some random variability to be expected. Substantial deviations from linearity in the normal $q-q$ plot suggest that the sample is drawn from a distribution that is not normal. The viewership for the episodes in season 6 appears to be drawn from an approximately normal distribution, while the distributions of viewers for the other displayed seasons appear not to be drawn from normal distributions. Viewership for season 4 appears to be drawn from a distribution which is right-skewed with a longer right tail and a shorter left tail than one would expect if the data were drawn from a normal distribution. From the $q-q$ plot for season 5 the outlier is apparent, and additionally the tails of the distribution sampled appear to be longer than one would expect

[12] The **MAD (median absolute deviation [from the median])** is defined as

$$c \cdot \text{med}_i |x_i - \text{med}_j\{x_j\}|,$$

where c is commonly set to 1.4826 so that the MAD is a consistent estimator of the population standard deviation at the normal.

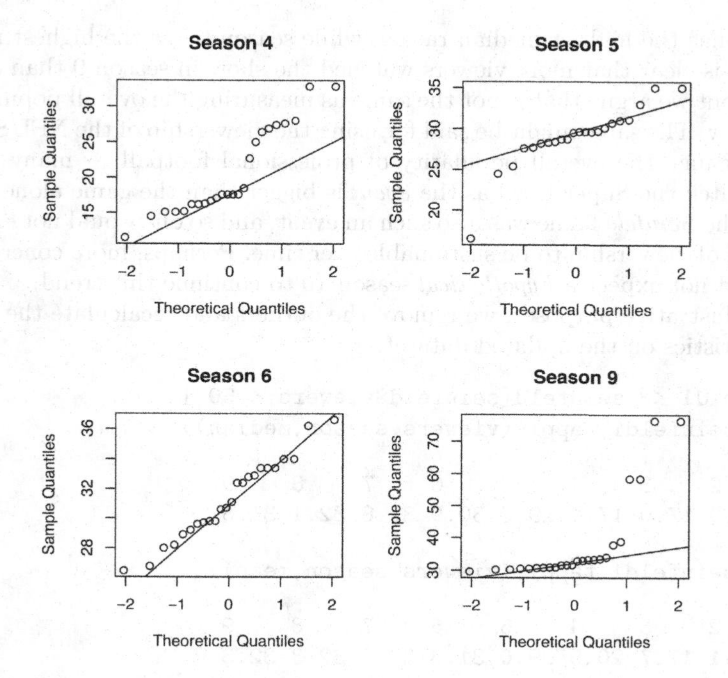

FIGURE 1.2
Normal $q-q$ plots for selected seasons.

if the data were drawn from a normal distribution. Based on the $q-q$ plot for season 9, outliers appear to be close to $2\times$ as large as would be expected, if the data were normally distributed.

Outliers, such as those in season 9, can drastically impact summary statistics. For example, suppose we were asked the question: when was the height of popularity (in terms of viewership[13]) of the show? To answer this question, we calculate the summary statistics consisting of the sample medians and means for each season. The following code segment computes these summary statistics for each of the 9 seasons.

```
> with(seinfeld,tapply(viewers,season,median))

   1    2    3    4    5    6    7    8    9
19.4 17.0 17.0 17.6 29.6 30.9 32.8 32.1 32.2

> with(seinfeld,tapply(viewers,season,mean))

   1    2    3    4    5    6    7    8    9
19.2 18.1 17.7 20.5 29.0 31.3 33.2 32.3 38.1
```

[13]In practice, one might use a different measurement such as *share*.

Season 7 has the highest median rating, while season 9 has the highest mean rating. It is clear that more viewers watched the show in season 9 than other seasons, but we argue that is not the same as measuring the overall popularity of the show. The same might be said for using the viewership of the NFL Super Bowl to gauge the overall popularity of professional football — many more people watch the Super Bowl as the *event* is bigger than the game alone. The night of the *Seinfeld* finale was also such an event, and so one would not expect that level of viewership to be sustainable over time. Perhaps, more concretely, one would not expect a *hypothetical* season 10 to continue this trend.

For illustrative purposes, we remove the outliers and recalculate the summary statistics on the updated dataset.

```
> seinfeld1 <- seinfeld[seinfeld$viewers < 50,]
> with(seinfeld1,tapply(viewers,season,median))

   1    2    3    4    5    6    7    8    9
19.4 17.0 17.0 17.6 29.6 30.9 32.8 32.1 31.6

> with(seinfeld1,tapply(viewers,season,mean))

   1    2    3    4    5    6    7    8    9
19.2 18.1 17.7 20.5 29.0 31.3 33.2 32.3 32.3
```

Now, based on these summary statistics, both the median and mean point to season 7 as the most popular. Also, both measures of center are similar.

Notice in this example how sensitive the sample mean is to the outliers. Its value for season 9 is 38.1 on the original data, but its value drops to 32.3 when the outliers are removed. On the other hand, the sample median is much less sensitive to the outliers. Its value is 32.2 on the original data and 31.6 on the data with the outliers removed. The sample median is not impaired by the outliers, while the sample mean is. We say that the sample median is a *robust* statistic, while the sample mean is not.

As we will see through out this book, outliers can have a substantial impact on traditional (normal theory) estimates as well as their associated inference. Some may advocate for the removal of such data points prior to running the analysis. While we did so in this example for illustrative purposes, in most cases we would discourage doing so in practice. Instead, we recommend using one of the *robust* methods discussed in this book.

Taking another look at comparison boxplots, this time with the outliers removed as shown in Figure 1.3, we can more clearly see that distributions tend to be more right-skewed in the early seasons (2–4) and that there are possibly a couple of additional outliers in season 9.

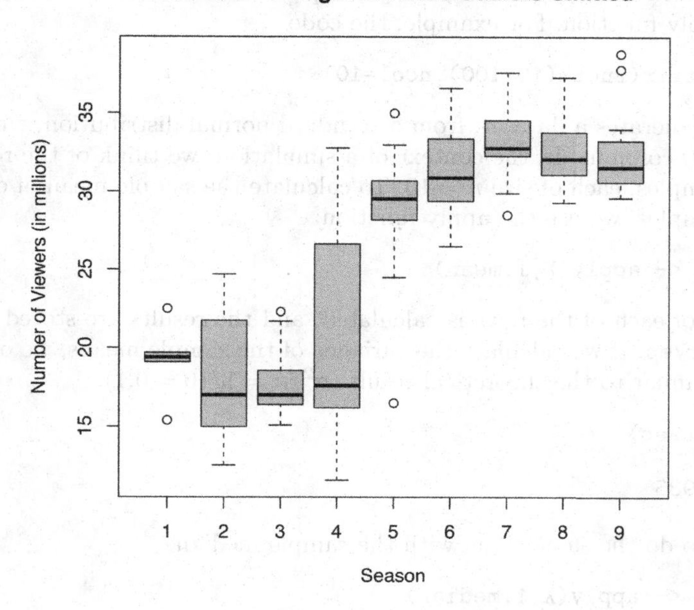

FIGURE 1.3
Comparison boxplots of the number of viewers versus season number with episodes airing the night of the series finale not shown.

1.3 Monte Carlo Simulation

Simulation is a powerful tool in modern statistics. Inferences for rank-based procedures discussed in this book are based, generally, on the asymptotic distribution of the estimators. Simulation studies allow us to examine their performance for small samples. Specifically, simulation is used to examine the empirical level (or power) of rank-based tests of hypotheses or the empirical coverage of their confidence intervals. Comparisons of estimators are often based on their empirical relative efficiencies (the ratio of the mean squared errors of the two estimators). Simulation is also used to examine the effect of violations of model assumptions on the validity of the rank-based inference. Another inference procedure used in this text is based on the bootstrap. This is a resampling technique, i.e., a Monte Carlo technique.

R software is an excellent tool for simulation studies, because a simple simulation may be carried out with only a few lines of code. One way to run a simulation in R is to generate many samples from a distribution and then use the apply function. For example, the code

```
> X <- matrix(rnorm(10*100),ncol=10)
```

randomly generates a dataset, from a standard normal distribution, with 100 rows and 10 columns. In the context of a simulation, we think of the rows as distinct samples, each of size $n = 10$. To calculate the sample mean of each of the 100 samples, we use the apply function:

```
> xbarVec <- apply(X,1,mean)
```

The mean of each of the rows is calculated, and the results are stored in the vector `xbarVec`. If we calculate the variance of the sample means, we observe that it is similar to the theoretical result ($\sigma^2/n = 1/10 = 0.1$).

```
> var(xbarVec)

[1] 0.090985
```

We can also do the same thing with the sample median

```
> xmedVec <- apply(X,1,median)
> var(xmedVec)

[1] 0.12146
```

Since the variability of the sample median is larger than the variability of the sample mean, we say that the sample mean is more efficient for these data. In the study of nonparametrics, relative efficiency is commonly used to compare two estimators at a given distribution or for a range of distributions. For example, Exercise 1.7.8 asks the reader to compare the efficiency of these two estimators of location when the data are drawn from a t_3 distribution. We will revisit relative efficiency throughout the text.

For the reproducibility of a random sequence, one can set the seed as illustrated in the following code segment.

```
> set.seed(99)
> sample.int(9,8)

[1] 1 4 6 7 5 3 2 8

> set.seed(99)
> sample.int(9,8)

[1] 1 4 6 7 5 3 2 8
```

To avoid always using the same sequence, a random seed could be selected via some *actual* random process. For example, one could use the system date in seconds.[14]

Distributions often used in this text for simulation studies include the normal, Student's t, and contaminated normal distributions. We assume that the reader is familiar with the normal and t-distributions. For a sample from a **contaminated normal distribution**, with a specified probability of ϵ a sample item is drawn from a $N(0, \sigma_c^2)$ distribution (*contaminated* observation) and with probability $(1 - \epsilon)$ a sample item is drawn from a $N(0, 1)$ distribution (*good* observation). For example, if $\epsilon = 0.05$ and $\sigma_c = 5$, then approximately 95% of the data sampled are from a $N(0, 1)$ distribution and approximately 5% of the data are sampled from a $N(0, 25)$ distribution. Such a distribution is denoted by $CN(0.05, 25)$. More details of the contaminated normal distribution are discussed in Exercise 1.7.7. In addition, we consider a specific type of contaminated normal where a small proportion (ϵ) of the data are multiplied by a value C (say, 100). This type of contamination can be thought of as a data entry error; i.e., dropping the decimal point (e.g., a value of \$1.00 is entered as \$100 or \$100.00 as \$10000)[15]. As we will see in the text, these types of errors, even when occurring at a frequency as low as $\epsilon = 0.001$ (0.1%), can cause traditional methods such as the sample mean to break down. We label this distribution as $CNX(\epsilon, C)$.

Normal q-q plots of a random sample of size $n = 10000$ from each of these four distributions are displayed in Figure 1.4. As to be expected, the sample drawn from the normal distribution roughly follows a straight line, while the normal q-q plots of the other distributions indicate that the samples were drawn from much heavier tailed distributions than the normal distribution.

1.3.1 Estimates of Center: Sample Mean and Sample Median

In the next code segment we run a simulation utilizing two of the distributions: $N(0, 1)$ and $CN(0.05, 25)$. The sample mean and sample median of each of the samples is then calculated.

```
> # declare constants
> nsim <- 5000 #number of sims
> n <- 33 #sample size for each sim
> set.seed(21791)
> # simulated data
> simMat_rnorm <- matrix(rnorm(nsim*n),nrow=nsim)
> simMat_rcn <- matrix(rcn_5_5(nsim*n),nrow=nsim)
> # calculate estimates
```

[14] `date +%s` on linux.

[15] The existence of term *fat-finger error*, in reference to larger than desired financial market orders being accidentally placed, suggests this type of error occurs with some measurable frequency.

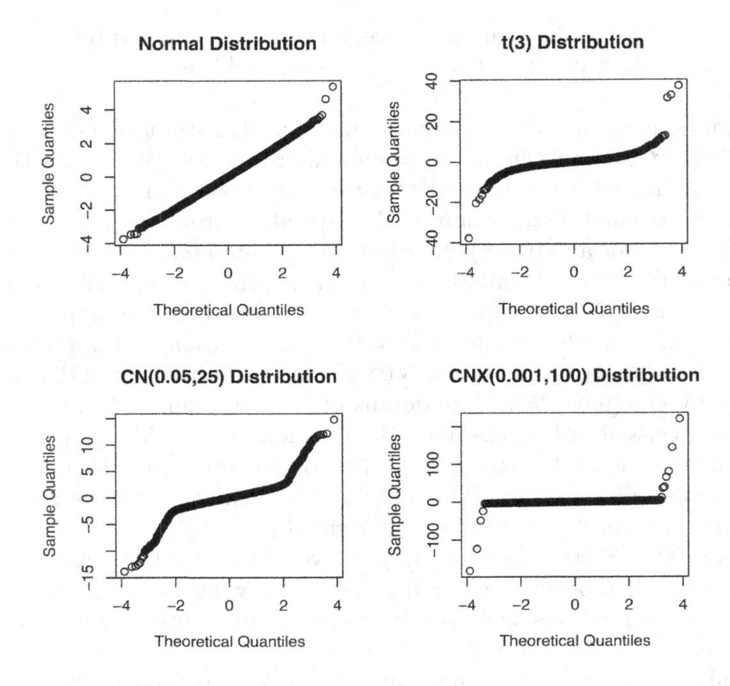

FIGURE 1.4
Normal q-q plots of random sample of size $n = 10000$ from various distributions. Note: while the xlim's of the plots are the same, the ylim's differ.

```
> xbarVec_rnorm <- apply(simMat_rnorm,1,mean)
> xdotVec_rnorm <- apply(simMat_rnorm,1,median)
> xbarVec_rcn <- apply(simMat_rcn,1,mean)
> xdotVec_rcn <- apply(simMat_rcn,1,median)
```

In Figure 1.5, histograms of the estimates from the 5000 simulation runs for each of the distributions are plotted. With a simulation size of 5000 the histograms provide a fairly good approximation of the sampling distribution of each of the two statistics under the two distributions. The histograms for the sample mean are plotted in the first row, while the histograms for the sample median are plotted in the second row. The left column contains the histograms when the samples are drawn from a $N(0, 1)$ distribution, while the right column contains the histograms when the samples are drawn from a $CN(0.05, 25)$ distribution. As can be seen when comparing the histograms of the sample mean, the variability increases when sampling from a CN distribution; however, there is much less of a change in the variability of the robust sample median. This increase in variability of the sample mean results in an increase in its standard error which has a negative impact on associated

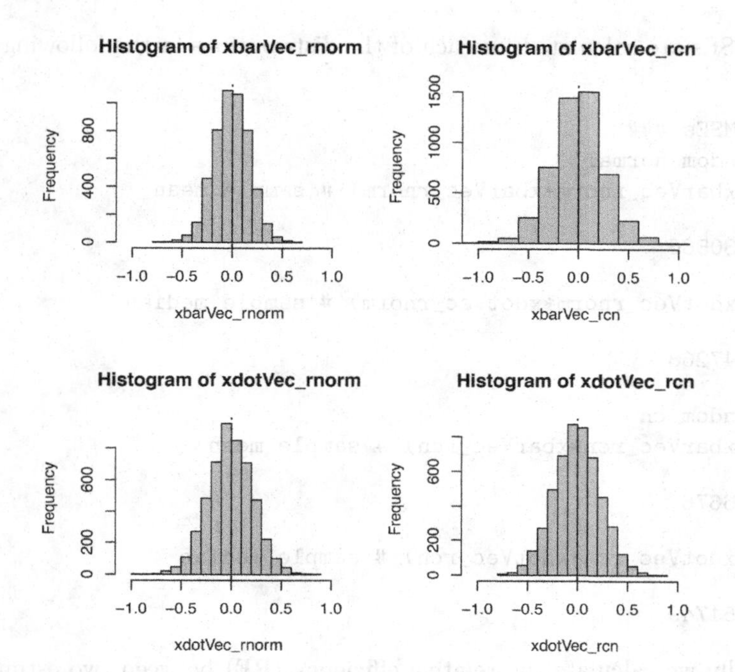

FIGURE 1.5
Simulation results of sample mean and sample median.

inference. We examine these differences between nonrobust traditional estimators and robust estimators in more detail throughout the text.

A useful metric is the **mean squared error (MSE)** of the estimator which describes how far, on average, an estimate is from the target. We calculate the **simulation estimated mean squared error (MSE)** based on results of our simulation. Suppose the target value of the estimator is θ and $\hat{\theta}_1, \ldots, \hat{\theta}_m$[16] are m estimated values based on m random samples. Then the simulated estimate of MSE is

$$\frac{1}{m}\sum_{i=1}^{m}(\hat{\theta}_i - \theta)^2.$$

In all cases considered in this section, the target θ is the center of a symmetric distribution. In this case, provided the mean of the population exists, both the mean and median of the population are the same with the common value θ, the center of the distribution. For the above simulations, each of the two distributions simulated (normal and contaminated normal) have center $\theta = 0$. So we have that $MSE(\bar{X}) = m^{-1}\sum_{i=1}^{m}\bar{x}_i^2$ and $MSE(\text{med}X) = m^{-1}\sum_{i=1}^{m}\hat{\theta}_i^2$.

[16]The notation $\hat{\theta}$ means the statistic $\hat{\theta}$ is an estimator of θ. It is pronounced theta-hat.

These MSEs are calculated for each of the distributions in the following code segment.

```
> #### MSEs ###
> ## random normal
> mean(xbarVec_rnorm*xbarVec_rnorm) # sample mean

[1] 0.030586

> mean(xdotVec_rnorm*xdotVec_rnorm) # sample median

[1] 0.047268

> ## random cn
> mean(xbarVec_rcn*xbarVec_rcn) # sample mean

[1] 0.06676

> mean(xdotVec_rcn*xdotVec_rcn) # sample median

[1] 0.051749
```

Usually we calculate the **relative efficiency (RE)** between two estimators as the ratio of their MSEs. For each distribution in our Monte Carlo study, we calculated these REs as the ratio of the MSE for the sample mean over the MSE of sample median. Hence, for a given distribution, if the RE is less than 1, then the sample mean is empirically more efficient than the sample median for that distribution. On the other hand, if the RE is greater than 1, then the sample median is empirically more efficient. A table of MSEs and REs from our simulation is output in the following R segment.

```
### Table of MSEs and REs ###

                 xbarMSE xmedMSE    RE
Normal             0.031   0.047 0.647
CN(0.05,25)        0.067   0.052 1.290
CNx(0.001,100)     0.307   0.045 6.748
```

Hence, for this Monte Carlo simulation, the sample mean is much more efficient than the sample median at the normal distribution, but the sample median is much more efficient than the sample mean for the heavier tailed distributions. In Chapter 2, we discuss a robust estimator that is almost as efficient as the sample mean at the normal distribution.

1.4 Functions

In this section, we present some basics for writing functions in R. Example uses of R functions include creating custom statistical analysis and aiding in a Monte Carlo simulation.

1.4.1 Single Line Functions

Single line functions can be used, for example, to specify a particular set of arguments to an existing function. These types of functions can be useful, for example, in simulation studies. As we have seen, the call `apply(datMat,1,f)` to the function `apply` will apply the function `f` to every row of a data matrix `datMat` with the row as the function's first argument. Many of the functions in R can be used for a variety of analyses or purposes based on the input or call to the function, so defining a set of arguments in a newly created function may simplify the simulation code. We will see examples of this use in the text.

For examples, consider the two single line R functions that obtain random samples from the contaminated normal distributions $CN(0.05, 25)$ and $CNX(0.001, 100)$ that are called `rcn_5_5` and `rcnx_01_100`, respectively. Both are single line functions, each of which calls a more general function with specific arguments. Both function definitions are provided in the following code segment as well as in `npsm`.

```
> rcn_5_5 <- function(n) rcn(n,0.05,5)
> rcnx_01_100 <- function(n) rcnx(n,0.001,100)
```

Both call a general random number generation function with specific arguments. The argument `n` of `rcn_5_5` is passed to `rcn` with values of the second and third argument specified. Similarly, the argument `n` of `rcn_01_100` is passed to `rcnx` with values of the second and third argument specified. Hence, the first function draws a random sample of size n from a $CN(0.05, 25)$ distribution, while the second obtains a sample of size n from a $CNX(0.001, 100)$ distribution.

1.4.2 Level and Power of a Statistical Test

Recall that the **level** (α) of a statistical test is defined as the probability that the test rejects the null hypothesis when in fact the null hypothesis is true. The **power** of a statistical test is defined as the probability that the test rejects the null hypothesis when it is in fact false.

For the following discussion, consider testing the hypotheses

$$H_0 : \mu = 0 \text{ versus } H_A : \mu \neq 0, \tag{1.1}$$

where μ is the population mean. In this case, the traditional normal theory *parametric* test is the one-sample t-test. In R, if the numeric vector `x` contains the sample, then the command `t.test(x)$p.value` returns the p-value of the two-sided in Hypothesis 1.1. A $\alpha = 0.05$ level test, of course, test rejects H_0, if the result of the call is less than or equal to 0.05.

The base R function `t.test` can be used for both one as well as two sample t-tests and may also be used to calculate confidence intervals depending on the arguments to the function call. In addition, the function is designed to return several outputs. To obtain a single output, to be used in simulations, for example, one may use a single line function. For instance, the function, `t.test_get_p.value`, defined below simply passes its arguments to `t.test` and returns the resulting p-value.

```
> t.test_get_p.value <- function(...) t.test(...)$p.value
```

This type of function is useful when only one result of the call to an existing function is needed. In the next code segment, we test it out on a simple simulated dataset. Note that the sample is from a standard normal distribution, so H_0, (1.1), is true.

```
> set.seed(20220202)
> n <- 30
> x <- rnorm(n)
> t.test_get_p.value(x)

[1] 0.61035
```

With the p-value 0.6103 the t-test fails to reject H_0 at level $\alpha = 0.05$; i.e., the sample does not support the alternative hypothesis in this case.

In the next segment we run a simple simulation of size 500 to estimate the level of the t-test for hypotheses (1.1), with $\alpha = 0.05$. The sampling is from a standard normal distribution, so H_0 is true.

```
> set.seed(1908)
> nsim <- 500
> sim_data <- matrix(rnorm(nsim*n),nrow=nsim)
> pvalVec <- apply(sim_data,1,t.test_get_p.value)
> mean(pvalVec <= 0.05)

[1] 0.052
```

The empirical level of significance[17] is close to the true value of 0.05. Exercise 1.7.6 asks the reader to approximate the power of the t-test under an alternative hypothesis.

[17]The values of the vector `pvalVec <= 0.05` are `TRUE` or `FALSE` depending on whether or not the p-values are less than 0.05. When the R function `mean` is applied, it converts logical values of `TRUE` and `FALSE` to the numerical values of 1 and 0, respectively.

1.4.3 Functions

When writing functions for general use, documentation on the function is important. For packages one can write `.Rd` files which are separate from the code itself. The package `roxygen2` (Wickham et al. 2020) uses a method where comments are added to the *header* of the function and then the documentation (`.Rd`) can be created with a call to `roxygenise`.[18]

Now we will write a function to compute MSE for a vector of estimates. As mentioned above, one could use `roxygen` style comments; however, for this simple two-liner, we just add a few notes on the input. Notice that for multiple line functions the lines are enclosed in braces (i.e., `{}`).

```
> mse <- function(thetahatVec,theta=0) {
+ # function to calculate sum (thetahat_i - theta)^2
+ # input :
+ #   thetahatVec - numeric vector
+ #   theta - numeric scalar
+ e <- thetahatVec - theta
+ mean(e*e)
+ }
```

The commands to define a function may be typed directly into an R session or copied and pasted from another file. Alternatively, the function may be sourced. If the function is in the text file `mse.r` in the working directory, this can be accomplished by the R command `source("mse.r")`. If the file is in another directory (or folder), then the path to it must be included; the path may be relative or absolute. For example, if `mse.r` is in the directory, `myFunctions` which is a subdirectory of the current working directory, then the command is `source("myFunctions/mse.r")`.

Once the function is sourced, we test it.

```
> round(c(mse(xbarVec_rcn),mse(xbarVec_rcnx100)),3)

[1] 0.067 0.307
```

1.5 Randomization

Many of the traditional nonparametric methods have only the assumption of a valid randomization; i.e., assignments of treatments to subjects or experimental units is random — so that each subject or experimental unit has an equal chance of being assigned each of the treatments. For example, suppose that

[18]See `cran.r-project.org/web/packages/roxygen2/vignettes/roxygen2.html` for more information.

an experimenter is comparing the effects that treatments A and B have on a response and that 10 patients have been drawn randomly from the reference population for the experiment with labels 1 through 10. Then assignment of treatment to patient must be performed by a valid randomization. A simple randomization is computed as

```
> set.seed(247468)
> rbind(1:10,sample(c("A","B"),10,replace=TRUE))

     [,1] [,2] [,3] [,4] [,5] [,6] [,7] [,8] [,9] [,10]
[1,] "1"  "2"  "3"  "4"  "5"  "6"  "7"  "8"  "9"  "10"
[2,] "B"  "A"  "B"  "A"  "A"  "B"  "B"  "A"  "B"  "A"
```

Hence patients 1, 2, 4, 6, 7, and 10 are assigned to receive treatment A, while the others are assigned to receive treatment B. The design is not perfectly balanced since 6 patients receive treatment A, while 4 receive treatment B. In terms of sample sizes, though, it can be shown that a balanced design is more powerful than an unbalanced design for the same total sample size.

In this section we include a short introduction to creating a *randomization list* or *rand list.* The goal is not to provide a comprehensive treatment of the topic, but rather to further illustrate and expand upon some of the concepts of this first chapter.

Example 1.5.1. Consider then a designed experiment with two possible treatment assignments denoted by A and B. We consider the case of a balanced design; i.e., an equal number of treatment assignment for each of A and B. The treatments are to be assigned at random to a set of n subjects or experimental units. One method to determine the sequence (or *randomization list*) is to use a permutation of an equal number of the treatment codes of each type. For illustration we take `n<-80` and generate such a list in the following code segment. We show the first 10 assignments and then use the function `table` to show the sample sizes for treatments A and B.

```
> set.seed(1969)
> rand_list_0 <- sample( rep_len(c('A','B'),n) )
> head(rand_list_0,10)

 [1] "A" "B" "A" "A" "B" "A" "A" "A" "B" "A"

> table(rand_list_0)

rand_list_0
 A  B
40 40
```

An alternative method is to use a **permuted block randomization** where blocks of an equal number of treatment codes are permuted. While both methods would result in the same proportions at the end of the study, a permuted block randomization would tend to stay closer to equal during the course of the study. This has the advantage of having closer to balanced numbers mid-study, for an interim analysis, for example. Block sizes need not be the same; for example, with two treatments in a balanced design, one may use a mixture of block sizes 2 and 4. In the following code segment we generate a randomization list using a permuted block randomization. A function is defined to create a permutation of size `n`, block sizes are then specified and permuted, then the function is applied to the vector of block sizes using the `lapply` function. The `unlist` function converts the list of lists to a single vector representing the randomization list.

```
> sample_trt_n <- function(n)
+    sample(rep_len(c('A','B'),n),n,replace=FALSE)
> block_sizes <- sample( rep(c(4,2),times=c(10,20)) )
> rand_list_PB <- unlist( lapply(block_sizes,sample_trt_n) )
```

So the block sizes are first permuted, and then given a block size the treatment codes within a block are permuted. As a check, we can view the number of blocks of each size as in the next code segment.

```
> table(block_sizes)

block_sizes
 2  4
20 10
```

The proportion of A's as a function of the total sample size (as the study would be enrolled) for both methods is provided in Figure 1.6. A function to compute the running proportion is defined in the next code segment which we use in creating the plots.

```
> propA <- function(x) cumsum(x == 'A')/seq_along(x)
```

The `cumsum` function returns a vector of the cumulative sum; i.e., has jth element $\sum_{i=1}^{j} x_i$; which, in this case, is counting the number of times the input is equal to the character `A`. The `seq_along` function returns a vector with j as the jth element. So the vector division returns the proportion assigned to treatment A after j subjects have been enrolled.

In this case, midway through enrollment, the proportion of subjects assigned to A for each of the methods is provided in the following code segment.

```
> propA(rand_list_0)[40:41]  # simple permutation

[1] 0.62500 0.63415

> propA(rand_list_PB)[40:41]  # permutation block

[1] 0.5000 0.4878
```

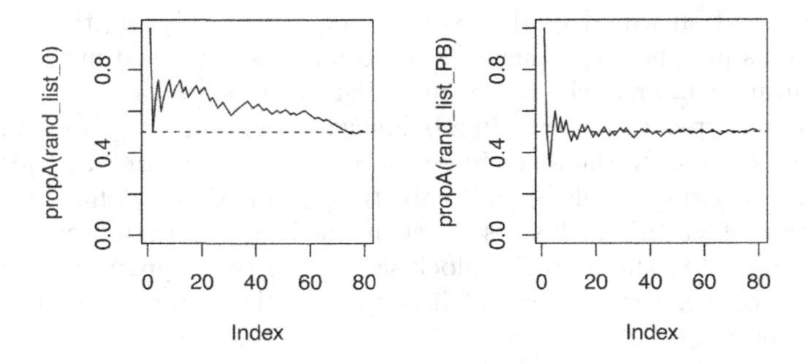

FIGURE 1.6
Two methods of creating a randomization list as discussed in Example 1.5.1. A simple permutation on the left and a permuted block on the right.

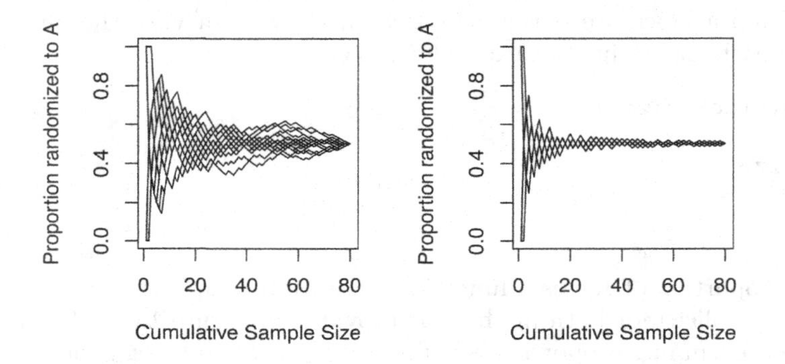

FIGURE 1.7
Proportion of A versus cumulative sample size for two methods — 20 of each type randomization list.

To provide a general sense, 20 lists of each type were generated; the proportion of A versus sample size is provided in Figure 1.7. ∎

Example 1.5.2. In a *stratified* randomization, subjects are first partitioned into subgroups or strata which are thought to have an impact on the outcome; i.e., in a clinical study are prognostic factors. Examples include demographics (e.g., sex) or disease severity. Randomization then occurs within a strata. Without knowing how many subjects will enroll in a study in each strata (assuming no cap), one approach is to use a sort of brute-force method and create a complete randomization list for each strata. The following code segment creates a complete randomization list for each strata level using such a method. There are two strata variables; the number of levels for each is defined in the code segment.

```
> set.seed(211212)
> # define constants
> num_strata_lev_1 <- 4  # number levels 1st strata
> num_strata_lev_2 <- 2  # number levels 2nd strata
> n <-88  # total sample size
> # create labels for strata
> p0 <- function(...) paste(...,sep=".")
> strata_levels <- as.vector(
+                         outer( paste0('S_1', 1:num_strata_lev_1),
+                         paste0('S_2', 1:num_strata_lev_2), p0 )
+                     )
> strata_levels

[1] "S_11.S_21" "S_12.S_21" "S_13.S_21" "S_14.S_21"
[5] "S_11.S_22" "S_12.S_22" "S_13.S_22" "S_14.S_22"

> sample_trt_n <- function(n)
+   sample(rep_len(c('A','B'),n),n,replace=FALSE)
> block_sizes <- rep_len(4,n/4)  # equal size blocks
> getPB <- function( ... )
+   unlist( lapply(block_sizes,sample_trt_n) )
> rand_list <- lapply(strata_levels, getPB)
> names(rand_list) <- strata_levels
```

`rand_list` is a list of lists keyed by the strata labels. For example, we view the first block of treatment assignments for the strata labeled `S_11.S_21` in the next code segment.

```
> rand_list$S_11.S_21[1:4]

[1] "B" "B" "A" "A"
```

In the next code segment we use alternative syntax to view the first block for the strata labeled `S_12.S_22`.

```
> rand_list[["S_12.S_22"]][1:4]

[1] "B" "A" "A" "B"
```

■

1.6 Density Estimation

Let $X_1, X_2, \ldots, X_n$ be a random sample on a random variable X that follows the location model

$$X_i = \theta + e_i, \quad i = 1, \ldots, n,$$

where the random errors e_i are iid with continuous pdf $f(x)$. Then the pdf of the observations X_i is $f_X(x) = f(x - \theta)$. Recall that, in general, the pdf satisfies the properties:

(i) The function $f_X(x)$ is nonnegative, $(f_X(x) \geq 0)$.

(ii) The total area under the curve $y = f_X(x)$ is 1, $(\int_{-\infty}^{\infty} f_X(x)\, dx = 1)$.

(iii) The probability that $a < X < b$ is the area under the curve $y = f_X(x)$ between a and b, $(P[a < X < b] = \int_a^b f_X(x)\, dx)$.

In Section 2.4, we discuss estimation of the center (effect), θ, of a distribution. In this section, we consider estimating the pdf $f_X(x)$. For illustration, consider the sample of obervations given in Table 1.1.

TABLE 1.1
Illustrative Sample

64	66	69	71	70	69	64	69	74	62

A simple estimator of the density, $f_X(x)$, is the *histogram*. Abutting intervals (groups), of equal length, are selected that cover the range of the observations $x_1, \ldots, x_n$. Suppose g intervals are selected, represented by $[a_0, a_1], (a_1, a_2], \ldots, (a_{g-1}, a_g]$. Let f_j denote the sample frequency of the interval j, i.e., the number of observations that fall in the interval $(a_{j-1}, a_j]$. A histogram displays a rectangle above interval j with height f_j. For computation, we use the histogram function, `hist`, in core R. By default, `hist` displays the frequencies; however, it may also display relative frequencies (or proportions) by using the option `probability=TRUE`.[19] Both these types of histograms are displayed in Figure 1.8 for the data in Table 1.1. By default `hist` produces a graphical display; however, this may be overridden to produce numeric summaries. In the next R code segment, the computed intervals, frequencies, relative frequencies, and mid-points of the intervals are provided.

```
> x <- c(64,66,69,71,70,69,64,69,74,62)
> hist(x,plot=FALSE)

$breaks
[1] 62 64 66 68 70 72 74
```

[19] See `help(hist)` for more information.

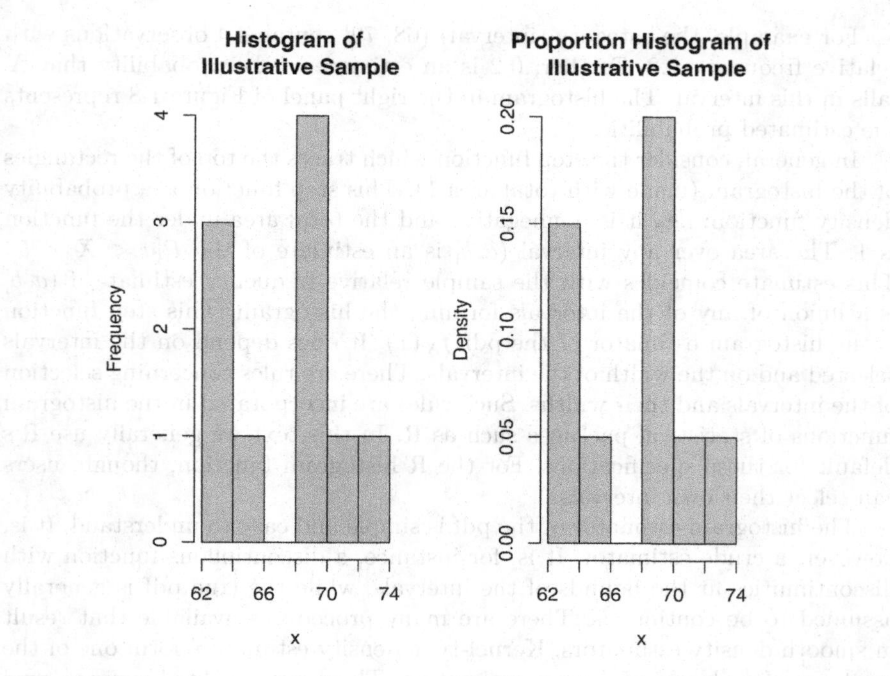

FIGURE 1.8
The left panel displays the histogram of the Illustrative Sample in Table 1.1. The right panel is a rescaled version which has total area 1.

```
$counts
[1] 3 1 0 4 1 1

$density
[1] 0.15 0.05 0.00 0.20 0.05 0.05

$mids
[1] 63 65 67 69 71 73

$xname
[1] "x"

$equidist
[1] TRUE

attr(,"class")
[1] "histogram"
```

For example, the category (interval) (68, 70] contains 4 observations with relative frequency 0.2. Further, 0.2 is an estimate of the probability that X falls in this interval. The histogram in the right panel of Figure 1.8 represents the estimated probabilities.

In general, consider the step function which traces the top of the rectangles of the histogram (made with total area 1). This step function is a probability density function; i.e., it is nonnegative and the total area under the function is 1. The area over any interval $(a, b]$ is an estimate of the $P[a < X < b]$. This estimate coincides with the sample relative frequency estimate if $(a, b]$ is a union of any of the intervals forming the histogram. This step function is the histogram estimator of the pdf $f_X(x)$. It does depend on the intervals selected and on the width of the intervals. There are rules concerning selection of the intervals and their widths. Such rules are incorporated in the histogram functions of statistical packages such as R. In this text we generally use R's default for these specifications. For the R histogram function, though, users can select their own intervals.

The histogram estimator of the pdf is simple and easy to understand. It is, however, a crude estimator. It is, for instance, a discontinuous function with discontinuities at the bounds of the intervals, while the true pdf is generally assumed to be continuous. There are many procedures available that result in smooth density estimators. Kernel-type density estimators form one of the most popular classes of density estimators. These are weighted average type estimators that are discussed in more detail in Chapter 4. The `density` function in base R is a kernel type estimator that we will use. Generally, density estimation requires a large sample as we illustrate in the next example.

Example 1.6.1. Assume X is normally distributed with mean 70 and standard deviation 4. We generate eight samples on X of various sample sizes. Plots of histograms are displayed in Figure 1.9. Each panel displays the histogram of the sample size indicated in the display title. Each histogram is next overlaid with a density estimate. The core R function `hist` is used followed by nested calls of `density` and `lines`; i.e., `lines(density(x))`.

This example illustrates the need for large datasets to accurately estimate the pdf. With smaller sample sizes, the shape of the estimated distribution is quite dissimilar to the true distribution. For the later sample sizes, though, the density estimate is unimodal. ∎

1.6.1 Some Details

Suppose $X_1, X_2, \ldots, X_n$ is a random sample from a population with the unknown probability density function (pdf) $f(x)$. In this section, we provide information on some of the details of an estimator of the pdf $f(x)$. Density estimation is similar to nonparametric regression which we discuss in a later chapter.

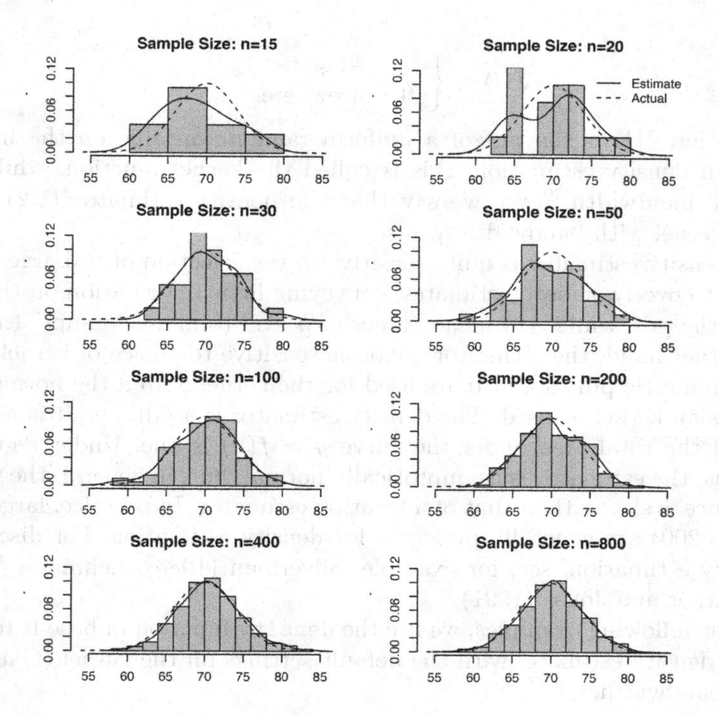

FIGURE 1.9
Samples of selected sizes from a $N(70, 4^2)$ distribution. Each panel displays a histogram as well as a density estimate (solid line) and actual pdf (dashed line). The sample size is indicated in the title of each plot.

Let x be in the support X_i, that is, the set of values where $f(x) > 0$. For any $h > 0$, the mean value theorem implies that

$$P(x - h < X_i < x + h) = f(\xi)h \doteq f(x)2h,$$

for some ξ between $x - h$ and $x + h$; see, for example, page 232 of Hogg et al. (2019). An estimate of the left side of this equation is the number of X_i's that fall in the interval $(x - h, x + h)$ divided by n, which in turn yields as an estimate of $f(x)$:

$$\hat{f}(x) = \frac{\#_i\{x - h < X_i < x + h\}}{2hn}.$$

Since $x - h < X_i < x + h$ is equivalent to $-1 < (x - X_i)/h < 1$, this estimate of $f(x)$ can be written as

$$\hat{f}(x) = \frac{1}{hn} \sum_{i=1}^{n} I\left(\frac{x - X_i}{h}\right), \tag{1.2}$$

where

$$I(t) = \begin{cases} \frac{1}{2} & -1 < t < 1 \\ 0 & \text{elsewhere.} \end{cases} \tag{1.3}$$

The function $I(t)$ is the pdf of a uniform random variable on the interval $(-1, 1)$. In density estimation, this is called the **kernel** function, while h is called the **bandwidth.**[20] So, we say that the density estimator (1.2) has a uniform kernel with bandwidth h.

The density estimator is quite sensitive to the selection of h. Large values of h lead to overly smooth estimates, conveying little information on the true shape of the pdf, while very small values of h lead to many "bumps" (nodes). On the other hand, the estimator is not as sensitive to choice of kernel. Generally, symmetric pdfs about 0 are used for the kernel. Often the normal pdf, the Gaussian kernel, is used. The density estimator is a pdf, i.e., it is nonnegative and the total area under the curve $y = \hat{f}(x)$ is one. Under regularity conditions, the estimator is asymptotically normal, but, in general, the rate of convergence is slower than that of a location estimator. In practice, large samples ($n > 200$) are generally preferred for density estimation. For discussion on density estimation, see, for example, Silverman (1986), Lehmann (1999), and Sheather and Jones (1991).

For the following examples, we use the `density` function in base R to compute the density estimate, with the default settings on the kernel (Gaussian) and the bandwidth.

Example 1.6.2 (Weights of Professional Baseball Players). Using the `baseball_players1000` dataset which contains information on 1000 professional baseball players. Included among the recorded variables is the weight of the baseball player. In the next R segment, a histograph is computed and drawn. Then a density estimate of the weights is computed and overlaid on the histogram. Note that the weight of one of the baseball players is missing, so we use the `na.rm=TRUE` option. In Figure 1.10 it is overlaid with the histogram of the weights.

```
> hist(baseball_players1000$weight,xlab="Weight (lbs)",
+       probability=TRUE,ylim=c(0,.02),
+       main="Histogram of Weight for 1000 Baseball Players")
> lines(density(baseball_players1000$weight,na.rm=TRUE))
```

There does appear to be some asymmetry in the distribution of weights. ■

Remark 1.6.1. Consider a situation where we are interested in the effect that a new treatment has on a response and we have considerable historical data on the response under the standard treatment. Often the difference between the new treatment and standard treatment effects is incremental. That is, if X_i and Y_j are responses under the standard and new treatments, respectively, then our model is $X_i = \theta + e_i$ and $Y_j = \theta + \Delta + e_j$, where the random errors e_i

[20] Bandwidth is sometimes referred to as *window width.*

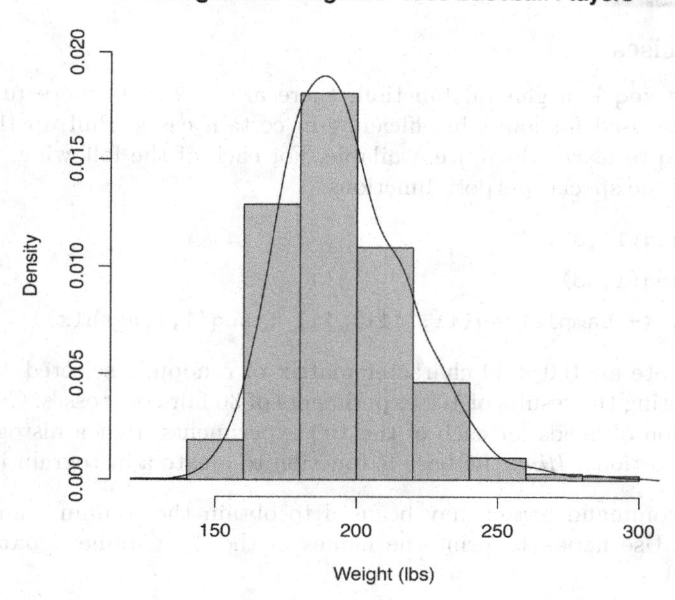

FIGURE 1.10
Histogram of weights of professional baseball players based on the sample in `baseball_players1000`. Density estimate overlaid.

and e_j have the same pdf[21]. Thus, Δ is the incremental effect due to the new treatment. In this case, the form of the pdf of Y_j is the same as that of X_i, (just shifted). Hence, density estimation based on historical data should prove useful. For instance, if responses appear to have a skewed distribution, then analyses appropriate for skewed data can be used. Such analyses are discussed in ensuing chapters.

Remark 1.6.2. Lehmann (1999) offers an informative summary of density estimation and a comparison of the theories for density estimation and for the estimation of effects, such as θ. In general, the asymptotic theory for $\hat{\theta}$ is of order $1/n^{1/2}$, where n is the sample size, whereas the asymptotic theory for the kernel estimator $\hat{f}(x)$ is of larger order $1/n^{2/5}$. Hence, density estimators converge at a slower rate. Also, the estimates of the effects discussed in this text are asymptotically unbiased (have asymptotic mean θ), whereas kernel density estimators are asymptotically biased. ■

[21]This is the location model of Chapter 3.

1.7 Exercises

1.7.1. While `seq` is a general function, there are special purpose functions which can be used for gains in efficiency in certain cases. Pull up the help menu for `seq` to learn which are available. For each of the following, rewrite using one of the special purpose functions.

(a) `seq(11,33)`

(b) `seq(1,33)`

(c) `x <- sample(seq(11,1111),11) ; seq(1,length(x))`

1.7.2. Generate an 100×30 character matrix of randomly selected `'H'` and `'T'` representing the results of 100 experiments of 30 fair coin tosses. Calculate the proportion of heads for each of the 100 experiments. Plot a histogram of the 100 proportions. *Hint*: the base R function to create a histogram is `hist`.

1.7.3. The command `names` may be used to obtain the column names of a data frame. Use `names` to print the names of the data frame `egData` from Section 1.1.3.

1.7.4. The command `names` may be used to assign column names to a data frame. Use `names` to assign the names `c('Char','Ind','X','Y')` to the data frame `egData` from Section 1.1.3.

1.7.5.

Use the data read in to R in Section 1.1.3 to complete the following.

(a) The variables `cr` and `i0` are categorical variables. Use the `table` command to create frequency distributions for each of the two variables.

(b) Create a boxplot and normal probability plot of the variable `x1`.

(c) Create a boxplot and normal probability plot of the variable `y`.

1.7.6. Estimate the power of a t-test of $H_0 : \mu = 0$ versus $H_A : \mu > 0$ when the true mean is $\mu = 0.5$. Assume a random sample of size $n = 25$ from a normal distribution with $\sigma = 1$. Assume $\alpha = 0.05$. Use a simulation size of 1000.

1.7.7. The contaminated normal distribution is frequently used in simulation studies. A standardized variable, X, having this distribution can be written as

$$X = (1 - I_\epsilon)Z + cI_\epsilon Z,$$

where $0 \leq \epsilon < 1$, I_ϵ has a binomial distribution with $n = 1$ and probability of success ϵ, Z has a standard normal distribution, $c > 1$, and I_ϵ and Z are independent random variables. When sampling from the distribution of X, $(1-\epsilon)100\%$ of the time the observations are drawn from a $N(0,1)$ distribution but $\epsilon 100\%$ of the time the observations are drawn from a $N(0,c^2)$. These latter observations are often outliers. The distribution of X is a mixture distribution; see, for example, Section 3.4.1 of Hogg et al. (2019). We say that X has a $CN(c,\epsilon)$ distribution.

(a) Using the R functions `rbinom` and `rnorm`, write an R function which obtains a random sample of size n from a contaminated normal distribution $CN(c,\epsilon)$.

(b) Obtain samples of size 100 from a $N(0,1)$ distribution and a $CN(16, 0.25)$ distribution. Form histograms and comparison boxplots of the samples. Discuss the results.

1.7.8. Using a t_3 distribution, estimate the relative efficiency of the sample median to the sample mean using a simulation size of 1000. Which estimator is more efficient for t_3 data? *Hint*: see `help(rt)`.

1.7.9. Using the `npsm` dataset `baseball_players1000` data, create comparison boxplots of the heights of pitchers and non-pitchers.

1.7.10. Using the `npsm` dataset `baseball_players1000` data, create a histogram of year of major league debut for this sample of players.

1.7.11. Let X be a random variable with the exponential distribution, i.e., X has pdf $f(x) = \exp\{-x\}$, $x > 0$. Suppose $Y = \log(X)$. In R, a sample of size n can be generated from the distribution of Y with the command `samp <- log(rexp(n)); plot(density(samp))`, where the second command plots the density estimate of the pdf of Y based on the sample `samp`. Run these commands for the sample sizes used in Figure 1.9. For each value of n, comment on the shape of the density estimate (symmetric, left-skewed, right-skewed, unimodal) and what you think the shape of the true pdf is.

1.7.12. In Exercise 1.7.11, we can show that the pdf of the random variable Y is given by

$$f(y) = \exp\{y - \exp(y)\}, \quad -\infty < y < \infty.$$

Obtain the plot of this pdf overlaid with the density estimate for $n = 100$ from Exercise 1.7.11. Is the estimate close to the actual pdf? The function $f(y)$ is the pdf of an **extreme-valued** distribution.

1.7.13. Johnson and Wichern (2007) discuss a dataset involving carapace measurements for female and male painted turtles. The `npsm` dataset `turtle` contains the length, width, and height carapace measurements for a sample of turtles. The fourth column is an indicator variable which is 1 for female and 2 for male. For the female turtles:

(a) Obtain the density plot for the measurement Length. Comment on the plot.

(b) Locate the sample median and mean on the plot.

1.7.14. Continuing Exercise 1.7.13, for the male turtles:

(a) Obtain the density plot for the measurement Length. Comment on the plot.

(b) For comparison purposes, obtain the density plots for female and male length measurements on the same set of axes. Comment on the comparison.

(c) Locate the male and female sample medians on your plot.

1.7.15. Create density comparison plots, male versus female, for the other two carapace measurements Width and Height in the `npsm` dataset `turtle`. Comment on the comparison.

2

One-Sample Problems

2.1 Introduction

In this chapter we present some basic nonparametric statistical procedures for one-sample problems. We begin with methods of inference for a proportion. Followed by classical nonparametric procedures including the sign and Wilcoxon tests. We also briefly discuss density estimation which was introduced in the previous chapter.

Our discussion includes a brief overview of the distribution-free sign test. For the one-sample problem for continuous data, we present the signed-rank Wilcoxon nonparametric procedure and review the parametric t procedure. Also we discuss inference based on bootstrapping (resampling).

Our discussion focuses on the computation of these methods via R. More details of these nonparametric procedures can be found in the books by Hettmansperger and McKean (2011), Higgins (2003), and Hollander and Wolfe (1999).

2.2 One-Sample Proportion Problems

In this section, we consider discrete random variables whose ranges consist of two categories which we generally label as failure (0) and success (1). Let X denote such a random variable. Let p denote the probability of success. Then we say that X has a Bernoulli distribution with the probability model

x	0	1
$P(X = x)$	$1 - p$	p

It is easy to show that the mean of X is p and that the variance of X is $p(1 - p)$.

Statistical problems consist of estimating p, forming confidence intervals for it, and testing hypotheses, e.g., of the form

$$H_0 : p = p_0 \text{ versus } H_A : p \neq p_0, \tag{2.1}$$

where p_0 is specified. One-sided hypotheses can be similarly formulated.

Let $X_1, \ldots, X_n$ be a random sample on X. Let S be the total number of successes in the sample of size n. Then S has a binomial distribution with the distribution

$$P(S = j) = \binom{n}{j} p^j (1-p)^{n-j}, \quad j = 0, 1, \ldots, n. \tag{2.2}$$

The estimate of p is the sample proportion of successes; i.e.,

$$\hat{p} = \frac{S}{n}. \tag{2.3}$$

Based on the asymptotic normality of S, an approximate $(1-\alpha)100\%$ confidence interval for p is

$$(\hat{p} - z_{\alpha/2}\sqrt{\hat{p}(1-\hat{p})/n}, \hat{p} + z_{\alpha/2}\sqrt{\hat{p}(1-\hat{p})/n}). \tag{2.4}$$

Example 2.2.1 (Squeaky Hip Replacements)**.** As a numerical example, Devore (2012), page 284, reports on a study of 143 subjects who have obtained ceramic hip replacements. Ten of the subjects in the study reported that their hip replacements squeaked. Consider patients who receive such a ceramic hip replacement and let p denote the true proportion of those whose replacement hips develop a squeak. Based on the data, we next compute[1] the estimate of p and a confidence interval for it.

```
> phat<-10/143
> zcv<-qnorm(0.975)
> phat+c(-1,1)*zcv*sqrt(phat*(1-phat)/143)

[1] 0.02813069 0.11172945
```

Hence, we estimate between roughly 3 and 11% of patients who receive ceramic hip replacements such as the ones in the study will report squeaky replacements. ■

Asymptotic tests of hypotheses involving proportions, such as (2.1), are often used. For hypotheses (2.1), the usual test is to reject H_0 in favor of H_A, if $|z|$ is large, where

$$z = \frac{\hat{p} - p_0}{\sqrt{p_0(1-p_0)/n}} \tag{2.5}$$

Note that z has an asymptotic $N(0,1)$ distribution under H_0, so an equivalent test statistic is based on $\chi^2 = z^2$. The p-value for a two-sided test is p-value $= P[\chi^2(1) > (\text{Observed } \chi^2)]$. This χ^2-formulation is the test and p-value computed by the R function `prop.test` with `correct=FALSE` indicating that a continuity correction not be applied. The two-sided hypothesis is the default, but one-sided hypotheses can be tested by specifying the `alternative` argument. The null value of p is set by the argument `p`.

[1]The base R function `prop.test` provides a confidence interval which is computed by inverting the score test.

Example 2.2.2 (Left-Handed Professional Ball Players). As an example of this test, consider testing whether the proportion of left-handed professional baseball players is the same as the proportion of left-handed people in the general population, which is about 0.15. For our sample we use the dataset `baseball` that consists of observations on 59 professional baseball players, including throwing `hand` ('L' or 'R'). The following R segment computes the test:

```
> ind<-with(baseball,throw=='L')
> prop.test(sum(ind),length(ind),p=0.15,correct=FALSE)

        1-sample proportions test without continuity correction

data:  sum(ind) out of length(ind), null probability 0.15
X-squared = 5.0279, df = 1, p-value = 0.02494
alternative hypothesis: true p is not equal to 0.15
95 percent confidence interval:
 0.1605598 0.3779614
sample estimates:
        p
0.2542373
```

Because the p-value of the test is 0.02494, H_0 is rejected at the 5% level. ■

The above inference is based on the asymptotic distribution of S, the number of successes in the sample. This statistic, though, has a binomial distribution, (2.2), and inference can be formulated based on it. This includes finite sample tests and confidence intervals. For a given level α, though, these confidence intervals are conservative; that is, their true confidence level is at least $1-\alpha$; see Section 4.3 of Hogg et al. (2019). These tests and confidence intervals are computed by the R function `binom.test`. We illustrate its computation for the baseball example.

```
> binom.test(sum(ind),59,p=.15)

        Exact binomial test

data:  sum(ind) and 59
number of successes = 15, number of trials = 59, p-value = 0.04192
alternative hypothesis: true probability of success is not equal to 0.15
95 percent confidence interval:
 0.1498208 0.3844241
sample estimates:
probability of success
             0.2542373
```

Note that the confidence interval traps $p = 0.15$, even though the two-sided test rejects H_0. This example illustrates the conservativeness of the finite sample confidence interval.

2.3 Sign Test

The sign test requires only the weakest assumptions of the data. For instance, in comparing two objects, the sign test only uses the information that one object is better in some sense than the other.

As an example, suppose that we are comparing two brands of ice cream, say Brand A and Brand B. A blindfolded taster is given the ice creams in a randomized order with a washout period between tastes. The taster's response is the preference of one ice cream over the other. For illustration, suppose that 12 tasters have been selected. Each taster is put through the blindfolded test. Suppose the results are such that Brand A is preferred by 10 of the tasters, Brand B by one of the tasters, and one taster has no preference. These data present pretty convincing evidence in favor of Brand A. How likely is such a result due to chance if the null hypothesis is true, i.e., no preference in the brands? As our sign test statistic, let S denote the number of tasters that prefer Brand A to Brand B. Then for our data $S = 10$. The null hypothesis is that there is no preference in brands; that is, one brand is selected over the other with probability 1/2. Under the null hypothesis, then, S has a binomial distribution with the probability of success of 1/2 and, in this case[2], $n = 11$ as the number of trials. A two-sided p-value can be calculated as follows.

```
> 2*(1-pbinom(9,11,1/2))

[1] 0.011719
```

On the basis of this p-value we would reject the null hypothesis at the 5% level. If we make a one-sided conclusion, we would say Brand A is preferred over Brand B.

The sign test is an example of a distribution-free (nonparametric) test. In the example, suppose we can measure numerically the goodness of taste. The distribution of the sign test under the null hypothesis does not depend on the distribution of the measure; hence, the term distribution-free. The distribution of the sign test, though, is not distribution-free under alternative hypotheses.

[2]The observation of no preference has been omitted from the analysis; see Section 2.5.

2.3.1 Power Simulation

The power of a statistical test of hypothesis was briefly introduced in Section 1.4.2 where we ran a small simulation to obtain an empirical power curve for the t-test at the normal distribution. In this section we run another small simulation to compare the t-test and the sign test at two distributions.

In `npsm` we provide a simple sign test function which computes the sign test for a one-sample problem and returns the p-value using the exact method described in the previous subsection.

In this section we test the following one-sided hypothesis,

$$H_0 : \theta = 0 \text{ versus } H_A : \theta > 0. \tag{2.6}$$

We generate data under the null hypothesis and a range of alternatives.

Before we get started, we create a couple of single line functions which return the p-value for the hypothesis defined in Equation 2.6.

```
> # define single line functions to obtain a p-value
> ttest_pval_g <- function(x)
+  t.test(x,alternative='greater')$p.value
> stest_pval_g <- function(x)
+   signtest_pvalue(x,alternative='greater')
```

In the next code segment we set some constants for our study.

```
> set.seed(1710) # seed for reproducibility
> nsim <- 5000   # number of simulations
> n <- 30        # sample size for each run
> alpha <- 0.05  # nominal level
```

The range of θ we consider for this study is defined next.

```
> thetaVec <- seq(0,1,by=0.25)
```

We specify the distributions that will be used as a list of functions in the following code segment,

```
> rdists <- list(rnorm,rcn_5_5)
```

so that the `kth` distribution may be extracted using `rdata <- rdists[[k]]` and a sample of size `n` for that distribution may be generated as `rdata(n)`. Matrices to hold the results are defined in the following code segment.

```
> power_t_test <- power_sign_test <-
+   matrix(nrow=length(thetaVec),ncol=length(rdists))
```

With the above in place, the entire simulation can be done with just a few lines of code as is done in the next code segment.

```
> set.seed(1122)
```

```
> for( k in 1:length(rdists) ) {
+ for( j in 1:length(thetaVec) ) {
+
+   rdata <- rdists[[k]]
+
+ # rows of sim_data represent realizations of the experiment
+   sim_data <- matrix(rdata(nsim*n),ncol=n)
+   sim_data <- sim_data + thetaVec[j]
+
+   p.values_t.test <- apply(sim_data, 1, ttest_pval_g)
+   power_t_test[j,k] <- mean(p.values_t.test < alpha)
+
+   p.values_sign.test <- apply(sim_data, 1, stest_pval_g)
+   power_sign_test[j,k] <- mean(p.values_sign.test < alpha)
+
+ }
+ }
```

The power curves are plotted in Figure 2.1. Notice the t-test dominates the sign test at the normal distribution. When contaminated samples are added, the tests are closer to having equal power.

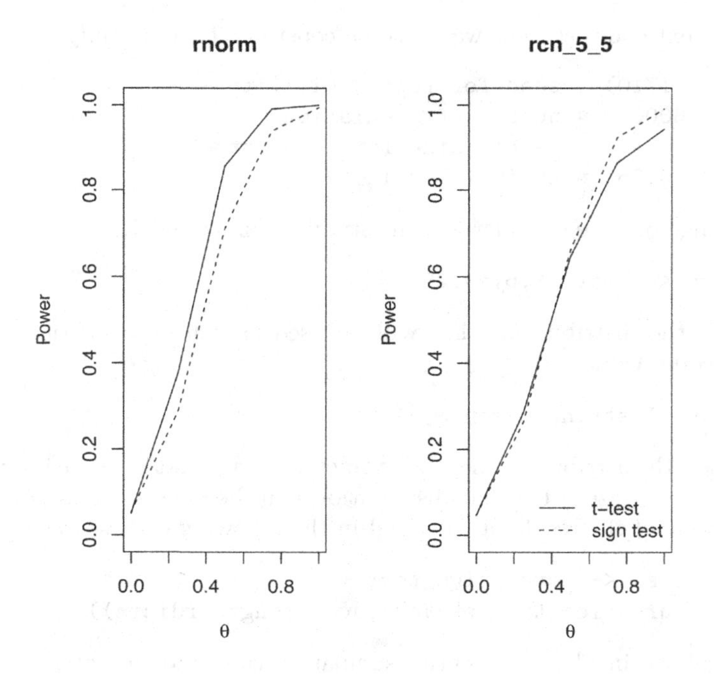

FIGURE 2.1
Power curves for the t-test and sign test for two distributions.

2.4 Signed-Rank Wilcoxon

The sign test discussed in the last section is a nonparametric procedure for the one-sample or paired problem. Although it requires few assumptions, the power can be low, for example relative to the t-test at the normal distribution. In this section, we present the **signed-rank Wilcoxon** procedure which is a nonparametric procedure that has power nearly that of the t-test for normal distributions, and it generally has power greater than that of the t-test for distributions with heavier tails than the normal distribution. More details for the Wilcoxon signed-rank test can be found in the references cited in Section 2.1. We discuss these two nonparametric procedures and the t-procedure for the one-sample location problem, showing their computation using R. For each procedure, we also discuss the R syntax for computing the associated estimate and confidence interval for the location effect.

We begin by defining a location model to set the stage for future discussions. Let $X_1, X_2, \ldots, X_n$ denote a random sample which follows the model

$$X_i = \theta + e_i, \tag{2.7}$$

where, to simplify discussion, we assume that the random errors, $e_1, \ldots, e_n$ are independent and identically distributed (iid) with a continuous probability density function $f(t)$ which is symmetric about 0. We call this model the **location model**. Under the assumption of symmetry, any location measure (parameter) of X_i, including the mean and median, is equal to θ. Suppose we are interested in testing the hypotheses

$$H_0 : \theta = 0 \text{ versus } H_A : \theta > 0. \tag{2.8}$$

The sign test of the last section is based on the test statistic

$$S = \sum_{i=1}^{n} \text{sign}(X_i),$$

where $\text{sign}(t) = -1, 0$, or 1 for $t < 0$, $t = 0$, or $t > 0$, respectively. Let

$$S^+ = \#_i\{X_i > 0\}. \tag{2.9}$$

Then $S = 2S^+ - n$. This assumes that none of the X_i's is equal to 0. In practice, generally observations with value 0 are omitted and the sample size is reduced accordingly. Note that under H_0, S^+ has a binomial distribution with n trials and probability of success 1/2. Hence, critical values of the test are obtained from the binomial distribution. Since the null distribution of S does not depend on $f(t)$, we say that the sign test is **distribution-free**. Let s^+ denote the observed (realized) value of S^+ when the sample is drawn. Then the p-value of the sign test for the hypotheses (2.8) is $P_{H_0}(S^+ \geq s^+) = 1 -$

$F_B(s^+ - 1; n, 0.5)$, where $F_B(t; n, p)$ denotes the cdf of a binomial distribution over n trials with probability of success p (`pbinom` is the R function which returns the cdf of a binomial distribution).

The traditional t-test of the hypotheses (2.8) is based on T, the sum of the observations.[3] The distribution of the statistic T depends on the population pdf $f(x)$. In particular, it is not distribution-free. The usual form of the test is the t-ratio

$$t = \frac{\overline{X}}{s/\sqrt{n}}, \tag{2.10}$$

where $\overline{X}$ and s are, respectively, the mean and standard deviation of the sample. If the population is normal, then t has a Student t-distribution with $n-1$ degrees of freedom. Let t_0 be the observed value of t. Then the p-value of the t-test for the hypotheses (2.8) is $P_{H_0}(t \geq t_0) = 1 - F_T(t_0; n-1)$, where $F_T(t; \nu)$ denotes the cdf of the Student t-distribution with ν degrees of freedom (`pt` is the R function which returns the cdf of a t-distribution). This is an exact p-value if the population is normal; otherwise, it is an approximation.

The difference between the t-test and the sign test is that the t-test statistic is a function of the distances of the sample items from 0 in addition to their signs. The **signed-rank Wilcoxon** test statistic, however, uses only the ranks of these distances. Let $R|X_i|$ denote the rank of $|X_i|$ among $|X_1|, \ldots, |X_n|$, from low to high.[4] Then the signed-rank Wilcoxon test statistic is

$$W = \sum_{i=1}^{n} \text{sign}(X_i) R|X_i|. \tag{2.11}$$

Unlike the t-test statistic, W is distribution-free under H_0. Its distribution, though, cannot be obtained in closed-form. There are iterated algorithms for its computation which are implemented in R (`psignrank`, `qsignrank`, etc.). Usually the statistic computed is the sum of the ranks of the positive items, W^+, which is

$$W^+ = \sum_{X_i > 0} R|X_i| = \frac{1}{2}W + \frac{n(n+1)}{4}. \tag{2.12}$$

The R function `psignrank` computes the cdf of W^+. Let w^+ be the observed value of W^+. Then, for the hypotheses (2.8), the p-value of the signed-rank Wilcoxon test is $P_{H_0}(W^+ \geq w^+) = 1 - F_{W^+}(w^+ - 1; n)$, where $F_{W^+}(x; n)$ denotes the cdf of the signed-rank Wilcoxon distribution for a sample of size n.

2.4.1 Estimation and Confidence Intervals

Each of these three tests has an associated estimate and confidence interval for the location effect θ of Model (2.7). They are based on inversions[5] of the

[3] For comparison purposes, T can be written as $T = \sum_{i=1}^{n} \text{sign}(X_i)|X_i|$.

[4] The ranking of tied observations is discussed in Section 2.5.

[5] See, for example, Chapter 1 of Hettmansperger and McKean (2011).

associated process. In this section we present the results and then show their computation in R. As in the last section, assume that we have modeled the sample $X_i, X_2, \ldots, X_n$ as the location model given in expression (2.7).

The confidence intervals discussed below, involve the order statistics of a sample. We denote the order statistics with the usual notation; that is, $X_{(1)}$ is the minimum of the sample, $X_{(2)}$ is the next smallest, $\ldots$, and $X_{(n)}$ is the maximum of the sample. Hence, $X_{(1)} < X_{(2)} < \cdots < X_{(n)}$. For example, if the sample results in $x_1 = 51, x_2 = 64, x_3 = 43$, then the ordered sample is given by $x_{(1)} = 43, x_{(2)} = 51, x_{(3)} = 64$.

The estimator of the location parameter θ associated with sign test is the sample median which we write as,

$$\hat{\theta} = \text{median}\{X_1, X_2, \ldots, X_n\}. \tag{2.13}$$

For $0 < \alpha < 1$, a corresponding confidence interval for θ of confidence $(1-\alpha)100\%$ is given by $(X_{(c_1+1)}, X_{(n-c_1)})$, where $X_{(i)}$ denotes the ith order statistic of the sample and c_1 is the $\alpha/2$ quantile of the binomial distribution, i.e., $F_B(c_1; n, 0.5) = \alpha/2$; see Section 1.3 of Hettmansperger and McKean (2011) for details. This confidence interval is distribution-free and, hence, has exact confidence $(1-\alpha)100\%$ for any random error distribution. Due to the discreteness of the binomial distribution, for each value of n, there is a limited number of values for α. Approximate interpolated confidence intervals for the median are presented in Section 1.10 of Hettmansperger and McKean (2011).

With regard to the t-test, the associated estimator of location is the sample mean $\overline{X}$. The usual confidence interval for θ is $(\overline{X} - t_{\alpha/2,n-1}[s/\sqrt{n}], \overline{X} + t_{\alpha/2,n-1}[s/\sqrt{n}])$, where $F_T(-t_{\alpha/2,n-1}; n-1) = \alpha/2$. This interval has the exact confidence of $(1-\alpha)100\%$ provided that the population is normal. If the population is not normal, then the confidence coefficient is approximately $(1-\alpha)100\%$. Note the t-procedures are not distribution-free.

For the signed-rank Wilcoxon, the estimator of location is the Hodges–Lehmann estimator which is given by

$$\hat{\theta}_W = \text{med}_{i \le j}\left\{\frac{X_i + X_j}{2}\right\}. \tag{2.14}$$

The pairwise averages, $A_{ij} = (X_i + X_j)/2$, $i \le j$, are called the Walsh averages of the sample. Let $A_{(1)} < \cdots < A_{(n(n+1)/2)}$ denote the ordered Walsh averages. Then a $(1-\alpha)100\%$ confidence interval for θ is

$$(A_{(c_2+1)}, A_{([n(n+1)/2]-c_2)}),$$

where c_2 is the $\alpha/2$ quantile of the signed-rank Wilcoxon distribution. Provided the random error pdf is symmetric, this is a distribution-free confidence interval which has exact confidence $(1-\alpha)100\%$. Note that the range of W^+ is the set $\{0, 1, \ldots n(n+1)/2\}$ which is of order n^2. So for moderate sample sizes the signed-rank Wilcoxon does not have the discreteness problems that the inference based on the sign test has; meaning α is close to the desired level.

2.4.2 Computation in R

The signed-rank Wilcoxon and t procedures can be computed by the intrinsic R functions `wilcox.test` and `t.test`, respectively. Suppose `x` is the R vector containing the sample items. Then for the two-sided signed-rank Wilcoxon test of $H_0 : \theta = 0$, the call is

```
wilcox.test(x,conf.int=TRUE).
```

This returns the value of the test statistic W^+, the p-value of the test, the Hodges–Lehmann estimate of θ and the distribution-free 95% confidence interval for θ. The `t.test` function has similar syntax. The default hypothesis is two-sided. For the one-sided hypothesis $H_A : \theta > 0$, use `alternative="greater"` as an argument. If we are interested in testing the null hypothesis $H_0 \ \theta = 5$, for example, use `mu=5` as an argument. For, say, a 90% confidence interval, use the argument `conf.level = .90`. For more information, see the help page (`help(wilcox.test)`). Although the sign procedure does not have an intrinsic R function, it is simple to code such a function. One such R-function is given in Exercise 2.11.16.

Example 2.4.1 (Nursery School Intervention). This dataset is drawn from a study discussed by Siegel (1956). It involves eight pairs of identical twins who are of nursery school age. In the study, for each pair, one is randomly selected to attend nursery school, while the other remains at home. At the end of the study period, all 16 children are given the same social awareness test. For each pair, the response of interest is the difference in the twins' scores, (Twin at School − Twin at Home). Let θ be the true median effect. As discussed in Remark 2.4.1, the random selection within a pair ensures that the response is symmetrically distributed under $H_0 : \theta = 0$. So the signed-rank Wilcoxon process is appropriate for this study. The following R session displays the results of the signed-rank Wilcoxon and the Student t-tests for one-sided tests of $H_0 : \theta = 0$ versus $H_A : \theta > 0$.

```
> school<-c(82,69,73,43,58,56,76,65)
> home<-c(63,42,74,37,51,43,80,62)
> response <- school - home
> wilcox.test(response,alternative="greater",conf.int=TRUE)

        Wilcoxon signed rank test


data:  response
V = 32, p-value = 0.02734
alternative hypothesis: true location is greater than 0
95 percent confidence interval:
   1 Inf
sample estimates:
(pseudo)median
          7.75
```

```
> t.test(response,alternative="greater",conf.int=TRUE)

        One Sample t-test

data:  response
t = 2.3791, df = 7, p-value = 0.02447
alternative hypothesis: true mean is greater than 0
95 percent confidence interval:
 1.781971      Inf
sample estimates:
mean of x
    8.75
```

Both procedures reject the null hypothesis at level 0.05. Note that the one-sided test option forces a one-sided confidence interval. To obtain a two-sided confidence interval, use the two-sided option. ■

Remark 2.4.1 (Randomly Paired Designs). The design used in the nursery school study is called a randomly paired design. For such a design, the experimental unit is a block of length two. In particular, in the nursery school study, the block was a set of identical twins. The factor of interest has two levels or there are two treatments. Within a block, the treatments are assigned at random, say, by a flip of a fair coin. Suppose H_0 is true; i.e., there is no treatment effect. If d is a response realization, then whether we observe d or $-d$ depends on whether the coin came up heads or tails. Hence, D and $-D$ have the same distribution; i.e., D is symmetrically distributed about 0. Thus the symmetry assumption for the signed-rank Wilcoxon test automatically holds. ■

As a last example, we present the results of a small simulation study.

Example 2.4.2. Which of the two tests, the signed-rank Wilcoxon or the t-test, is the more powerful? The answer depends on the distribution of the random errors. Discussions of the asymptotic power of these two tests can be found in the references cited at the beginning of this chapter. In this example, however, we compare the powers of these two tests empirically for a particular situation. Consider the situation where the random errors of Model (2.7) have a t-distribution with 2 degrees of freedom. Note that it suffices to use a standardized distribution such as this because the tests and their associated estimators are equivariant to location and scale changes. We are interested in the two-sided test of $H_0 : \theta = 0$ versus $H_A : \theta \neq 0$ at level $\alpha = 0.05$. The R code below obtains 10,000 samples from this situation. For each sample, it records the p-values of the two tests. Then the empirical power of a test is the proportion of times its p-values is less than or equal to 0.05.

```
n = 30; df = 2; nsims = 10000; mu = .5; collwil = rep(0,nsims)
collt = rep(0,nsims)
for(i in 1:nsims){
```

```
    x = rt(n,df) + mu
    wil = wilcox.test(x)
    collwil[i] = wil$p.value
    ttest = t.test(x)
    collt[i] = ttest$p.value
}
powwil = rep(0,nsims); powwil[collwil <= .05] = 1
powerwil = sum(powwil)/nsims
powt = rep(0,nsims); powt[collt <= .05] = 1
powert = sum(powt)/nsims
```

We ran this code for the three situations: $\theta = 0$ (null situation) and the two alternative situations with $\theta = 0.5$ and $\theta = 1$. The empirical powers of the tests are:

Test	$\theta = 0$	$\theta = 0.5$	$\theta = 1$
Wilcoxon	0.0503	0.4647	0.9203
t	0.0307	0.2919	0.6947

The empirical α level of the signed-rank Wilcoxon test is close to the nominal value of 0.05, which is not surprising because it is a distribution-free test. On the other hand, the t-test is somewhat conservative. In terms of power, the signed-rank Wilcoxon test is much more powerful than the t-test. So in this situation, the signed-rank Wilcoxon is the preferred test. ∎

2.4.3 Density Estimation Revisited

In Example 1.6.1, Section 1.6, we considered estimating the pdf $f_X(x)$. In this example, we assumed that X was normally distributed with mean 70 and standard deviation 4. We generated eight samples on the random variable X of selected sample sizes. We now replot the histograms of these samples in Figure 2.2. As before, the title of each panel displays the sample size, and each histogram is overlaid with a density estimate. In addition, for each sample, we have added the Hodges–Lehmann estimate and 95% confidence interval for the true median θ which is 70.

Suppose our inference is concerned with estimating the effect θ, using the Hodges–Lehmann estimator with the associated distribution-free confidence interval. For reference, the Hodges–Lehmann estimate and exact 95% confidence interval are provided in Table 2.1. The estimates are generally close to 70 and all but one of the 95% confidence intervals were successful in trapping the true center, $\theta = 70$. Further, notice how the confidence intervals become more precise (shorter length) as n increases. The figure shows the slowness of convergence of density estimation relative to the speed of convergence of the estimation of the effect θ.

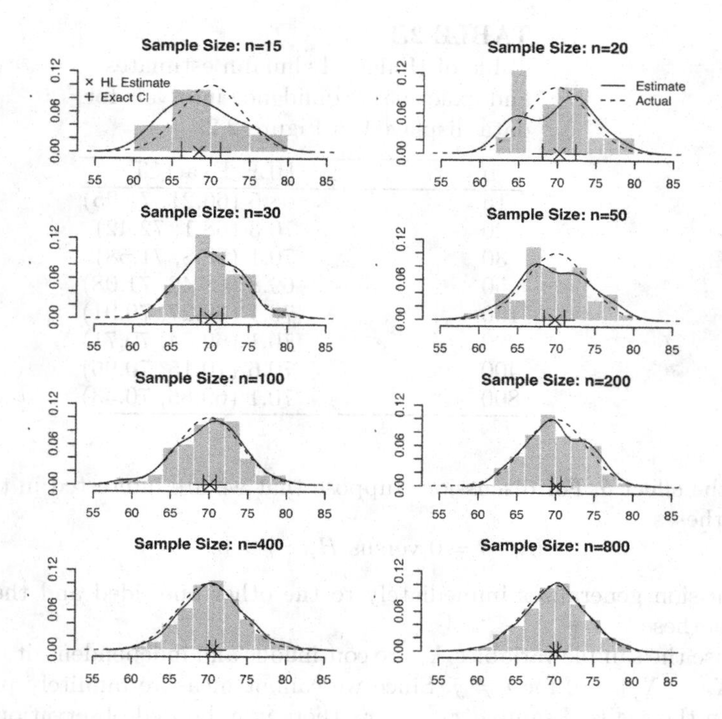

FIGURE 2.2
Samples of selected sizes from a $N(70, 4^2)$ distribution. Each panel displays a histogram as well as a density estimate (solid line) and actual pdf (dashed line). The Hodges–Lehmann estimate is indicated by a $\times$ symbol. The lower and upper bounds of the exact 95% confidence interval are given by the pair of $+$ symbols. The sample size is indicated in the title of each plot.

2.5 Adjustments for Ties

In this chapter we have presented two distribution-free tests for the median in a one-sample location model, namely the sign test and signed-rank Wilcoxon test. In this section, we discuss adjustments to these tests, if there are **ties** in the sample observations.

For notation, let $X_1, \ldots, X_n$ be a random sample on a random variable X. Assume that X_i follows the location model

$$X_i = \theta + e_i, \quad 1 \leq i \leq n,$$

where the random errors $e_1, \ldots, e_n$ are iid with continuous pdf $f(t)$ and cdf $F(t)$ and the true median of e_i is 0. For the signed-rank Wilcoxon discussion, additionally assume that the pdf $f(t)$ is symmetric about 0. Hence, the median

TABLE 2.1
Table of Hodges–Lehmann estimates and exact 95% confidence intervals for data displayed in Figure 2.2.

n	HLE_ExactCI
15	68.5 (66.21, 71.25)
20	70.3 (68.1, 72.42)
30	70.1 (68.8, 71.58)
50	69.8 (68.42, 71.08)
100	70.1 (69.27, 70.91)
200	70.1 (69.54, 70.71)
400	70.6 (70.15, 70.96)
800	70.1 (69.86, 70.43)

of X_i is the effect θ. For discussion, suppose that we are interested in testing the hypotheses

$$H_0 : \theta = 0 \text{ versus } H_A : \theta > 0. \tag{2.15}$$

Our discussion generalizes immediately to the other one-sided and the two-sided hypotheses.

Because the sample variables X_i are continuous and independent, it follows that $P(X_i = X_j) = 0$ for $i \neq j$. Since we cannot measure infinitely precise, though, in the realized sample $x_1, \ldots, x_n$ there may be tied observations. We begin with modifications to the sign test.

2.5.1 Sign Test

Recall that the sign test statistic is $S^+ = \sum_{i=1}^n I(X_i > 0)$. The *tie issue* that may occur with the sign test concerns the handling of sample items that are 0; as they are neither positive nor negative. In practice, usually these values are excluded from the analysis and the effective sample size is reduced accordingly. As an illustration, suppose the realized sample results in the values displayed in the first row of Table 2.2. Excluding the observation with value 0 leaves 11 observations. The value of the sign test statistic is $S^+ = 7$. Hence, the p-value is $P_{H_0}(S^+ \geq 7) =$ `1-pbinom(6,11,.5)` $= 0.2744$.

TABLE 2.2
The top row contains the realized values of the illustrative sample, while the bottom row contains the ranks of the absolute values.

x_i	2.2	4.1	−3.9	−2.6	7.7	−3.1	0.0	3.9	7.7	−3.9	3.3	7.3
$R\|x_i\|$	1.0	8.0	6.0	2.0	10.5	3.0	NA	6.0	10.5	6.0	4.0	9.0

For the corresponding estimation and confidence intervals for θ, though, we generally retain the 0 values; i.e., no changes are made to the sample. Recall that the estimate of θ corresponding to the sign test is the sample median (as illustrated in the following code segment).

```
> x <- c(2.2,4.1,-3.9,-2.6,7.7,-3.1,0.0,3.9,7.7,-3.9,3.3,7.3)
> m <- median(x)
> m

[1] 2.75
```

Hence, for the illustrative sample $\hat{\theta} = 2.75$.

2.5.2 Signed-Rank Wilcoxon

Recall that the signed-rank Wilcoxon test statistic for the hypotheses (2.15) is given by

$$W^+ = \sum_{X_i > 0} R|X_i|. \tag{2.16}$$

Two issues with ties may occur when using the Wilcoxon test with real data: sample items with values 0 and groups of sample items that have the same absolute value (tied). As with the sign test, sample items having the value 0 are excluded from the calculation, and the sample size is reduced accordingly. There are several ways to rank tied observations. We use the method of **average ranks** where the rank given to each item in a group of tied observations is the average value of the integer ranks which would have been assigned if ties had been ignored (or broken at random). For instance, consider the illustrative sample given in the top row of Table 2.2. The smallest sample item in absolute value is 2.2, and it is assigned a rank of 1. Next, the sample items with absolute values 2.6, 3.1, and 3.3 receive the respective ranks 2, 3, and 4. Next, each member of the group of the three sample items with absolute value of 3.9 is assigned the average of the ranks 5, 6, and 7; i.e., each receives rank 6. The next item in absolute value is 4.1 which receives rank 8. Continuing in this way, the rank assigned to each member of the group of two sample items with absolute value 7.7 is the average of the ranks 10 and 11. The bottom row of Table 2.2 contains the final assigned ranks. Notice that the sample item with value 0 has been excluded from the ranking. Adding the ranks of the positive observations, we get $W^+ = 49$. The average ranking method is the default tie procedure in the base R function `wilcox.test`

In the case of ties, the null distribution of the signed-rank Wilcoxon test statistic, W^+, is still distribution-free; however, its distribution additionally depends on the number of tied groups and how many items there are within such a group. The distribution can be tabulated, but this becomes an increasing laborious task as the sample size increases. Instead, many users prefer to use the asymptotic version of the test which depends on the null mean and variance of W^+. Even in the case of ties, the null mean of the signed-rank Wilcoxon test statistic remains at $n(n+1)/4$. The null variance, however, is now given by

$$\text{Var}(W^+) = \frac{n(n+1)(2n+1)}{24} - \frac{1}{2}\sum_{j=1}^{g} t_j(t_j-1)(t_j+1), \tag{2.17}$$

where g is the number of groups with observations tied with the same absolute value and t_j is the number of observations in group j.

For example, in the illustrative sample displayed in Table 2.2, there are two groups of observations tied with the same absolute value. Group 1 contains three observations tied with the absolute value 3.9 and group 2 contains two observations tied with the absolute value of 7.7. Since there are $n = 11$ nonzero items, the mean of W^+ is $11(12)/4 = 33$ and the variance of W^+ is

$$\text{Var}(W^+) = \frac{11(12)(23)}{24} - \frac{1}{2}[3(2)(4) + 2(1)(3)] = 126.5 - 15 = 111.5. \tag{2.18}$$

Since $W^+ = 49$, the asymptotic p-value, using the continuity correction, is

$$P_{H_0}(W^+ \geq 49) \doteq P\left[Z \geq \frac{48.5 - 33}{\sqrt{111.5}}\right] = P(Z \geq 1.4679) = 0.07107.$$

Notice in expression (2.18) that the second term is much smaller than the first. In general, this occurs unless there is an abundance of tied observations. So some practitioners use the uncorrected variance instead of the corrected variance. For the illustrative example, this leads to $Z = (48.5 - 33)/\sqrt{126.5} = 1.378118$ with the p-value of 0.08408, which is close to base R's p-value of 0.08356 computed by the function `wilcox.test`. Since the tie adjustment to the variance is positive, the value of the standardized statistic without the tie corrected variance is always smaller than the standardized statistic with the tie corrected variance. Hence, the error in using the standardized statistic without the tie corrected variance is in the conservative direction.

For the estimation and confidence intervals for θ using the signed-rank Wilcoxon process, no adjustments for ties are made; that is, the original sample is used. We also recommend retaining the 0 values. Base R, though, does exclude 0 values for the estimation and confidence intervals computed by its R function `wilcox.test`.

2.6 Confidence Interval Based Estimates of Standard Errors

In Section 2.4, we presented the Wilcoxon analysis and Hodges–Lehmann estimator for the location parameter θ in Model (2.7). We also presented distribution-free confidence intervals for θ for the Wilcoxon and sign procedures. In practice, though, **standard errors** (SE) of the estimates are often reported along with the estimator. A standard error for an estimator $\hat{\theta}$ of θ is an estimate of the square-root of the variance of $\hat{\theta}$. That is,

$$\text{se}(\hat{\theta}) = \sqrt{\text{var}(\hat{\theta})}.$$

In this section we discuss estimates of standard error for the Hodges–Lehmann estimate of location. Also mentioned are standard error estimates for the sample mean and sample median. Asymptotic theory for the sample median and the Hodges–Lehmann estimator can be found, for example, in Chapter 1 of Hettmansperger and McKean (2011).

Assume that the location model (2.7) holds with location parameter θ. Let $\hat{\theta}_W$ denote the Hodges–Lehmann estimator of location given in expression (2.14). Under the assumptions of the model, it follows that $\hat{\theta}_W$ has an approximate $N(\theta, \tau_W^2/n)$ distribution, where the parameter τ_W is a scale parameter that depends on the actual (unknown) error distribution; see Remark 2.6.1.

A confidence interval based estimator of the standard error is given by

$$\text{se}(\hat{\theta}) = \frac{\hat{\tau}_W}{\sqrt{n}} = \frac{A_{([n(n+1)/2]-c_2)} - A_{(c_2+1)}}{2z_{\alpha/2}} \tag{2.19}$$

where $A.$ are the Walsh averages (defined in Subsection 2.4.1). As noted below, we recommend a 95% confidence interval.

Other estimates of τ_W have been considered; see Remark 2.6.1.

Example 2.6.1 (Nursery School Intervention, Continued). The following lines of code computes the estimate of τ_W based on (2.19) for the twins data (Example 2.4.1) of the last section.

```
> school<-c(82,69,73,43,58,56,76,65)
> home<-c(63,42,74,37,51,43,80,62)
> response <- school - home
> n <- length(response) ; zcv <- qnorm(0.975)
> CI <- wilcox.test(response,conf.int=TRUE)$conf.int
> sqrt(n)*diff(CI)/(2*zcv)

[1] 14.07
```

■

Example 2.6.2. Consider the case where X_i has a $N(\theta, \sigma^2)$ distribution. Then it follows that $\tau_W^2 = (\pi/3)\sigma^2$. Presented in Figure 2.3 are comparison boxplots of confidence interval estimates of the scale parameter τ_W for a selected set of sample sizes. For each of the sample sizes, 20 simulated datasets were generated with $\theta = 0$ and $\sigma = 1$. The horizontal line on the plot is at $\tau_W = \sqrt{\pi/3}$. As the sample size increases, the estimate becomes more and more accurate; i.e., the estimate is a consistent estimate of τ_W.

■

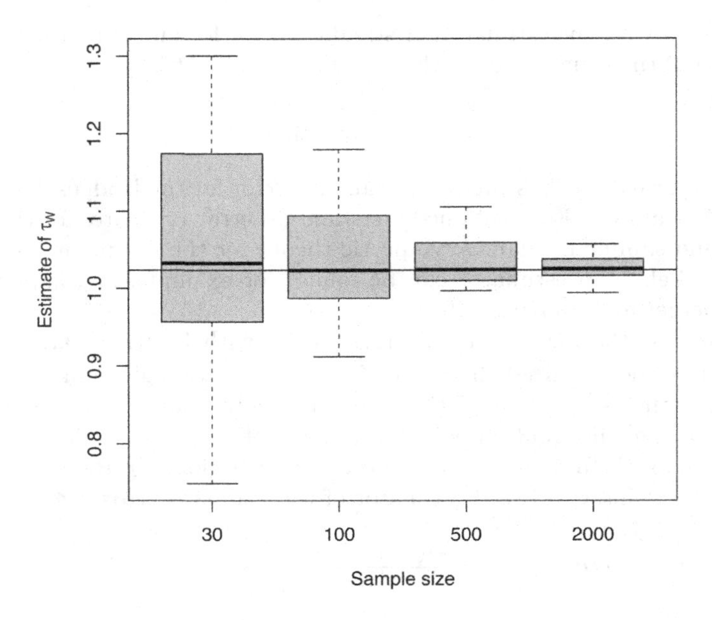

FIGURE 2.3
Estimates of τ_W for simulated datasets of select sizes from $N(0,1)$ distributions. For each sample size, 20 simulated datasets were generated.

Next consider the sample median $\hat{\theta}$. It can be shown that $\hat{\theta}$ has an approximate $N(\theta, \tau_S^2/n)$ distribution. The parameter τ_S is a scale parameter that depends on the actual (unknown) error distribution; see Remark 2.6.2. Hence, the SE of the median requires an estimate of τ_S. In a Monte Carlo study of estimators of τ_S, McKean and Schrader (1984) found that the estimator which is proportional to the length of a distribution-free confidence interval for the median performed well in terms of empirical validity, power, and efficiency when compared to other estimators of τ_S that were included in the study. Based on their study, they recommended 95% as the confidence percentage to be used in the estimator. From the last section, this interval is $(X_{(c+1)}, X_{(n-c)})$, where the asymptotic cut point is $c = (n/2) - 1.96(\sqrt{n/2}) - (1/2)$. The estimator of τ_S is then given by

$$\hat{\tau}_S = \sqrt{\frac{n}{n-1}} \frac{\sqrt{n}(X_{(n-c)} - X_{(c+1)})}{2(1.96)}. \tag{2.20}$$

The first factor on the right-side of expression (2.20) is a degree of freedom correction. So the SE of the sample median is $\hat{\tau}_S/\sqrt{n}$. The `Rfit` function `taustar` computes $\hat{\tau}_S$.

Suppose we use the sample mean $\bar{X}$ to estimate θ. Assuming that X_i has finite variance σ^2, the distribution of $\bar{X}$ is approximately $N(\theta, \sigma^2/n)$. If s denotes the sample standard deviation, then the standard error of $\bar{X}$ is $s/\sqrt{n}$.

An estimator $\hat{\theta}$ and its SE offers a brief but informative summary of the data. The estimator is an estimate of the center of the distribution, while the SE^2 is an estimate of the variability in the data. The SE is also an estimate of the precision of $\hat{\theta}$. Further, the quickly formed interval $\hat{\theta} \pm 2 * SE$ provides an approximation to a 95% confidence interval for θ.

Example 2.6.3 (Simulated Example — Fat-Finger Error). In this example a dataset is generated from a normal distribution with median $\theta = 50$ except there is a fat-finger error where the last observation is replaced by 10 times its original value. Estimates and standard errors are calculated for two methods: sample mean as well as the Wilcoxon–Hodges–Lehmann.

```
> set.seed(928364)
> n <- 15
> theta <- 50
> x <- round(rnorm(n,theta),1)
> x[n] <- 10*x[n]
> x

 [1]  50.0  48.1  48.0  50.7  50.3  50.3  50.5  50.1  50.1
[10]  49.5  50.5  51.0  49.4  51.2 490.0

> ## sample mean method ##
> xbar <- mean(x)
> se_xbar <- sd(x)/sqrt(n)
> ## Wilcoxon/Hodges--Lehmann method ##
> fit <- wilcox.test(x,conf.int=TRUE)
> xhl <- fit$est
> se_xhl <- (fit$conf.int[2]-fit$conf.int[1])/(2*1.96)
```

The results are given in the following output.

```
      est   se
mean 79.3 29.3
HL   50.2  0.3
```

Using the quick method discussed above, we can see that both the confidence intervals trap the true sample median of 50. However, the Hodges–Lehmann method is much more precise than the sample mean method in this case. ■

Remark 2.6.1 (τ_W). The parameter τ_W is the scale parameter defined by

$$\tau_W^2 = \frac{1}{12[\int f^2(x)\, dx]^2}. \tag{2.21}$$

As noted above, for a normal distribution with standard deviation σ, $\tau_W = \sqrt{\pi/3}\sigma$. Another estimator of τ_W is the **KSM** estimator developed by Koul et al. (1987) that is discussed in Chapter 4. The KSM estimator is a consistent estimator of τ_W for asymmetric pdf's $f(x)$ as well as symmetric pdfs. The `Rfit` function `gettau` computes $\hat{\tau}_W$. ∎

Remark 2.6.2 (τ_S)**.** The parameter τ_S is the scale parameter defined by

$$\tau_S^2 = \frac{1}{4f^2(\theta)}. \tag{2.22}$$

For example, if $f_X(x)$ is the $N(\theta, \sigma^2)$ pdf, then it follows that $\tau_S = \sqrt{\pi/2}\sigma$. ∎

Remark 2.6.3 (Asymmetrical Distributions)**.** For notation, let $X_1, X_2, \ldots, X_n$ be a random sample from a population with pdf $f_X(x)$ and cdf $F_X(x) = P[X \leq x]$. Suppose that the pdf $f_X(x)$ is asymmetrical. In this case, the three location estimators discussed in this section may all be estimating different parameters. The sample mean is a consistent estimator of the population mean $\mu = E(X)$; the sample median is a consistent estimator of the population median, $\theta = F_X^{-1}(1/2)$; and the Hodges–Lehmann estimator is a consistent estimator of the pseudo-median given by $\text{med}[(X_1 + X_2)/2]$ (see, for example, Høyland (1965)). So for asymmetrical distributions, comparisons among the three estimators are not recommended. Of the three, the population median is the most descriptive; for instance, half of the time an observation from the population is less than θ, and half of the time it is greater than θ. The mean, on the other hand, is a centroid which is not as descriptive. The confidence intervals and standard errors of the sample mean and median remain the same as in the symmetric case but those of the Hodges–Lehmann estimator differ from their symmetric form; (see, for example, Høyland (1965)). ∎

2.7 Asymptotic Tests

The signed-rank and sign tests presented in the two previous Sections (2.3 and 2.4) are based on the null distributions of the test statistics, and, hence, they are often referred to as **exact tests**. There are also asymptotic versions of these tests that are useful, especially the asymptotic Wilcoxon signed-rank test which we describe first. For notation, let $X_1, X_2, \ldots, X_n$ be a random sample which follows the location model, (2.7), with true median θ. For the discussion, we consider the one-sided hypotheses

$$H_0 : \theta = 0 \text{ versus } H_A : \theta > 0. \tag{2.23}$$

2.7.1 Signed-Rank Wilcoxon Test

Recall that the signed-rank Wilcoxon test statistic W^+ is the sum of the ranks of the absolute values of the positive sample items that is formerly defined in expression (2.11). Note that the average of the ranks, ($\{1, 2, \ldots n\}$), assigned is $(n+1)/2$. Under H_0, we expect that half of the sample items will be positive and half negative. So intuitively, the null mean of W^+ is $\mu_W = (n/2)(n+1)/2 = n(n+1)/4$. Although not intuitively obvious, the null variance of W^+ is $V_W = n(n+1)(2n+1)/24$. A Central Limit theorem shows that W^+ has an asymptotic normal distribution. Hence, the asymptotic α-level test is given by

$$\text{Reject } H_0 \text{ in favor of } H_A \text{ if } Z_w \geq z_\alpha, \tag{2.24}$$

where $Z_W = (W^+ - \mu_W)/\sqrt{V_W}$.

We generally test using p-values, so a few remarks are in order. The distribution of W^+ is discrete with range consisting of the integers from 0 to $n(n+1)/2$. So we recommend using the **continuity correction** when calculating the p-value of the asymptotic test. For a simple illustration, suppose that we are testing the hypotheses (2.23) and that the realized value of W^+ is w^+. Then the p-value of the exact signed-rank Wilcoxon test is $P_{H_0}(W^+ \geq w^+)$. When approximating this probability with the continuous normal distribution, though, this excludes the area over the interval $(w^+ - 0.5, w^+)$. Hence, the asymptotic p-value that we recommend is the continuity corrected p-value which is the probability on the right-side of the following expression:

$$P_{H_0}(W^+ \geq w^+) = P_{H_0}(W^+ \geq w^+ - 0.5) \doteq P\left(Z \geq \frac{w^+ - 0.5 - \mu_W}{\sqrt{V_W}}\right), \tag{2.25}$$

where the random variable Z has a standard normal distribution. If the alternative is $H_A : \theta < 0$ and w^+ is the realized value, then the asymptotic p-value is obtained from the standardizing of $P_{H_0}(W^+ \leq w^+ + 0.5)$. Generally, the p-values of the exact and asymptotic tests are close even for small sample sizes. The next example offers an illustration of the asymptotic test.

Example 2.7.1 (Asymptotic Test on Generated Data). Consider testing the hypotheses $H_0 : \theta = 50$ versus $H_A : \theta > 50$. The following code obtains a random sample from a normal distribution with $\theta = 53$. Hence, we know that H_A is true. The realized (rounded) sample is

```
> set.seed(2328283)
> x <- round(rnorm(10,53,5),digits=2); x

 [1] 53.09 58.14 55.15 57.95 59.03 51.91 47.23 55.34 51.92
[10] 48.00
```

In the next segment, we compute the signed-rank Wilcoxon test statistic.

```
> rs <- rank(abs(x-50))                # ranks of abs. values
> wplus <- sum(rs[x-50 > 0]); wplus    # signed-rank test stat.

[1] 48
```

The value of w^+ exceeds the null mean $\mu_W = 39$ of the signed-rank statistic by 9 units. The asymptotic p-value is

```
> n <- length(x);  muw <- n*(n+1)/4;  vw <- n*(n+1)*(2*n+1)/24
> pval <- 1 - pnorm((wplus - .5 - muw)/sqrt(vw));  pval

[1] 0.020746
```

Hence, H_0 is rejected at level $\alpha = 0.05$. For comparison, here is the p-value for the exact test:

```
> wilcox.test(x,mu=50,alternative="greater")$p.value

[1] 0.018555
```

The p-values of the asymptotic and exact tests are close, differing by about 0.0022. ∎

2.7.2 Sign Test

From Section 2.3, the sign test statistic for the hypotheses (2.23) is $S^+ = \#_i\{X_i > 0\}$. For the sign test, the distribution of X_i need not be symmetric about θ. The range of S^+ is the set $\{0, 1, \ldots, n\}$. Recall that under H_0, S^+ has a binomial distribution with n trials and probability of success 1/2. The null mean and variance of S^+ are, respectively, $n/2$ and $n/4$. As does the signed-rank Wilcoxon test statistic, S^+ has an asymptotic normal distribution. The discussion concerning asymptotic p-values for the signed-rank Wilcoxon test follows in the same way for the sign test as illustrated in the next example.

Example 2.7.2 (Asymptotic Test on Generated Data, Continued). Consider the dataset discussed in Example 2.7.1. Eight of the ten observations exceed 50; hence, the sign test statistic has value 8. The asymptotic and exact p-values for the sign test are:

```
> n<-length(x); mus<-n/2; vs <- n/4; splus <- sum(x-50>0); splus

[1] 8

> asypval <- 1 - pnorm((splus - .5 - mus)/sqrt(vs));
> exactpval <- 1 -pbinom(7,10,0.5)
> c(asypval,exactpval)

[1] 0.056923 0.054688
```

For this situation, the exact and asymptotic p-values are close. At $\alpha = 0.05$, the sign test fails to reject H_0. So for this example, at $\alpha = 0.05$, the signed-rank Wilcoxon and sign tests lead to different decisions. ∎

Note that the range of the sign test statistic S^+ is of order n, while the range of the signed-rank Wilcoxon test statistic W^+ is of order n^2. The range of W^+ is much more dense than that of the S^+. In general, while asymptotic p-values are useful for the signed-rank Wilcoxon for small sample sizes, we recommend that exact p-values be used for the sign test.

The difference in decisions for the sign and signed-rank Wilcoxon tests in the example is not surprising. The sample was drawn from a normal distribution. For normal distributions, the Student t-test is the most powerful test. In terms of efficiency, as discussed in Chapter 3, the signed-rank Wilcoxon test is approximately 96% as efficient as the Student t-test for normal populations. On the other hand, the sign test is only 64% as efficient as the Student t-test for normal populations. So for normal populations, the signed-rank Wilcoxon test is much more powerful than the sign test.

2.8 Bootstrap

As computers have become more powerful, the bootstrap, as well as resampling procedures in general, has gained widespread use. The bootstrap is a general tool that is used to measure the error in an estimate or the significance of a test of hypothesis. In this book we demonstrate the bootstrap for a variety of problems, though we still only scratch the surface; the reader interested in a thorough treatment is referred to Efron and Tibshirani (1993) or Davison and Hinkley (1997). In this section we illustrate estimation of confidence intervals and p-values for the one-sample and paired location problems.

To fix ideas, recall a histogram of the sample is often used to provide context of the distribution of the random variable (e.g., location, variability, shape). One way to think of the bootstrap is that it is a procedure to provide some context for the sampling distribution of a statistic. A bootstrap sample is simply a sample from the original sample taken with replacement. The idea is that if the sample is representative of the population, or more concretely, the histogram of the sample resembles the pdf of the random variable, then sampling from the sample is representative of sampling from the population. Doing so repeatedly will yield an estimate of the sampling distribution of the statistic.

R offers a number of capabilities for implementing the bootstrap. We begin with an example which illustrates the bootstrap computed by the R function `sample`. Using `sample` is useful for illustration; however, in practice, one will likely want to implement one of R's internal functions and so the library `boot` (Canty and Ripley 2013) is also discussed.

Example 2.8.1. To illustrate the use of the bootstrap, first generate a sample of size 25 from a normal distribution with mean 30 and standard deviation 5.

```
> x<-rnorm(25,30,5)
```

In the following code segment, we obtain 1000 bootstrap samples, and for each sample we calculate the sample mean. The resulting vector `xbar` contains the 1000 sample means. Figure 2.4 contains a histogram of the 1000 estimates. We have also included a plot of the true pdf of the sampling distribution of $\bar{X}$; i.e., a $N(30, 5^2/25)$.

```
> B<-1000 # number of bootstrap samples to obtain
> xbar<-rep(0,B)
> for( i in 1:B ) {
+ xbs<-sample(x,length(x),replace=TRUE)
+ xbar[i]<-mean(xbs)
+ }
```

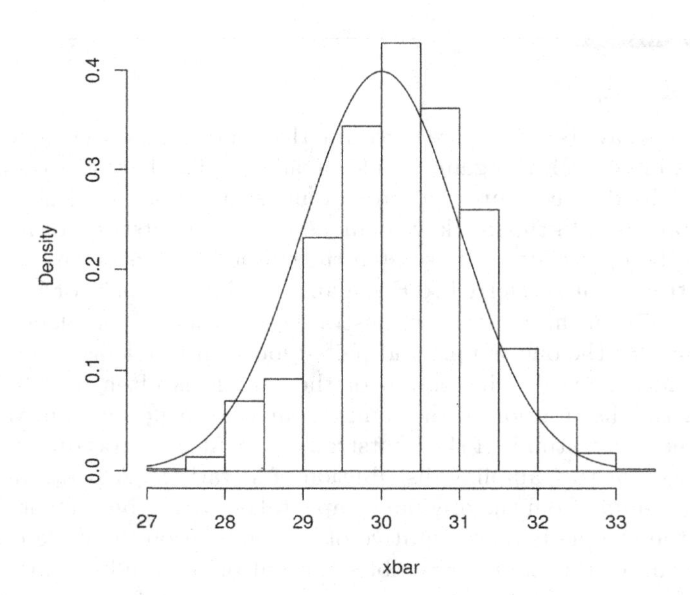

FIGURE 2.4
Histogram of 1000 bootstrap estimates of the sample mean based on a sample of size $n = 25$ from a $N(30, 5^2)$ distribution. The pdf of a $N(30, 1)$ is overlaid.

The standard deviation of the bootstrap sampling distribution may serve as an estimate of the standard error of the estimate.

```
> se.xbar<-sd(xbar)
> se.xbar

[1] 0.9568816
```

The estimated standard error may then be used for inference. For example, as we know the distribution of the sample mean is normally distributed, we can calculate an approximate 95% confidence interval using t-critical values as follows. We have included the usual t-interval for comparison.

```
> tcv<-qt(0.975,length(x)-1)
> mean(x)+c(-1,1)*tcv*se.xbar

[1] 28.12227 31.87773

> mean(x)+c(-1,1)*tcv*sd(x)/sqrt(length(x))

[1] 29.89236 30.10764
```

■

2.8.1 Percentile Bootstrap Confidence Intervals

In Example 2.8.1 we presented a simple confidence interval based on a bootstrap estimate of standard error. Such an estimate requires assumptions on the sampling distribution of the estimate; for example, that the sampling distribution is symmetric and that the use of t-critical values is appropriate. In this section we present an alternative, the percentile bootstrap confidence interval, which is free of such assumptions. Let $\hat{\theta}$ be any location estimator.

Let $\boldsymbol{x} = [x_1, \ldots, x_n]^T$ denote a vector of observations observed from the distribution F. Let $\hat{\theta}$ denote the estimate of θ based on this sample. Define the empirical cumulative distribution function of the sample by

$$\hat{F}_n(t) = \frac{1}{n}\sum_{i=1}^{n} I(x_i \leq t). \tag{2.26}$$

Then a **bootstrap sample** is a sample taken with replacement from $\hat{F}_n$; i.e., $x_1^*, \ldots x_n^*$ are iid $\hat{F}_n$. Denote this sample by $\boldsymbol{x}^* = [x_1^*, \ldots, x_n^*]^T$. Let $\hat{\theta} = T(\boldsymbol{x})$ be the estimate based on the original sample. Similarly $\hat{\theta}^* = T(\boldsymbol{x}^*)$ is the estimate based on the bootstrap sample. The bootstrap process is repeated a large number of times, say B, from which we obtain $\hat{\theta}_1^*, \ldots, \hat{\theta}_B^*$. Since the empirical distribution of the bootstrap estimates approximates the sampling distribution of $\hat{\theta}$, we may use it to obtain an estimate of certainty in our estimate $\hat{\theta}$. To obtain our confidence interval, we order the estimates $\hat{\theta}_{(1)}^* \leq \hat{\theta}_{(2)}^* \leq, \ldots, \leq \hat{\theta}_{(B)}^*$. Let $m = [\alpha/2 * B]$ then $(\hat{\theta}_{(m)}^*, \hat{\theta}_{(B-m)}^*)$ is an approximate $(1-\alpha) * 100\%$ confidence interval for θ. That is, the end points of the **percentile bootstrap**

confidence interval are the $\alpha/2$ and $1-\alpha/2$ percentiles of the empirical distribution of $\hat{\theta}_1^*, \ldots, \hat{\theta}_B^*$.

Returning again to our example, let $T(\boldsymbol{x}) = \frac{1}{n}\sum_{i=1}^n x_i$. The following code segment obtains a 95% bootstrap percentile confidence interval.

```
> quantile(xbar,probs=c(0.025,0.975),type=1)

    2.5%    97.5%
28.30090 32.14894

> m<-0.025*1000
> sort(xbar)[c(m,B-m)]

[1] 28.30090 32.14894
```

Next we illustrate the use of the `boot` library to obtain a similar result.

```
> bsxbar<-boot(x,function(x,indices) mean(x[indices]), B)
> boot.ci(bsxbar)

BOOTSTRAP CONFIDENCE INTERVAL CALCULATIONS
Based on 1000 bootstrap replicates

CALL :
boot.ci(boot.out = bsxbar)

Intervals :
Level      Normal              Basic
95%   (29.89, 30.11 )   (29.89, 30.11 )

Level     Percentile            BCa
95%   (29.89, 30.11 )   (29.89, 30.12 )
Calculations and Intervals on Original Scale

> quantile(bsxbar$t,probs=c(0.025,0.975),type=1)

    2.5%    97.5%
29.88688 30.11383
```

2.8.2 Bootstrap Tests of Hypotheses

In bootstrap testing, the resampling is conducted under conditions that ensure the null hypothesis, H_0, is true. This allows the formulation of a bootstrap p-value. In this section we illustrate the use of the bootstrap testing procedure by applying it to the paired and one-sample problems discussed in Sections 2.3–2.4.

Consider a randomly paired design (c.f., Remark 2.4.1) where pairs of subjects are sampled from a population and assigned, at random, to one of two treatments, say A and B. Let θ denote the median of the distribution of differences: $D = X_A - X_B$. To test if treatment A tends to yield higher responses than treatment B, one may test the hypothesis $H_0 : \theta = 0$ versus $H_A : \theta > 0$. Let $d_1, \ldots, d_n$ denote the paired differences: $x_{A,i} - x_{B,i}$ for $i = 1, \ldots, n$. Let T be the selected test statistic, where large values of T indicate rejection of H_0.

The bootstrap testing procedure for a paired sample problem is as follows. First, sample, with replacement, from the set of pairs; then treatment A is assigned at random to one of member of the pair, while the other member is assigned treatment B. Notice this preserves the correlation of the paired design. Further, notice this sampling method is equvient to sampling, with replacement, from the set $d_1, \ldots, d_n, -d_1, \ldots, -d_n$ a sample of size n. Hence, the null hypothesis is true for each bootstrap sample.

Let $T_1^*, \ldots, T_B^*$ be the test statistics based on the B bootstrap samples. These form an estimate of the null distribution of the test statistic T. The bootstrap p-value is then calculated as

$$p\text{-value} = \frac{\#\{T_i^* \geq T\}}{B}. \tag{2.27}$$

Example 2.8.2 (Nursery School Intervention Continued). There is more than one way to implement the bootstrap testing procedure for the paired problem, the following is one which utilizes the set of differences.

```
> d<-school-home
> dpm<-c(d,-d)
```

Then `dpm` contains all the $2n$ possible differences. Obtaining bootstrap samples from this vector ensures the null hypothesis is true. In the following, we first obtain $B = 5000$ bootstrap samples and store them in the matrix `dbs`.

```
> n<-length(d)
> B<-5000
> dbs<-matrix(sample(dpm,n*B,replace=TRUE),ncol=n)
```

In the next code segment, we use the `apply` function to obtain the Wilcoxon test statistic for each bootstrap sample. First, we define a function which will return the value of the test statistic which is used in our call to `apply`. The last line calculates the p-value based on (2.27).

```
> wilcox.teststat<-function(x) wilcox.test(x)$statistic
> bs.teststat<-apply(dbs,1,wilcox.teststat)
> mean(bs.teststat>=wilcox.teststat(d))

[1] 0.0238
```

Hence, the p-value $= 0.0238$ and is significant at the 5% level. ■

For the second problem, consider the one-sample location problem where the goal is to test the hypothesis

$$H_0 : \theta = \theta_0 \text{ versus } \theta > \theta_0.$$

Let $x_1, \ldots, x_n$ be a random sample from a population with location θ. Let $\hat{\theta} = T(\boldsymbol{x})$ be an estimate of θ.

To ensure the null hypothesis is true, that we are sampling from a distribution with location θ_0, we take our bootstrap samples from

$$x_1 - \hat{\theta} + \theta_0, \ldots, x_n - \hat{\theta} + \theta_0. \tag{2.28}$$

Denote the bootstrap estimates as $\hat{\theta}_1^*, \ldots, \hat{\theta}_B^*$. Then the bootstrap p-value is given by

$$\frac{\#\{\hat{\theta}_i^* \geq \hat{\theta}\}}{B}.$$

We illustrate this bootstrap test with the following example.

Example 2.8.3 (Bootstrap test for sample mean). In the following code segment, we first take a sample from a $N(1.5, 1)$ distribution. Then we illustrate a test of the hypothesis

$$H_0 : \theta = 1 \text{ versus } H_A : \theta > 1.$$

The sample size is $n = 25$.

```
> x<-rnorm(25,1.5,1)
> thetahat<-mean(x)
> x0<-x-thetahat+1 #theta0 is 1
> mean(x0) # notice H0 is true

[1] 1

> B<-5000
> xbar<-rep(0,B)
> for( i in 1:B ) {
+ xbs<-sample(x0,length(x),replace=TRUE)
+ xbar[i]<-mean(xbs)
+ }
> mean(xbar>=thetahat)

[1] 0.02
```

In this case, the p-value $= 0.02$ is significant at the 5% level. ■

2.9 Robustness*

In this section, we briefly discuss the robustness properties of the three estimators discussed so far in this chapter, namely, the mean, the median, and the Hodges–Lehmann estimate. Three of the main concepts in robustness are efficiency, influence, and breakdown. In Chapter 3, we touch on efficiency, while in this section we briefly explore the other two concepts for the three estimators.

The finite sample version of the influence function of an estimator is its sensitivity curve. It measures the change in an estimator when an outlier is added to the sample. More formally, let the vector $\boldsymbol{x}_n = (x_1, x_2, \ldots, x_n)^T$ denote a sample of size n. Let $\hat{\theta}_n = \hat{\theta}_n(\boldsymbol{x}_n)$ denote an estimator. Suppose we add a value x to the sample to form the new sample $\boldsymbol{x}_{n+1} = (x_1, x_2, \ldots, x_n, x)^T$ of size $n+1$. Then the **sensitivity curve** for the estimator is defined by

$$S(x; \hat{\theta}) = \frac{\hat{\theta}_{n+1} - \hat{\theta}_n}{1/(n+1)}. \tag{2.29}$$

The value $S(x; \hat{\theta})$ measures the rate of change of the estimator at the outlier x.

As an illustration consider the sample

$$\{1.85, 2.35, -3.85, -5.25, -0.15, 2.15, 0.15, -0.25, -0.55, 2.65\}.$$

The sample mean of this dataset is -0.09 while the median and Hodges–Lehmann estimates are both 0.0. The top panel of Figure 2.5 shows the sensitivity curves of the three estimators, the mean, median, and Hodges–Lehmann for this sample when x is in the interval $(-20, 20)$. Note that the sensitivity curve for the mean is unbounded; i.e., as the outlier becomes larger the rate of change of the mean becomes larger; i.e., the curve is unbounded. On the other hand, the median and the Hodges–Lehmann estimators change slightly as x changes sign, but their changes soon become constant no matter how large $|x|$ is. The sensitivity curves for the median and Hodges–Lehmann estimates are bounded.

While intuitive, a sensitivity curve depends on the sample items. Its theoretical analog is the influence function which measures rate of change of the functional of the estimator at the probability distribution, $F(t)$, of the random errors of the location model. We say an estimator is **robust** if its influence function is bounded. Down to a constant of proportionality and center, the influence functions of the mean, median, and Hodges–Lehmann estimators at a point x are respectively x, $\text{sign}(x)$, and $F(x) - 0.5$. Hence, the median and the Hodges–Lehmann estimators are robust, while the mean is not. The influence functions of the three estimators are displayed in the lower panel of Figure 2.5 for a normal probability model. For the median and Hodges–Lehmann estimators, they are smooth versions of their respective sensitivity curves.

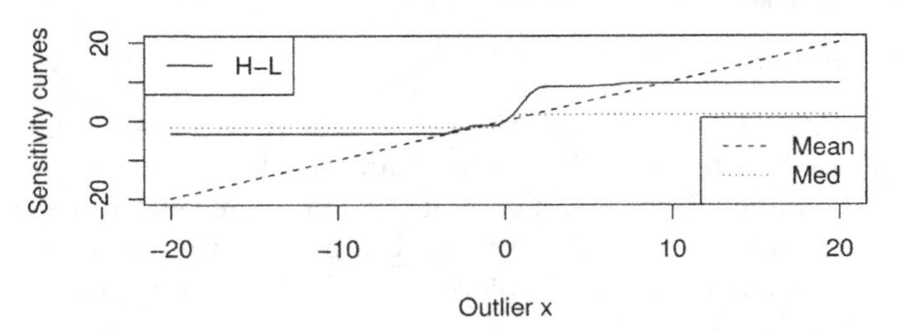

Influence Functions

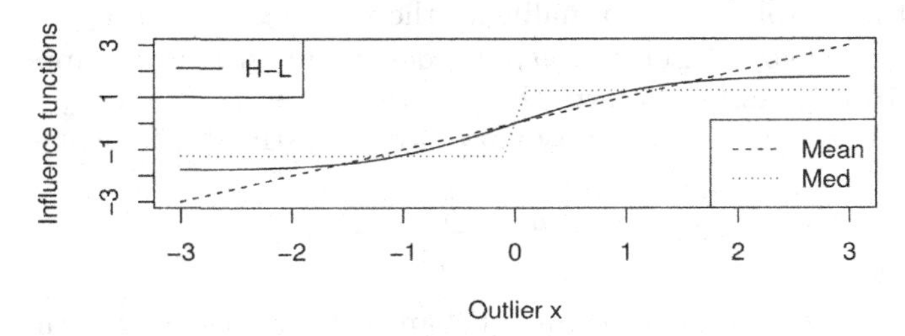

FIGURE 2.5
The top panel shows the sensitivity curves for the mean, median, and Hodges–Lehmann estimators for the sample given in the text. The bottom panel displays the influence function of the three estimators.

To briefly define the breakdown point of an estimator, consider again a sample $\boldsymbol{x}_n = (x_1, x_2, \ldots, x_n)^T$ from a location model with parameter θ. Let $\hat{\theta} = \hat{\theta}(\boldsymbol{x}_n)$ be an estimator of θ. Suppose we contaminate m points in the sample, so that the sample becomes

$$\boldsymbol{x}_n^* = (x_1^*, \ldots, x_m^*, x_{m+1}, \ldots, x_n)^T,$$

where $x_1^*, \ldots, x_m^*$ are the contaminated points. Think of the contaminated points as very large (nearly ∞) in absolute value. The smallest value of m so that the value of the estimator $\hat{\theta}(\boldsymbol{x}_n^*)$ becomes meaningless is the breaking point of the estimator and the ratio m/n is called the finite sample breakdown point of $\hat{\theta}$. If this ratio converges to a finite value, we call this value the **breakdown point** of the estimator. Notice for the sample mean that one point of contamination suffices to make the mean meaningless (as $x_1^* \to \infty$, $\overline{x} \to \infty$). Hence, the breakdown point of the mean is 0. On the other hand, the sample median can tolerate almost half of the data being contaminated. In the limit, its ratio converges to 0.50. So we say that median has 50% breakdown. The Hodges–Lehmann estimate has breakdown point 0.29; see, for instance, Chapter 1 of Hettmansperger and McKean (2011).

In summary, the sample median and the Hodges–Lehmann estimator are robust, with positive breakdown points. The mean is not robust and has breakdown point 0. Of the two, the sample median and the Hodges–Lehmann estimator, because of its higher breakdown, it would seem that the median is preferred. This, however, ignores efficiency between estimators which is discussed in Chapter 3. Efficiency generally reverses this preference.

2.10 Power and Sample Size Determination*

In this chapter, we have discussed rank-based tests of hypothesis and location parameter estimation for the one-sample location model. In this section, we discuss in more detail the behavior of rank-based tests under alternatives to the null hypothesis — we explore the power of these tests. This leads to sample size determination which plays a crucial role in the planning of one-sample designs, including randomized paired designs. Power was introduced, using simulations, in Section 1.4.2 and Section 2.3.1. Power was estimated for the Wilcoxon test and t-test in Example 2.4.2 under two alternatives in addition to the level of the test (i.e., the power under the null hypothesis).

In this section, we consider location models, in our discussion of power and sample size calculations. For further discussion, as well as sample size under more general conditions, see Hollander and Wolfe (1999).

2.10.1 Power of One-Sample Rank-Based Tests

Let $X_1, \ldots, X_n$ be a random sample that follows the one-sample location model

$$X_i = \theta + e_i, \quad 1 \leq i \leq n, \tag{2.30}$$

where $e_1, \ldots, e_n$ are iid with continuous pdf $f(t)$ which is symmetric about 0. Hence, the median of X_i is θ and X_i is symmetrically distributed about θ. Due to symmetry, θ is also the mean of X_i. For discussion, our hypotheses are

$$H_0 : \theta = 0 \text{ versus } H_A : \theta > 0. \tag{2.31}$$

There is no loss in generality in assuming that the null hypothesis is $\theta = 0$, for if the null hypothesis is $\theta = \theta_0$, for a specified θ_0, then the sample $X_i' = X_i - \theta_0$, $i = 1, \ldots, n$, is used instead of the original sample for the analysis.

As our test statistic, we consider the signed-rank Wilcoxon statistic discussed in Section 2.4. We also discuss the sign test later in Section 2.10.4. For convenience, the signed-rank Wilcoxon statistic is given by

$$W^+(0) = \sum_{X_i > 0} R|X_i|, \tag{2.32}$$

where $R|X_i|$ is the rank of $|X_i|$ among $|X_1|, \ldots, |X_n|$. Recall that the null mean and variance of $W^+(0)$ are given by $n(n+1)/4$ and $n(n+1)(2n+1)/24$, respectively. The decision rule for a level α test is:

$$\text{Reject } H_0 \text{ in favor of } H_A, \text{ if } W^+(0) \geq c_\alpha, \tag{2.33}$$

where the critical value c_α is defined by $P_0[W^+(0) \geq c_\alpha] = \alpha$.

The test statistic $W^+(0)$ is distribution-free under H_0, but it is not distribution-free under alternatives. To study the behavior of $W^+(0)$ under alternatives, consider its **power function** given by

$$\gamma_n(\theta) = P_\theta[W^+(0) \geq c_\alpha], \quad \theta \geq 0. \tag{2.34}$$

The power function at θ is the probability that the test rejects H_0 when θ is the true median. This power function $\gamma_n(\theta)$ has the properties:

1. $\gamma_n(0) = \alpha$;
2. $\gamma_n(\theta)$ is monotonically increasing for all $-\infty < \theta < \infty$;
3. The test is **resolving**: $\gamma_n(\theta) \to 1$ as $\theta \to \infty$;
4. The test is **consistent**: For all $\theta > 0$, $\gamma_n(\theta) \to 1$ as $n \to \infty$.

The power function satisfies the **Asymptotic Power Lemma** given by,

$$\gamma_n(\theta_n) \sim 1 - \Phi(z_\alpha - (\sqrt{n}\theta_n)/\tau_W), \tag{2.35}$$

where $\theta_n = \theta^*/\sqrt{n}$, for $\theta^* > 0$, $\Phi(t)$ is the cdf of a standard normal distribution, and τ_W is the scale parameter discussed in Section 2.6.[6]

In this section, we use the (KSM) estimate of τ_W that was proposed by Koul et al. (1987). This estimate is discussed in Chapter 3. This estimate is computed by the `Rfit` function `gettau`. The following generated dataset from a normal distribution with mean 50 and standard deviation 10 serves as an illustration:

```
> library(Rfit)
> set.seed(8266382)
> x <- round(rnorm(12,50,10)); x

 [1] 49 36 60 43 57 32 50 51 63 65 51 61

> sd(x); gettau(x)

[1] 10.484

[1] 13.787
```

[6]These properties are proved in Chapter 1 of Hettmansperger and McKean (2011).

Note, if the null hypothesis is $H_0 : \theta = \theta_0$, then expression (2.35) becomes

$$\gamma_n(\theta_n) \sim 1 - \Phi(z_\alpha - (\sqrt{n}\theta_n)/\tau_W), \tag{2.36}$$

where $\theta_n = (\theta^* - \theta_0)/\sqrt{n}$, for $\theta^* > \theta_0$.

For $\theta > 0$, the power function depends on the pdf $f(t)$ of the random errors as well as the signs and absolute ranks of the observations. So in general, the power function cannot be obtained in closed form. As seen in the next example, for a given situation, it is easy to estimate $\gamma_n(\theta)$ by simulation.

Example 2.10.1 (Simulation of Power at θ_A). Consider the situation where the distribution of X_i is $N(\theta, 8^2)$ and we are interested in testing $H_0 : \theta = 100$ versus $H_A : \theta > 100$ at level 0.05 using the sample size $n = 30$. The following code simulates the power of the signed-rank Wilcoxon to detect the alternative $\theta_A = 104$.

```
> set.seed(9856718)
> nsims<-10000 # number of simulations
> n<-30 # sample size
> alpha <- 0.05 # significance level
> theta0 <- 100
> thetaA <- 104
> sigma <- 8
> ic <- 0
> for(i in 1:nsims){
+   x <- rnorm(n,thetaA,sigma)
+   pval <- wilcox.test(x,mu=theta0,alternative="greater")$p.value
+   if(pval <= alpha){ic <- ic+1}
+ }
> ic/nsims # estimated power

[1] 0.8345
```

Using expression (2.36), we can approximate the power for this situation. Recall from Section 2.6 that the ratio of asymptotic variances between LS and the Wilcoxon processes at the normal distribution is $\sigma^2/\tau_W^2 = 3/\pi = 0.955$. For this situation $\sigma = 8$, leading to $\tau_W = 8\sqrt{\pi/3} = 8.19$. Hence, the approximation of power is:

```
> tauW <- 8*sqrt(pi/3)
> theta1 <- thetaA - theta0
> za <- qnorm(1-alpha)
> 1 - pnorm(za - sqrt(n)*theta1/tauW) # approximate power

[1] 0.84881
```

The estimated power and the approximate power are similar. ■

Remark 2.10.1 (Approximating Power). Suppose an experiment is being planned and a signed-rank Wilcoxon analysis is designated as the primary analysis. Further, suppose, a preliminary estimate of τ_W is available, say $\hat{\tau}_W = 8.24$. Expressions (2.35) and (2.36) can be used to approximate the power the experimental design will have to detect an alternative of interest. This involves substituting the estimate $\hat{\tau}_W$ for τ_W in these expressions, so it should be viewed as an approximation. As an illustration, suppose the hypotheses are $H_0 : \theta = 10$ versus $H_A : \theta > 10$ and the significance level is $\alpha = 0.05$.

Now, suppose $\theta_A = 12$ is an important alternative to detect. If a sample size of $n = 25$ is arbitrarily chosen, what is the approximate power to detect an alternative of $\theta = 12$? The following code chunk performs the calculation.

```
> theta1 <- 12-10; za <- qnorm(1-0.05)
> 1 - pnorm(za - sqrt(25)*theta1/8.24)

[1] 0.33314
```

Hence, the approximate power is only 33%, and the experiment would be considered *underpowered.* If $\theta_A = 12$ had been an important alternative to detect, then a larger sample size should have been used. Sample size determination is the topic of the next section. Generally, experiments are powered at a minimum of 80% for the primary analysis. ∎

2.10.2 Sample Size Determination

Consider the planning of the experiment to obtain a sample. Assume that the reference population and response variable, X, are designated. Suppose that we are interested in the hypotheses $H_0 : \theta = 0$ versus $H_A : \theta > 0$, where θ is the true median of X, and that we intend to use the Wilcoxon signed-rank analysis. Suppose we have selected α as the desired level of significance and we have determined a practical alternative θ_A that we want to detect, along with the probability β of detecting θ_A. Often in practice $\beta = 0.80$ is used. In order to conduct the experiment, we need to determine the sample size. With the information that we have, $(\gamma(\theta_A) \doteq \beta)$, it follows from expression (2.35) that the necessary (approximate) sample size is

$$n_W \doteq \left(\frac{z_\alpha - z_\beta}{\theta_A} \right)^2 \tau_W^2, \tag{2.37}$$

where τ_W is the Wilcoxon scale parameter discussed in Section 2.6, and z_β is the normal critical value with β as the area to the right of z_β.

In general, the unknown on the right-side of this equation is τ_W. Often we can make an educated guess of the value of τ_W. For instance, suppose we can assume that the distribution of X is close to a normal distribution. Then if we have an educated guess of σ, the standard deviation of X, we can use $\sqrt{\pi/3}\sigma$ as an educated guess of τ_W. As a second instance, suppose the experiment concerns a change in a process and we have data on the unchanged (standard) process. Let the R vector `xold` contain this standard data. Then the `Rfit` command `gettau(xold)` may serve as an adequate approximation to τ_W. Finally perhaps a small pilot study can be run to obtain an estimate of τ_W. We illustrate this discussion in Example 2.10.2 after a remark on other hypotheses.

Remark 2.10.2 (Other hypotheses). For the other one-sided hypotheses H_0 : $\theta = 0$ versus $H_A : \theta < 0$, the decision rule for a level α test is

$$\text{Reject } H_0 \text{ in favor of } H_A, \text{ if } W(0) \leq c'_\alpha, \tag{2.38}$$

where the critical value c'_α is defined by $P_0[W(0) \leq c'_\alpha] = \alpha$. Denote the power function for this test by $\gamma_2(\theta)$. It can be shown that, $\gamma_2(-\theta) = \gamma(\theta)$, for all θ, where $\gamma(\theta)$ is the power function given in expression (2.34) for the "greater than" alternative. Hence, for sample size determination at the alternative $-\theta_A$, expression (2.37) can be used with θ_A.

For the two-sided hypotheses $H_0 : \theta = 0$ versus $H_A : \theta \neq 0$, the decision rule for a level α test is

$$\text{Reject } H_0 \text{ in favor of } H_A, \text{ if } W(0) \leq c'_\alpha/2 \text{ or } W(0) \geq c_\alpha/2. \tag{2.39}$$

Let $\gamma_3(\theta)$ be the power function of this test. Then it follows that for all θ

$$\gamma_3(\theta) = \gamma_2(\theta) + \gamma(\theta). \tag{2.40}$$

Suppose we want to determine the sample size for the alternative $\theta_A > 0$. In general, for $\theta > 0$, $\gamma_2(\theta)$ is decreasing and $\gamma_2(\theta) < \alpha/2$, hence, the contribution of $\gamma_2(\theta)$ to the power is negligible. Hence, for sample size determination at the alternative θ_A, expression (2.37) can be used with θ_A and with α replaced by $\alpha/2$.

Note if the null hypothesis is $H_0 : \theta = \theta_0$, then expression (2.37) becomes

$$n_W \doteq \left(\frac{z_\alpha - z_\beta}{\theta_A - \theta_0}\right)^2 \tau_W^2. \tag{2.41}$$

As with the approximate power, it is the increment from the null that plays a role in sample size determination. ∎

Example 2.10.2 (Determination of Sample Size for the One-Sample Problem). Consider a random variable X that is approximately normally distributed with median θ. Suppose that we are interested in testing $H_0 : \theta = 50$ versus $H_A : \theta > 50$ at level $\alpha = 0.05$. In particular, we want to determine the sample size so that the power to detect the alternative $\theta = 54$ is 80%. It is thought that the range of X is about 35 units. Interpreting the range as the interval $\theta \pm 2\sigma$ which traps about 95% of the observations, $\sigma \doteq 35/4$. Recall for a normal distribution that $\tau_W = \sqrt{\pi/3}\sigma$; hence, in this situation, $\tau_W \doteq 8.95$. Hence, using expression (2.41), the desired sample size is

$$n_W \doteq \left(\frac{1.645 - \text{qnorm}(.20)}{54 - 50} \right)^2 8.95^2 = 30.96.$$

So, we would use $n_W = 31$ as the sample size. ■

Expressions similar to (2.37) and (2.41) hold for the LS analysis based on the sample mean. The only difference is that τ_W is replaced by σ, the standard deviation of X.

2.10.3 Randomized Paired Design

The **randomized paired design** (RPD) is frequently used in practice. Suppose that we have two treatments T_1 and T_2 and we want to decide which, if either, is more effective in terms of a response of interest. In the RPD, the experimental unit is a block of length 2. Each block consists of a pair of subjects, (each member of the pair may be the same subject), whose responses are generally positively, highly correlated. Recall the nursery school data discussed in Example 2.4.1 in which a block was a set of identical twins. For another example, when it is possible, T_1 and T_2 are applied to the same subject with a *washout* period between the applications. In this case each member of the block is the same subject.

For the RPD, n blocks are randomly selected, and within a block T_1 is randomly assigned to one member of the block and then T_2 is assigned to the other member of the block. For the *ith* block, let X_i denote the response for the member of the block that received treatment T_1 and let Y_i denote the response for the member of the block that received treatment T_2. Then the sample of interest consists of the paired differences $D_i = Y_i - X_i$, for $i = 1, \ldots, n$. Let θ be the true median of D_i. Suppose the hypotheses of interest are $H_0 : \theta = 0$ versus $H_A : \theta > 0$; i.e., the responses under treatment T_2 are generally larger than those under treatment T_1. Recall that within a pair treatments are assigned at random. Then the null distribution of D_i is symmetric about 0. So, an appropriate test statistic, is the signed-rank Wilcoxon statistic given by

$$W^+(0) = \sum_{D_i > 0} R|D_i|, \tag{2.42}$$

where $R|D_i|$ is the rank of $|D_i|$ among $|D_1|, \ldots, |D_n|$. The decision rule for a level α test is:

$$\text{Reject } H_0 \text{ in favor of } H_A, \text{ if } W^+(0) \geq c_\alpha, \tag{2.43}$$

where the critical value c_α is defined by $P_0[W^+(0) \geq c_\alpha] = \alpha$.

To conduct the paired design experiment, we need to determine the number of pairs (blocks) to select. Suppose θ_A is a practical alternative of interest, β

is selected as the power of the test to detect θ_A, the level of test is set at α and the signed-rank Wilcoxon analysis is selected as the method of analysis. Then the number of pairs to select is given by expression (2.37). Of course, for the scale parameter τ_W, either a prior estimate or an educated guess of it is required. Note in this situation that τ_W is the scale parameter for the paired differences D_i. For educated guessing, suppose Y_i and X_i have the common variance σ^2. Then

$$\mathrm{Var}(D_i) = \mathrm{Var}(Y_i - X_i) = 2\sigma^2 - 2\sigma^2\rho = 2\sigma^2(1-\rho), \tag{2.44}$$

where ρ is the correlation coefficient between Y_i and X_i. So besides a guess of the standard deviation, a guess of the correlation coefficient is needed.

Example 2.10.3 (Sample Size Determination for a RPD). Suppose a randomized paired design on two treatments T_1 and T_2 is in the planning stage. There is interest in testing at level 0.05 the hypotheses $H_0 : \theta = 0$ versus $H_A : \theta > 0$, where θ is the median of the paired differences. Further, the specific alternative of interest is $\theta_A = 5$ and the number of pairs n should be sufficiently large so that the power to detect θ_A is 0.90. In order to select the number of pairs (blocks of length 2), the following data from a small pilot study is available.

```
> t1 <- c(80.2,30.0,29.0,30.1,25.2,21.9,39.1)
> t2 <- c(53.6,35.7,37.9,34.6,21.3,30.5,42.5)
```

Based on the pilot study, the following R segment obtains an estimate of τ_W.

```
> d <- t2-t1; d

[1] -26.6   5.7   8.9   4.5  -3.9   8.6   3.4

> tauhatW <- gettau(d,0)
> tauhatW

[1] 9.3633
```

Using $\hat{\tau}_W$ in expression (2.37), the next R segment computes the necessary sample size for the RPD.

```
> nw <- ceiling( ((qnorm(.95) - qnorm(.10))/5)^2*tauhatW^2)
> nw

[1] 31
```

Hence, based on the estimate of τ_W, 31 pairs of subjects should be selected for the design. For comparison, if the standard deviation of the vector **d** is used as a guess of τ_W, the estimated sample size is:

```
> sigmahat <- sd(d)
> sigmahat
```

```
[1] 12.518

> nt <- ceiling( ((qnorm(.95) - qnorm(.10))/5)^2*sigmahat^2 )
> nt

[1] 54
```

Using the standard deviation, 54 pairs are required. ∎

2.10.4 Sign Test

The other robust test for θ discussed in this chapter is the sign test. Using the same notation as in Section 2.10.1, let $X_1, \ldots, X_n$ be a random sample that follows the one-sample location model given in expression (2.30). Consider the hypotheses (2.31); i.e., $H_0 : \theta = 0$ versus $H_A : \theta > 0$. Recall, that the sign test statistic is the number of positive X_i's, i.e.,

$$S^+(0) = \#_i\{X_i > 0\}. \tag{2.45}$$

Under H_0, $S^+(0)$ has the binomial distribution with n trials and the probability of success at $p = 1/2$, $b(n, 1/2)$. The level α decision rule is

$$\text{Reject } H_0 \text{ in favor of } H_A \text{ if } S^+(0) \geq c_{S,\alpha}, \tag{2.46}$$

where $c_{S,\alpha}$ is the upper critical value of the $b(n, 1/2)$ distribution, i.e., $\alpha = P_0[S^+(0) \geq c_{S,\alpha}]$.

Let $\gamma_{S,n}(\theta) = P_\theta[S^+(0) \geq c_{S,\alpha}]$ denote the power function of the sign test. The power function $\gamma_{S,n}(\theta)$ has the same properties as that of the power function (2.34) of the signed-rank Wilcoxon test. So it is monotonically increasing, resolving, and consistent. It has an asymptotic power lemma similar to that of the signed-rank Wilcoxon test except τ_W is replaced by τ_S where τ_S is the scale parameter $\tau_S = [2f(0)]^{-1}$ where $f(t)$ is the pdf of the random errors, (2.30); see the discussion in Section 2.6. As discussed in Section 2.6, $\tau_S/\sqrt{n}$ is the asymptotic standard deviation of the sample median.

Suppose we want to determine the sample size n_S for a level α sign test to detect the alternative θ_A with power β. By the above discussion, n_S satisfies:

$$n_S \doteq \left(\frac{z_\alpha - z_\beta}{\theta_A}\right)^2 \tau_S^2. \tag{2.47}$$

As with the signed-rank Wilcoxon, the ratio of the sample size for a LS t-test to n_S is the ratio of the asymptotic variances between the sample mean and the sample median Recall at the normal distribution, $\sigma^2/\tau_S^2 = 2/\pi = 0.634$. So, if the random errors are approximately normal, then $n_s \doteq n_{LS}/0.634 = 1.57 n_{LS}$. Thus, for approximately normal errors, an experimenter would be ill-advised to use n_S for the sample size. In such cases, we would recommend the signed-rank analysis. Recall, that $n_W \doteq (\pi/3) n_{LS} = 1.05 n_{LS}$.

2.10.5 Sample Size Determination for Estimation

Consider the one-sample location model of expression (2.30). Suppose we are interested in the estimation of the median θ. Specifically, we want to determine the sample size so that we are fairly confident that the estimation is within a preset bound. We consider the Hodges–Lehmann estimator and the sample median.

Recall that the Hodges–Lehmann estimator of θ is the median of the Walsh averages; i.e.,

$$\hat{\theta}_{HL} = \text{median}\left\{\frac{X_i + X_j}{2} : 1 \leq i \leq j \leq n\right\} \tag{2.48}$$

In Section 2.6, we discussed the associated $(1-\alpha)100\%$ symmetric confidence interval for θ that is given by

$$\hat{\theta}_{HL} \pm t_{\alpha/2,n-1}\frac{\hat{\tau}_W}{\sqrt{n}}. \tag{2.49}$$

The term following the $\pm$ sign is the error part of the confidence interval which is an estimate of $t_{\alpha/2,n-1}\tau_W/\sqrt{n}$.

The formulation of sample size determination is as follows. Suppose we want to determine the sample size n so that the estimation of θ is within B units with confidence $(1-\alpha)100\%$; that is, determine n so that

$$z_{\alpha/2}\frac{\tau_W}{\sqrt{n}} \leq B.$$

Note that we replaced the t-critical value by the corresponding z-critical that does not depend on n. Solving for n, we have

$$n \geq \left(\frac{z_{\alpha/2}\tau_W}{B}\right)^2. \tag{2.50}$$

Generally, the unknown on the right-side is τ_W. The discussion concerning the estimation or educated guess of τ_W for the signed-rank Wilcoxon test in Section 2.10.2 pertains to this situation, also.

Example 2.10.4 (Sample Size Determination for a RPD, Continued). Consider the randomized paired design concerning two treatments T_1 and T_2 discussed in Example 2.10.3. Let θ be the median of the paired differences. Suppose we want to estimate θ within 3.5 units and with confidence 95%. The results of the small pilot study discussed in Example 2.10.3 provided 10.35 as an estimate of τ_W. Substituting this in expression (2.50), we have

```
> nhl <- ceiling( ((qnorm(.975)*10.35)/3.5)^2)
> nhl

[1] 34
```

Hence, the recommended number of pairs is 34. ∎

Sample size determination for the sample median is the same as for the Hodges–Lehmann estimate except that in the expression (2.50) the scale parameter τ_S replaces τ_W.

As we illustrate throughout the book, the most efficient estimate — the one which requires the fewest samples — depends on the underlying error distribution, f. These concepts are discussed more, in the context of the two-sample problem, in Section 3.7.

2.11 Exercises

2.11.1. Using the `baseball_player1000` data, estimate the proportion of baseball players who are pitchers. Provide a point estimate as well as a 95% confidence interval using Wald's method.

2.11.2. Plot a histogram of estimated proportions from simulated binomial experiements with $n = 30$ and $p = 0.33$. Use a simulation size of 1000. Overlay a normal density with mean p and variance $p(1-p)/n$.

2.11.3. Repeat the previous exercise with a simulation size of 5000.

2.11.4. In the case of rare events, a sample from a binomial process may result in zero events. The Wald interval for p is undefined in such cases. One solution used in practice is to use the **rule of three** where the interval is defined as $[0, 3/n)$.

A Wald interval for p may have poor accuracy, and one way to improve its accuracy is to add $0.5z^2_{\alpha/2}$ S and the same number of F in the calculation of $\hat{p}$. Using this estimate of $\hat{p}$ in the calculation of the Wald interval, conduct a simulation with sample sizes $n = 30, 100, 250$ and probability of success $p = 0.05, 0.1$ and calculate the empirical confidence level for each combination. Also, calculate the Wald interval without the adjustment, using the rule of three in the case of zero events.

2.11.5. Conduct a simulation to compare the three methods for calculating a confidence interval for p. Estimate the empirical confidence level for each test for each combination of the following:

- sample sizes of $n = 30, 150, 750$
- probability of success $p = 0.05, 0.15, 0.45$

Use a large simulation size (say, at least 10000) to achieve a high degree of accuracy in the estimates of the empirical confidence levels. Discuss.

2.11.6. *Sample size calculation for a confidence interval for a proportion.* Suppose the primary analysis of a study is to estimate a proportion using a Wald interval with maximum length L. Without knowledge of the true proportion p, one can assume $p = 0.5$ and obtain a *conservative* value for n. Using the code below as a basis, fill in the values of a vector `nvec` corresponding to the values of `Lvec`, which represent several possible maximum lengths. Assume a confidence level of 0.95. Since sample sizes should be rounded up, use the `ceiling` function.

```
> Lvec <- seq(0.05,0.2,by=0.05)
```

Hint: using the formula for a Wald interval for p, set the difference between the upper and lower bounds of the confidence interval equal to L and solve for n.

2.11.7. Repeat the previous exercise with $p = 0.75$.

2.11.8. In the following code chunk, a function is defined which uses a `switch` statement to determine the p-value to calculate. Rewrite the simulation code in Subsection 2.3.1 to use the below function so that in place of a list of distributions `rdists` a vector of `types` is used. *Hint*: the `apply` will pass additional arguments to the function `FUN` via the `...` argument.

```
> get_pval <- function(x,type='t',...) {
+ # x - sample of size n from a numeric distribution
+ # type - either 't' or 's' for t-test or sign test
+
+ # check input
+ type.options <- c('t','s')
+ if( !(type %in% type.opts) ) stop("invalid argument for type")
+
+
+ switch(type,
+        t=ttest_pval_g,
+        s=stest_pval_g
+     )
+
+ }
```

2.11.9. Create a plot similar to Figure 2.1 for data from a $t(3)$ distribution and a $CNx(0.001, 100)$ distribution.

2.11.10. Verify, via simulation, the level of the `wilcox.test` when sampling from a standard normal distribution. Use $n = 30$ and levels of $\alpha = 0.1, 0.05, 0.01$. Based on the resulting estimate of α, the empirical level, obtain a 95% confidence interval for α.

2.11.11. Redo Exercise 2.11.10 for a t-distribution using 1,2,3,5,10 degrees of freedom.

2.11.12. Redo Example 2.8.1 without a `for` loop and using the `apply` function.

2.11.13. Redo Example 2.4.2 without a `for` loop and using the `apply` function.

2.11.14. Suppose in a poll of 500 registered voters, 269 responded that they would vote for candidate P. Obtain a 90% percentile bootstrap confidence interval for the true proportion of registered voters who plan to vote for P.

2.11.15. For Example 2.4.1 obtain a 90% two-sided confidence interval for the treatment effect.

2.11.16. Write an R function which computes the sign analysis. For example, the following commands compute the statistic S^+, assuming that the sample is in the vector `x`.

```
xt <- x[x!=0];  nt <- length(xt);  ind <- rep(0,nt);
ind[xt > 0] <-1;  splus <- sum(ind)
```

2.11.17. Calculate the sign test for the nursery school example, Example 2.4.1. Show that the p-value for the one-sided sign test is 0.1445.

2.11.18. The data for the nursery school study were drawn from page 79 of Siegel (1956). In the data table, there is an obvious typographical error. In the 8th set of twins, the score for the twin that stayed at home is typed as 82 when it should be 62. Rerun the signed-rank Wilcoxon and t-analyses using the typographical error value of 82.

2.11.19. Perform the simulation study of Example 2.4.2 when the population has a $CN(16, 0.25)$ distribution. For the alternatives, select values of θ so the spread in empirical powers of the signed-rank Wilcoxon test ranges from approximately 0.05 to 0.90.

2.11.20. The ratio of the expected squared lengths of confidence intervals is a measure of efficiency between two estimators. Based on a simulation of size 10,000, estimate this ratio between the Hodges–Lehmann and the sample mean for $n = 30$ when the population has a standard normal distribution. Use 95% confidence intervals. Repeat the study when the population has a t-distribution with 2 degrees of freedom.

2.11.21. Suppose the cure rate for the standard treatment of a disease is 0.60. A new drug has been developed for the disease, and it is thought that the cure rate for patients using it will exceed 0.60. In a small clinical trial, 48 patients having the disease were treated with the new drug and 34 were cured.

(a) Let p be the probability that a patient having the disease is cured by the new drug. Write the hypotheses of interest in terms of p.

(b) Determine the p-value for the clinical study. What is the decision for a α-level of 0.05?

2.11.22. Let p be the probability of success. Suppose it is of interest to test

$$H_0: \; p = 0.30 \text{ versus } H_A: \; p < 0.30.$$

Let S be the number of successes out of 75 trials. Suppose we reject H_0, if $S \leq 16$.

(a) Determine the significance level of the test.

(b) Determine the power of the test if the true p is 0.25.

(c) Determine the power function for the test for the sequence of the probabilities of success in the set $\{0.02, 0.03, \ldots, 0.35\}$. Then obtain a plot of the power curve.

2.11.23. Conduct a Monte Carlo simulation to approximate the power of the test discussed in Example 2.8.3 when the true $\theta = 1.5$.

2.11.24. Let $0 < \alpha < 1$. Suppose I_1 and I_2 are respective confidence intervals for two parameters θ_1 and θ_2 both with confidence coefficient $1 - (\alpha/2)$; that is,

$$P_{\theta_i}[\theta_i \in I_i] = 1 - \frac{\alpha}{2}, \quad i = 1, 2.$$

Show that the simultaneous confidence for both intervals is at least $1 - \alpha$, i.e.,

$$P_{\theta_1,\theta_2}[\{\theta_1 \in I_1\} \cap \{\theta_2 \in I_2\}] \geq 1 - \alpha.$$

Hint: Use the method of complements and Boole's inequality, $P[A \cup B] \leq P(a) + P(B)$. Extend the argument to m intervals each with confidence coefficient $1 - (\alpha/2)$ to obtain a set of m simultaneous Bonferroni confidence intervals.

2.11.25. The code below generates a sample of size n from the logistic distribution and puts it in the variable `f1`:

```
y<-runif(n);f1<-log(y/(1-y))+5
```

(a) Using the code, generate samples of size 5, 10, 20, 30, 100, and 200. Obtain a 3×2 page of the density plots of your samples. Comment on the shape of each plot. Do your comments differ as the sample size increases?

(b) Obtain the Hodges–Lehmann estimator of the location parameter θ for each sample. Obtain the associated distribution-free confidence interval for θ too. Table these with column headings: n, $\hat{\theta}_W$, confidence interval, and the length of the confidence interval. Comment on the table. What is happening to the last column as n increases?

(c) If the true θ is 5, how many of your confidence intervals are successful?

2.11.26. In Section 2.4.3 a simulated sample of selected sizes was generated from a common distribution. A confidence interval was calculated for the population median for each of the samples. What proportion of the confidence intervals did not capture the true value? Is the result surprising? Why or why not?

2.11.27. Consider the following sorted sample of observations on a continuous random variable X.

```
> sort(x)

 [1]  62  76  79  83  84  85  94  99  99  99 100 103 108 116
[15] 119 119 120
```

Suppose the hypotheses of interest are

$$H_0 : \theta = 85 \text{ versus } H_A : \theta > 85,$$

where θ is the true median of X.

(a) Using the tie adjustments of Section 2.5, compute the signed-rank test statistic W^+.

(b) Compute the null mean and variance (adjusted for ties).

(c) Using the normal approximation, obtain the p-value and conclude in terms of the data at $\alpha = 0.05$.

(d) Compare part (c) with the p-value computed by the R function `wilcox.test`.

2.11.28. For Exercise 2.11.27, test the hypotheses using the sign test. Compare the results with those of the signed-rank Wilcoxon test.

2.11.29. For Exercise 2.11.27, obtain the Hodges–Lehmann estimate of θ along with the associated 95% confidence interval for θ. Repeat using the sample median as the estimate along with its associated 95% confidence interval for θ. Compare the results in terms of precision.

2.11.30. For the sample in Exercise 2.11.27,

(a) Obtain the p-value using the one-sample t-test.

(b) Suppose, in recording the measurements, a simple typographical error occurred resulting in the data:

```
> sort(x)

 [1]   6  76  79  83  84  85  94  99  99  99 100 103 108
[14] 116 119 119 120
```

Obtain the p-values of the signed-rank and one-sample t-tests for these data. Compare the results with those obtained on data of Exercise 2.11.27. Which of the tests appears to be robust?

2.11.31. For the twin data of Example 2.6.1, obtain the estimates and standard errors for the sample mean, sample median, and Hodges–Lehmann methods. Discuss.

2.11.32. For the Hodges–Lehmann procedure, the following code runs a simulation of the SE confidence interval versus the distribution-free confidence interval. It collects the confidence intervals in the R matrix `coll`. Note that the error distribution is $N(0, 1)$ so the true θ is 0. Also, the sample size is set at 30, the number of simulations is set at 100, and the true confidence level is 0.95. All these quantities are easy to change.

```
> library(Rfit)
> n <-30; nsims <- 100; tc <- qt(.975,n-1)
> inse <- 0; innp <- 0
> for(i in 1:nsims){
+   x<-rnorm(n)
+   fit <- wilcox.test(x,conf.int=T);
+   hlest <- fit$est
+   tauhat <- gettau(x)
+   lbse <- hlest - tc*(tauhat/sqrt(n))
+   ubse <- hlest + tc*(tauhat/sqrt(n))
+   lbnp <- fit$conf.int[1]; ubnp <- fit$conf.int[2]
+   if(lbse*ubse < 0){inse <- inse +1}
+   if(lbnp*ubnp < 0){innp <- innp +1}
+ }
> inse/nsims;innp/nsims
```

(a) Explain why the final line (`inse/nsims;innp/nsims`) contains the estimates of the confidence levels, based on 100 simulations, for the SE confidence interval and the distribution-free confidence interval, respectively.

(b) Now change the number of simulations to 1000 and then run the code. Check the validities of the confidence levels (the proportions `inse/nsims` and `innp/nsims` should be close to 0.95).

(c) Find 95% confidence intervals for the proportions `inse/nsims` and `innp/nsims`. If an interval traps 0.95, it implies statistical significance of validity at level 0.05.

2.11.33. Change the code in Exercise 2.11.32 so that the simulation collects squared-lengths of the confidence intervals. The ratio of the means of these squared-lengths is a measure of efficiency, (shorter intervals are more precise). Determine the efficiency for 1000 simulations. Which method is better? Why?

2.11.34. Repeat Exercise 2.11.32 if the normal distribution is replaced by a t-distribution with 3 degrees of freedom.

2.11.35. Repeat Exercise 2.11.33 if the normal distribution is replaced by a t-distribution with 3 degrees of freedom.

2.11.36. Consider the situation described in Remark 2.10.1. Determine the sample size for the signed-rank Wilcoxon analysis to detect the alternative $\theta_A = 12$ with 80% power.

2.11.37. The distribution simulated in Example 2.10.1 was the normal distribution. For this exercise keep all of the parameters of the simulation the same except replace the normal distribution with a t-distribution having 3 degrees of freedom. Simulate the power of the signed-rank Wilcoxon to detect the alternative $\theta_A = 104$ for this situation. Discuss the results.

2.11.38. For the situations described in Example 2.10.1 and Exercise 2.11.37, estimate the powers of the t-test. Compare the signed-rank Wilcoxon and t-test simulated powers.

2.11.39. Consider the data from a small randomized pair design over two methods M_1 and M_2 and with 8 subjects that is given below. The experimenters are interested in the hypotheses $H_0 : \theta = 0$ versus $H_A : \theta > 0$, where θ is the true median of the paired differences ($M_2 - M_1$).

Subject	1	2	3	4	5	6	7	8
M_1	118	104	126	78	108	108	86	76
M_2	144	132	92	114	102	99	88	98

(a) Use the signed-rank Wilcoxon test at level 0.05 to test these hypotheses.

(b) The experimenters were not happy with the result in (a). They decide to run a larger experiment. Using $\alpha = 0.05$ as the level of the test, how many pairs of subjects are needed if the experimenters want to detect the alternative $\theta_A = 10$ with power 80%. Use the small study to approximate τ_W.

(c) Suppose instead of testing the hypotheses, the experimenters want to estimate θ within 12 units of precision at confidence 95%. Determine the number of pairs needed to achieve this. Use the small study to approximate τ_W.

3

Two-Sample Problems

In this chapter, we consider two-sample problems. These are simple but often used models in practice. Even in more complicated designs, contrasts of interest often involve two levels. So the ideas discussed here carry over to these designs.

In Sections 3.1–3.2, we discuss the two-sample Wilcoxon procedure. This includes the usual distribution-free rank test as well as the associated rank-based estimation, standard errors of estimates, and confidence intervals for the shift in locations. Wilcoxon procedures are based on the linear score function. Section 3.2.2 discusses rank-based procedures based on normal scores which are optimal if the underlying populations have normally distributed errors. We extend this discussion to general rank scores with a brief consideration of efficiency in Section 3.7. A Hogg-type adaptive scheme for rank score selection is presented in Section 3.8. For all of these generalizations, as shown, the computation of both testing and estimation (including standard errors and confidence intervals) is easily carried out by the R functions in the packages `Rfit` and `npsm`.

In Section 3.5, rank-based procedures for the two-sample scale problem are discussed. The Fligner–Kileen rank-based procedure for testing and estimation in this setting is optimal under normality and, unlike the traditional F-test based on variances, it possesses both robustness of validity and power for nonnormal situations. Rank-based procedures for the related Behrens–Fisher problem (Section 3.6) are also considered. As in other chapters, the focus is on the R computation of these rank-based procedures.

3.1 Introductory Example

In this section, we discuss the Wilcoxon rank-based analysis for the two-sample location problem in context of a real example. In this problem, there are two populations which we want to compare. From each population, we have a sample and based on these two samples we want to infer whether or not there is a difference in location between the populations and, if possible, to measure, with standard error, the difference (size of the effect) between the populations. The test component of the analysis is the Wilcoxon two-sample rank test,

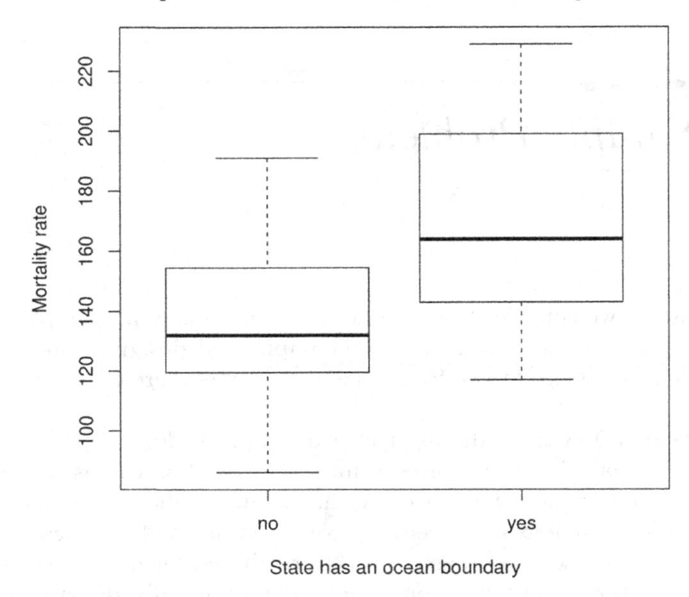

FIGURE 3.1
Mortality rates for white males due to malignant melanoma in the United States.

and the estimation procedure is based on the associated Hodges–Lehmann estimate. We discuss the R computation of the analyses.

Example 3.1.1 (Malignant Melanoma)**.** The dataset `USmelanoma`, from the package `HSAUR2` (Everitt and Hothorn 2014), contains, for each of the 50 states, variables for mortality rate for death due to malignant melanoma of white males, latitude, longitude, and an indicator variable denoting whether the state has an ocean boundary. The primary research questions of interest are: first, is there a difference in mortality for ocean states and non-ocean states, and second, does the difference prevail after adjusting for latitude and longitude? We address the first question in this chapter and the second question on adjustment in Chapter 5; see Exercise 5.9.19.

In Figure 3.1 we present comparison boxplots for the states that are ocean states (yes) and for the states that are non-ocean states (no). There appears to be an increase in the mortality rate due to malignant melanoma for the ocean states. The difference in the medians is an estimate of the "shift" in mortality rate between the non-ocean and ocean states.

In Table 3.1, results of the tests and the associated estimates of the size effect are displayed for the traditional analysis (two-sample t) and the rank-based analysis (two-sample Wilcoxon). The p-value of the Wilcoxon analysis indicates that there is a significant difference in mortality rates, (from malignant melanoma), between the non-ocean and ocean states. The mortality rate

TABLE 3.1
Estimates of increase in mortality due to malignant melanoma in white males in the United States.

	Test	p-value	Estimate	Std Error
Least Squares	3.60	0.00	31.49	8.55
Wilcoxon	3.27	0.00	31.00	9.26

is significantly higher in the ocean states. The Wilcoxon estimate of the shift from non-ocean states to ocean states is about 31 units with standard error 9 units. The results of the traditional LS analysis are similar. ∎

In the next two subsections, we present the Wilcoxon rank-based analysis. We first discuss testing and then estimation.

3.2 Rank-Based Analyses

For the most part, in this book, we are concerned with the rank-based fitting of models and testing hypotheses defined in terms of parameters. The roots of nonparametric methods, though, are in distribution-free tests, that work well for both the location model, (3.3), as well as for nonparametric settings. In this section, we first briefly discuss these procedures in the general nonparametric setting of stochastic ordering for which these tests are consistent and then discuss them in terms of the location model.

3.2.1 Wilcoxon Test for Stochastic Ordering of Alternatives

Let $X_1, \ldots, X_{n_1}$ denote a random sample from a distribution with the cdf F. Similarly, let $Y_1, \ldots, Y_{n_2}$ denote a random sample from a distribution with the cdf G. Assume that the samples are independent of one another. Let $n = n_1 + n_2$ denote the total sample size. We begin with the general case which only requires that the response variables be measured on at least an ordinal scale.

The hypotheses of interest are given by

$$H_0 : F(t) = G(t) \text{ vs. } H_A : F(t) \leq G(t), \tag{3.1}$$

where the inequality is strict for at least some t, (in the common support of both X and Y). For the alternative, we often say that X is **stochastically larger** than Y or that X tends to beat Y. This concept is illustrated by the cdfs in the left top panel of Figure 3.2. The bottom panels show the corresponding pdfs. The right panels show a location shift model, which forms a subfamily of stochastic ordering.

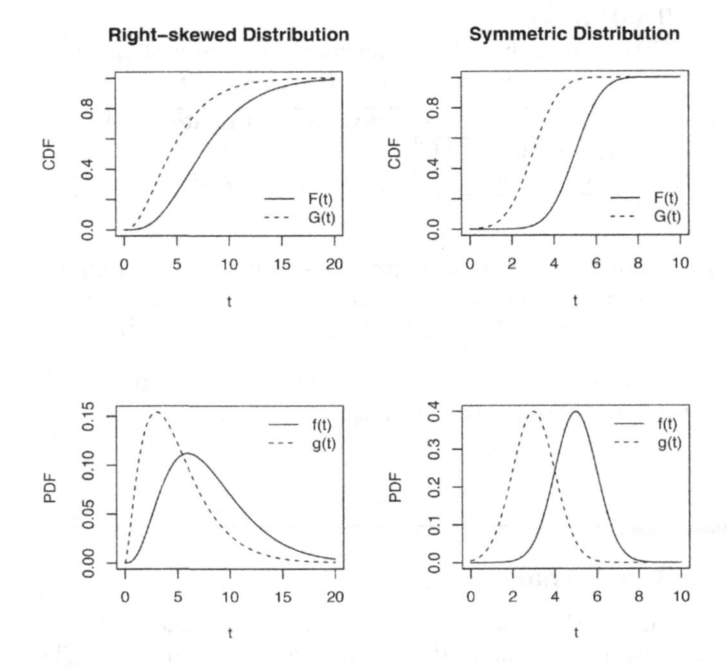

FIGURE 3.2
Plots illustrating stochastic ordering. $X \sim f, F$ is stochastically larger than $Y \sim g, G$.

For the discussion in this section, we consider the test based on Wilcoxon scores. For this test, the samples are combined into one sample and then ranked from 1 to n, low to high. Let $R(Y_j)$ denote the rank of Y_j in this combined ranking[1]. Then the Wilcoxon test statistic is

$$T = \sum_{j=1}^{n_2} R(Y_j). \tag{3.2}$$

The command `wilcox.test` is available in base R, and we highlight its use in this section. For the hypotheses (3.1), the Wilcoxon test rejects H_0 for small values of T. Under H_0 the two samples are from the same population; hence, any subset of ranks is equilikely as any other subset of the same size. For example, the probability that a subset of n_2 rankings is assigned to the Y's is $\binom{n}{n_2}^{-1}$. Thus the null distribution of T is free of the population distribution, and we say that the Wilcoxon test is **distribution-free**. Therefore, the p-value of the Wilcoxon test can be based on the exact distribution of T (by calculating the distribution of the ranks) or by using a large sample approximation.

[1]For adjustments to the rankings due to tied observations, see Section 3.4.

Example 3.2.1 ((O)esophageal Cancer). This example is based on the case-control study of esophageal cancer in Ile-et-Vilaine, France (Breslow et al. 1980). These data are available in the `datasets` package.[2] We test the hypothesis that alcohol consumption is the same in the two groups, using as our dataset a sample of cases and controls.

Figure 3.3 displays a stacked bar chart of ordinal data in this study. The following code segment illustrates the creation of the graphic. Note: we have plotted the highest category of alcohol consumption in the lowest bar.

```
> library(datasets)
> x<-rep(esoph$alcgp,esoph$ncases)      # responses for cases
> y<-rep(esoph$alcgp,esoph$ncontrols) # responses for controls
> z<-c(x,y)                             # combined vector
> w<-c(rep(1,length(x)),rep(0,length(y))) # indicator for cases
> tab1 <- table(z,w)
> tab1

           w
z             0   1
  0-39g/day 386  29
  40-79     280  75
  80-119     87  51
  120+       22  45

> tab <- prop.table(tab1,margin=2)
> tab <- tab[nrow(tab):1,]
> tab

           w
z                  0        1
  120+      0.028387 0.225000
  80-119    0.112258 0.255000
  40-79     0.361290 0.375000
  0-39g/day 0.498065 0.145000

> barplot(tab,names.arg=c('Controls','Cases'),
+   legend.text=rownames(tab),ylim=c(0,1.5),axes=FALSE)
> axis(2,at=seq(0,1,by=0.2))
```

Below is the output for the Wilcoxon procedure for testing if the case group tends to have higher levels of alcohol consumption than the control group. In this case we use the default settings of a large sample p-value with a continuity correction. Note that we have converted the data from the original ordered

[2]See `help(esoph)` which indicates the data are from "...a case-control study of (o)esophageal cancer...".

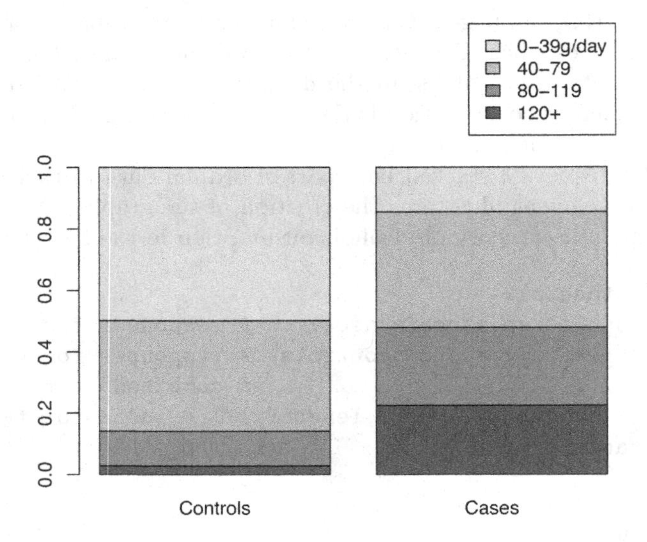

FIGURE 3.3
Stacked bar chart of the esophageal cancer data contained in the `datasets` dataset `esoph`.

categorical (or ordinal) data to numeric, as required by the base R function `wilcox.test`.

```
> xn<-as.numeric(x)
> yn<-as.numeric(y)
> wilcox.test(xn,yn,alt='greater')

        Wilcoxon rank sum test with continuity correction

data:  xn and yn
W = 115612, p-value <2e-16
alternative hypothesis: true location shift is greater than 0
```

The result is highly significant: the cases tend to consume more alcohol. This result is not surprising based on the bar chart. ■

3.2.2 Analyses for a Shift in Location

We next discuss the Wilcoxon rank-based analysis for the two-sample location problem. Let X and Y be continuous random variables. Let $F(t)$ and $f(t)$ respectively denote the cdf and pdf of X, and $G(t)$ and $g(t)$ respectively denote the cdf and pdf of Y. Then we say X and Y follow a **location model**, if for some parameter Δ, $-\infty < \Delta < \infty$,

$$G(t) = F(t - \Delta) \text{ and } g(t) = f(t - \Delta). \tag{3.3}$$

The parameter Δ is the **shift** in location between the random variables Y and X; for instance, it is the difference in medians or in means (provided that the means exist). We can write $\mathcal{L}(Y) = \mathcal{L}(X + \Delta)$, where $\mathcal{L}$ means distribution of (i.e., in law). The location model assumes, in particular, that the scale parameters of X and Y are the same.

Consider two independent samples drawn from the location model. Let $X_1, \ldots, X_{n_1}$ be a random sample on a random variable X with cdf and pdf $F(t)$ and $f(t)$, respectively, and let $Y_1, \ldots, Y_{n_2}$ be a random sample on a random variable Y with cdf and pdf $F(t - \Delta)$ and $f(t - \Delta)$, respectively. Let $n = n_1 + n_2$ denote the total sample size. Assume that the random samples are independent of one another. The hypotheses of interest are

$$H_0 : \Delta = 0 \text{ versus } H_A : \Delta \neq 0. \tag{3.4}$$

One-sided alternatives can also be used. In addition to tests of hypotheses, we also discuss estimates of the shift parameter Δ; both point estimates and confidence intervals.

Let $R(Y_i)$ denote the rank of Y_i among the combined samples, i.e., among $X_1, \ldots, X_{n_1}, Y_1, \ldots, Y_{n_2}$. Then the Wilcoxon test statistic is

$$T = \sum_{i=1}^{n_2} R(Y_i). \tag{3.5}$$

In the case of tied observations (ties) in the data, one usually assigns the average of the ranks that is allotted to these tied observations.[3]

There is a second formulation of the Wilcoxon test statistic that is often used. Consider the set of all $n_1 n_2$ differences $\{Y_j - X_i\}$ and let T^+ denote the number of positive differences, i.e.,

$$T^+ = \#_{i,j}\{Y_j - X_i > 0\}. \tag{3.6}$$

We then have the identity

$$T^+ = T - \frac{n_2(n_2 + 1)}{2}. \tag{3.7}$$

The form (3.6) is usually referred to as the Mann–Whitney test statistic.

For the two-sided hypothesis (3.4), the test rejects H_0 for small or large values of T. As discussed in the last section, T is distribution-free under H_0 and, hence, exact critical values can be determined from its distribution. If the samples are placed in the R vectors `x` and `y`, respectively, the R function `wilcox.test(y,x)` computes T^+ and returns the p-value of the test. By

[3]Handling of ties is discussed further in Section 3.4.

default, for small sample sizes ($n < 50$) and no ties, the p-value is based on the exact distribution; otherwise, it is based on an asymptotic approximation which we discuss below. The argument `exact=FALSE` results in the asymptotic distribution being used and `exact=TRUE` results in the exact distribution being used if no ties are in the data. When obtaining asymptotic p-values, use of a continuity correction may be overridden by setting the argument `correct=FALSE`. The default hypothesis is the two-sided hypothesis. The argument `alternative="less"` (`alternative="greater"`) results in testing the alternative $H_A : \Delta < 0$ ($H_A : \Delta > 0$).

Approximate p-values are based on the asymptotic distribution of T (or T^+). It can be shown that the mean and variance of T under the null hypothesis are given by $n_2(n+1)/2$ and $n_1 n_2(n+1)/12$, respectively. Furthermore, T has an asymptotic normal distribution. Hence, the standardized test statistic is given by

$$z = \frac{T - [n_2(n+1)/2]}{\sqrt{n_1 n_2(n+1)/12}}. \tag{3.8}$$

In this formulation, p-values are obtained by using the standard normal distribution. In the package `npsm` we have provided the R function `rank.test` which computes this standardized form. To set these ideas and illustrate the R computation, consider the following simple example.

Example 3.2.2 (Generated t_5-Data). The following are two samples generated from a t-distribution with 5 degrees of freedom. The true shift parameter Δ was set at the value 8. The following code is used to generate the samples.

```
> x<-round(rt(11,5)*10+42,1)
> y<-round(rt(9,5)*10+50,1)
```

The sorted data are

```
> sort(x)

 [1] 29.1 31.1 32.4 34.9 41.0 42.6 45.0 47.9 52.5 59.3 76.6

> sort(y)

[1] 40.1 45.8 47.2 48.1 49.5 58.3 58.7 62.0 64.8
```

We first obtain the analysis based on the `wilcox.test` function in base R.

```
> wilcox.test(x,y,exact=TRUE)

        Wilcoxon rank sum exact test

data:  x and y
W = 27, p-value = 0.095
alternative hypothesis: true location shift is not equal to 0
```

The function `wilcox.test` uses the T^+ (3.6) formulation of the Wilcoxon test statistic, which for these data has the value 27 with p-value 0.0952. Next we obtain the analysis based on the `rank.test` function in `npsm`.

```
> rank.test(x,y)

statistic =  1.7094 , p-value =  0.087375
```

The results of the exact and asymptotic analyses are quite similar; neither would reject the null hypothesis at a 5% level, while both would be considered significant evidence against the null at a 10% level. Note that `wilcox.test(x,y,exact=FALSE,correct=FALSE)` provides the same p-value as `rank.test(x,y)`. ∎

Often an estimate of the shift parameter Δ is desired. The estimator of Δ associated with the Wilcoxon analysis is the Hodges–Lehmann estimator, which is the median of all pairwise differences:

$$\hat{\Delta}_W = \text{med}_{i,j}\{Y_j - X_i\}. \tag{3.9}$$

This estimate is obtained by inverting the Wilcoxon test as we discuss in the next section. Note, there are $N_d = n_1 n_2$ pairwise differences between the sample of Ys and the sample of Xs. A distribution-free confidence interval can be constructed from the differences. Let $D_{(1)} < D_{(2)} < \cdots < D_{(N_d)}$ be the ordered differences. If the confidence level is $1-\alpha$, take c to be the lower $\alpha/2$ critical point of the null distribution of T^+, i.e.,

$$\alpha/2 = P_{H_0}[T^+ \leq c].$$

Then the interval $(D_{(c+1)}, D_{(n_1 n_2 - c)})$ is a $(1-\alpha)100\%$ confidence interval for Δ. The asymptotic value for c is given by

$$c = \frac{n_1 n_2}{2} - \frac{1}{2} - z_{\alpha/2}\sqrt{\frac{n_1 n_2 (n+1)}{12}},$$

which is rounded to nearest integer. It is generally quite close to the actual value. The R function `wilcox.test` computes the estimate and confidence interval. As an illustration, reconsider Example 3.2.2. The following code segment illustrates the how to obtain a confidence interval for Δ.

```
> wilcox.test(y,x,conf.int=TRUE)

        Wilcoxon rank sum exact test

data:  y and x
W = 72, p-value = 0.095
alternative hypothesis: true location shift is not equal to 0
95 percent confidence interval:
```

```
 -1.0 18.4
sample estimates:
difference in location
                  10.4
```

Note that the confidence interval $(-1, 18.4)$ includes the true Δ which is 8. The point estimate is $\hat{\Delta} = 10.4$. Different confidence levels can be set by changing the value of the input variable `conf.level`. For example, use `conf.level=.99` for a 99% confidence interval.

Note that we can perform the Wilcoxon test equivalently using the following formulation of the Wilcoxon test statistic:

$$T_W = \sum_{i=1}^{n_2} a[R(Y_i)],$$

where $a(i) = \varphi_W[i/(n+1)]$ and $\varphi(u) = \sqrt{12}[u-(1/2)]$. The following identity is easy to obtain:

$$T_W = \frac{\sqrt{12}}{n+1}\left[T - \frac{n_2(n+1)}{2}\right].$$

Hence, the statistic T_W is a linear function of the ranks and, for instance, its z score formulation is exactly the same as that of T. We call the function $\varphi_W(u)$ a score function and, more specifically, the **Wilcoxon score function**. We also call the related scores, $a_W(i)$, the **Wilcoxon scores**.

Normal Scores

Certainly functions other than the linear function can be used as score functions. For certain error distributions rank-based methods based on scores other than the Wilcoxon scores may be more efficient than Wilcoxon procedures. In Section 3.7 we discuss efficiency, offering an optimality result. Later, in Section 3.8, we present an adaptive scheme which automatically selects an appropriate score. In this section, we discuss the analysis based on the scores which are fully asymptotically efficient if the underlying populations are normally distributed. These scores are called the **normal scores.**

The normal scores are generated by the function.

$$\varphi_{ns}(u) = \Phi^{-1}(u), \tag{3.10}$$

where $\Phi^{-1}(u)$ denotes the inverse of the standard normal cumulative distribution function. Letting $a_{ns}(i) = \varphi_{ns}[i/(n+1)]$, the associated test statistic is

$$T_{ns} = \sum_{i=1}^{n_2} a_{ns}[R(Y_i)].$$

Under H_0 this test statistic is distribution-free. Usually, the associated z-test statistic is used. The mean and variance of T_{ns} under the null hypothesis are

0 and

$$\mathrm{Var}_{H_0}(T_{ns}) = \frac{n_1 n_2}{n-1} \sum_{i=1}^{n} a_{ns}^2(i),$$

respectively. Using these moments, an asymptotic level α test of the hypotheses (3.4) is

$$\text{Reject } H_0 \text{ in favor of } H_A, \text{ if } |z_{ns}| \geq z_{\alpha/2}, \tag{3.11}$$

where

$$z_{ns} = \frac{T_{ns}}{\sqrt{\mathrm{Var}_{H_0}(T_{ns})}}.$$

The z-test formulation is computed by the `npsm` function `rank.test`. The following code segment displays the results of the normal scores test for the data of Example 3.2.2. Notice that the normal scores are called by the argument `scores=nscores`.

```
> rank.test(x,y,scores=nscores)

statistic =  1.6067 , p-value =  0.10811
```

The standardized test statistic has the value 1.61 with p-value 0.1081.

Estimates and confidence intervals are also available as with `rank.test`. The next section discusses the details of the computation, but we illustrate the use of `rank.test` to obtain these values.

```
> rank.test(x,y,scores=nscores,conf.int=TRUE)

statistic =  1.6067 , p-value =  0.10811
  percent confidence interval:
-2.6893 21.689
Estimate: 9.5
```

3.2.3 Analyses Based on General Score Functions

Recall in Section 3.2.2 that besides the Wilcoxon scores, we introduced the normal scores. In this section, we define scores in general and discuss them in terms of efficiency. General scores are discussed in Section 2.5 of Hettmansperger and McKean (2011). A set of rank-based scores is generated by a function $\varphi(u)$ defined on the interval $(0,1)$. We assume that $\varphi(u)$ is nondecreasing, square-integrable, and, without loss of generality, standardize as

$$\int_0^1 \varphi(u)\,du = 0 \text{ and } \int_0^1 \varphi^2(u)\,du = 1. \tag{3.12}$$

The generated scores are then $a_\varphi(i) = \varphi[i/(n+1)]$. Because $\int_0^1 \varphi(u)\,du = 0$, we also may assume that the scores sum to 0, i.e., $\sum_{i=1}^n a[i] = 0$. The Wilcoxon and normal scores functions are given respectively by $\varphi_W(u) = \sqrt{12}[u-(1/2)]$

and $\varphi_{ns}(u) = \Phi^{-1}(u)$ where $\Phi^{-1}(u)$ is the inverse of the standard normal cumulative distribution function.

For general scores, the associated process is

$$S_\varphi(\Delta) = \sum_{j=1}^{n_2} a_\varphi[R(Y_j - \Delta)], \tag{3.13}$$

where $R(Y_j - \Delta)$ denotes the rank of $Y_j - \Delta$ among $X_1, \ldots, X_{n_1}$ and $Y_1 - \Delta, \ldots, Y_{n_2} - \Delta$. A test statistic for the hypotheses

$$H_0 : \Delta = 0 \text{ versus } H_A : \Delta > 0 \tag{3.14}$$

is $S_\varphi = S_\varphi(0)$. Under H_0, the Xs and Ys are identically distributed. It then follows that S_φ is distribution-free. Although exact null distributions of S_φ can be numerically generated, usually the null asymptotic distribution is used. The mean and variance of $S_\varphi(0)$ under the null hypothesis are

$$E_{H_0}[S_\varphi(0)] = 0 \text{ and } \sigma_\varphi^2 = V_{H_0} = \tfrac{n_1 n_2}{n(n-1)} \sum_{i=1}^{n} a_\varphi^2(i). \tag{3.15}$$

Furthermore, the null distribution of $S_\varphi(0)$ is asymptotically normal. Hence, our standardized test statistic is

$$z_\varphi = \frac{S_\varphi(0)}{\sigma_\varphi}. \tag{3.16}$$

For the hypotheses (3.14), an asymptotic level α test is to reject H_0 if $z_\varphi \geq z_\alpha$, where z_α is the $(1-\alpha)$ quantile of the standard normal distribution. The two-sided and the other one-sided hypotheses are handled similarly.

The `npsm` function `rank.test` computes this asymptotic test statistic z_φ along with the corresponding p-value for all the intrinsic scores found in `Rfit`. The arguments are the samples, the specified score function, and the value of `alternative` which is `greater`, `two.sided`, or `less` for the respective alternatives $\Delta > 0$, $\Delta \neq 0$, or $\Delta < 0$. For example, the call `rank.test(x,y,scores=nscores)` computes the normal scores asymptotic test for a two-sided alternative. Other examples are given below.

For a general score function $\varphi(u)$, the corresponding estimator $\hat{\Delta}_\varphi$ of the shift parameter Δ solves the equation

$$S_\varphi(\Delta) \doteq 0. \tag{3.17}$$

It can be shown that the function $S_\varphi(\Delta)$ is a step function of Δ which steps down at each difference $Y_j - X_i$; hence, the estimator is easily computed numerically. The asymptotic distribution of $\widehat{\Delta}_\varphi$ is given by

$$\widehat{\Delta}_\varphi \text{ has an approximate } N\left(\Delta, \tau_\varphi^2 \sqrt{\frac{1}{n_1} + \frac{1}{n_2}}\right), \tag{3.18}$$

where τ_φ is the scale parameter

$$\tau_\varphi^{-1} = \int_0^1 \varphi(u)\varphi_f(u)\,du \tag{3.19}$$

and

$$\varphi_f(u) = -\frac{f'(F^{-1}(u))}{f(F^{-1}(u))}. \tag{3.20}$$

In practice, τ_φ must be estimated based on the two samples. We recommend the Koul, Sievers, and McKean (1987) estimator which is implemented in `Rfit`. We denote this estimate by $\hat{\tau}_\varphi$. Thus the standard error of the estimate is

$$\mathrm{SE}(\hat{\Delta}_\varphi) = \hat{\tau}_\varphi\sqrt{\frac{1}{n_1}+\frac{1}{n_2}}. \tag{3.21}$$

Based on (3.18), an approximate $(1-\alpha)100\%$ confidence interval for Δ is

$$\widehat{\Delta}_\varphi \pm t_{\alpha/2,n-2}\widehat{\tau_\varphi}\sqrt{\frac{1}{n_1}+\frac{1}{n_2}}, \tag{3.22}$$

where $t_{\alpha/2,n-2}$ is the upper $\alpha/2$ critical value of a t-distribution[4] with $n-2$ degrees of freedom. Standard errors are easily computed using the summary of the regression fit as discussed next.

3.2.4 Linear Regression Model

Looking ahead to the next chapter, we frame the two-sample location problem as a regression problem. We begin by continuing our discussion of the normal scores rank-based analysis of the previous section. In particular, we next discuss the associated estimate of the shift parameter Δ. The `rfit` function of the R package `Rfit` has the built-in capability to fit regression models with a number of known score functions, including the normal scores. So, it is worthwhile to take a few moments to set up the two-sample location model as a regression model. Let $\boldsymbol{Z} = (X_1,\ldots,X_{n_1},Y_1,\ldots,Y_{n_2})^T$. Let $\mathbf{c}$ be a $n\times 1$ vector whose *ith* component is 0 for $1 \le i \le n_1$ and 1 for $n_1+1 \le i \le n = n_1+n_2$. Then we can write the location model as

$$Z_i = \alpha + c_i\Delta + e_i, \tag{3.23}$$

where $e_1,\ldots,e_n$ are iid with pdf $f(t)$ and true median 0. Hence, we can estimate Δ by fitting this regression model. If we use the method of least squares, then the estimate is $\overline{Y}-\overline{X}$. If we use the rank-based fit with Wilcoxon scores, then the estimate of Δ is the median of the pairwise differences. If instead normal scores are called for, then the estimate is the estimate associated with the

[4]For a discussion on the appropriateness of t-critical values, see McKean and Sheather (1991).

normal scores test. The following R segment obtains the rank-based normal scores fit for the data of Example 3.2.2. Notice that `scores=nscores` calls for the normal scores rank-based fit. If this call is deleted, the default Wilcoxon fit is computed instead.

```
> z <- c(x,y); ci <- c(rep(0,length(x)),rep(1,length(y)))
> fitns <- rfit(z ~ ci,scores=nscores)
> coef(summary(fitns))

            Estimate Std. Error t.value    p.value
(Intercept)     41.8     4.4224  9.4519 2.1111e-08
ci               9.5     5.8019  1.6374 1.1891e-01
```

Hence, the normal scores estimate of Δ is 9.5. Notice that the summary table includes the standard error of the estimate, namely 5.8. Using this, an approximate 95% confidence interval for Δ, based on the upper 0.025 t-critical with 18 degrees of freedom, is calculated in the next code segment.

```
> deltahat <- coef(summary(fitns))[2,1]
> se_deltahat <- coef(summary(fitns))[2,2]
> tcv <- qt(0.025,18,lower.tail=FALSE)
> # 95% CI
> round(deltahat + c(-1,1)*tcv*se_deltahat,2)

[1] -2.69 21.69

>
```

We close this section with the Wilcoxon and normal scores analyses of a dataset taken from a preclinical study.

Example 3.2.3 (Quail Data)**.** The data for this problem were extracted from a preclinical study on low density lipids (LDL). Essentially, for this example, the study consisted of assigning 10 quail to a diet containing an active drug compound (treated group), which hopefully reduces LDL, while 20 other quail were assigned to a diet containing a placebo (control group); see Section 2.3 of Hettmansperger and McKean (2011) for a discussion of this preclinical experiment. At the end of a specified time, the LDL levels of all 30 quail were obtained. The data are available in the `quail2` dataset. A comparison boxplot of the data is given in Figure 3.4.

The boxplots clearly show an outlier in the treated group. Further, the plot indicates that the treated group has generally lower LDL levels than the placebo group. Using the `rfit` function, we compute the Wilcoxon and normal scores estimates of Δ and their standard errors for the Wilcoxon and normal scores analyses. For comparison purposes, we also present these results for the least squares (LS) analysis.

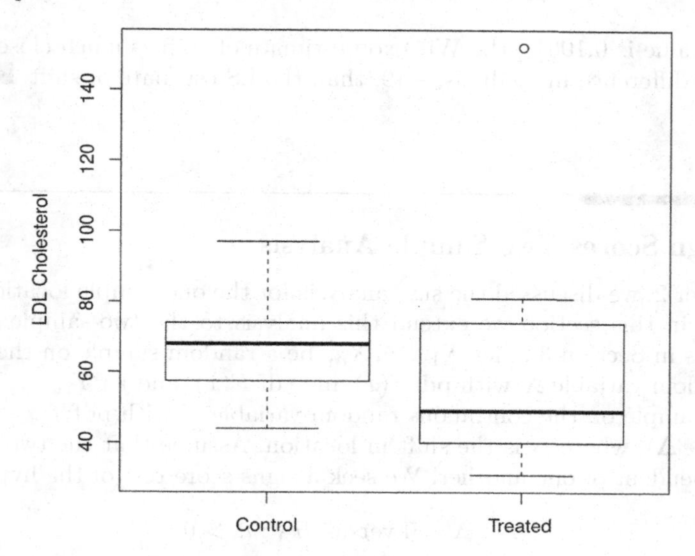

FIGURE 3.4
Comparison boxplots of low density lipids cholesterol.

```
> fit<-rfit(ldl~treat,data=quail2)
> coef(summary(fit))

            Estimate Std. Error t.value    p.value
(Intercept)       65     4.2690 15.2261 4.4871e-15
treat            -14     8.3847 -1.6697 1.0612e-01

> fitns<-rfit(ldl~treat,data=quail2,scores=nscores)
> coef(summary(fitns))

            Estimate Std. Error t.value    p.value
(Intercept)       65     4.1169 15.7885 1.8005e-15
treat            -13     7.6699 -1.6949 1.0118e-01

> fitls<-lm(ldl~treat,data=quail2)
> coef(summary(fitls))

            Estimate Std. Error  t value   Pr(>|t|)
(Intercept)     67.2     5.1756 12.98393 2.2702e-13
treat           -5.0     8.9645 -0.55776 5.8144e-01
```

The Wilcoxon and normal scores estimates and standard errors are similar. The LS estimate, though, differs substantially from the robust fits. The LS analysis is severely affected by the outlier. Although not significant at the 5%

level, (p-value is 0.1061), the Wilcoxon estimate of shift is much closer to the boxplots' difference in medians, -19, than the LS estimate of shift. ∎

3.3 Sign Scores Two-Sample Analysis

In Chapter 2, we discussed the sign analysis for the one-sample location problem, and in this section we extend this analysis to the two-sample location model. As in Section 3.2, let $X_1, \ldots, X_{n_1}$ be a random sample on the continuous random variable X with pdf $f(x)$ and cdf $F(x)$, and let $Y_1, \ldots, Y_{n_2}$ be a random sample on the continuous random variable Y with pdf $f(x - \Delta)$ and cdf $F(x - \Delta)$, where Δ is the shift in location. Assume that the two samples are independent of one another. We seek a signs score test of the hypotheses

$$H_0 : \Delta = 0 \text{ versus } H_A : \Delta > 0. \tag{3.24}$$

As we note later, the other one-sided ($\Delta < 0$) and the two-sided ($\Delta \neq 0$) tests are similar. We also want to determine the associated point estimate of Δ along with a confidence interval for Δ.

3.3.1 Mood's Median Test

The sign-score function is given by $\varphi_s(u) = \text{sign}[u - (1/2)]$ where the function $\text{sign}[v]$ is defined by

$$\text{sign}[v] = \begin{cases} 1 & v > 0 \\ 0 & v = 0 \\ -1 & v < 0. \end{cases} \tag{3.25}$$

Then following the discussion on general scores in Section 3.2.3, by expression (3.13), the sign process is given by

$$S_s(\Delta) = \sum_{j=1}^{n_2} \text{sign}\left[\frac{R(Y_j - \Delta)}{n+1} - \frac{1}{2}\right] = \sum_{j=1}^{n_2} \text{sign}\left[R(Y_j - \Delta) - \frac{n+1}{2}\right], \tag{3.26}$$

where $R(Y_j - \Delta)$ denotes the rank of $Y_j - \Delta$ among $X_1, \ldots, X_{n_1}$ and $Y_1 - \Delta, \ldots, Y_{n_2} - \Delta$ and the second equality follows from the definition of the sign function, (3.25).

The sign test statistic for the hypotheses (3.24) is $S_s(0)$, i.e.,

$$S_s(0) = \sum_{j=1}^{n_2} \text{sign}\left[R(Y_j) - \frac{n+1}{2}\right]. \tag{3.27}$$

Note that $(n + 1)/2$ is the midrank of the ranks $1, 2, \ldots, n$. Under H_0, the distributions of the X_i's and Y_j's are the same; hence, it is equilikely for the

$R(Y_j)$ to be below the midrank as above it; i.e., sign $\left[R(Y_j) - \frac{n+1}{2}\right]$ is -1 or 1 each with probability 1/2. Thus $S_s(0)$ is distribution-free under H_0 and its mean is 0. On the other-hand, under H_A, the $R(Y_j)$'s are more likely to be above the midrank than below it. This is the intuition behind the decision rule:

$$\text{Reject } H_0 : \Delta = 0 \text{ in favor of } H_A : \Delta > 0 \text{ at level } \alpha \text{ if } S_s(0) \geq c_\alpha.$$

To determine the critical point c_α we need the null distribution of $S_s(0)$. We first describe the asymptotic test. A simple equivalence, however, leads to the exact null distribution of $S_s(0)$ that we discuss secondly.

For the asymptotic test, as discussed above $E_{H_0}[S_s(0)] = 0$ and it follows from expression (3.15) that $V_{H_0}(S_s(0)) = n_1 n_2/(n-1)$. Since $S_s(0)$ is asymptotically normal under H_0, the standardized test statistic $Z_s = S_s(0)/\sqrt{n_1 n_2/(n-1)}$ can be used to test hypotheses (3.24). Thus, the decision rule for the asymptotic level α test is:

$$\text{Reject } H_0 : \Delta = 0 \text{ in favor of } H_A : \Delta > 0 \text{ if } Z_s = \frac{S_s(0)}{\sqrt{n_1 n_2/(n-1)}} \geq z_\alpha, \quad (3.28)$$

where z_α is the upper α-quantile of a standard $N(0,1)$ distribution.

Next, to determine the exact null distribution of the sign test statistic, denote the combined sample median by $\hat{M} = \text{med}\{X_1, \ldots, X_{n_1}, Y_1, \ldots, Y_{n_2}\}$. Then $R(Y_j) > (n+1)/2$ if and only if $Y_j > \hat{M}$. Hence, $S_s(0)$ is counting the number of Y_j's above $\hat{M}$ minus the number below. This leads[5] to

$$S_s(0) \doteq 2\#_{1 \leq j \leq n_2}\{Y_j > \hat{M}\} - n_2 = 2M_s(0) - n_2,$$

where

$$M_s(0) = \#_{1 \leq j \leq n_2}\{Y_j > \hat{M}\}. \quad (3.29)$$

Thus $M_s(0) = [S_s(0) + n_2]/2$, is an equivalent test statistic to $S_s(0)$. The null mean of $M_s(0)$ is $n_2/2$. The test based on $M_s(0)$ is called **Mood's median test**; see Mood (1950). Note further that an observation either exceeds $\hat{M}$ or is less than or equal to it. Hence, we can summarize Mood's test in a 2×2 contingency table. There are two cases depending on whether n is even or odd. Consider the even case first and let $r = n/2$. The contingency table[6] is given by

	$\# > \hat{M}$	$\# \leq \hat{M}$	
Y_j's	$M_s(0)$	$n_2 - M_s(0)$	n_2
X_i's	$r - M_s(0)$	$r - (n_2 - M_s(0))$	n_1
	r	r	n

[5]We have counted a Y_j equal to $\hat{M}$ as a -1 in $S_s(0)$; hence, the approximate equality.

[6]The table includes observations which are equal to $\hat{M}$. Some authors exclude these items, so their results may differ slightly.

Note that both table margins are fixed. Next, we derive the exact distribution of $M_s(0)$. Note that there are $\binom{n}{r}$ ways to select the r combined sample items that exceed $\hat{M}$ and under H_0 these are each equilikely. Now, consider the event $\{M_s(0) = k\}$. There are $\binom{n_2}{k}$ ways to select the $Y's$ which exceed $\hat{M}$ and, hence, $\binom{n_1}{r-k}$ ways to select the X's which exceed $\hat{M}$. Hence,

$$P[M_s(0) = k] = \frac{\binom{n_2}{k}\binom{n_1}{r-k}}{\binom{n}{r}}, \quad k = 0, \ldots, n_2, \tag{3.30}$$

where we adopt the usual convention that $\binom{s}{t}$ is zero if $t > s$. Thus, $M_s(0)$ has a hypergeometric distribution. If n is odd, then use $r = (n-1)/2$ in constructing the contingency table and deriving the hypergeometric distribution for $M_s(0)$.

The hypergeometric version of the test is called Fisher's exact test. The base R function `fisher.test` can be used to carry out Fisher's exact test on the contingency table, as we illustrate in the next example. The asymptotic version of the test, (3.28), is computed by the usual `chisq.test` for association on the contingency table; see Exercise 3.12.8.

Example 3.3.1 (Plant Growth). Consider the dataset `PlantGrowth` in base R. Dried weights of plants were obtained under control (`ctrl`) conditions and two treatments (`trt1` and `trt2`). As shown in the code chunk below, there are 10 observations in each sample.

```
> with(PlantGrowth,table(group))

group
ctrl trt1 trt2
  10   10   10
```

In this example, we examine `trt2` versus `ctrl`.

```
> ind <- PlantGrowth$group %in% c('ctrl','trt2')
> PlantGrowth_c2 <- PlantGrowth[ind,]
> PlantGrowth_c2$group <- factor(PlantGrowth_c2$group) # tidying up
```

Let Δ be the shift in location from the control to the treated responses. We are interested in testing the hypotheses $H_0 : \Delta = 0$ versus $H_A : \Delta > 0$, using Mood's test. Figure 3.5 shows a comparison boxplot of the two samples. It appears that the plants grown under the treatment weighed more than those grown under control conditions. In the next R segment, we obtain the samples and then find the combined sample median.

```
> mhat <- median(PlantGrowth_c2$weight)
> mhat

[1] 5.275

> with(PlantGrowth_c2,tapply(weight,group,sort))
```

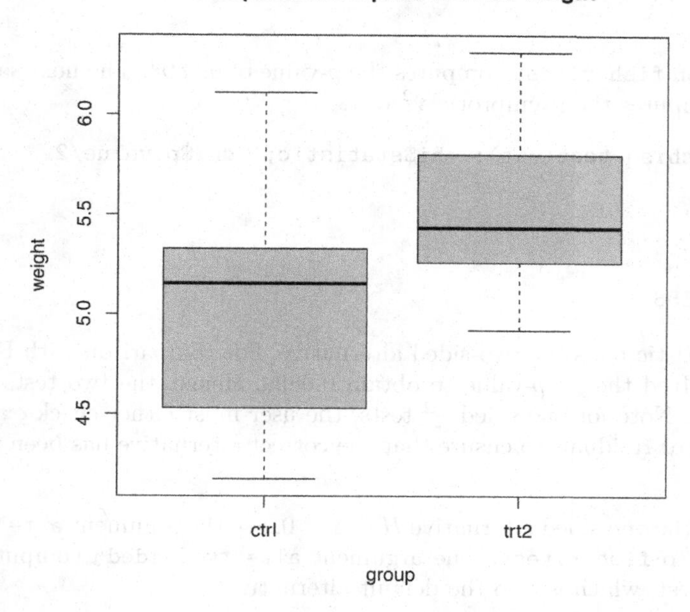

FIGURE 3.5
Comparison boxplots of the weights of the plants grown under control conditions (`ctrl`) versus the weights of the plants grown under Treatment 2 conditions (`trt2`).

```
$ctrl
 [1] 4.17 4.50 4.53 4.61 5.14 5.17 5.18 5.33 5.58 6.11

$trt2
 [1] 4.92 5.12 5.26 5.29 5.37 5.50 5.54 5.80 6.15 6.31
```

Based on the ordered data, it is easy to see by inspection that $M_s(0) = 7$ and, from that, to construct the contingency table which, in the next R segment, is entered in the matrix `mat`. We then compute Fisher's exact test.

```
> r1 <- c(7,3);r2<-c(3,7); mat <- rbind(r1,r2); mat

   [,1] [,2]
r1    7    3
r2    3    7

> ft <- fisher.test(mat,alt="greater");  ft$p.value
```

```
      r1
0.089448
```

The function `fisher.test` computes the p-value of 0.0894. The next segment of code computes the asymptotic χ^2 test.

```
> chi <- chisq.test(mat); chi$statistic;  chi$p.value/2

X-squared
      1.8

[1] 0.089856
```

The χ^2 statistic tests the two-sided alternative. For comparison with Fisher's test, we halved the χ^2 p-value to obtain 0.0899. Hence, the two tests essentially agree. Note for one-sided χ^2 tests, the user must either check expected frequencies or residuals to ensure that the correct alternative has been tested. ■

To test the one-sided alternative $H_A; \Delta < 0$, use the argument `alt="less"` in the call to `fisher.test`. The argument `alt="two.sided"` computes the two-sided test, which is also the default alternative.

3.3.2 Estimation of the Shift in Location

Next, we consider the signed score estimator of the shift in location Δ. Let $\hat{\theta}_1 = \text{med}\{X_1, \ldots, X_{n_1}\}$ and $\hat{\theta}_2 = \text{med}\{Y_1, \ldots, Y_{n_2}\}$. According to expression (3.17), the estimator of Δ for the signed score procedure solves the equation,

$$S_s(\Delta) = \sum_{j=1}^{n_2} \text{sign}\left[R(Y_j - \Delta) - \frac{n+1}{2}\right] \doteq 0, \tag{3.31}$$

where the $R(Y_j - \Delta)$ is the rank of $Y_j - \Delta$ among $\{X_1, \ldots, X_{n_1}, Y_1 - \Delta, \ldots, Y_{n_2} - \Delta\}$. Because the rankings are invariant to an additive constant, these rankings are the same as the rankings of this set: $\{X_1 - \hat{\theta}_1, \ldots, X_{n_1} - \hat{\theta}_1, Y_1 - \hat{\theta}_1 - \Delta, \ldots, Y_{n_2} - \hat{\theta}_1 - \Delta\}$. The estimator of Δ must be chosen so half of the ranks of the Y part is below $(n+1)/2$ (the mid-rank) and half above. Note in this last set, half of the $X_i - \hat{\theta}_1$'s are negative and half are positive. If we take $\hat{\Delta} = \hat{\theta}_2 - \hat{\theta}_1$ then also half of the $Y_j - \hat{\theta}_1 - \hat{\Delta} = Y_j - \hat{\theta}_2$ will be negative and half will be positive. Hence, the rank-based estimator of Δ based on sign scores is the difference in sample medians, i.e., $\hat{\Delta}_S = \hat{\theta}_2 - \hat{\theta}_1$.

For $i = 1, 2$, recall from Chapter 2 that $\hat{\theta}_i$ has an asymptotic $N(\theta_i, \tau_S^2/n_i)$ distribution, where the scale parameter τ_S is defined in expression (2.22). Since the samples are independent, we have

$$\hat{\Delta} \text{ has an approximate } N\left(\Delta, \tau_S^2\left(\tfrac{1}{n_1} + \tfrac{1}{n_2}\right)\right) \text{ distribution.} \tag{3.32}$$

For $i = 1, 2$, let $\hat{\tau}_{S,i}$ be the estimator of τ_S defined in expression (2.20). Under the assumptions of the model, the distribution of X and Y differ by at most the shift in location Δ. In particular, both $\hat{\tau}_{S,1}$ and $\hat{\tau}_{S,2}$ are estimating the same scale parameter τ_S. In this case it seems reasonable to use the **pooled estimator**

$$\hat{\tau}_{Sp}^2 = \frac{(n_1 - 1)\hat{\tau}_{S,1}^2 + (n_2 - 1)\hat{\tau}_{S,2}^2}{n_1 + n + 2 - 2} \tag{3.33}$$

as the estimator of τ_S^2. Hence, the standard error of $\hat{\Delta}_S$ is

$$SE(\hat{\Delta}_S) = \hat{\tau}_{Sp}\sqrt{\frac{1}{n_1} + \frac{1}{n_2}}. \tag{3.34}$$

The associated approximate $(1-\alpha)100\%$ Wald type confidence interval for Δ, is

$$\hat{\Delta}_S \pm t_{\alpha/2, n_1+n_2-2}\hat{\tau}_{Sp}\sqrt{\frac{1}{n_1} + \frac{1}{n_2}}. \tag{3.35}$$

The `npsm` function `mood.ci` calculates the estimate $\hat{\Delta}_S$, its standard error, and the associated confidence interval.

Example 3.3.2 (Quail Data, Continued). Consider, again, the quail dataset discussed in Example 3.2.3. Recall that 10 quail were assigned to a diet containing a drug compound (treated group) which was developed as a treatment to reduce low density lipids (LDL), while 20 quail were assigned to the control group which had the same diet but without the drug compound. At the end of the study period, the response measured was the LDL levels of the quail. Let Δ denote the shift in location from the control group to the treated group. Figure 3.4 displays comparison boxplots of the responses. The data are in the `npsm` dataset `quail2`. The following R segment computes the sign score estimate $\hat{\Delta}_S$, its corresponding standard error, and a 95% approximate confidence interval for Δ.

```
> x <- quail2$ldl[quail2$treat==0]
> y <- quail2$ldl[quail2$treat==1]
> fit <- mood.ci(y,x,var.equal=TRUE)
> fit

95 percent confidence interval:
[1] -37.83479  -0.16521

> fit$estimate ; fit$stderr

[1] -19

[1] 9.1948
```

Thus $\hat{\Delta}_S = -19.0$ with the SE of 9.19. The 95% confidence interval does not contain 0, so a significant decrease in LDL level between the treated and control groups at the 5% level has been detected. ■

We now have discussed in detail two rank-based procedures for the two-sample location model: the Wilcoxon analysis based on the linear score function and Mood's median analysis based on the sign score function. One way of comparing these analyses is in terms of their efficiency that is discussed in Section 3.7. In general, unless the error distribution has very heavy tails, the Wilcoxon is more efficient. So of the two analyses, we generally recommend the Wilcoxon analysis.

3.4 Adjustments for Ties

In this section, we discuss adjustments for ties for the two-sample distribution-free tests we have considered previously. We consider the most general model. Let $X_1, \ldots, X_{n_1}$ be a random sample on a continuous random variable X with pdf $f(t)$ and cdf $F(t)$ and let $Y_1, \ldots, Y_{n_2}$ be a random sample on a continuous random variable Y with pdf $g(t)$ and cdf $G(t)$. Denote the combined sample size by $n = n_1 + n_2$. Suppose that we are interested in testing the hypotheses that X tends to be larger than Y; i.e.,

$$H_0 : F(t) = G(t) \text{ versus } H_A : F(t) \leq G(t), \tag{3.36}$$

where the inequality is strict for some value of t. For our discussion, we consider the Wilcoxon test statistic

$$T = \sum_{j=1}^{n_2} R(Y_j), \tag{3.37}$$

where $R(Y_j)$ denotes the rank of Y_j among $X_1, \ldots, X_{n_1}$ and $Y_1, \ldots, Y_{n_2}$. Because the sample variables X_i and Y_j are continuous and independent, $P(X_i = Y_j) = 0$ for $i \neq j$. In practice, though, we cannot measure infinitely precise, so there may be ties in the realized samples.

A common method of ranking when ties occur is the method of average ranks that was discussed in Section 2.5 for the one-sample problem. Recall that if a group of observations are tied at the same value, then each observation in the group is assigned as its rank the average value of the integer ranks which would have been assigned to the group of observations if ties had been ignored. As a simple illustration, consider the combined samples of observations in the first row of Table 3.2. For the group of three observations tied at the value 51, each receives the rank 6 which is the average of ranks 5, 6, and 7. The combined rankings of the samples are displayed in the second row of the table.

TABLE 3.2
The samples on X and Y are displayed in the first row while their assigned ranks are given in the second row.

	x_i						y_j							
Data	33	35	49	51	56	74	49	51	51	53	58	60	61	62
Rank	1	2	3.5	6	9	14	3.5	6	6	8	10	11	12	13

While the null distribution of the Wilcoxon test statistic T is still distribution-free under tied observations, it differs from the null distribution when there are no ties. As in the one-sample problem, this distribution depends on how many groups of tied observations there are in the combined samples and how many observations there are in each group. In practice, usually the asymptotic null distribution is used. The null mean of T remains at $n_2(n+1)/2$, but the null variance differs. It is given by

$$\text{Var}_0(T) = \frac{n_1 n_2 (n+1)}{12} - \left[\frac{n_1 n_2}{12n(n-1)} \sum_{j=1}^{g} t_j(t_j - 1)(t_j + 1) \right], \tag{3.38}$$

where g is the number of groups of tied observations and t_j is the number of observations in group j. The standardized test statistic is

$$Z = \frac{T - \frac{n_2(n+1)}{2}}{\sqrt{\text{Var}_0(T)}}.$$

For the hypotheses (3.36), the nominal α-level, asymptotic test rejects H_0 in favor of H_A, if $Z \geq z_\alpha$.

For the example dataset displayed in Table 3.2, an easy calculation shows: $T = 69.5$, the null mean is $8(15)/2 = 60$, and the null variance is $60 - 0.6593 = 59.3407$. Using the continuity correction, the p-value is then given by

$$P_0(T \geq 69.5 - 0.5) \cdot = P\left(Z \geq \frac{69}{\sqrt{59.3407)}} \right) = 0.1213.$$

Remark 3.4.1. Upon omitting the adjustment due to ties in the variance, $Z = 1.161895$ with p-value 0.1226. There is little difference in the results because the tie adjustment is small relative to the unadjusted variance. Note that in omitting the adjustment due to ties in the variance, the test is conservative compared to the test using tie adjusted variance. This is true because the term in the brackets of expression (3.38) is nonnegative.

For the tie adjusted Mann–Whitney version of the Wilcoxon test statistic, define the indicator

$$I(x) = \begin{cases} 1 & \text{if } x > 0 \\ \frac{1}{2} & \text{if } x = 0 \\ 0 & \text{if } x < 0. \end{cases} \tag{3.39}$$

Then the tie adjusted Mann–Whitney version of the test statistic is given by

$$T^+ = \sum_{i=1}^{n_1} \sum_{j=1}^{n_2} I(Y_j - X_i). \tag{3.40}$$

The identity between the test statistics T and T^+ still holds for the tie adjusted versions; that is, $T^+ = T - n_2(n_2+1)/2$. This seems odd at first because T^+ ignores tie groups which contain observations from only one sample. It is easy to see for such a group, however, that if the integer values of the ranks assigned to the group are used instead of the average ranks the contribution from the group to T remains the same. Hence, only the ties between Y_j and X_i count in the statistic T.

For the illustrative example, this identity yields $T^+ = 69.5 - 36 = 33.5$. Recall that the core R function `wilcox.test` calculates the Mann–Whitney version of the test. Exercise 3.12.4 shows `wilcox.test` calculates T^+ to be 33.5, also.

For general score functions, the R function `rank.test` in the library `npsm` handles ties with average ranks.

As in the one-sample problem, for the location model, the estimation of the shift parameter Δ and the associated confidence intervals no adjustments are made for ties.

3.5 Scale Problem

Besides differences in location, we are often interested in the difference between scales for populations. Let $X_1, \ldots, X_{n_1}$ be a random sample with the common pdf

$$\frac{1}{\sigma_1} f\left(\frac{t-\theta_1}{\sigma_1}\right)$$

and $Y_1, \ldots, Y_{n_2}$ be a random sample with the common pdf

$$\frac{1}{\sigma_2} f\left(\frac{t-\theta_2}{\sigma_2}\right),$$

where $f(t)$ is a pdf that is symmetric about 0 and $\sigma_1, \sigma_2 > 0$. In this section our hypotheses of interest are

$$H_0: \eta = 1 \text{ versus } H_A: \eta \neq 1, \tag{3.41}$$

where $\eta = \sigma_2/\sigma_1$. Besides discussing rank-based tests for these hypotheses, we also consider the associated estimation of η, along with a confidence interval for η. So here the location parameters θ_1 and θ_2 are nuisance parameters.

As discussed in Section 2.10 of Hettmansperger and McKean (2011), there are asymptotically distribution-free rank-based procedures for this problem. We discuss the Fligner–Killeen procedure based on folded, aligned samples. The aligned samples are defined by

$$\begin{aligned} X_i^* &= X_i - \text{med}\{X_l\}, \quad i = 1, \ldots, n_1 \\ Y_j^* &= Y_j - \text{med}\{Y_l\}, \quad j = 1, \ldots, n_2. \end{aligned} \tag{3.42}$$

Next, the folded samples are $|X_1^*|, \ldots, |X_{n_1}^*|, |Y_1^*|, \ldots, |Y_{n_2}^*|$. The folded samples consist of positive items and their logs, essentially, differ by a location parameter, i.e., $\Delta = \log(\eta)$. This suggests the following log-linear model. Define Z_i by

$$Z_i = \begin{cases} \log |X_i^*| & i = 1, \ldots, n_1 \\ \log |Y_{i-n_1}^*| & i = n_1 + 1, \ldots, n_1 + n_2. \end{cases}$$

Let $\boldsymbol{c}$ be the indicator vector with its first n_1 entries set at 0 and its last n_2 entries set at 1. Then the log-linear model for the aligned, folded sample is

$$Z_i = \Delta c_i + e_i, \quad i = 1, 2, \ldots, n. \tag{3.43}$$

Our rank-based procedure is clear. Select an appropriate score function $\varphi(u)$ and generate the scores $a_\varphi(i) = \varphi[i/(n+1)]$. Obtain the rank-based fit of Model (3.43) and, hence, the estimator $\hat{\Delta}_\varphi$ of Δ. The estimator of η is then $\hat{\eta}_\varphi = \exp\{\hat{\Delta}_\varphi\}$. For specified $0 < \alpha < 1$, denote by (L_φ, U_φ) the confidence interval for Δ based on the fit; i.e.,

$$L_\varphi = \hat{\Delta}_\varphi - t_{\alpha/2}\hat{\tau}_\varphi\sqrt{\tfrac{1}{n_1} + \tfrac{1}{n_2}} \text{ and } U_\varphi = \hat{\Delta}_\varphi + t_{\alpha/2}\hat{\tau}_\varphi\sqrt{\tfrac{1}{n_1} + \tfrac{1}{n_2}}.$$

An approximate $(1-\alpha)100\%$ confidence interval for η is $(\exp\{L_\varphi\}, \exp\{U_\varphi\})$. Similar to the estimator $\hat{\eta}_\varphi$, an attractive property of this confidence interval is that its endpoints are always positive.

This confidence interval for η can be used to test the hypotheses (3.41); however, the gradient test is often used in practice. Because the log function is strictly increasing, the gradient test statistic is given by

$$S_\varphi = \sum_{j=1}^{n_2} a_\varphi[R(\log(|Y_j^*|)] = \sum_{j=1}^{n_2} a_\varphi[R|Y_j^*|], \tag{3.44}$$

where the ranks are over the combined folded, aligned samples. The standardized test statistic is $z = (S_\varphi - \mu_\varphi)/\sigma_\varphi$, where

$$\mu_\varphi = \tfrac{1}{n}\textstyle\sum_{i=1}^n a(i) = \overline{a} \text{ and } \sigma_\varphi^2 = \tfrac{n_1 n_2}{n(n-1)}\textstyle\sum_{i=1}^n (a(i) - \overline{a})^2. \tag{3.45}$$

What scores are appropriate? The case of most interest in applications is when the underlying distribution of the random errors in Model (3.43) is normal. In this case the optimal score[7] function is given by

---

[7]See Section 2.10 of Hettmansperger and McKean (2011).

$$\varphi_{FK} = \left(\Phi^{-1}\left(\frac{u+1}{2}\right)\right)^2. \tag{3.46}$$

Hence, the scores are of the form **squared-normal scores**. Note that these are light-tail score functions, which is not surprising because the scores are optimal for random variables which are distributed as $\log(|W|)$ where W has a normal distribution. Usually, the test statistic is written as

$$S_{FK} = \sum_{j=1}^{n_2} \left(\Phi^{-1}\left(\frac{R|Y_j^*|}{2(n+1)} + \frac{1}{2}\right)\right)^2. \tag{3.47}$$

This test statistic is discussed in Fligner and Killeen (1976) and Section 2.10 of Hettmansperger and McKean (2011). The scores generated by the score function (3.46) are in `npsm` under `fkscores`. Using these scores, straightforward code leads to the computation of the Fligner–Killeen procedure. We have assembled the code in the R function `fk.test` which has similar arguments as other standard two-sample procedures.

```
> args(fk.test)

function (x, y, alternative = c("two.sided", "less", "greater"),
    conf.level = 0.95)
NULL
```

In the call, `x` and `y` are vectors containing the original samples (not the folded, aligned samples); the argument `alternative` sets the hypothesis (default is two-sided); and `conf.level` sets the confidence coefficient of the confidence interval. The following example illustrates the Fligner–Killeen procedure and its computation.

Example 3.5.1 (Effect of Ozone on Weight of Rats)**.** Draper and Smith (1966) present an experiment on the effect of ozone on the weight gain of rats. Two groups of rats were selected. The control group ($n_1 = 21$) lived in an ozone-free environment for 7 days, while the experimental group ($n_2 = 22$) lived in an ozone environment for 7 days. At the end of the 7 days, the weight gain of each rat was taken to be the response. The comparison boxplots of the data, Figure 3.6, show a disparity in scale between the two groups.

For this example, the following code segment computes the Fligner–Killeen procedure for a two-sided alternative. The data are in the dataset `sievers`. We first split the groups into two vectors `x` and `y` as input to the function `fk.test`.

```
> x <- with(sievers,weight.gain[group=='Control'])
> y <- with(sievers,weight.gain[group=='Ozone'])
> fk.test(x,y)

statistic =  2.096 , p-value =  0.036084
```

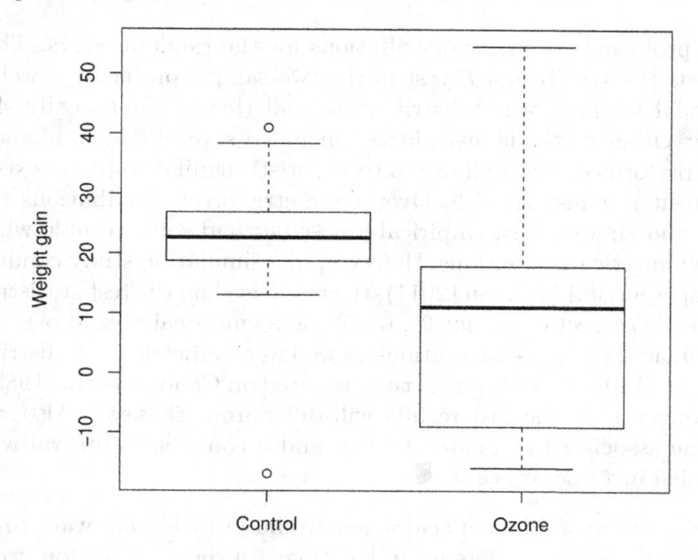

FIGURE 3.6
Comparison boxplots of weight gain in $n_1 = 21$ controls and $n_2 = 22$ ozone treated rats.

```
95 percent confidence interval:
1.0025 5.6365
Estimate: 2.377
```

Hence, the rank-based estimate of η is 2.377 with the 95% confidence interval of $(1.002, 5.636)$. The standardized Fligner–Killeen test statistic has value 2.09 with the p-value 0.0361 for a two-sided test. Thus, there is evidence that rats exposed to ozone have larger variability in their weight gains than nonexposed rats. ∎

The score function (3.46) is optimal for scale if the original samples are from normal populations. Several other score functions are discussed in Hettmansperger and McKean (2011). For example, the Wilcoxon scores are optimal for scale if $|X|$ follows a $F(2, 2)$-distribution.

It is well known that the traditional F-test based on the ratio of sample variances is generally invalid for nonnormal populations. This can be shown theoretically as in Section 2.10.2 of Hettmansperger and McKean (2011). On the other hand, the Fligner–Killeen test is asymptotically distribution-free over all symmetric error pdfs $f(x)$, and, as the next remark discusses, appears to be valid for skewed-contaminated normal distributions. We discuss several pertinent simulation studies next.

Remark 3.5.1 (Simulation Studies Concerning the FK-Test). Conover et al. (1983) discuss the results of a large simulation study of tests for scale in the

k-sample problem over many distributions for the random errors. The traditional Bartlett's test (usual F-test in the two-sample problem) is well known to be invalid for nonnormal distributions and this is dramatically shown in this study. Other methods investigated included several folded, aligned tests. One that performed very well uses a test statistic similar to (3.47) except that the exponent is 1 instead of 2. Over symmetric error distributions that test was valid and showed high empirical power but had some trouble with validity for asymmetric distributions. However, in a simulation study conducted by Hettmansperger and McKean (2011), the test based on the test statistic (3.47), (i.e., with the correct exponent 2), (3.47), was empirically valid over a family of contaminated skewed distributions as well as symmetric error distributions.

The base R function `fligner.test` is based on Conover et al. (1983) which used exponent 1, so the test results will differ from `fk.test`. Also, `fk.test` obtains the associated estimate of effect and a confidence interval which are not available in `fligner.test`. ■

In the presence of different scales, one would probably not want to perform the usual analysis for a difference in locations. In the next section, we discuss a rank-based analysis using placements, which is appropriate.

3.6 Placement Test for the Behrens–Fisher Problem

Suppose that we have two populations which differ by location and scale, and we are interested in testing that the locations are the same. To be specific, assume for this section that

$$\begin{aligned} X_1, X_2, \ldots, X_{n_1} &\quad \text{is a random sample on } X \text{ with cdf } F(x) \\ Y_1, Y_2, \ldots, Y_{n_2} &\quad \text{is a random sample on } Y \text{ with cdf } G(x). \end{aligned} \tag{3.48}$$

Let θ_X and θ_Y denote the medians of the distributions (populations) $F(x)$ and $G(y)$, respectively. Our two-sided hypothesis of interest is

$$H_0: \theta_X = \theta_Y \text{ versus } H_A\, \theta_X \neq \theta_Y. \tag{3.49}$$

This is called the Behrens–Fisher problem and the traditional test in this situation is the two-sample t-test which uses a t-statistic with the Satterthwaite degrees of freedom correction. This is the default test in the R function `t.test(x,y)`, where `x` and `y` are the R vectors containing the samples. In this section, we discuss a version of the two-sample Mann–Whitney–Wilcoxon test which serves as a robust alternative to this approximate t-test.

The nonparametric procedure that we consider was proposed by Fligner and Policello (1981); see also, Section 2.11 of Hettmansperger and McKean (2011) and Section 4.4 of Hollander and Wolfe (1999). It is a modified

Mann–Whitney–Wilcoxon test. For its underlying theory, we further assume that the cdfs $F(x)$ and $G(y)$ are symmetric; i.e., symmetric about θ_X and θ_Y, respectively. Provided we have sufficient data, an important diagnostic for checking symmetry is the comparison boxplot of the samples which offers a graphical check of difference in scales and the symmetry of the distributions.

The test statistic is the Mann–Whitney–Wilcoxon statistic defined in expression (3.6), which we rewrite here:

$$T^+ = \#_{i,j}\{Y_j - X_i > 0\} = \sum_{j=1}^{n_2} R(Y_j) - \frac{n_2(n_2+1)}{2}.$$

Under the symmetry assumption, $E_{H_0}(T^+) = n_1 n_2/2$. Thus the null expectation of T^+ is the same as in the location problem. The null variance, though, differs from that of the location problem. It is most easily seen in terms of what are known as the **placements.**

Let $P_1, \ldots, P_{n_1}$ denote the placements of the X_is in terms of the Y-sample, which is defined as

$$P_i = \#_j\{Y_j < X_i\}; \quad i = 1, \ldots, n_1. \tag{3.50}$$

In the same way, define the placements of the Y_js in terms of the X-sample as $Q_1, \ldots, Q_{n_2}$ where

$$Q_j = \#_i\{X_i < Y_j\}; \quad j = 1, \ldots, n_2. \tag{3.51}$$

Placements are ranks within a sample, but to avoid confusion with the ranks on the combined samples, the term placement is used. Define

$$\overline{P} = \frac{1}{n_1}\sum_{i=1}^{n_1} P_i \qquad \overline{Q} = \frac{1}{n_2}\sum_{i=1}^{n_2} Q_j$$

$$V_1 = \sum_{i=1}^{n_1}(P_i - \overline{P})^2 \qquad V_2 = \sum_{j=1}^{n_2}(Q_j - \overline{Q})^2. \tag{3.52}$$

Then the standardized test statistic is

$$Z_{fp} = \frac{T^+ - n_1 n_2/2}{(V_1 + V_2 + \overline{P}\overline{Q})^{1/2}}. \tag{3.53}$$

Under H_0, Z_{fp} has an asymptotic standard normal distribution. So, the Fligner–Policello test of the hypothesis (3.49) is

$$\text{Reject } H_0 : \theta_X = \theta_Y \text{ in favor of } H_A : \theta_X \neq \theta_Y \text{ if } Z_{fp} \geq z_{\alpha/2}. \tag{3.54}$$

This test has asymptotically level α.

For computation of this test **npsm** provides the R function `fp.test`. Its use is illustrated in the following example.

Example 3.6.1 (Geese). On page 136 of Hollander and Wolfe (1999) a study of healthy and lead-poisoned geese is discussed. The study involved 7 healthy geese and 8 lead-poisoned geese. The response of interest was the amount of plasma glucose in the geese in mg/100 ml of plasma. As discussed in Hollander and Wolfe (1999), the hypotheses of interest are:

$$H_0 : \theta_L = \theta_H \text{ vs. } H_A : \theta_L > \theta_H, \tag{3.55}$$

where θ_L and θ_H denote the true median plasma glucose values of lead-poisoned geese and healthy geese, respectively. The data are listed below. As shown, the R vectors `lg` and `hg` contain respectively the responses of the lead-poisoned and healthy geese. The sample sizes are too small for comparison boxplots, so, instead, the comparison dotplots found in Figure 3.7 are presented. Using Y as the plasma level of the lead-poisoned geese, the test statistic $T^+ = 40$ and its mean under the null hypothesis is 28. So there is some indication that the lead-poisoned geese have higher levels of plasma glucose. The following code segment computes the Fligner–Policello test of the hypotheses, (3.55), based on **npsm** function `fp.test`:

```
> lg = c(293,291,289,430,510,353,318)
> hg = c(227,250,277,290,297,325,337,340)
> fp.test(hg,lg,alternative='greater')

statistic =  1.467599 , p-value =  0.07979545
```

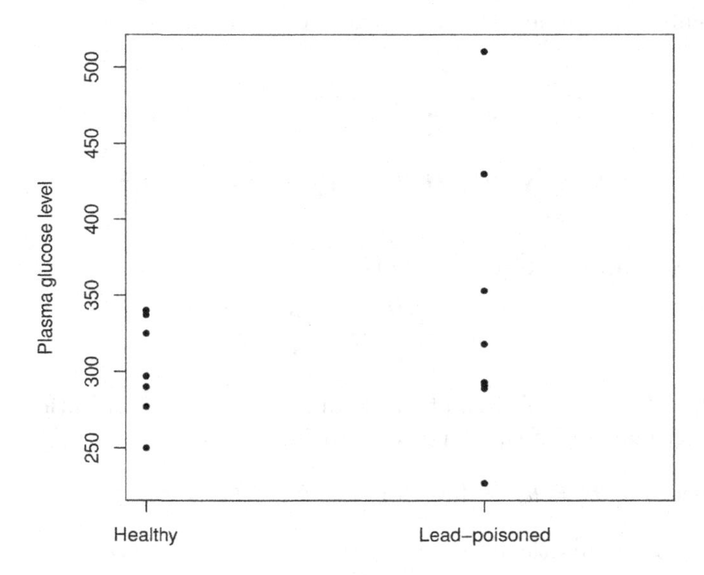

FIGURE 3.7
Comparison dotplots of plasma glucose in healthy and toxic geese.

Hence, for this study, the Fligner–Policello test finds that lead-poisoned geese have marginally higher plasma glucose levels than healthy geese. ∎

Discussion

In a two-sample problem, when testing for a difference in scale, we recommend the Fligner–Killeen procedure discussed in the last section. Unlike the traditional F-test for a difference in scale, the Fligner–Killeen procedure possesses robustness of validity. Besides the test for scales being the same, the Fligner–Killeen procedure offers an estimate of the ratio of scales and a corresponding confidence interval for this ratio. This gives the user a robust disparity measure (with confidence) for the scales difference in the data, which may be more useful information than an estimate of the difference in locations. For example, based on this information, a user may want to estimate the shift in location after a transformation or to use a weighted estimate of shift based on robust estimates of scale. If, though, a user wants to test for location differences in the presence of difference in scales, then we recommend the robust test offered by the Fligner–Policello procedure.

3.6.1 Estimation of a Shift Parameter, Δ

For this section, we assume the same model as in testing, Model (3.48), including the symmetry of the two pdfs $f(x)$ and $g(y)$ with their respective medians θ_X and θ_Y. The parameter of interest is the shift $\Delta = \theta_Y - \theta_X$ in location. The following two robust estimators of Δ are readily formulated: the difference in sample medians and the difference in the one-sample Hodges–Lehmann estimators. We begin with the difference in sample medians.

Define the sample medians and the estimate of Δ by

$$\begin{aligned} \hat{\theta}_{SX} &= \text{med}\{X_1, \ldots, X_{n_1}\} \\ \hat{\theta}_{SY} &= \text{med}\{Y_1, \ldots, Y_{n_2}\} \\ \hat{\Delta}_S &= \hat{\theta}_{SY} - \hat{\theta}_{SX}. \end{aligned} \tag{3.56}$$

Recall that the sample medians are asymptotically normal. Then because the samples are independent, it follows that

$$\hat{\Delta}_S \text{ has an approximate } N\left(\Delta, \tfrac{\tau_{SX}^2}{n_1} + \tfrac{\tau_{SY}^2}{n_2}\right), \tag{3.57}$$

where $\tau_{SX} = [2f(\theta_X)]^{-1}$ and $\tau_{SY} = [2g(\theta_Y)]^{-1}$. The estimators $\hat{\tau}_{SX}$ and $\hat{\tau}_{SY}$ of these scale parameters are given in expression (2.20). The `Rfit` function `taustar` computes them as follows. If the sample is in the R vector `x`, then the command `taustar(x,0)` computes $\hat{\tau}_{XS}$. The standard error of $\hat{\Delta}_S$ is given by

$$\text{SE}(\hat{\Delta}_S) = \sqrt{\frac{\hat{\tau}_{SX}^2}{n_1} + \frac{\hat{\tau}_{SY}^2}{n_2}}, \tag{3.58}$$

from which an approximate $(1-\alpha)100\%$ Wald-type confidence interval for Δ is given by $\hat{\Delta}_S \pm t_{\alpha/2,n_1+n_2-2}\mathrm{SE}(\hat{\Delta}_S)$.

In much the same way, the estimator of Δ based on the difference of the one-sample Hodges–Lehmann estimators is defined. Let

$$\begin{aligned}\hat{\theta}_{HX} &= \mathrm{med}\left\{\frac{X_i + X_j}{2}\right\}, \quad 1 \le i \le j \le n_1, \\ \hat{\theta}_{HY} &= \mathrm{med}\left\{\frac{Y_i + Y_j}{2}\right\}, \quad 1 \le i \le j \le n_2, \\ \hat{\Delta}_H &= \hat{\theta}_{HY} - \hat{\theta}_{HX}. \end{aligned} \tag{3.59}$$

Based on Section 2.6 and the independence of the samples, the approximate distribution of $\hat{\Delta}_H$ is:

$$\hat{\Delta}_H \text{ has an approximate } N\left(\Delta, \tfrac{\tau_{HX}^2}{n_1} + \tfrac{\tau_{HY}^2}{n_2}\right), \tag{3.60}$$

where $\tau_{HX} = [\sqrt{12}\int f^2(x)\,dx]^{-1}$ and $\tau_{HY} = [\sqrt{12}\int g^2(y)\,dy]^{-1}$. The estimators $\hat{\tau}_{HX}$ and $\hat{\tau}_{HY}$ of these scale parameters are the KSM estimators discussed in Section 3.2. The `Rfit` function `gettau` computes them as follows. If the sample is in the R vector `x`, then the command `gettau(x)` computes $\hat{\tau}_{XH}$. The standard error of $\hat{\Delta}_H$ is

$$\mathrm{SE}(\hat{\Delta}_H) = \sqrt{\frac{\hat{\tau}_{HX}^2}{n_1} + \frac{\hat{\tau}_{HY}^2}{n_2}} \tag{3.61}$$

which leads to an approximate $(1-\alpha)100\%$ confidence interval for Δ given by $\hat{\Delta}_H \pm t_{\alpha/2,n_1+n_2-2}\mathrm{SE}(\hat{\Delta}_H)$.

We have constructed two R functions that compute the respective analyses based on differences in medians and differences in one-sample Hodges–Lehmann estimates. The function for medians is `mood.ci`. For this analysis, we want the nonpooled SE, (3.58), that is obtained by the argument `var.equal=FALSE` which is the default. The use of the function `hodges_lehmann.ci` is similar to that of `mood.ci`. We illustrate these R-functions in Example 3.6.2.

As discussed in Section 3.2, the Wilcoxon estimator is another robust estimator of the shift parameter Δ. Recall that it is the median of the sample differences. For Model (3.48), it has an asymptotic normal distribution centered at Δ, but its asymptotic variance is more complicated than that of $\hat{\Delta}_H$; see Høyland (1965). We recommend bootstrapping its variance.

3.6.2 Classical Behrens–Fisher Model

The classical Behrens–Fisher model is a sub model of Model (3.48) in which the pdfs of X and Y have the same form. That is the pdfs of X and Y are respectively

$$f(x) = h\left[\tfrac{x-\theta_X}{\sigma_X}\right] \text{ and } g(y) = h\left[\tfrac{y-\theta_Y}{\sigma_Y}\right], \tag{3.62}$$

where $h(z)$ is a pdf which is symmetric about 0 and σ_X^2 and σ_Y^2 are the respective variances[8] of X and Y. Let $\Delta = \theta_Y - \theta_X$ be the shift in locations and let $\eta = \sigma_Y/\sigma_X$ be the ratio of scale parameters. This is also the model of Section 3.5, the two-sample scale problem.

For this model, a complete rank-based analysis can be performed. Tests concerning Δ can be conducted using the placement test while tests concerning η can be conducted using the Fligner–Killeen test. In terms of estimation, the estimator $\hat{\Delta}_H$ can be used to estimate the shift in locations Δ along with the SE($\hat{\Delta}_H$) and/or a $(1-\alpha)100\%$ confidence interval for Δ. The Fligner–Killeen procedure can be used to estimate η and to obtain $(1-\alpha)100\%$ confidence interval for η. We illustrate this analysis in Example 3.6.2.

Under the Behrens–Fisher model, a few points on efficiency between the analyses are in order. Denote the difference in sample means estimator by $\hat{\Delta}_{LS} = \bar{Y} - \bar{X}$. Under this model, the efficiency[9] of $\hat{\Delta}_H$ relative to $\hat{\Delta}_{LS}$ is the same as the efficiency in the one-sample location model; see, for example, Høyland (1965). In particular, if the pdf $h(z)$ is the normal pdf, then this efficiency is 0.955. For heavy tailed-error distributions, generally, $\hat{\Delta}_H$ is much more efficient than $\hat{\Delta}_{LS}$. Hence, $\hat{\Delta}_H$ is a highly efficient robust estimator. The efficiency of $\hat{\Delta}_S$ (difference in sample medians) relative to $\hat{\Delta}_{LS}$ is the same as that of the sample median relative to the sample mean in the one-sample problem. At the normal, this efficiency is only 0.637. So, $\hat{\Delta}_S$, while robust, is not highly efficient. For this Behrens–Fisher model, the efficiency of the two-sample Wilcoxon estimator, $\hat{\Delta}_W$, is a function of η and, hence, is more complex; see Høyland (1965).

Example 3.6.2 (Song Length). Verzani (2014) discussed a dataset concerning the lengths of songs on albums recorded by the band U2.[10] In this example, we considered the albums *Zooropa* and *Rattle & Hum*. For these albums, the responses (length of songs in seconds) are given by:

Zooropa									
390	258	298	255	324	240	416	319	225	284
Rattle & Hum									
187	186	179	269	382	264	353	38	350	267
229	383	254	302	195	43	337	390		

Figure 3.8 displays the comparison boxplot of the song lengths for these albums. Based on this plot, the sample median song length of the *Zooropa* album is larger than that of the *Rattle & Hum* album, while the song lengths of the *Rattle & Hum* show more variation than that of the songs on the *Zooropa* album. There is, however, considerable variability in the data. To estimate

[8] If the reader does not want to assume the existence of variances, a robust scale functional can be used instead of the standard deviation.

[9] Technically the asymptotic relative efficiency; see Section 3.7.

[10] The complete dataset is `u2` in the package `UsingR` by Verzani (2014).

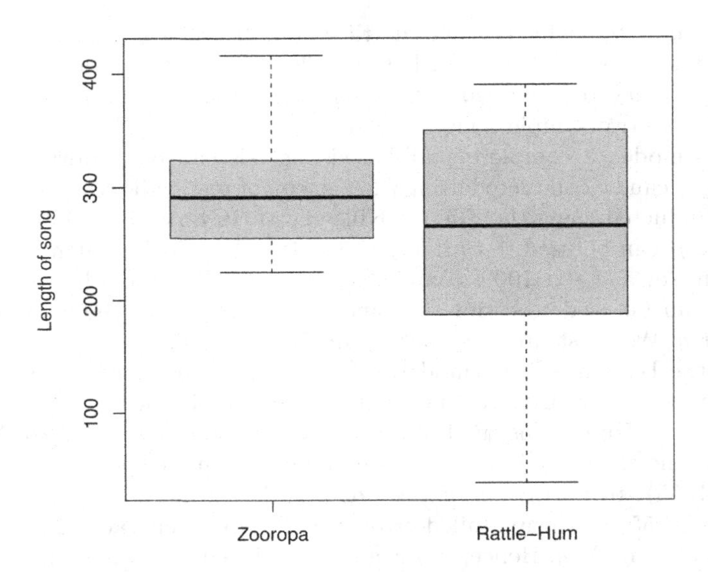

FIGURE 3.8
Comparison boxplots of the lengths of songs in seconds on the albums *Zooropa* and *Rattle & Hum* by the band U2.

the effect Δ along with a 95% confidence interval for it, we selected the analysis based on the differences of the one-sample Hodges–Lehmann estimates. The computations are in the next R segment, where the R vectors `zoo` and `rh` contain the song lengths for the respective albums *Zooropa* and *Rattle & Hum*.

```
> fitW <- hodges_lehmann.ci(zoo,rh)
> fitW

95 percent confidence interval:
[1] -45.908  95.908

> fitS <- mood.ci(zoo,rh)
> fitS

95 percent confidence interval:
[1] -95.371 146.371

> precision <- fitS$stderr^2/fitW$stderr^2
> precision

[1] 2.9057
```

```
> with(fitW,print(c(estimate,stderr)))

[1] 25.000 34.496
```

The Hodges–Lehmann estimate of Δ is 25.0 seconds with a standard error of 34.5 seconds. The confidence interval contains 0; so, the effect is not statistically significant. We also computed the analysis based on the difference of sample medians. This analysis is much less precise than that based on the difference of the Hodges–Lehmann estimates. Using the ratio of squared standard errors as our estimate of precision, the Hodges–Lehmann analysis is almost 3 times more precise than the analysis based on sample medians. Notice that these results on precision confirm the above discussion on efficiency. These data contain no outliers, suggesting that the tail weight is moderately heavy at most and that the Hodges–Lehmann analysis will be more efficient (precise) than the median analysis.

In the last R segment, we compute the Fligner–Killeen analysis of the ratio of variances, $\eta^2 = \sigma^2_{\text{rh}}/\sigma^2_{\text{zoo}}$. The estimate of η is 1.78, but, since 1.0 is in the confidence interval it is not significant.

```
> fk.test(zoo,rh)

statistic =  1.2476 , p-value =  0.21217
95  percent confidence interval:
0.82256 3.8532
Estimate: 1.7803
```

■

Discussion

In this section, we presented rank-based analyses for the shift parameter Δ in the two-sample Behrens–Fisher model. In this model, besides a shift in location, the two population scale parameters may also differ. Our analysis includes the Fligner–Policello placements test for hypotheses concerning Δ that is a Wilcoxon type test. We also presented two estimation procedures for Δ, including, for each, associated standard errors and confidence intervals for Δ. One is based on the difference of sample medians, $(\hat{\Delta}_S)$, while the other is based on the difference of one-sample Hodges–Lehmann estimators, $(\hat{\Delta}_H)$. Of the two, we recommend $\hat{\Delta}_H$ because not only is it robust, but it has high efficiency relative to the difference in sample means. This efficiency is 0.955 when the distribution of the random errors of the model is normal.

In practice, an experimenter usually is not sure whether the model is a Behrens–Fisher model or a shift in location model as discussed in Section 3.2. In terms of recommendation of a highly efficient robust analysis, should the experimenter use the difference of one-sample Hodges–Lehmann estimators, $(\hat{\Delta}_H)$, or the Wilcoxon estimator, the median of the differences between the samples, $(\hat{\Delta}_W)$? Subject matter may be present to help in the decision; for example, a knowledge of past, similar type problems. Or perhaps it is thought that the change in responses will be small, i.e., of first order. In which case, $\hat{\Delta}_W$ would be selected. A useful diagnostic here is the comparison boxplot between the samples. The sample interquartile range (length of the box) is a difference of robust estimates, so a comparison of the lengths of the boxes is based on robust estimates.

Let $\eta = \sigma_X/\sigma_Y$. Often in practice a pretest of the hypotheses $H_0 : \eta = 1$ versus $H_A : \eta \neq 1$ is conducted first. Then if H_0 is rejected, an analysis based on the Behrens–Fisher model is conducted; else an analysis based on the two-sample location model of Section 3.2 is conducted. The parametric F-test of variances is often used as the pretest. We do not recommend this test because it lacks robustness of validity, as discussed in Section 3.5. Instead, we recommend the use of the Fligner–Killeen test discussed in Section 3.5. It possesses both robustness of validity and robustness of power. Its analysis, also includes a robust estimator of η along with an associated confidence interval for η.

If this analysis is used for the situation in Example 3.6.2 with $\alpha = 0.05$ for the pretest (Fligner–Killeen), then, in terms of estimation of Δ, the two-sample location analysis based on $\hat{\Delta}_W$ would have been selected. Its computation is:

```
> wilcox.test(zoo,rh,conf.int=TRUE)

        Wilcoxon rank sum test with continuity correction

data:  zoo and rh
W = 110, p-value = 0.34
alternative hypothesis: true location shift is not equal to 0
95 percent confidence interval:
 -39 112
sample estimates:
difference in location
                 36.09
```

For the two analyses, the estimates of shift are $\hat{\Delta}_H = 25.0$ and $\hat{\Delta}_W = 36.1$. Based on their respective confidence intervals each containing 0, though, neither estimate is significant at $\alpha = 0.05$.

3.7 Efficiency and Optimal Scores*

In this section, we return to the two-sample location problem, (3.23). We discuss the optimal score function in the advent that the user knows the form of the population distributions. In a first reading, this section can be skipped, but even a brief browsing serves as motivation for the next section on adaptive score selection schemes. Recently Doksum (2013) developed an optimality result for the Hodges–Lehmann estimate based on a rank-based likelihood.

3.7.1 Efficiency

In this subsection, we briefly discuss the robustness and efficiency of the rank-based estimates. This leads to a discussion of optimizing the analysis by the suitable choice of a score function.

When evaluating estimators, an essential property is their robustness. In Section 2.9, we discussed the influence function of an estimator. Recall that it is a measure of the sensitivity of the estimator to an outlier. We say an estimator is **robust** if its influence function is bounded for all possible outliers.

Consider the rank-based estimator $\widehat{\Delta}_\varphi$ based on the score function $\varphi(u)$. Recall that its standard deviation is $\tau_\varphi/\sqrt{n}$; see expression (3.19). Then for an outlier at z, $\widehat{\Delta}_\varphi$ has the influence function

$$\mathrm{IF}(z;\widehat{\Delta}_\varphi) = -I_X(z)\left(\tau_\varphi \frac{n}{n_1}\right)\varphi[(F(z)] + I_Y(z)\left(\tau_\varphi \frac{n}{n_1}\right)\varphi[(F(z)], \tag{3.63}$$

where $I_X(z)$ is 1 or 0 depending on whether z is thought as being from the sample of Xs or Ys, respectively.[11] The indicator $I_Y(z)$ is defined similarly. For a rank-based estimator, provided its score function $\varphi(u)$ is bounded, its influence function is a bounded function of z; hence, the rank-based estimator is robust. There are a few scores used occasionally in practice, which have unbounded score functions. In particular, the normal scores, discussed in Section 3.7.1, belong to this category. In practice, though, generally scores with bounded score functions are used.

In contrast, the influence function for the LS estimator $\overline{Y} - \overline{X}$ of $\widehat{\Delta}$ is given by

$$\mathrm{IF}(z;\widehat{\Delta}_\varphi) = -I_X(z)\left(\frac{n}{n_1}\right)z + I_Y(z)\left(\frac{n}{n_2}\right)z; \tag{3.64}$$

which is an unbounded function of z. Hence, the LS estimator is not robust.

We next briefly consider relative efficiency of estimators of Δ for the two-sample location model (3.23). Suppose we have two estimators, $\widehat{\Delta}_1$ and $\widehat{\Delta}_2$, such that, for $i = 1, 2$, $\sqrt{n}(\widehat{\Delta}_i - \Delta)$ converges in distribution to the $N(0, \sigma_i^2)$ distribution. Then the asymptotic relative efficiency (ARE) between the estimators is the ratio of their asymptotic variances; i.e.,

$$\mathrm{ARE}(\widehat{\Delta}_1, \widehat{\Delta}_2) = \frac{\sigma_2^2}{\sigma_1^2}. \tag{3.65}$$

[11] See Chapter 2 of Hettmansperger and McKean (2011).

Note that $\widehat{\Delta}_2$ is more efficient than $\widehat{\Delta}_1$ if this ratio is less than 1. Provided that the estimators are location and scale equivariant, this ratio is invariant to location and scale transformations, so for its computation only the form of the pdf needs to be known. Because they minimize norms, all rank-based estimates and the LS estimate are location and scale equivariant. Also, all these estimators have an asymptotic variance of the form $\kappa^2\{(1/n_1)+(1/n_2)\}$. The scale parameter κ is often called the constant of proportionality. Hence, in the ARE ratio, the design part (the part in braces) of the asymptotic variance cancels out, and the ratio simplifies to the ratio of the parameters of proportionality; i.e., the κ^2s. For addtional details regarding the ensuing discussion of AREs, see Hettmansperger and McKean (2011) and Hollander and Wolfe (1999).

First, consider the ARE between a rank-based estimator and the LS estimator. Assume that the random errors in Model (3.23) have variance σ^2 and pdf $f(t)$. Then, from the above discussion, the ARE between the rank-based estimator with score generating function $\varphi(u)$ and the LS estimator is

$$\text{ARE}(\widehat{\Delta}_\varphi, \widehat{\Delta}_{LS}) = \frac{\sigma^2}{\tau_\varphi^2}, \tag{3.66}$$

where τ_φ is defined in expression (3.19).

If the Wilcoxon scores are chosen, then $\tau_W = [\sqrt{12}\int f^2(t)\,dt]^{-1}$. Hence, the ARE between the Wilcoxon and the LS estimators is

$$\text{ARE}(\widehat{\Delta}_W, \widehat{\Delta}_{LS}) = 12\sigma^2\left[\int f^2(t)\,dt\right]^2. \tag{3.67}$$

If the error distribution is normal, then this ratio is 0.955; see, for example, Hettmansperger and McKean (2011). Thus, the Wilcoxon estimator loses less than 5% efficiency to the LS estimator if the errors have a normal distribution.

In general, the Wilcoxon estimator has a substantial gain in efficiency over the LS estimator for error distributions with heavier tails than the normal distribution. To see this, consider a family of contaminated normal distributions with cdfs

$$F(x) = (1-\epsilon)\Phi(x) + \epsilon\Phi(x/\sigma_c), \tag{3.68}$$

where $0 < \epsilon < 0.5$ and $\sigma_c > 1$. If we are sampling from this cdf, then $(1-\epsilon)100\%$ of the time we are sampling from a $N(0,1)$ distribution while $\epsilon 100\%$ of the time we are sampling from a heavier tailed $N(0, \sigma_c^2)$ distribution. For illustration, consider a contaminated normal with $\sigma_c = 3$. In the first row of Table 3.3 are the AREs between the Wilcoxon and the LS estimators for a sequence of increasing contamination. Note that even if $\epsilon = 0.01$, i.e., only 1% contamination, the Wilcoxon is more efficient than LS.

If sign scores are selected, then $\tau_S = [2f(0)]^{-1}$. Thus, the ARE between the sign and the LS estimators is $4\sigma^2 f^2(0)$. This is only 0.64 at the normal distribution, so medians are much less efficient than means if the errors have a normal distribution. Notice from Table 3.3 that for this mildly contaminated

TABLE 3.3
AREs among the Wilcoxon (W), sign (S), and LS estimators when the errors have a contaminated normal distribution with $\sigma_c = 3$ and proportion of contamination ϵ.

	ϵ, (Proportion of Contamination)							
	0.00	0.01	0.02	0.03	0.05	0.10	0.15	0.25
ARE(W, LS)	0.955	1.009	1.060	1.108	1.196	1.373	1.497	1.616
ARE(S, LS)	0.637	0.678	0.719	0.758	0.833	0.998	1.134	1.326
ARE(W, S)	1.500	1.487	1.474	1.461	1.436	1.376	1.319	1.218

normal, the proportion of contamination must exceed 10% for the sign estimator to be more efficient than the LS estimator. The third row of the table contains the AREs between the Wilcoxon and sign estimators. In all of these situations of the contaminated normal distribution, the Wilcoxon estimator is more efficient than the sign estimator.

If the true pdf of the errors is $f(t)$, then the optimal score function is $\varphi_f(u)$ defined in expression (3.20).[12] These scores are asymptotically efficient; i.e., achieve the Rao–Cramer lower bound asymptotically similar to maximum likelihood estimates. We, next, discuss optimal scores and corresponding AREs for three distributions: normal, Laplace, and logistic.

Suppose $f(t)$ is a normal pdf. Then the rank-based estimator, $\widehat{\Delta}_{ns}$, based on normal scores is optimal. As we noted earlier, however, the influence function of $\widehat{\Delta}_{ns}$ is unbounded so, $\widehat{\Delta}_{ns}$ is not robust. Further, as Huber (1981) notes, for heavy-tailed contaminated distributions, the asymptotic variance of $\widehat{\Delta}_{ns}$ generally increases rapidly as the rate of contamination increases. This is not true for the robust estimates, such as the rank-based estimator, $\widehat{\Delta}_W$, based on the Wilcoxon scores.

Next, consider the logistic distribution whose pdf is given by,

$$f(x) = \frac{1}{b}\frac{\exp\{-[x-a]/b\}}{(1+\exp\{-[x-a]/b\})^2}, \quad -\infty < x < \infty, \tag{3.69}$$

The logistic pdf is similar to the normal pdf, but with slightly heavier tails than the normal pdf. The Wilcoxon scores are optimal for errors with a logistic distribution. The ARE between the Wilcoxon and LS estimators is 1.097 at the logistic distribution; hence, the Wilcoxon estimator is slightly more efficient than LS for logistically distributed random errors.

The Laplace (double exponential) pdf is given by

$$f(x) = \frac{1}{2}e^{-|x|}, \quad -\infty < x < \infty. \tag{3.70}$$

This pdf has heavier tails than the normal and logistic pdfs. The sign scores are optimal if the errors have a Laplace distribution. The ARE between the

[12]See Chapter 2 of Hettmansperger and McKean (2011).

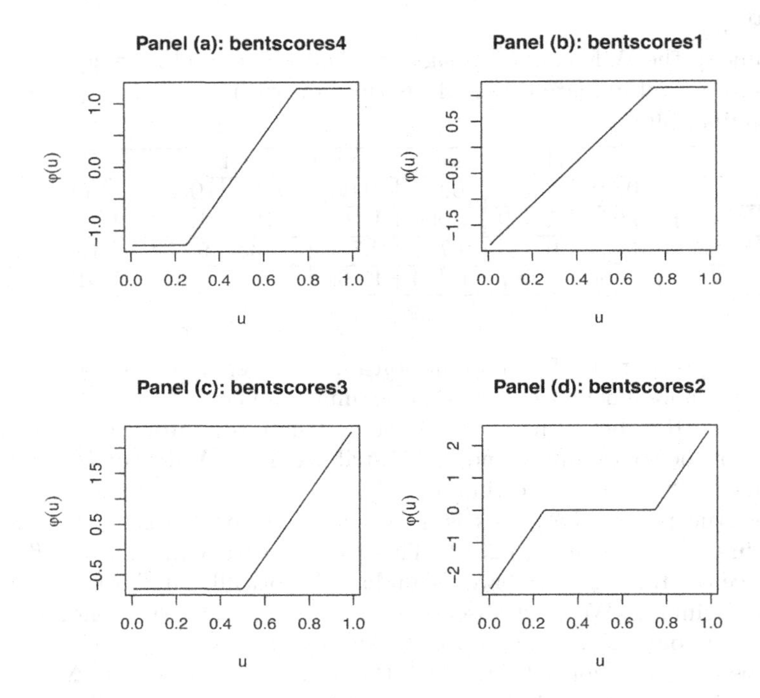

FIGURE 3.9
Plots of four bent score functions available in `Rfit`.

sign scores estimator and LS estimator is 2.0 at the Laplace distribution. The ARE between the sign scores estimator and the Wilcoxon estimator is 4/3 at the Laplace distribution, while it is 3/4 at the logistic distribution.

This range of efficiencies between the optimal scores for the logistic and Laplace distributions suggests a simple family of score functions called the Winsorized Wilcoxons. These scores are generated by a nondecreasing piecewise continuous function defined on $(0, 1)$ which is flat at both ends and linear in the middle. As discussed in McKean, Vidmar, and Sievers (1989), these scores are optimal for densities with "logistic" middles and "Laplace" tails. Four typical score functions from this family are displayed in Figure 3.9. If the score function is odd about 1/2, as in Panel (a), then it is optimal for a symmetric distribution; otherwise, it is optimal for a skewed distribution. Those in Panels (b) and (c) are optimal for distributions which are skewed right and skewed left, respectively. The type in Panel (d) is optimal for light-tailed distributions (lighter than the normal distribution), which occasionally are of use in practice. In `Rfit`, these scores are used via the options: `scores=bentscores4` for the type in Panel (a); `scores=bentscores1` for the type in Panel (b); `scores=bentscores3` for the type in Panel (c); and `scores=bentscores2` for the type in Panel (d).

In the two-sample location model, as well as in all fixed effects linear models, the ARE between two estimators summarizes the difference in analyses based on the estimators. For example, the asymptotic ratio of the squares of the lengths of the confidence intervals based on the two estimators is their ARE. Once we have data, we call the estimated ARE between two estimators the **precision coefficient** or the estimated precision of one analysis over the other. For instance, if we use the two rank-based estimators of Δ based on their respective score functions φ_1 and φ_2, then

$$\text{Precision}(\text{Analysis based on } \hat{\Delta}_1, \text{Analysis based on } \hat{\Delta}_2) = \frac{\hat{\tau}^2_{\varphi_2}}{\hat{\tau}^2_{\varphi_1}}. \tag{3.71}$$

For a summary example of this section, we reconsider the quail data. This time we select an appropriate score function which results in an increase in precision (empirical efficiency) over the Wilcoxon analysis.

Example 3.7.1 (Quail Data, Continued). The data discussed in Example 3.2.3 were drawn from a large study involving many datasets, which was discussed in some detail in McKean et al. (1989). Upon examination of the residuals from many of these models, it appeared that the underlying error distribution is skewed with heavy right-tails, although outliers in the left tail were not usual. Hence, scores of the type `bentscore1` were considered. The final recommended score was of this type with a bend at $u = 3/4$, which is the score graphed in Panel (b) of the above plot. The `Rfit` of the data, using these scores, computes the estimate of shift in the following code segment:

```
> mybentscores = bentscores1
> mybentscores@param<-c(0.75,-2,1)
> fit <- rfit(ldl~treat, scores=mybentscores, data=quail2)
> summary(fit)

Call:
rfit.default(formula = ldl ~ treat, data = quail2, scores
   = mybentscores)

Coefficients:
            Estimate Std. Error t.value p.value
(Intercept)    66.00       4.12   16.04 1.2e-15 ***
treat         -16.00       7.66   -2.09   0.046 *
---
Signif. codes:
0 '***' 0.001 '**' 0.01 '*' 0.05 '.' 0.1 ' ' 1

Multiple R-squared (Robust): 0.12813
Reduction in Dispersion Test: 4.115 p-value: 0.05211
```

The estimate of τ for the bentscores is

```
> fit$tauhat

[1] 19.79
```

while the estimate of τ for the Wilcoxon scores (default in `Rfit`) is

```
> rfit(ldl~treat, data=quail2)$tauhat

[1] 21.649
```

In summary, using these bent scores, the rank-based estimate of shift is $\widehat{\Delta} = -16$ with standard error 7.66, which is significant at the 5% level for a two-sided test. Note that the estimate of shift is closer to the estimate of shift based on the boxplots than the estimate based on the Wilcoxon score; see Example 3.2.3. From the code segment, the estimate of the relative precision (3.71) of the bent scores analysis versus the Wilcoxon analysis is $19.79/21.649 = 0.836$. Hence, the Wilcoxon analysis is about 16% less precise than the bentscore analysis for this dataset. ∎

The use of Winsorized Wilcoxon scores for the quail data was based on an extensive investigation of many datasets. Estimation of the score function based on the rank-based residuals is discussed in Naranjo and McKean (1987). Similar to density estimation, though, large sample sizes are needed for these score function estimators. Adaptive procedures for score selection are discussed in the next section.

As we discussed above, if no assumptions about the distribution can be reasonably made, then we recommend using Wilcoxon scores. The resulting rank-based inference is robust and is highly efficient relative to least squares based inference. Wilcoxon scores are the default in `Rfit` and are used for most of the examples in this book.

3.8 Adaptive Rank Scores Tests

As discussed in Section 3.7.1, the optimal choice for score function depends on the underlying error distribution. However, in practice, the true distribution is not known and we advise against making any strong distributional assumptions about its exact form. That said, there are times, after working with prior similar data perhaps, that we are able to choose a score function which is appropriate for the data at hand. More often, though, the analyst has a single dataset with no prior knowledge. In such cases, if hypothesis testing is the main inference desired, then the adaptive procedure introduced by Hogg (1974) allows the analyst to first look at the (combined) data to choose an appropriate score function and then allows him to use the same data to test a statistical hypothesis, without inflating the type I error rate.

In this section, we present the version of Hogg adaptive scheme for the two-sample location problem that is discussed in Section 10.6 of Hogg, McKean, and Craig (2019). Consider a situation where the error distributions are either symmetric or right-skewed with tail weights that vary from light-tailed to heavy-tailed. For this situation, we have chosen a scheme that is based on selecting one of the four score functions in Figure 3.10.

Suppose each of the four tests from which the adaptive test is chosen has the level of significance α. Then under H_0, because the selected tests are distribution-free and the selector statistics are functions of the combined sample order statistics, the adaptive test has exact size α[13]; i.e., the adaptive test has the same level of significance as the tests from which it is selected.

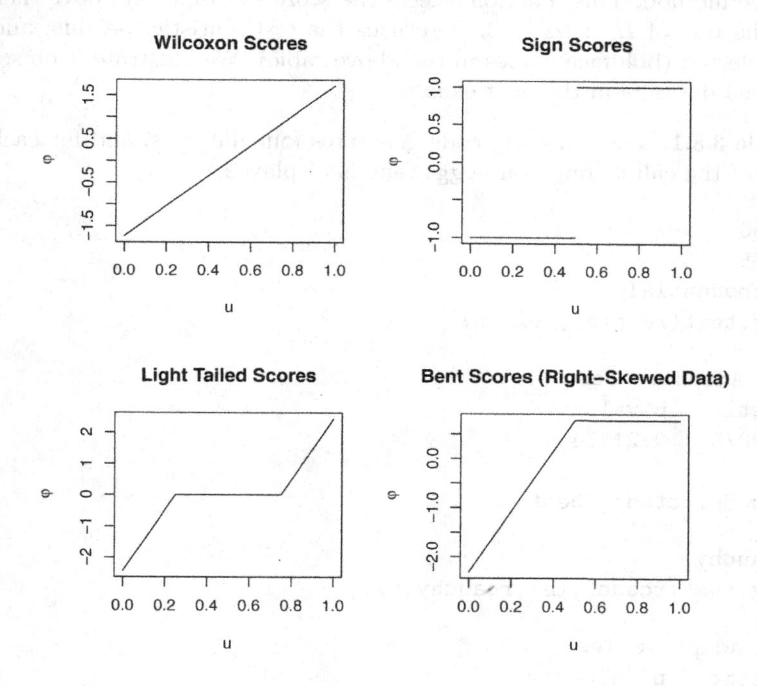

FIGURE 3.10
Score functions used in Hogg's adaptive scheme.

The choice of the score function is made by selector statistics which are based on the order statistics of the combined data. For our scheme, we have chosen two which we label Q_1 and Q_2, which are

$$Q_1 = \frac{\bar{U}_{0.05} - \bar{M}_{0.5}}{\bar{M}_{0.5} - \bar{L}_{0.05}} \quad \text{and} \quad Q_2 = \frac{\bar{U}_{0.05} - \bar{L}_{0.05}}{\bar{U}_{0.5} - \bar{L}_{0.5}}, \tag{3.72}$$

[13]See Section 10.6 of Hogg et al. (2019).

where $U_{0.05}$ is the mean of the Upper 5%, $M_{0.5}$ is the mean of the Middle 50%, and $L_{0.05}$ is the mean of the Lower 5% of the combined sample. Note $Q1$ is a measure of skewness and $Q2$ is a measure of tail heaviness. Benchmarks for the selectors must also be chosen. The following table shows the benchmarks that we used for our scheme.

Benchmark	Score Selected
$Q_2 > 7$	**Sign** Scores
$Q_1 > 1$ & $Q_2 < 7$	**Bent** Score for Right-Skewed Data
$Q_1 \leq 1$ & $Q_2 \leq 2$	**Light**-Tailed Scores
else	**Wilcoxon**

The R function `hogg.test` is available in `npsm` which implements this adaptive method. This function selects the score as discussed above then obtains the test of $H_0 : \Delta = 0$. It returns the test statistic, p-value, and the score selected (boldface names in the above table). We illustrate it on several generated datasets in the next example.

Example 3.8.1. The following code generates four datasets, and for each the results of the call to function `hogg.test` is displayed.

```
> m<-50
> n<-55
> # Exponential
> hogg.test(rexp(m),rexp(n))

Hogg's Adaptive Test
Statistic   p.value
  4.32075   0.21434

Scores Selected:  bent

> # Cauchy
> hogg.test(rcauchy(m),rcauchy(n))

Hogg's Adaptive Test
Statistic   p.value
  3.56604   0.31145

Scores Selected:  bent

> # Normal
> hogg.test(rnorm(m),rnorm(n))

Hogg's Adaptive Test
Statistic   p.value
  2.84318   0.57671
```

```
Scores Selected:  Wilcoxon

> # Uniform
> hogg.test(runif(m,-1,1),runif(n,-1,1))

Hogg's Adaptive Test
Statistic   p.value
  -2.1321    0.2982

Scores Selected:  light
```

■

In applying the adaptive scheme to a particular application, other score functions can be selected for the scheme which might be more suitable for the problem at hand. The number of score functions can also vary.

Remark 3.8.1. In an investigation of microarray data, Kloke et al. (2010) successfully used Hogg's adaptive scheme. The idea is that each gene may have a different underlying error distribution. Using the same score function, for example the Wilcoxon, for each of the genes may result in low power for any number of genes, perhaps ones which would have a much higher power with a more appropriately chosen score function. As it is impossible to examine the distribution of each of the genes in a microarray experiment an automated adaptive approach seems appropriate.

Al-Shomrani (2003) extended Hogg's adaptive scheme to estimation and testing in a linear model. It is based on an initial set of residuals and, hence, its tests are not exact level α. However, for the situations covered in Al-Shomrani's extensive Monte Carlo investigation, the overall liberalness of the adaptive tests was slight. We discuss Al-Shomrani's scheme in Chapter 8. Hogg's adaptive strategy was recently extended to mixed linear models by Okyere (2011). ■

3.9 Power and Sample Size Determination for Rank-Based Tests*

The model for this section, is the two-sample location model. Briefly, let X and Y be continuous independent random variables. Let $f_X(t)$ denote the pdf of X. Then the pdf of Y is $f_Y(t) = f_X(t-\Delta)$, where the parameter Δ is the shift in location from the distribution of X to the distribution of Y. For example, Δ is the difference in the medians of Y and X or equivalently the difference in means; i.e., $\Delta = \theta_Y - \theta_X = \mu_Y - \mu_X$. Let $X_1, \ldots, X_{n_1}$ be a random sample on X and $Y_1, \ldots, Y_{n_2}$ be a random sample on Y. Assume that the samples are independent of one another. In previous sections, we discussed rank-based tests for the shift parameter Δ. We discussed the distribution of the test statistics under the null hypothesis that $\Delta = 0$. In this section, we discuss their behavior under alternative hypotheses. In particular, we explore the power of the tests to detect alternatives of interest. In practice, if possible, before the samples are drawn, we want to plan the experimental design. This includes determining appropriate sample sizes. Sample size determination is the second topic of this section, and it is directly linked with the power of the test to detect alternatives of interest.

3.9.1 Power of Rank-Based Tests

For the discussion on the power of rank-based tests, consider the one-sided hypotheses given in expression

$$H_0 : \Delta = 0 \text{ versus } H_A : \Delta > 0. \tag{3.73}$$

The "less than" alternative and the two-sided alternative are discussed briefly in Section 3.9.2. Our discussion holds for general score functions $\varphi(u)$. So consider the distribution-free test statistic $S_\varphi(0)$ of Section 3.2.3; i.e.,

$$S_\varphi(0) = \sum_{j=1}^{n_2} a_\varphi[R(Y_j)], \tag{3.74}$$

where $R(Y_j)$ is the rank of Y_j in the combined sample $X_1, \ldots, X_{n_1}, Y_1, \ldots, Y_{n_2}$ and $a_\varphi(j) = \varphi[j/(n+1)]$. For a given significance level α, let c_{α,n_1,n_2} be the critical value of $S_\varphi(0)$. Thus the decision rule is

$$\text{Reject } H_0 \text{ at level } \alpha, \text{ if } S_\varphi(0) \geq c_{\alpha,n_1,n_2}. \tag{3.75}$$

The **power function** of the test is given by

$$\gamma_n(\Delta) = P_\Delta[S_\varphi(0) \geq c_{\alpha,n_1,n_2}], \quad \Delta > 0, \tag{3.76}$$

where the subscript Δ on P means that the probability is determined when Δ is the true shift parameter. Thus, $\gamma(\Delta)$ is the probability that the test rejects H_0 when the alternative Δ is true.

The null distribution of the test statistic $S_\varphi(0)$ is distribution-free, but it is not distribution-free under alternatives. Hence, for $\Delta > 0$ the power depends on the pdf $f_X(t)$ and Δ as well as the ranks and sample sizes. Thus, we cannot obtain the power function in closed form. The following theoretical results, though, are informative about the nature of the power function. For their proofs, see Chapter 2 of Hettmansperger and McKean (2011). For the power function $\gamma(\Delta)$, (3.76), and the one-sided alternative, (3.73), we have the following properties:

1. First, $\gamma_n(0) = \alpha$; hence, the level of the test is the power of the test when $\Delta = 0$.
2. The power function $\gamma_n(\Delta)$ is a monotonically increasing function of Δ; hence, the power of the test increases as Δ increases. Because $\gamma_n(0) = \alpha$, this implies that $\gamma_n(\Delta) \geq \alpha$ for all Δ; i.e., the power of the test never falls below the level, (we say that the test is **unbiased**).
3. For a fixed alternative Δ, $\gamma_n(\Delta) \to 1$ as $n \to \infty$. We say that the test is **consistent** for any alternative.
4. For fixed n, $\gamma_n(\Delta) \to 1$ as $\Delta \to \infty$. We say that the test is **resolving**.
5. For n large and all $\Delta_n > 0$,

$$\gamma_n(\Delta_n) \sim 1 - \Phi\left(z_\alpha - \sqrt{\frac{n_1 n_2}{n}} \tau_\varphi^{-1} \Delta_n\right), \tag{3.77}$$

where τ_φ is the scale parameter defined in expression (3.19), $\Delta_n = \delta/\sqrt{n}$i ($\delta > 0$), and $\Phi(t)$ is the cdf of a standard normal random variable. The right-side is the limit of the left-side as $n \to \infty$. This is often called the **Asymptotic Power Lemma**.

Although in general we cannot obtain $\gamma_n(\Delta)$ in closed form, as the following example shows, it is straightforward to simulate the power function for a specified situation.

Example 3.9.1 (Simulation of the Power Function). Although for this example, we have selected a specific situation, as we note later, it is easy to change the code for other situations. Assume that the sample sizes are $n_1 = 15$ and $n_2 = 20$ and that the samples are generated as a linear transformation of a t-distribution with 2 degrees of freedom; that is,

$$\begin{aligned} X_i &= \theta + bT_i, \quad i = 1, \ldots, n_1 \\ Y_j &= \Delta + \theta + bT_{n_1+j}, \quad j = 1, \ldots, n_2, \end{aligned}$$

where $T_1, T_2, \ldots T_{n_1+n_2}$ are iid with a common t-distribution having 2 degrees of freedom, θ is the median of X, $b > 0$ is a scale parameter, and Δ is the shift parameter. Our test statistics are location and scale equivariant, so it does not matter what values we assign to θ and b, but for this example we set $\theta = 25$ and $b = 10$. As specific alternatives, we have selected the values of Δ as $0, 4, 8, 12, 16, 20$, and 24. For statistical methods, we have chosen the Wilcoxon two-sample test statistic and the two-sample t-test (LS) and we use the base R functions `wilcox.test` and `t.test` for their respective computations. We set the level for both tests at $\alpha = 0.05$. Hence, a test statistic rejects H_0 if its p-value satisfies $p \leq 0.05$. The following code runs 1000 simulations at each value of Δ. For each simulation, the test statistics are computed, and the number of times the null hypothesis is rejected is counted for each of the two methods. At each Δ, the empirical powers are just the number of these rejections divided by the number of simulations run at a value of Δ. These empirical powers are collected for each test, and the results are displayed graphically in Figure 3.11. These graphs represent estimates of the true power functions under the conditions of the simulation.

```
> set.seed(947828)
> n1<-15; n2<-20
> df<-2
> alpha<-0.05
> scale<-10; theta<-25
> nsims <- 1000
> deltaVec <- c(0, 4, 8, 12, 16, 20, 24)
> power_w <- power_t <- rep(NA,length(deltaVec))
> for(j in 1:length(deltaVec)){
+   delta <- deltaVec[j]
+   nw<-0; nt<-0
+   for(i in 1:nsims){
+     x <- rt(n1,df)*scale + theta
+     y <- rt(n2,df)*scale + theta + delta
+     wt<-wilcox.test(y,x,alt="greater")
+     tt <- t.test(y,x,alt="greater")
+     if(wt$p.value <= alpha){nw<-nw+1}
+     if(tt$p.value <= alpha){nt<-nt+1}
+   }
+   power_w[j] <- nw/nsims
+   power_t[j] <- nt/nsims
+ }
> plot(c(min(deltaVec),max(deltaVec)),c(0,1),pch='',
+   xlab='Delta',ylab='Empirical Power')
> title("Wilcoxon and t-test Empirical Power Functions")
> lines(power_w~deltaVec,lty=1)
> lines(power_t~deltaVec,lty=2)
> legend('topleft',c("Wilcoxon test","t-test"),lty=c(1,2),bty='n')
```

Note that both empirical power functions are monotonically increasing, verifying the property of monotonicity.

Below we print the results of the simulation.

```
     deltaVec power_w power_t
[1,]        0   0.046   0.044
[2,]        4   0.192   0.137
[3,]        8   0.485   0.325
[4,]       12   0.775   0.542
[5,]       16   0.912   0.678
[6,]       20   0.979   0.790
[7,]       24   0.990   0.856
```

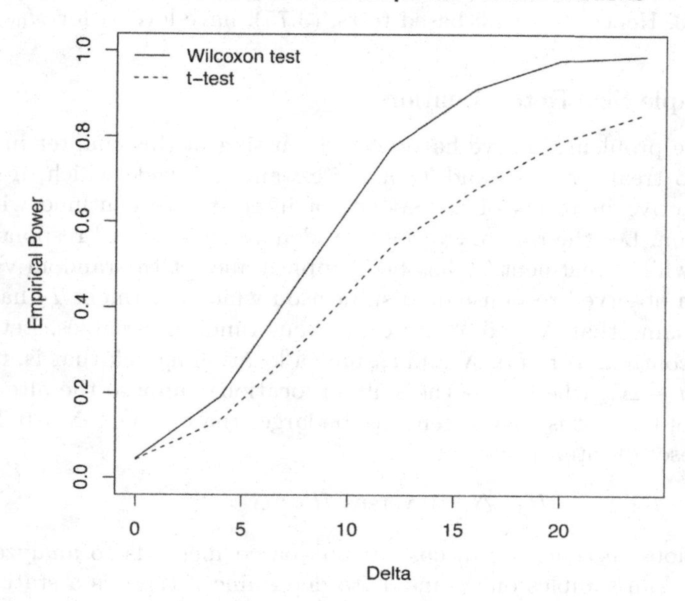

FIGURE 3.11
Empirical power functions for the Wilcoxon and t-test (LS) for the situation described in Example 3.9.1.

The empirical levels are 0.046 and 0.044, respectively for the Wilcoxon and t-test. Notice the empirical levels are close to 0.05 and because the levels are about the same, we can compare the empirical powers. The empirical powers of the Wilcoxon test are *greater than* the respective powers of the LS test. At $\Delta = 12$, the gap between the tests' powers is over 20 percentage points. Even at $\Delta = 24$, the gap is 15 percentage points. For this error distribution, the Wilcoxon test has definitely out performed the LS test. Note that from $\Delta = 4$ to $\Delta = 24$, the power of the Wilcoxon test increases from 0.192 to 0.990 which is a verification of the resolving property. ■

By making changes to the first two lines of code in Example 3.9.1, it is easy to change the situation simulated. For different random error distributions, replace `rt` with the appropriate R function; for example, if normal random errors are desired use `rnorm`. As we have noted, the tests are location and scale equivariant, so only the form of the pdf is important. In all these changes, though, the alternatives need to change accordingly. With modern computing, this simulation is quick to compute. So, for example, a range of suitable alternatives can be determined quickly by trial and error.

Suppose that we change the null hypothesis to $H_{02} : \Delta \leq 0$. Then the level of the test is defined to be the maximum of $\gamma(\Delta)$ for $\Delta \leq 0$. Because the power function of the rank-based test is increasing for all Δ, $\gamma(\Delta) \leq \gamma(0) = \alpha$ for all $\Delta \leq 0$. Hence, the rank-based tests, (3.75), have level α for H_{02}, also.

3.9.2 Sample Size Determination

Consider the problem that we have been discussing in this chapter in which we have two treatments T_1 and T_2 and we want to decide which, if either, is more effective in terms of a response of interest. We continue with the same notation. Let the random variable X denote an observed response on a subject on which treatment T_1 has been applied and let the random variable Y denote an observed response on a subject on which treatment T_2 has been applied. Assume that X and Y are continuous random variables. Let $f_X(t)$ denote the continuous pdf of X and assume a location model; that is, the pdf of Y is $f_X(t - \Delta)$, where Δ is the shift in location. Suppose the alternative hypothesis of interest is that Y tends to be larger than X; i.e., $\Delta > 0$. Hence, the hypotheses of interest are:

$$H_0 : \Delta = 0 \text{ versus } H_A : \Delta > 0. \tag{3.78}$$

In the previous sections, we discussed rank-based methods to analyze independent random samples on X and Y to determine if there is a statistically significant difference between two treatments. In this section, we discuss how to determine sample size so that the study is adequately powered. We focus on the **completely randomized design** (CRD) — a commonly used experimental design for this problem. In Section 2.10, we considered randomized paired (block) designs for this setup.

In a CRD, n subjects are selected at random from a reference population. Then n_1 of these subjects are randomly assigned to T_1 while the remaining $n_2 = n - n_1$ subjects are assigned to T_2. Let $X_1, \ldots, X_{n_1}$ denote the respective responses of the subjects that are assigned to T_1 and $Y_1, \ldots, Y_{n_2}$ denote the respective responses of the subjects that are assigned to T_2. An important part of the planning of the experiment is to make sure that the experiment is run under controlled conditions, so that, in particular, the assumptions (i) $X_1, \ldots, X_{n_1}$ are independent and identically distributed (iid), (ii) $Y_1, \ldots, Y_{n_2}$ are iid, and (iii) the two samples are independent of one another are reasonable assumptions.

To carry out the study, the sample sizes need to be determined that are of the appropriate size. If the sample sizes are too small, the statistical tests will tend to lack the necessary power to detect the alternatives of interest. On the other hand, if the sample sizes are too large, we may have had adequate power to detect alternatives of interest at smaller sample sizes — thereby conserving resources.

Suppose we have chosen to use the Wilcoxon rank-based test for the analysis to assess the hypothesis of interest. Except as noted, the discussion is the same for any other score function. Assume that the level of significance α has been selected and an alternative $\Delta_A > 0$ has been chosen that is of scientific interest to be able to detect. This is generally a value which makes a practical difference, a value which may lead in practice to the use of treatment T_2 instead of treatment T_1. Further, the power β to detect Δ_A has been chosen; i.e., Δ_A solves $\gamma(\Delta_A) = \beta$. Adequately powered studies will use a power of 80% or higher.

In terms of sample sizes n_1 and n_2, it can be shown that the most powerful design is the balanced design, i.e., $n_1 = n_2$. Denote the common sample size as $n_W^* = n_1 = n_2$. Using the fifth property of the power function given in Section 3.9.1, the approximate sample size n^* to achieve power β at the alternative Δ_A is

$$n_W^* \doteq \left(\frac{z_\alpha - z_\beta}{\Delta_A}\right)^2 2\tau_W^2, \tag{3.79}$$

where τ_W is the scale parameter defined in Equation (3.19) and z_β is the standard normal critical value with area to the right β; see Chapter 2 of Hettmansperger and McKean (2011) for details. For a general score function $\varphi(u)$, the derivation of expression (3.79) remains the same except τ_φ replaces τ_W.

As an illustration, suppose α and β are set respectively at 0.05 and 0.80, $\Delta_A = 5$, and $\tau_W = 10$. Then $z_\alpha =$ `qnorm(.05,lower.tail=FALSE)`$\doteq 1.645$ and $z_\beta =$ `qnorm(.80,lower.tail=FALSE)`$\doteq -0.842$. Substituting these values into expression (3.79) yields

$$n_W^* = \left(\frac{1.645 - (-0.842)}{5}\right)^2 2 * 10^2 = 49.48 \doteq 50.$$

Hence, for the design we would set $n_1 = n_2 = 50$.

Notice that the quantities in expression (3.79) make sense intuitively. If a higher power is desired, then z_β decreases and, hence, the quantity $z_\alpha - z_\beta$ increases resulting in a larger sample size. If a lower alternative $\Delta_{A2} < \Delta_A$ is selected then, since this term is in the denominator, n_W^* increases. If the scale of the measurements increases, then so does τ_W; hence, n_W^* increases.

For the sample size determination of the LS procedure, (the two-sample t-test), an analogous derivation results in

$$n_{LS}^* \doteq \left(\frac{z_\alpha - z_\beta}{\Delta_A}\right)^2 2\sigma_e^2, \tag{3.80}$$

where σ_e^2 is the variance of the random errors. Notice that the ratio $n_{LS}^*/n_W^* = \sigma_e^2/\tau_W^2$ is just the ARE of the Wilcoxon estimator (median of the sample differences) of Δ to the LS estimator (difference of sample means) that was discussed in Section 3.7. This ratio is also called the **asymptotic relative efficiency** (ARE) of the Wilcoxon two-sample test to the LS two-sample test. Recall that if the random errors are normally distributed, this ratio is $3/\pi = 0.955$. In this case, the sample size for the LS test is approximately 95% of the sample size for the Wilcoxon test. On the other hand, if the random errors have a Laplace (double exponential) distribution, this ratio is 1.5. So for this heavy-tailed error distribution, the required sample size for the t-test (LS analysis) is 50% larger than the sample size required for the Wilcoxon test to achieve the same power.

In practice, generally, the unknown value in expression (3.79) is the scale parameter τ_W. The analogous problem for the LS procedure is that σ_e is unknown. If the experimenter thinks that the errors are about normally distributed then, using the ARE between the Wilcoxon and LS at the normal distribution, $\tau_W = \sqrt{\pi/3}\sigma_e \doteq 1.02\sigma_e$ can be used, provided that σ_e is known. In this case an educated guess at the standard deviation σ_e provides an educated guess of τ_W. The assumption of normality, though, is often not warranted, especially if robust procedures are desired for the analysis. One situation which occurs frequently is when the treatment T_1 is the control, the standard treatment. In this situation, there is often historical data to use as a sample on X. If the R vector `xold` contains such a sample, then the Rfit call `gettau(xold)` provides an estimate of τ_W. If none of the above options are possible, a solution is to run a small pilot study, which provides an estimate of τ_W. In the next example, we simulate this situation.

Example 3.9.2 (Pilot Study). Suppose we are planning a two-sample design where a test of the hypotheses (3.73) is of primary interest. We have chosen $\alpha = 0.05$ as our level of significance and we are interested in detecting the practical alternative $\Delta = 8$ with power $\beta = 0.80$. We have selected the two-sample Wilcoxon procedure as our method of analysis. In order to determine the sample size for the experiment, we need a reasonable guess (estimate) of τ_W. It is further known that the distributions of the samples are heavy-tailed (prone to outliers), so the normal approximation to τ_W based on the standard deviation of the random errors is not warranted. In order to determine the sample size, we decide to do a pilot study with $n_1 = n_2 = 5$. In practice, we would then obtain 10 subjects for the study (independent of one another) and randomly assign 5 of them to treatment T_1 and the other 5 to treatment T_2. The following code simulates this process, resulting in the two samples.

For the simulation, the random errors are distributed as $e = 8T$ where T has the t-distribution with 2 degrees of freedom, the median of X is 20, and the shift parameter Δ is 8.

```
> set.seed(3647431)
> n1 <- 5; n2 <- 5
> err<- rt(10,2)
> x <- 20 + 8*err[1:n1]
> y <- 8 + 20 + 8*err[(n1+1):(n1+n2)]
> x; y

[1] 21.975 33.099 26.412 25.665 14.977

[1] 13.008 77.073 29.250 37.538 28.487
```

The following code computes the Wilcoxon analysis of the samples using the regression model which provides the pilot's estimate of τ_W. We also computed the LS estimate of σ.

```
> ind <- c(rep(0,n1),rep(1,n2))
> zed <- c(x,y)
> fitw <- rfit(zed~ind); tauhat <- fitw$tauhat
> fitls <- lm(zed~ind); sfitls <- summary(fitls)
> sigmahat<-sfitls$sigma
> tauhat

[1] 13.362

> sigmahat

[1] 17.643
```

As Exercise 3.12.19 shows, the Wilcoxon test is not significant at the 5% level for the pilot study data. Using expression (3.79), we next compute the sample size to achieve 80% power at the alternative $\Delta = 8$ for the Wilcoxon analysis.

```
> alpha <- 0.05; beta <- 0.80; delta <- 8
> za <- qnorm(alpha,lower.tail=FALSE)
> zb <- qnorm(beta,lower.tail=FALSE)
> nw <- ceiling( (((za-zb)/delta)^2)*(2*tauhat^2) )
> nls <- ceiling( (((za-zb)/delta)^2)*(2*sigmahat^2) )
> nw  # sample size based on Wilcoxon

[1] 35

> nls # sample size based on Least Squares

[1] 61
```

The sample size for the Wilcoxon analysis is $n_1 = n_2 = 35$. Hence, the sampling plan for the CRD is to randomly select 70 subjects and randomly assign 35 of these to treatment T_1 and the remaining 35 to treatment T_2. Notice that if the LS analysis is selected, then $n_1 = n_2 = 61$ is the sample size so, instead of 70 subjects, 122 would be required.

In the next R segment we simulate the experiment using $n_1 = n_2 = 35$ subjects. The random generation model remains the same as in the pilot study. Based on the generated samples, we then compute the Wilcoxon test of the hypotheses (3.78).

```
> set.seed(4764731)
> n1 <-35;n2<-35
> err<- rt(n1+n2,2)
```

```
> x<-20 + 8*err[1:n1]; y<-8 + 20 + 8*err[(n1+1):(n1+n2)]
> wilcox.test(y,x,alt="greater")

        Wilcoxon rank sum exact test

data:  y and x
W = 883, p-value = 0.00063
alternative hypothesis: true location shift is greater than 0
```

With p-value 0.0006, the null hypothesis is rejected in favor of the alternative $\Delta > 0$. The next R segment, obtains the rank-based fit of the corresponding linear model using Wilcoxon scores.

```
> ind <- c(rep(0,n1),rep(1,n2)); zed <- c(x,y)
> fitw <- rfit(zed~ind); sfitw <- summary(fitw); sfitw

Call:
rfit.default(formula = zed ~ ind)

Coefficients:
            Estimate Std. Error t.value p.value
(Intercept)    18.08       1.68   10.79  <2e-16 ***
ind             8.71       2.69    3.24  0.0018 **
---
Signif. codes:
0 '***' 0.001 '**' 0.01 '*' 0.05 '.' 0.1 ' ' 1

Multiple R-squared (Robust): 0.14542
Reduction in Dispersion Test: 11.571 p-value: 0.00113

> deltahat <- sfitw$coef[2,1]
> se <- sfitw$coef[2,2]
> tc <- qt(0.975,68)
> c(deltahat-tc*se,deltahat+tc*se)

[1]  3.3451 14.0684
```

Using the t-ratio test for Δ based on the fit, the conclusion is the same as for the distribution-free Wilcoxon test. The estimate of Δ is 8.71 with standard error 2.69. Note: the true shift $\Delta = 8$ is in the 95% confidence interval. ■

Sample Size Determination for Other Hypotheses

Sample size determination for the other two hypotheses follows from the above discussion. First, consider the one-sided hypotheses,

$$H_0 : \Delta = 0 \text{ versus } H_A : \Delta < 0. \tag{3.81}$$

For level α, let $c_{1-\alpha,n_1,n_2}$ be the lower critical value of $S_\varphi(0)$; i.e., $P_{\Delta=0}[S_\varphi(0) \leq c_{1-\alpha,n_1,n_2}] = \alpha$. Then the decision rule is

$$\text{reject } H_0 \text{ at level } \alpha, \text{ if } S_\varphi(0) \leq c_{1-\alpha,n_1,n_2}. \tag{3.82}$$

Let $\gamma_2(\Delta)$ denote the power function of this test. It follows that $\gamma_2(-\Delta) = \gamma(\Delta)$, where $\gamma(\Delta)$ is the power function of the greater than alternative defined in expression (3.76); see Chapter 2 of Hettmansperger and McKean (2011). Hence, γ_2 is a monotonically decreasing function. To determine the sample size for the alternative $-\Delta_A < 0$, just use Δ_A in expression (3.79).

Next, suppose we are testing the two-sided alternative,

$$H_0 : \Delta = 0 \text{ versus } H_A : \Delta \neq 0. \tag{3.83}$$

Then the decision rule is

$$\text{reject } H_0 \text{ at level } \alpha, \text{ if } S_\varphi(0) \leq c_{1-\alpha/2,n_1,n_2} \text{ or if } S_\varphi(0) \geq c_{\alpha/2,n_1,n_2}. \tag{3.84}$$

The power function of this test is $\gamma_3(\Delta) = \gamma_2(\Delta) + \gamma(\Delta)$ with significance level $\alpha/2$ for both one-sided hypotheses. Suppose we want to determine the sample size for the alternative $\Delta_A > 0$. Because $\Delta_A > 0$, the contribution to the overall power from the *less than* alternative, $\gamma_2(\Delta)$, is negligible.[14] Hence, use Δ_A in expression (3.79) with α replaced by $\alpha/2$ to determine the sample size.

Sample Size Determination for the Estimation of Δ

A related problem is determining the sample size for the estimation of the shift parameter Δ. As in the last section, consider the two-sample location model

$$\begin{aligned} X_i &= \theta + e_i, \quad 1 \leq i \leq n_1 \\ Y_i &= \theta + \Delta + e_{n_1+i}, \quad 1 \leq i \leq n_2, \end{aligned} \tag{3.85}$$

where the random errors e_i are iid with continuous pdf $f(t)$.

Suppose we have chosen an appropriate score function $\varphi(u)$ for our rank-based analysis. Sample size determination for estimation follows from the confidence interval for Δ. Recall that the $(1-\alpha)100\%$ confidence interval (Wald type) for Δ is:

$$\hat{\Delta} \pm t_{\alpha/2,n_1+n_2-2}\hat{\tau}_\varphi\sqrt{\frac{1}{n_1} + \frac{1}{n_2}}. \tag{3.86}$$

For sample size determination, suppose we want to estimate Δ within B units and with confidence level $(1-\alpha)$. As discussed in the last section, the most

[14]For $\Delta_A > 0$, $\gamma_2(\Delta) < \alpha/2$ and it is decreasing.

powerful design based on n_1 and n_2 is the design with $n_1 = n_2$. Let $n^* = n_1 = n_2$. Then, using $z_{\alpha/2}$ for $t_{\alpha/2,n_1+n_2-2}$ we want to determine n^* so that

$$z_{\alpha/2}\tau_\varphi\sqrt{\frac{2}{n^*}} \leq B.$$

Solving for n^*, we obtain the sample size as

$$n^* \geq \left(\frac{z_{\alpha/2}}{B}\right)^2 2\tau_\varphi^2. \tag{3.87}$$

The rule makes sense intuitively. If the level of confidence increases, then $z_{\alpha/2}$ increases; hence, n^* increases. If sampling noise increases, then τ_φ^2 increases; hence, n^* increases. Finally if B decreases (more precision), then n^* increases. As with testing, generally the unknown on the right-side is τ_φ. The discussion concerning estimating and/or guessing of τ_φ in the testing section applies to the estimation case, too.

As an illustration, consider the situation described in Example 3.9.2. Suppose we want to estimate Δ within $B = 6$ units with 95% confidence. From the pilot study, $\hat{\tau}_W = 13.36$. Hence, the sample size is

$$n^* = \left(\frac{1.96}{6}\right)^2 2 * 13.36^2 = 38.09.$$

So to estimate Δ within 6 units of precision, we would recommend using 39 as the common sample size.

3.10 k-Nearest Neighbors and Cross-Validation

We introduce classification via the k-nearest neighbors (k-NN) classifier. We also introduce the concept of model selection. Classification is one of the main points of focus in what is commonly referred to as machine learning today. For a more in-depth discussion of classification and machine learning in general, we refer the reader to James et al. (2013), Hastie et al. (2017), as well as Kuhn and Johnson (2013).

Classification is used to make decisions on new observations based on past data. A classifier is *trained* based on a random sample of observations from the population of interest; i.e., based on a set of covariates[15] upon which a response variable[16] is to be classified. Based on this learned classifier, new observations are then classified into one of two (or more) groups. Some examples of classifiers include:

[15] also called predictor variables, explanatory variables, or features
[16] also referred to as outcome variable, class label, or label

- e-mail classification as spam or regular e-mail
- loan or credit applications based on likelihood of a default
- precision medicine to choose an optimal treatment from a set based on a patient's individual characteristics, symptoms, or other medical information

with many more applications currently being developed.

Denote the training dataset as $(\boldsymbol{x}_1^T, y_1), (\boldsymbol{x}_2^T, y_2), \ldots, (\boldsymbol{x}_n^T, y_n)$. Where $\boldsymbol{x}_i$ is the set of covariates and y_i is the response variable (or class label) for the ith experimental unit. For simplicity assume that the class labels are 0 and 1; however, many of the classifiers used in practice may be used to classify observations into several (i.e., > 2) classes. The goal is then to classify new observations as one of the classes based on the set of covariates; i.e., $\boldsymbol{x}_{\text{new1}}, \boldsymbol{x}_{\text{new2}}, \ldots$.

For classification, k-nearest neighbors (k-NN) is a basic method of classification, that, as the name suggests, looks at the class labels of the *nearest* data points to classify an observation. Often Euclidean distance is used; however, other metrics are possible. The value k is a parameter that needs to be selected; often cross-validation is used as we will illustrate. So, for example, to classify the data point $\boldsymbol{x}_{\text{new1}}$, one would look at the class labels of the k nearest data points and make the prediction based on majority rules.

Often in classifier training or, more generally, model building, the data are split into training and test datasets.

Remark 3.10.1. At times, given sufficient sample size, a validation set is also used in addition to the training and test datasets. The validation dataset can be used, for example, in place of cross-validation. There does not appear to be a consensus on guidelines for splits. In choosing the size of the splits, there are several things to consider including: total samples size, number of class labels, number of predictor variables. One would want a large enough sample size for the classifier to provide accurate predictions and, if using one, a test set which is representative enough to provide an adequate assessment of the model performance for new data. Every problem is different, though we offer some general guidelines for consideration. For small (for classification) datasets (say, n in the 100s) there may not be sufficient data for a test set, so cross-validation may be the best choice. For medium (for classification) datasets (say, n in the 1000s), a fraction of the data may be set aside for a test and/or validation sets; for larger n, a smaller fraction can provide the same precision in the performance metric. For large (for classification) datasets (say, n in the 10000sa,) 5–20% of the data could be set aside for validation and test sets, which for many problems, would provide sufficient data for model building and accurate assessment of model performance on an independent test set. For big data (see Chapter 11), multiple validation and/or test sets are possible, perhaps, even allowing one to estimate the distribution or error of the performance metric based on multiple test sets.

The classifier is trained based on the training set, and the test set is then used to assess the performance of the classifier on *new* data which provides an estimate as to how the classifier will perform in practice. As with all training or model building, there is always the danger of *overfitting* in which the model fits the sample used to build the model (or classifier), but does not generalize well to new data. The test set allows one to see if model performance metrics are reproduced with the new dataset. See references noted above for more discussion training and test sets.

To fix ideas, consider the example dataset `sim_class2`. In this dataset, there are 1000 rows and 4 columns. The first several rows are displayed in the following code segment.

```
> head(sim_class2,4)

  train     x1     x2 y
1     0 -0.729 -0.265 1
2     1  0.056 -0.077 0
3     0  0.094 -0.295 0
4     1 -0.403  0.239 0
```

The variable `train` is an indicator variable representing if the row should be used in the training (`1`) or test(`0`) sets. The variable `y` is the binary response. The `x`'s are used to predict or classify the observation as `0` or `1`.

We first split the dataset into training and test sets based on the variable `train`. In this case, we use the package `dplyr` (Wickham et al. 2023) — including `magrittr`[17] (Bache and Wickham 2022) — for our data wrangling.

```
> library(dplyr)
> train_set <- sim_class2 %>% filter(train==1) %>% select(-train)
> dim(train_set)

[1] 800   3

> test_set <- sim_class2 %>% filter(train!=1) %>% select(-train)
> dim(test_set)

[1] 200   3
```

A scatterplot of the training data is displayed in the top left panel of Figure 3.12 with the two class labels displayed by the two colors.

The function `knn` in the library `class` (Venables and Ripley 2002) implements the k-NN classifier using Euclidean distance. The function `knn` has three required inputs: the design matrix for the training set, the design matrix for the test set (for which predictions are desired), and vector of class labels

[17]The `magrittr` operator, `%>%`, passes the output of the function on the left as the first argument to the function on the right.

for the training set. The default is to use $k = 1$ neighbors. To obtain the predictions for the training set, one may input the training set design matrix as the test design matrix as is illustrated in the following code segment. In this case we specify that $k = 3$ neighbors are to be used.

```
> library(class)
> xtrain <- select(train_set,-y)
> ytrain <- train_set[,'y']
> fit1 <- knn(xtrain,xtrain,ytrain,k=3)
> head(fit1)

[1] 0 0 0 1 0 1
Levels: 0 1
```

The output for **knn** is the predicted class labels.

To obtain the predictions for the test set, we would instead include the matrix of predictors for the test set as the second argument in the **knn** call.

```
> xtest <- select(test_set,-y)
> ytest <- test_set[,'y']
> fit2 <- knn(xtrain,xtest,ytrain,k=3)
> head(fit2)

[1] 1 0 0 0 1 0
Levels: 0 1
```

A simple metric to evaluate the quality of the fit is a **confusion matrix**. A confusion matrix tabulates the agreement between the observed and predicted values.

```
> # confusion matrix for test set based on k=3 nearest neighbors
> table(observed=ytest,predicted=fit2)

        predicted
observed   0   1
       0 100   6
       1   9  85
```

Out of the 200 observations $6 + 9 = 15$ were misclassified. i.e. the **misclassification rate** is 0.075.

To choose the number of nearest neighbors, k, for the classifier, often cross-validation is used. In cross-validation, the data used to train the classifier are split into a number of folds, k.[18] Typically, 5-fold, 10-fold, or leave-one-out (a.k.a., n-fold) cross-validation is used.

[18] It is somewhat of an unfortunate situation that k is used both for the the number of nearest neighbors in the classifier and the number of chunks in the cross-validation, however that is the tradition.

For example, in 5-fold cross-validation, the data is split, at random, into 5 chunks. The first chunk is held out and the model is fit on the remaining 4 chunks for each of the values of the parameter under consideration (e.g. $k = 1, 3, 5, \ldots 33$ nearest-neighbors); the error rate is then calculated on the held out set for each of the fitted models. This is repeated for each of the 5 chunks so that 5 error rates are obtained for each of the candidate models. The 5 error rates are then averaged for each of the candidate number of nearest-neighbors; the one with the lowest error rate being chosen. For more information, see James et al. (2013), Hastie et al. (2017), and Kuhn and Johnson (2013).

The function **knn_cv** in the package **npsm** performs k-fold cross-validation on a training set to choose the number of nearest neighbors. The function assumes the data frame contains the design matrix in the first set of columns and the vector of class labels in the last column. The use is illustrated in the following code segment.

```
> set.seed(19180511)
> fit_cv <- knn_cv(train_set,k.cv=10)
> fit_cv

 10-fold Cross-Validation for kNN
Best k=5, CV error rate=0.06375
```

Using plot(fit_cv) produces a plot of the misclassification error rate versus the value of k. This plot, based on the training set, is displayed in the lower left panel of Figure 3.12; we see the that lowest cross-validation occurs when k =5. Also, displayed are grids of predicted values for the case when $k = 1$ (top right) and when k is chosen by cross-validation (bottom right).

The following code segment produces a confusion matrix based on a k-NN fit with $k = 1$ and k chosen by cross-validation.

```
> xtest <- select(test_set,-y)
> ytest <- test_set[,'y']
> yhat1 <- knn(xtrain,xtest,ytrain,k=1)
> # confusion matrix for test set based on k=1 nearest neighbors
> table(observed=ytest,predicted=yhat1)

        predicted
observed  0  1
       0 97  9
       1 11 83

> yhatcv <- knn(xtrain,xtest,ytrain,k=fit_cv$k.best)
> # confusion matrix for test set based with k chosen via
   cross-validation
> table(observed=ytest,predicted=yhatcv)

        predicted
```

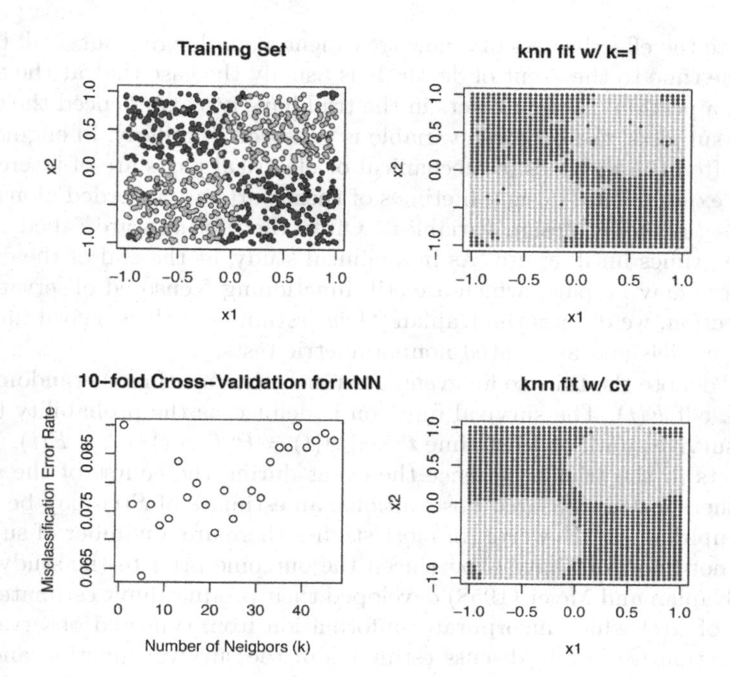

FIGURE 3.12
Top Left: Scatterplot of the training set from `sim_class2`. The two colors represent the two levels of the class variable (`y`). Bottom Left: Results of `plot(fit_cv)` for the training data. Top Right: Grid of predicted values based on training the classifier with $k = 1$. Bottom Right: Grid of predicted values based on training the classifier with k chosen by cross-validation.

```
observed   0  1
       0 100  6
       1   9 85
```

The misclassification rate for $k = 1$ is 0.1 and 0.075 when k is chosen via cross-validation. It is often the case that k-NN with $k = 1$ results in *overfitting* to the training data so that the result does not generalize well to new data.

3.11 Kaplan–Meier and Log-Rank Test

In survival or reliability analysis, the investigator is interested in time to an event of interest as the outcome variable. Often in a clinical trial, the goal is to evaluate the effectiveness of a new treatment at prolonging survival; i.e., to extend the time to the event of death. It is usually the case that at the end of followup, a portion of the subjects in the trial have not experienced the event; for these subjects, the outcome variable is **censored**. Similarly, in engineering studies, often the lifetimes of mechanical or electrical parts are of interest. In a typical experimental design, lifetimes of these parts are recorded along with covariates (including design variables). Often the lifetimes are called failure times, i.e., times until failure. As in a clinical study, at the end of the experiment, there may be parts which are still functioning (censored observations). In this section, we discuss the Kaplan–Meier estimate of the survival function for these models and associated nonparametric tests.

Let T denote the time to an event. Assume T is a continuous random variable with cdf $F(t)$. The survival function is defined as the probability that a subject survives until at least time t; i.e., $S(t) = P(T > t) = 1 - F(t)$. When all subjects in the trial experience the event during the course of the study, so that there are no censored observations, an estimate of $S(t)$ may be based on the empirical cdf. However, in most studies there are a number of subjects who are not known to have experienced the outcome prior to the study completion. Kaplan and Meier (1958) developed their product-limit estimate as an estimate of $S(t)$ which incorporates information from censored observations. In this section we briefly discuss estimates of the survival function and also illustrate them via small samples. The focus, however, is on the R syntax for analysis. We describe how to store time to event data and censoring in R, as well as computation of the Kaplan–Meier estimate and the log-rank test – which is a standard test for comparing two survival distributions.

We begin with a brief overview of survival data as well as simple examples which illustrate the calculation of the Kaplan–Meier estimate.

Example 3.11.1 (Treatment of Pulmonary Metastasis). In a study of the treatment of pulmonary metastasis arising from osteosarcoma, survival time was collected; the data are provided in Table 3.4.
As there are no censored observation an estimate of the survival function at time t is

$$\hat{S}(t) = \frac{\#\{t_i > t\}}{n} \tag{3.88}$$

which is based on the empirical cdf. Because of the low number of distinct time points, the estimate (3.88) is easily calculated by hand which we briefly

TABLE 3.4
Survival Times (in months) for Treatment of Pulmonary Metastasis.

11	13	13	13	13	13	14	14	15	15	17

illustrate next. Since $n = 11$, the result is

$$\hat{S}(t) = \begin{cases} 1 & 0 \le t < 11 \\ \frac{10}{11} & 11 \le t < 13 \\ \frac{5}{11} & 13 \le t < 14 \\ \frac{3}{11} & 14 \le t < 15 \\ \frac{1}{11} & 15 \le t < 17 \\ 0 & t \ge 17. \end{cases}$$

The estimated survival function is plotted in Figure 3.13. ∎

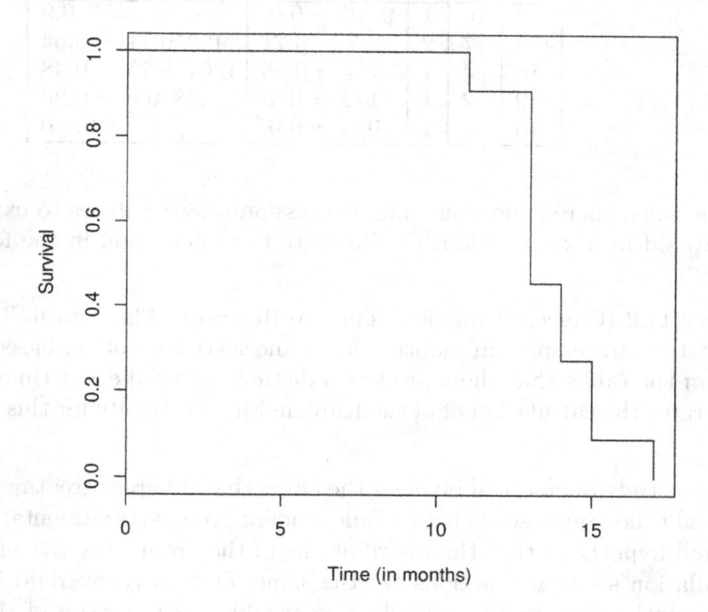

FIGURE 3.13
Estimated survival curve ($\hat{S}(t)$).

Though expression (3.88) aids in the understanding of survival functions, it is not often useful in practice. In most clinical studies, at the end of followup, there are subjects who have yet to experience the event being studied. In this case, the Kaplan–Meier product limit estimate is used which we describe briefly next. Suppose n experimental units are put on test. Let $t_{(1)} < \ldots < t_{(k)}$ denote the ordered distinct event times. If there are censored responses, then $k < n$. Let n_i = #subjects at risk at the beginning of time $t_{(i)}$ and d_i = #events occurring at time $t_{(i)}$ (i.e., during that day, month, etc.). The **Kaplan–Meier** estimate of the survival function is defined as

$$\hat{S}(t) = \prod_{t_{(i)} \le t} \left(1 - \frac{d_i}{n_i}\right). \tag{3.89}$$

TABLE 3.5
Time in remission (in months) in cancer study.

Relapse	3	6.5	6.5	10	12	15
Lost to followup	8.4					
Alive and in remission at the end of study	4	5.7	10			

TABLE 3.6
Illustration of the Kaplan–Meier estimate.

t	n	d	$1 - d/n$	$S(t)$
3	10	1	9/10 = 0.9	0.9
6.5	7	2	5/7 = 0.71	0.9*0.71 = 0.64
10	4	1	3/4 = 0.75	0.64*0.75 = 0.48
12	2	1	1/2 = 0.5	0.48*0.5 = 0.24
15	1	1	0/1 = 0.0	0

Note that when there is no censoring, expression (3.89) reduces to expression (3.88). To aid in interpretation, we illustrate the calculation in the following example.

Example 3.11.2 (Cancer Remission: Time to Relapse)**.** The data in Table 3.5 represent time to relapse (in months) in a cancer study. Notice, based on the top row of the table, that there are $k = 5$ distinct survival event times. Table 3.6 illustrates the calculation of of the Kaplan–Meier estimate for this dataset. ∎

Often a study on survival involves the effect that different treatments have on survival time. Suppose we have r independent groups (treatments). Let H_0 be the null hypothesis that the distributions of the groups are the same; i.e., the population survival functions are the same. Obviously, overlaid Kaplan–Meier survival curves provide an effective graphical comparison of the times until failure of the different treatment groups. A nonparametric test that is often used to test for a difference in group survival times is the log-rank test. This test is complicated, and complete details can be found, for example, in Kalbfleisch and Prentice (2002). Briefly, as above, let $t_1 < t_2 < \cdots < t_k$ be the distinct failure times of the combined samples. Then at each time point t_j, it can be shown that the number of failures in Group i conditioned on the total number of failures has a distribution-free hypergeometric distribution under H_0. Based on this, a goodness-of-fit type test statistic (called the log-rank test) can be formulated which has a χ^2-distribution with $r - 1$ degrees of freedom under H_0. The next example illustrates this discussion for the time until relapse of two groups of patients who had survived a lobar intracerebral hemorrhage.

Example 3.11.3 (Hemorrhage Data)**.** For demonstration we use the hemorrhage data discussed in Chapter 6 of Dupont (2002). The study population consisted of patients who had survived a lobar intracerebral hemorrhage and whose genotype was known. The outcome variable was the time until recurrence of lobar intracerebral hemorrhage. The investigators were interested in examining the genetic effect on recurrence, as there were three common alleles $e2, e3$, and $e4$. The analysis was focused on the effect of homozygous $e3/e3$ (Group 1) versus at least one $e2$ or $e4$ (Group 2). The data are available at the author's website. The following code segment illustrates reading the data into `R` and converting it to a survival dataset which includes censoring information. Many of the functions for survival data are available in the `R` package `survival` (Therneau 2013).

```
> library(survival)

> with(hemorrhage,Surv(round(time,2),recur))

 [1]  0.23   1.05+  1.22   1.38+  1.41   1.51+  1.58+  1.58   3.06   3.32
[11]  3.52   3.55   4.04+  4.63+  4.76   8.08+  8.44+  9.53  10.61+ 10.68+
[21] 11.86+ 12.32  13.27+ 13.60+ 14.69+ 15.57  16.72+ 17.84+ 18.04+ 18.46+
[31] 18.46+ 18.46+ 18.66+ 19.15  19.55+ 19.75+ 20.11+ 20.27+ 20.47+ 24.77
[41] 24.87  25.56+ 25.63+ 26.32+ 26.81+ 28.09  30.52+ 32.95+ 33.05+ 33.61
[51] 34.99+ 35.06+ 36.24+ 37.03+ 37.52  37.75+ 38.54+ 38.97+ 39.16+ 40.61+
[61] 42.22+ 42.41+ 42.78+ 42.87  43.27+ 44.65+ 45.24+ 46.29+ 46.88+ 47.57+
[71] 53.88+
```

In the output are survival times (in months) for 71 subjects. However, one subject's genotype information is missing and is excluded from analysis. Of the remaining 70 subjects, 32 are in Group 1 and 38 are in Group 2. A + sign indicates a censored observation; meaning that at that point in time the subject had yet to report recurrence. The study could have ended or the subject could have been lost to followup. Kaplan–Meier estimates are available through the command `survfit`. The resulting estimates may then be plotted, as is usually the case for Kaplan–Meier estimates, as the following code illustrates. If confidence bands are desired, one may use the `conf.type` option to survfit. Setting `conf.type='plain'` returns the usual Greenwood (1926) estimates.

```
> fit<-with(hemorrhage, survfit(Surv(time,recur)~genotype))

> plot(fit,lty=1:2,
+          ylab='Probability of Hemorrhage-Free Survival',
+          xlab='Time (in Months)'
+ )
> legend('bottomleft',c('Group 1', 'Group 2'),lty=1:2,bty='n')
```

As illustrated in Figure 3.14, patients that were homozygous $e3/e3$ (Group 1) seem to have significantly greater survival.

```
> with(hemorrhage, survdiff(Surv(time,recur)~genotype))
```

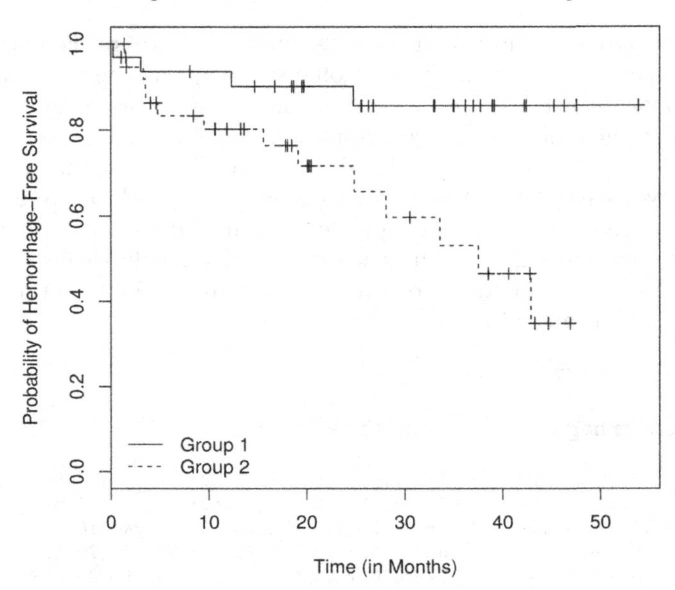

FIGURE 3.14
Plots of Kaplan–Meier estimated survival distributions.

```
Call:
survdiff(formula = Surv(time, recur) ~ genotype)

n=70, 1 observation deleted due to missingness.

             N Observed Expected (O-E)^2/E (O-E)^2/V
genotype=0 32        4     9.28      3.00      6.28
genotype=1 38       14     8.72      3.19      6.28

 Chisq= 6.3  on 1 degrees of freedom, p= 0.0122
```

Note that the log-rank test statistic is 6.3 with p-value 0.0122 based on a null χ^2-distribution with 1 degree of freedom. Thus the log-rank test confirms the difference in survival time of the two groups. ■

3.11.1 Gehan's Test

Gehan's test, (see Higgins (2003)), sometimes referred to as the Gehan–Wilcoxon test, is an alternative to the log-rank test. Gehan's method is a generalization of the Wilcoxon procedure discussed in Chapter 3. Suppose in a randomized controlled trial, subjects are randomized to one of two treatments, say, with survival times represented by X and Y. Represent the sample as $X_1, \ldots, X_{n_1}$ and $Y_1, \ldots, Y_{n_2}$ with a censored observation denoted with a

TABLE 3.7
Survival Times (in days) for Undergoing Standard Treatment (S) and a New Treatment (N).

S	94	180+	741	1133	1261	382	567+	988	1355+
N	155	375	951+	1198	175	521	683+	1216+	

plus sign, X_i^+, for example. Only unambiguous pairs of observations are used. Not used are ambiguous observations such as when an observed X is greater than a censored Y ($X_i > Y_j^+$) or when both observations are censored. The test statistic is defined as the number of times each of the X clearly beats Y minus the number of times Y clearly beats X. Let S_1 denote the set of uncensored observations, S_2 denote the set of observations for which X is censored and Y is uncensored, and S_3 denote the set where Y is censored and X is uncensored. Then Gehan's test statistic can be represented as

$$U = (\#_{S_1}\{X_i > Y_j\} + \#_{S_2}\{X_i^+ \geq Y_j\}) - (\#_{S_1}\{Y_j > X_i\} + \#_{S_3}\{Y_j^+ \geq X_i\}).$$

Example 3.11.4 (Higgins' Cancer Data). Example 7.3.1 of Higgins (2003) describes an experiment to assess the effect of a new treatment relative to a standard. The data are in the dataset `cancertrt`, but for convenience the data are also given in Table 3.7. We illustrate the computation of Gehan's test based on the `npsm` function, `gehan.test`. There are three required arguments to the function: the survival time, an indicator variable indicating that the survival time corresponds to an event (and is not censored), and a dichotomous variable representing one of two treatments; see the output of the `args` function below.

```
> args(gehan.test)

function (time, event, trt)
NULL
```

We use the function `gehan.test` next on the `cancertrt` dataset.

```
> with(cancertrt,gehan.test(time,event,trt))

statistic =  -0.6071557 , p-value =  0.5437476
```

The results agree with those in Higgins. The two-sided p-value = 0.5437 which is not significant. As a final note, using the `survdiff` function with `rho=1` gives the Peto–Peto modification of the Gehan test. ■

3.11.2 Composite Outcomes

At times a study may use a **composite outcome** which consists of two or more important events ideally having a *similar* severity. An example from cardiovascular reseach is MACE (major adverse cardiac event) which is commonly defined as a composite of cardiovascular death, nonfatal myocardial infarction, and nonfatal stroke. An example from oncology is progression-free survival, where both time to death and time to disease progression are both considered in the analysis. See Mao and Kim (2021) for a recent review of methods for composite outcomes. In this subsection, we briefly introduce the concept and refer readers to the references on the topic provided for additional information.

For illustration, consider the case of two composite outcomes, say, $t^{(1)}$ and $t^{(2)}$, representing the (possibly censored) times to events (1) and (2), where the occurrence of event (1) is considered to be of similar but slightly greater severity than event (2). Denote censoring with the indicator variables $c^{(1)}$ and $c^{(2)}$ for the two events. For illustration, consider a random subset of the data from Lee (1992) provided in Table 3.8.

TABLE 3.8
Random subset of data with composite outcomes. Notation: t denotes time and c is a censoring indicator; r and s represent the two composite events.

subjid	trt	t_r	c_r	t_s	c_s
4	1	2.3	0	5.4	0
5	1	6.4	0	19.5	0
7	1	1.8	0	7.9	0
8	1	5.5	0	16.9	1
10	1	33.7	1	33.7	1
11	1	17.1	1	17.1	1
12	2	4.3	0	8.0	0
20	2	23.0	1	23.0	1
25	2	14.9	0	24.4	0
26	2	4.5	0	7.7	0

The null hypothesis is no treatment difference on either outcome measure. The goal is to compute a *score* for each subject by comparing their set of outcomes to the set of every other subject in the study. Consider the comparison of subject i to subject j. First, the outcomes are compared on event (1) and if there is not a clear *winner*, then the outcomes are compared on event (2). For each of the event types, there are 4 possible censoring patterns for any pair of subjects. For our illustration, assume there are no ties. Let's consider each of the cases:

$c_i^{(1)} = 0, c_j^{(1)} = 0$:
if $t_i^{(1)} > t_j^{(1)}$, then $u_{ij} = +1$
if $t_i^{(1)} < t_j^{(1)}$, then $u_{ij} = -1$

$c_i^{(1)} = 0, c_j^{(1)} = 1$:

if $t_i^{(1)} < t_j^{(1)}$, then $u_{ij} = -1$
if $t_i^{(1)} > t_j^{(1)}$, then compare on event (2)

$c_i^{(1)} = 1, c_j^{(1)} = 0$:
if $t_i^{(1)} > t_j^{(1)}$, then $u_{ij} = +1$
if $t_i^{(1)} < t_j^{(1)}$, then compare on event (2)

$c_i^{(1)} = 1, c_j^{(1)} = 1$:
then compare on event (2)

For inconclusive comparisons after comparing on all components of the composite, then $u_{ij} = 0$. The total *score* for subject i is then $u_{i\cdot} = \sum_{j \neq i} u_{ij}$. The test statistics purposed by Finkelstein and Schoenfeld (1999) is

$$t^* = \sum_{i=1}^{N} u_{i\cdot} D_i \tag{3.90}$$

where D_i is an indicator for treatment. The test rejects the null hypothesis if there is a '...substantial treatment difference on either of the measures...' (Finkelstein and Schoenfeld 1999).

R packages available for computation include `WR` (Mao and Wang 2023) and `WINS` (Cui and Huang 2023). The Win-Ratio (Pocock et al. 2011) methodology tests the hypothesis that the probability of a win is equal to the probability of a loss.

3.12 Exercises

3.12.1. Using the data discussed in Example 3.11.4:

(a) Obtain a plot of the Kaplan–Meier estimates for the two treatment groups.

(b) Obtain the p-value based on the log-rank statistic.

(c) Obtain the p-value based on the Peto–Peto modification of the Gehan statistic.

3.12.2. For the dataset `hodgkins`, plot Kaplan–Meier estimated survival curves for both treatments. Note treatment code 1 denotes radiation of affected node, and treatment code 2 denotes total nodal radiation.

3.12.3. To simulate survival data, often it is useful to simulate multiple time points. For example, the time to event and the time to end of study. Then, events occurring after the time to end of study are censored. Suppose the time to the event of interest follows an exponential distribution with mean 5 years, and the time to the end of the study follows an exponential distribution with a mean of 1.8 years. For a sample size $n = 100$, simulate survival times from this model. Plot the Kaplan–Meier estimate.

3.12.4. For the illustrative example of Section 3.4, use the core R function `wilcox.test` to calculate the test statistic T^+.

3.12.5. Conover (1980) reports on a study involving men and women concerning which room temperature is most comfortable for each gender. The data are

Men	74	72	77	76	76	73	75	73	74	75
Women	75	77	78	79	77	73	78	79	78	80

Let the variables X and Y denote the most comfortable room temperatures for a man and woman, respectively. Assume a location model where X has pdf $f(t)$, Y has pdf $f(t - \Delta)$, and Δ is the shift in location. Consider the hypotheses $H_0 : \Delta = 0$ versus $H_A : \Delta > 0$.

(a) Obtain and comment on the comparison boxplots of the data.

(b) Using average ranks to break ties, obtain the joint ranking of the observations. Determine the value of the Wilcoxon test statistic T along with its null mean and variance.

(c) Calculate the asymptotic Wilcoxon test for H_0 versus H_A, concluding in terms of the data using the nominal level $\alpha = 0.05$.

3.12.6. Set up the linear model for the data in Exercise 3.12.5. Using `rfit` and Wilcoxon scores, obtain the fit of the model. In terms of the problem, discuss the estimate of Δ and its associated 95% confidence interval based on the standard error of $\hat{\Delta}$.

3.12.7. Continuing with Exercise 3.12.6, obtain the rank-based estimate of Δ using normal scores. Likewise, compute the least squares fit.

(a) Table the three (Wilcoxon, normal scores, and LS) estimates along with their 95% confidence intervals. Compare the results.

(b) Repeat part (a), if the data were reported as:

Men	74	72	77	76	76	73	75	73	74	75
Women	75	77	78	79	77	37	78	79	78	80

3.12.8. Consider the χ^2 version of Mood's test.

(a) Determine the expected frequencies for the cells of the contingency table found above expression (3.30).

(b) Show that the χ^2 statistic test statistic simplifies to

$$\chi^2 = \frac{\left(M_S(0) - \frac{n_2}{2}\right)^2}{\frac{n_1 n_2}{4n}}.$$

3.12.9. Consider the observations for `trt1` and `trt2` in the dataset discussed in Example 3.3.1. Let Δ be the shift in locations from `trt1` to `trt2`.

(a) Obtain comparison boxplots between the samples and comment on the plot in terms of the weight of the plants for each sample.

(b) Use Fisher's exact test to test $\Delta = 0$ versus $\Delta > 0$. Comment at level 0.05 in terms of the problem.

(c) Carry out the test in Part (b) using the χ^2 contingency table test. Compare the results with Part (b).

(d) Carry out the test in Part (b) using the Wilcoxon rank test. Compare the results with Part (b).

3.12.10. Consider the base R dataset `Seatbelts`. This dataset contains several variables recorded over time in Great Britain. The data were recorded by month from Jan 1969 to Dec 1984. This is a time series dataset. The proper analysis would be a time series analysis[19] but as an exercise for this section consider the number of drivers killed and whether or not the seatbelt law was in effect. As the first sample, consider the number killed before the law took effect, and as the second sample, the number killed after the law took effect. The samples can be obtained by:

```
nkilled <- Seatbelts[,1]; law <- Seatbelts[,8]
x <- nkilled[law==0]; y <- nkilled[law==1]
```

Let Δ be the shift in the number killed before the law took effect to after the law took effect. Use Fisher's exact test to test $H_0 : \Delta = 0$ versus $H_A : \Delta < 0$. Next, compute the approximate χ^2-test. Compare the results.

3.12.11. Consider the generated samples:

x:	54	60	62	66	70	70	71	74	80	80		
y:	13	35	37	37	38	53	53	62	63	63	92	94

Let $\Delta = \theta_x - \theta_y$ be the shift in locations. We are interested in testing $H_0 : \Delta = 0$ versus $H_A : \Delta \neq 0$ at level $\alpha = 0.05$.

(a) Obtain comparison boxplots between the samples. Comment on the plot and on the hypotheses in terms of the plot.

(b) Test the hypotheses using the Wilcoxon test. Using the p-value, conclude in terms of the problem.

[19] See the original source of the data, Harvey (1989).

(c) Test the hypotheses using the t-test. Use the argument `var.equal=T` in the call to `t.test`. Using the p-value, conclude in terms of the problem.

(d) Why do the results differ in Parts (b) and (c).

3.12.12. Consider Example 3.3.1. Determine the estimate $\hat{\Delta}_S$ of Δ, its standard error, and a 95% confidence interval for Δ. Summarize your results in words of the example.

3.12.13. Write an R-function that computes the estimate of Δ, its standard error, and confidence interval at a specified confidence level for the Wilcoxon procedure. As an example, use your function to compute these quantities for the two samples in Example 3.3.1.
Hint: If the samples are in `x` and `y`, then the following code obtains the standard error of the Wilcoxon estimate of Δ:

```
z <- c(x,y); ind <- c(rep(0,length(x)),rep(1,length(y)))
summary(rfit(z,ind))$coef[2,2]
```

3.12.14. For the data of Example 3.3.1, find 95% confidence intervals based on the differences in medians for all 3 comparisons (between `trt1` and `trt2` and each treatment with control). Table your results with columns: Estimate of Δ, Standard Error, and Confidence Interval. Is one treatment the "best" (heavier weight is better)? Is one treatment the "worst"? Discuss your findings.

3.12.15. Consider the following two samples:

x													
56	43	41	30	71	44	56	60	45	53	44	55		
y													
66	86	97	87	99	130	76	75	85	66	55	68	126	60

(a) Obtain comparison boxplots of the samples. Discuss the plot including the difference in medians and the ratio of the interquartile ranges.

(b) Use the pretest procedure on η with $\alpha = 0.10$ described at the end of Section 3.6.1. Then based on the result, obtain an estimate of Δ, along with the standard error of the estimate and a 90% confidence interval for Δ.

(c) Compare the estimate of Δ with the difference in medians and the estimate of η with the ratio of the interquartile ranges.

3.12.16. In Example 3.6.2 we considered the length of songs on two albums recorded by the band U2; see Verzani (2014) for the dataset. For this exercise, consider the albums *Joshua Tree* and *Boy*. For these albums, the responses (length of songs in seconds) are given by:

Joshua Tree										
337	277	296	272	258	292	177	212	323	253	314
Boy										
216	381	253	94	274	276	262	113	182	194	288

(a) Obtain comparison boxplots of the samples. Discuss the plot including the difference in medians and the ratio of the interquartile ranges.

(b)] Use the pretest procedure on η with $\alpha = 0.10$ described at the end of Section 3.6.1. Then based on the result, obtain an estimate of Δ, along with the standard error of the estimate and a 90% confidence interval for Δ. Discuss these results.

3.12.17. Consider the power study in Example 3.9.1. Obtain a similar power study including the plot of empirical power functions when the random errors have a t-distribution with 5 degrees of freedom; i.e., just change `df<-2` to `df<-5`. The set of alternatives may have to be changed to obtain a nice spread of power.

3.12.18. Repeat Exercise 3.12.17 when the random errors have a normal distribution.

3.12.19. Consider the pilot dataset in Example 3.9.2.

(a) Obtain comparison boxplots of the samples. Comment on the plot.

(b) Obtain the Wilcoxon two-sample test of the hypotheses (3.73). Use the p-value to conclude at the 5% level.

(c) Next, test the hypotheses using the LS test (two-sample t). Compare the results.

3.12.20. Consider Example 3.9.2.

(a) Determine the sample size for the Wilcoxon analysis to achieve 90% power to detect $\Delta_A = 8$.

(b) Determine the sample size for the Wilcoxon analysis to achieve 70% power to detect $\Delta_A = 8$.

(c) Discuss these results along with the 80% power result in the example. Guess the trend in sample sizes as power increases.

3.12.21. In Example 3.9.2 the alternative of interest is $\Delta_A = 8$. Determine the sample size for the Wilcoxon analysis to detect the alternatives $\Delta_A = 5$ with 80% power. Likewise, determine the sample size if $\Delta_A = 10$. Guess the trend in sample sizes as the alternative increases.

3.12.22. Suppose the focus of a completely randomized design for two treatments T_1 and T_2 on the response variable of interest, is the estimation of Δ the shift in location. The experimenters want to estimate Δ within 7 units of precision at 95% confidence, using the Hodges–Lehmann (median sample differences) estimate of Δ. In order to determine the sample size, they have at their disposal the following data from a pilot study:

T_1	8	49	46	56	57	39
T_2	67	55	52	45	64	47

Using the pilot study, determine the necessary sample sizes to estimate Δ within 7 units of precision at 95% confidence.

3.12.23. Using the pilot study data in Exercise 3.12.22, compute the Hodges–Lehmann (median sample differences) estimate of Δ along with the distribution-free 90% confidence interval. What is the precision of the estimate?

3.12.24. In Example 3.3.1 we use the `factor` function after subsetting the dataset. What is the purpose of this function call?

3.12.25. Compute the test statistics proposed by Finkelstein and Schoenfeld (1999) for the dataset in Table 3.8.

4

Regression

4.1 Introduction

In this chapter, a nonparametric, rank-based (R) approach to regression modeling is presented. Our primary goal is the estimation of parameters in a linear regression model and associated inferences. As with the previous chapters, we generally discuss Wilcoxon analyses, while general scores are discussed in Section 7.1. These analyses generalize the rank-based approach for the two-sample location problem discussed in the previous chapter. Focus is on the use of the R package `Rfit` (Kloke and McKean 2012). We illustrate the estimation, diagnostics, and inference including confidence intervals and test for general linear hypotheses. We assume that the reader is familiar with the general concepts of regression analysis.

Rank-based (R) estimates for linear regression models were first considered by Jurečková (1971) and Jaeckel (1972). The geometry of the rank-based approach is similar to that of least squares as shown by McKean and Schrader (1980). In this chapter, the `Rfit` implementation for simple and multiple linear regression is discussed. The first two sections of this chapter present illustrative examples. We present a short introduction to the more technical aspects of rank-based regression in Section 7.1. The reader interested in a more detailed introduction is referred to Chapter 3 of Hettmansperger and McKean (2011).

In Section 4.2 correlation is presented, including the two commonly used nonparametric measures of association of Kendall and Spearman. Using the bootstrap for rank regression is conceptually the same as other types of regression and Section 4.5 demonstrates the `Rfit` implementation. Nonparametric smoothers for regression models are considered in Section 4.6. In succeeding chapters we present rank-based ANOVA analyses and extend rank-based fitting to more general models.

4.2 Correlation

In a simple linear regression problem involving a response variable Y and a predictor variable X, the fit of the model is of main interest. In particular, we are often interested in predicting the random variable Y in terms of x and we treat x as nonstochastic. In certain settings, though, the pair (X, Y) is taken to be a random vector, and we are interested in measuring the strength of a relationship or association between the random variables X and Y. By no association, we generally mean that the random variables X and Y are independent, so the basic hypotheses of interest in this section are:

$$H_0 : X \text{ and } Y \text{ are independent versus } H_A : X \text{ and } Y \text{ are dependent.} \quad (4.1)$$

For this section, assume that (X, Y) is a continuous random vector with joint cdf and pdf $F(x, y)$ and $f(x, y)$, respectively. Recall that X and Y are independent random variables if their joint cdf factors into the product of the marginal cdfs; i.e., $F(x, y) = F_X(x)F_Y(y)$ where $F_X(x)$ and $F_Y(y)$ are the marginal cdfs of X and Y, respectively. In Section 6.2 we discussed a χ^2 goodness-of-fit test for independence when X and Y are discrete random variables. For the discrete case, independence is equivalent to the statement $P(X = x, Y = y) = P(X = x)P(Y = y)$ for all x and y. In fact, the null expected frequencies of the χ^2 goodness-of-fit test statistic are based on this statement.[1] In the continuous case, we consider **measures of association** between X and Y. We discuss the traditional Pearson's measure of association (the correlation coefficient ρ) and two popular nonparametric measures (Kendall's τ_K and Spearman's ρ_S).

Let $(X_1, Y_1), (X_2, Y_2), \ldots, (X_n, Y_n)$ denote a random sample of size n on the random vector (X, Y). Using this notation, we discuss the estimation, associated inference, and R computation for these measures. Our discussion is brief, and many of the facts that we state are discussed in detail in most introductory mathematical statistics texts; see, for example, Chapters 9 and 10 of Hogg et al. (2019).

4.2.1 Pearson's Correlation Coefficient

The traditional correlation coefficient between X and Y is the ratio of the covariance between X and Y to the product of their standard deviations, i.e.,

$$\rho = \frac{E[(X - \mu_X)(Y - \mu_Y)]}{\sigma_X \sigma_Y}, \quad (4.2)$$

where μ_X, σ_X and μ_Y, σ_Y are the respective means and standard deviations of X and Y. The parameter ρ requires, of course, the assumption of finite variance

[1] For the continuous case, this statement is simply $0 = 0$, so the χ^2 goodness-of-fit test is not an option.

for both X and Y. It is a measure of linear association between X and Y. It can be shown that it satisfies the properties: $-1 \leq \rho \leq 1$; $\rho = \pm 1$ if and only if Y is a linear function of X (with probability 1); and $\rho > (<)\, 0$ is associated with a positive (negative) linear relationship between Y and X. Note that if X and Y are independent, then $\rho = 0$. In general, the converse is not true. The contrapositive, though, is true; i.e., $\rho \neq 0 \Rightarrow X$ and Y are dependent.

Usually ρ is estimated by the nonparametric estimator. The numerator is estimated by the sample covariance, $n^{-1}\sum_{i=1}^{n}(X_i - \overline{X})(Y_i - \overline{Y})$, while the denominator is estimated by the product of the sample standard deviations (with n, not $n-1$, as divisors of the sample variances). This simplifies to the **sample correlation coefficient** given by

$$r = \frac{\sum_{i=1}^{n}(X_i - \overline{X})(Y_i - \overline{Y})}{\sqrt{\sum_{i=1}^{n}(X_i - \overline{X})^2 \cdot \sum_{i=1}^{n}(Y_i - \overline{Y})^2}}. \tag{4.3}$$

Similarly, it can be shown that r satisfies the properties: $-1 \leq r \leq 1$; $r = \pm 1$ if there is a deterministic linear relationship for the sample (X_i, Y_i); and $r > (<)\, 0$ is associated with a positive (negative) linear relationship between Y_i and X_i. The estimate of the correlation coefficient is directly related to simple least squares regression. Let $\hat{\sigma}_x$ and $\hat{\sigma}_y$ denote the respective sample standard deviations of X and Y. Then we have the relationship

$$r = \frac{\hat{\sigma}_x}{\hat{\sigma}_y}\hat{\beta}, \tag{4.4}$$

where $\hat{\beta}$ is the least squares estimate of the slope in the simple regression of Y_i on X_i. It can be shown that, under the null hypothesis, $\sqrt{n}r$ is asymptotically $N(0,1)$. Inference for ρ can be based on this asymptotic result, but usually the t-approximation discussed next is used.

If we make the much stronger assumption that the random vector (X, Y) has a bivariate normal distribution, then the estimator r is the maximum likelihood estimate (MLE) of ρ. Based on expression (4.4) and the usual t-ratio in regression, under H_0, the statistic

$$t = \frac{\sqrt{n-2}\, r}{\sqrt{1 - r^2}} \tag{4.5}$$

has t-distribution with $n-2$ degrees of freedom; see Section 9.7 of Hogg et al. (2019). Thus a level α test of the hypotheses (4.1) is to reject H_0 in favor of H_A if $|t| > t_{\alpha/2, n-2}$. Furthermore, for general ρ, it can be shown that $\log[(1+r)/(1-r)]$ is approximately normal with mean $\log[(1+\rho)/(1-\rho)]$. Based on this, approximate confidence intervals for ρ can be constructed. In practice, usually the strong assumption of bivariate normality cannot be made. In this case, the t-test and confidence interval are approximate. For computation in R, assume that the R vectors `x` and `y` contain the samples $X_1, \ldots, X_n$ and $Y_1, \ldots, Y_n$, respectively. Then the R function `cor.test` computes this analysis; see Example 4.2.1 below. If inference is not needed, the function `cor` may be used to just obtain the estimate.

4.2.2 Kendall's τ_K

Kendall's τ_K is the first nonparametric measure of association that we discuss. As above, let (X, Y) denote a jointly continuous random vector. Kendall's τ_K is a measure of **monotonicity** between X and Y. Let the two pairs of random variables (X_1, Y_1) and (X_2, Y_2) be independent random vectors with the same distribution as (X, Y). We say that the pairs (X_1, Y_1) and (X_2, Y_2) are **concordant** or **discordant** if

$$\text{sign}\{(X_1 - X_2)(Y_1 - Y_2)\} = 1 \text{ or } \text{sign}\{(X_1 - X_2)(Y_1 - Y_2)\} = -1,$$

respectively. Concordant pairs are indicative of increasing monotonicity between X and Y, while discordant pairs indicate decreasing monotonicity. Kendall's τ_K measures this monotonicity in a probability sense. It is defined by

$$\tau_K = P[\text{sign}\{(X_1 - X_2)(Y_1 - Y_2)\} = 1] - P[\text{sign}\{(X_1 - X_2)(Y_1 - Y_2)\} = -1]. \tag{4.6}$$

It can be shown that $-1 \leq \tau_K \leq 1$; $\tau_K > 0$ indicates increasing monotonicity; $\tau_K < 0$ indicates decreasing monotonicity; and $\tau_K = 0$ reflects neither monotonicity. It follows that if X and Y are independent, then $\tau_K = 0$. As with ρ, the converse is not true, but the contrapositive is true; i.e., $\tau_K \neq 0 \Rightarrow X$ and Y are dependent.

Using the random sample $(X_1, Y_1), (X_2, Y_2), \ldots, (X_n, Y_n)$, a straightforward estimate of τ_K is simply to count the number of concordant pairs in the sample and subtract from that the number of discordant pairs. Standardization of this statistic leads to

$$\hat{\tau}_K = \binom{n}{2}^{-1} \sum_{i<j} \text{sign}\{(X_i - X_j)(Y_i - Y_j)\} \tag{4.7}$$

as our estimate of τ_K. Since the statistic $\hat{\tau}_K$ is a Kendall's τ_K based on the empirical sample distribution, it shares the same properties; i.e., $\hat{\tau}_K$ is between -1 and 1, positive values of $\hat{\tau}_K$ reflect increasing monotonicity, and negative values reflect decreasing monotonicity. It can be shown that $\hat{\tau}_K$ is an unbiased estimate of τ_K. Further, under the assumption that X and Y are independent, the statistic $\hat{\tau}_K$ is distribution-free with mean 0 and variance $2(2n + 5)/[9n(n - 1)]$. Tests of the hypotheses (4.1) can be based on the exact finite sample distribution. Tie corrections for the test are available. Furthermore, distribution-free confidence intervals[2] for τ_K exist. R computation of the inference for Kendall's τ_K is obtained by the function `cor.test` with `method="kendall"`; see Example 4.2.1. Although this R function does not compute a confidence interval for τ_K, in Section 4.2.4 we provide an R function to compute the percentile bootstrap confidence interval for τ_K.

---

[2]See Chapter 8 of Hollander and Wolfe (1999).

4.2.3 Spearman's ρ_S

In defining Spearman's ρ_S, it is easier to begin with its estimator. Consider the random sample $(X_1, Y_1), (X_2, Y_2), \ldots, (X_n, Y_n)$. Denote by $R(X_i)$ the rank of X_i among $X_1, X_2, \ldots, X_n$ and likewise define $R(Y_i)$ as the rank of Y_i among $Y_1, Y_2, \ldots, Y_n$. The estimate of ρ_S is simply the sample correlation coefficient with X_i and Y_i replaced, respectively, by $R(X_i)$ and $R(Y_i)$. Let r_S denote this correlation coefficient. Note that the denominator of r_S is a constant and that the sample mean of the ranks is $(n+1)/2$. Simplification leads to the formula

$$r_S = \frac{\sum_{i=1}^{n}(R(X_i) - [(n+1)/2])(R(Y_i) - [(n+1)/2])}{n(n^2-1)/12}. \tag{4.8}$$

This statistic is a correlation coefficient, so it is between ± 1. It is ± 1 if there is a strictly increasing (decreasing) relation between X_i and Y_i; hence, similar to Kendall's $\hat{\tau}_K$, it estimates monotonicity between the samples. It can be shown that

$$E(r_S) = \frac{3}{n+1}[\tau_K + (n-2)(2\gamma - 1)],$$

where $\gamma = P[(X_2 - X_1)(Y_3 - Y_1) > 0]$. The parameter that r_S is estimating is not as easy to interpret as the parameter τ_K.

If X and Y are independent, it follows that r_S is a distribution-free statistic with mean 0 and variance $(n-1)^{-1}$. We accept H_A : X and Y are dependent for large values of $|r_S|$. This test can be carried out using the exact distribution or approximated using the z-statistic $\sqrt{n-1}r_S$. In applications, however, similar to expression (4.5), the t-approximation[3] is often used, where

$$t = \frac{\sqrt{n-2}r_S}{\sqrt{1-r_S^2}}. \tag{4.9}$$

There are distribution-free confidence intervals for ρ_S and tie corrections[4] are available. The R command `cor.test` with `method="spearman"` returns the analysis based on Spearman's r_S. This computes the test statistic and the p-value, but not a confidence interval for ρ_S. Although the parameter ρ_S is difficult to interpret, confidence intervals are important as they give a sense of the strength (effect size) of the estimate. As with Kendall's τ_K, in Section 4.2.4, we provide an R function to compute the percentile bootstrap confidence interval for ρ_S.

Remark 4.2.1 (Hypothesis Testing for Associations). In general, let ρ_G denote any of the measures of association discussed in this section. If $\rho_G \neq 0$ then X and Y are dependent. Hence, if the statistical test rejects $\rho_G = 0$, then we can statistically accept H_A that X and Y are dependent. On the other hand, if $\rho_G = 0$, then X and Y are not necessarily independent. Hence, if the test

[3]See, for example, page 347 of Huitema (2011).
[4]See Hollander and Wolfe (1999).

fails to accept H_A, then we should not conclude independence between X and Y. In this case, we should conclude that there is not significant evidence to refute $\rho_G = 0$. ∎

4.2.4 Computation and Examples

We illustrate the R function `cor.test` in the following example.

Example 4.2.1 (Baseball Data, 2010 Season). Datasets of major league baseball statistics can be downloaded at the site `baseballguru.com`. For this example, we investigate the relationship between the batting average of a full-time player and the number of home runs that he hits. By full-time we mean that the batter had at least 450 official at bats during the season. These data are in the `npsm` dataset `bb2010`. Figure 4.1 displays the scatterplot of home run production versus batting average for full-time players. Based on this plot, there is an increasing monotone relationship between batting average and home run production, although the relationship is not very strong.

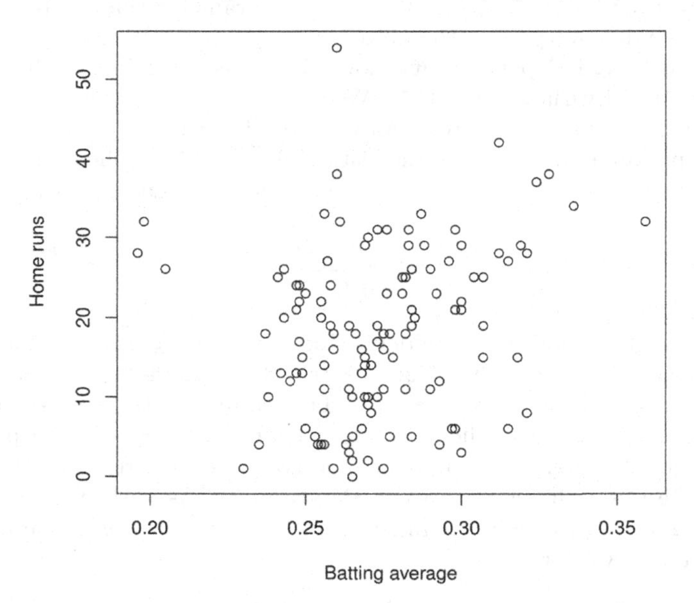

FIGURE 4.1
Scatterplot of home runs versus batting average for players who have at least 450 at bats during the 2010 Major League Baseball Season.

In the next code segment, the R analyses (based on `cor.test`) of Pearson's, Spearman's, and Kendall's measures of association are displayed.

```
> with(bb2010,cor.test(ave,hr))

        Pearson's product-moment correlation

data:  ave and hr
t = 2.2719, df = 120, p-value = 0.02487
alternative hypothesis: true correlation is not equal to 0
95 percent confidence interval:
 0.02625972 0.36756513
sample estimates:
      cor
0.2030727

> with(bb2010,cor.test(ave,hr,method="spearman"))

        Spearman's rank correlation rho

data:  ave and hr
S = 234500, p-value = 0.01267
alternative hypothesis: true rho is not equal to 0
sample estimates:
      rho
0.2251035

> with(bb2010,cor.test(ave,hr,method="kendall"))

        Kendall's rank correlation tau

data:  ave and hr
z = 2.5319, p-value = 0.01134
alternative hypothesis: true tau is not equal to 0
sample estimates:
      tau
0.1578534
```

For each of the methods, the output contains a test statistic and associated p-value as well as the point estimate of the measure of association. Pearson's also contains the estimated confidence interval (95% by default). For example, the results from Pearson's analysis give $r = 0.203$ and a p-value of 0.025. While all three methods show a significant positive association between home run production and batting average, the results for Spearman's and Kendall's procedures are somewhat stronger than that of Pearson's. Based on the scatterplot, Figure 4.1, there are several outliers in the dataset which may have impaired Pearson's r. On the other hand, Spearman's r_S and Kendall's $\hat{\tau}_K$ are robust to the effects of the outliers.

The output for Spearman's method results in the value of r_S and the p-value of the test. It also computes the statistic

$$S = \sum_{i=1}^{n} [R(X_i) - R(Y_i)]^2.$$

Although it can be shown that

$$r_S = 1 - \frac{6S}{n^2 - n};$$

the statistic S does not readily show the strength of the association, let alone the sign of the monotonicity. Hence, in addition, we advocate forming the z statistic or the t-approximation of expression (4.9). The latter gives the value of 2.53 with an approximate p-value of 0.0127. This p-value agrees with the p-value calculated by `cor.test` and the value of the standardized test statistic is readily interpreted. See Exercise 4.7.13.

As with the R output for Spearman's procedure, the output for Kendall's procedure includes $\hat{\tau}_K$ and the p-value of the associated test. The results of the analysis based on Kendall's procedure indicate that there is a significant monotone increasing relationship between batting average and home run production, similar to the results for Spearman's procedure. The estimate of association is smaller than that of Spearman's, but recall that they are estimating different parameters. Instead of a z-test statistic, R computes the test statistic T which is the number of pairs that are monotonically increasing. It is related to $\hat{\tau}_K$ by the expression

$$T = \left\{ \binom{n}{2} [1 + \hat{\tau}_K] \right\} / 2.$$

The statistic T does not lend itself easily to interpretation of the test. Even the sign of monotonicity is missing. As with the Spearman's procedure, we recommend also computing the standardized test statistic; see Exercise 4.7.14.
■

In general, a confidence interval yields a sense of the strength of the relationship. For example, a *quick* standard error is the length of a 95% confidence interval divided by 4. The function `cor.test` does not compute confidence intervals for Spearman's and Kendall's methods. We have written an R function, `cor.boot.ci`, which obtains a percentile bootstrap confidence interval for each of the three measures of association discussed in this section. Let $\boldsymbol{B} = [\boldsymbol{X}, \boldsymbol{Y}]$ be the matrix with the samples of X_i's in the first column and the samples of Y_i's in the second column. Then the bootstrap scheme resamples the rows of $\boldsymbol{B}$ with replacement to obtain a bootstrap sample of size n. This is performed n_{BS} times. For each bootstrap sample, the estimate of the measure of association is obtained. These bootstrap estimates are collected and the $\alpha/2$ and $(1 - \alpha/2)$ percentiles of this collection form the confidence interval. The default arguments of the function are:

```
> args(cor.boot.ci)

function (x, y, method = "spearman", conf = 0.95, nbs = 3000)
NULL
```

Besides Spearman's procedure, bootstrap percentile confidence intervals are computed for ρ and τ_K by using respectively the arguments `method="pearson"` and `method="kendall"`. Note that $(1-\alpha)$ is the confidence level and the default number of bootstrap samples is set at 3000. We illustrate this function in the next example.

Example 4.2.2 (Continuation of Example 4.2.1). The code segment below obtains a 95% percentile bootstrap confidence interval for Spearman's ρ_S.

```
> library(boot)
> with(bb2010,cor.boot.ci(ave,hr))

      2.5%      97.5%
0.05020961 0.39888150
```

The following code segment computes percentile bootstrap confidence intervals for Pearson's and Kendall's methods.

```
> with(bb2010,cor.boot.ci(ave,hr,method='pearson'))

       2.5%       97.5%
0.005060283 0.400104126

> with(bb2010,cor.boot.ci(ave,hr,method='kendall'))

      2.5%      97.5%
0.02816001 0.28729659
```

To show the robustness of Spearman's and Kendall's procedures, we changed the home run production of the 87*th* batter from 32 to 320; i.e., a typographical error. Table 4.1 compares the results for all three procedures on the original and changed data.[5]

Note that the results for Spearman's and Kendall's procedures are essentially the same on the original dataset and the dataset with the outlier. For Pearson's procedure, though, the estimate changes from 0.20 to 0.11. Also, the confidence interval has been affected. ∎

[5]These analyses were run in a separate step, so they may differ slightly from those already reported.

TABLE 4.1
Estimates and Confidence Intervals for the Three Methods. The first three columns contain the results for the original data, while the last three columns contain the results for the changed data.

	Original Data			Outlier Data		
	Est	LBCI	UBCI	Est2	LBCI2	UBCI2
Pearson's	0.20	0.00	0.40	0.11	0.04	0.36
Spearman's	0.23	0.04	0.40	0.23	0.05	0.41
Kendall's	0.16	0.03	0.29	0.16	0.04	0.29

4.3 Simple Linear Regression

In this section we present an example of how to utilize `Rfit` to obtain rank-based estimates of the parameters in a simple linear regression problem. Write the simple linear regression model as

$$Y_i = \alpha + x_i\beta + e_i, \quad \text{for } i = 1, \dots n \tag{4.10}$$

where Y_i is a continuous response variable for the ith subject or experimental unit, x_i is the corresponding value of an explanatory variable, e_i is the error term, α is an intercept parameter, and β is the slope parameter. We are interested in inference for the slope parameter β. The errors are assumed to be iid with pdf $f(t)$. Closed-form solutions exist for the least squares (LS) estimates of (4.10). However, in general, this is not true for rank-based (R) estimation.

For the rest of this section, we work with an example which highlights the use of `Rfit`. The dataset involved is `engel` which is in the package `quantreg` (Koenker 2013). The data are a sample of 235 Belgian working-class households. The response variable is food expenditure in Belgian francs, and the explanatory variable is the annual household income in Belgian francs.

A scatterplot of the data is presented in Figure 4.2 where the rank-based (R) and least squares (LS) fits are overlaid. Several outliers are present, which affect the LS fit. The data also appear to be heteroscedastic — scale appears to increase as annual income increases. The following code segment illustrates the creation of the graphic.

```
> library(quantreg)
> data(engel)
> plot(engel)
> abline(rfit(foodexp~income,data=engel))
> abline(lm(foodexp~income,data=engel),lty=2)
> legend("topleft",c('R','LS'),lty=c(1,2),bty='n')
```

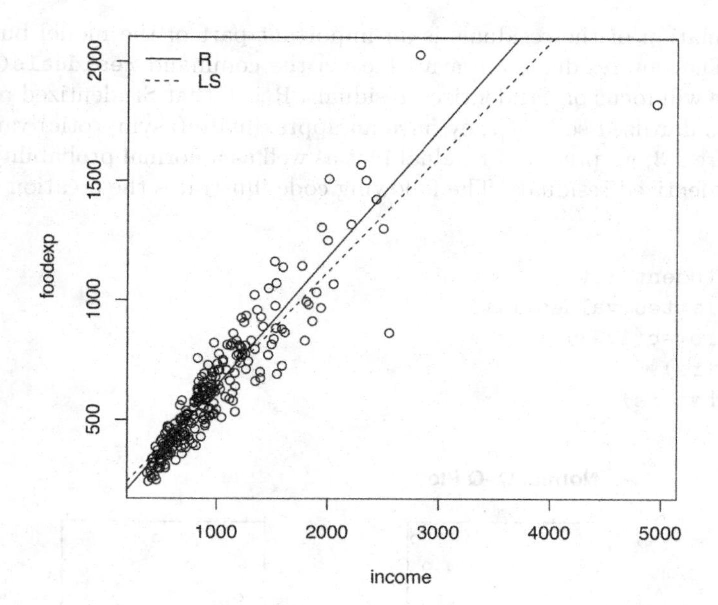

FIGURE 4.2
Scatterplot of Engel data with overlaid regression lines.

The command `rfit` obtains robust R estimates for the linear regression models, for example (4.10). To examine the coefficients of the fit, use the `summary` command. Critical values and p-values based on Student's t-distribution with $n - 2$ degrees of freedom are recommended for inference. For this example, `Rfit` used the t-distribution with 233 degrees of freedom to obtain the p-value.

```
> library(Rfit)
> fit<-rfit(foodexp~income,data=engel)
> coef(summary(fit))

              Estimate Std. Error t.value     p.value
(Intercept) 103.64804  12.774615  8.1136  2.8181e-14
income        0.53777   0.011507 46.7328 2.4311e-120
```

Readers with experience modeling in R will recognize that the syntax is similar to using `lm` to obtain a least squares analysis. The fitted regression equation is

$$\widehat{\text{foodexp}} = 103.648 + 0.538 * \text{income}.$$

A 95% confidence interval for the slope parameter (β) is calculated as $0.538 \pm 1.97 * 0.012 = 0.538 \pm 0.023$ or $(0.515, 0.56)$.

Examination of the residuals is an important part of the model building process. The raw residuals are available via the command **residuals(fit)**, though we will focus on Studentized residuals. Recall that Studentized residuals are standardized so that they have an approximate (asymptotic) variance 1. In Figure 4.3, we present a residual plot as well as a normal probability plot of the Studentized residuals. The following code illustrates the creation of the graphic.

```
> rs<-rstudent(fit)
> yhat<-fitted.values(fit)
> par(mfrow=c(1,2))
> qqnorm(rs)
> plot(yhat,rs)
```

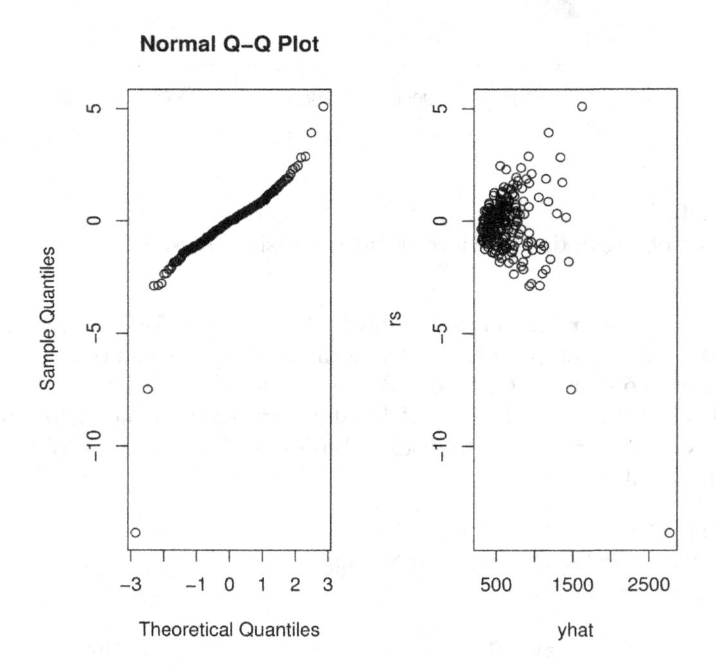

FIGURE 4.3
Diagnostic plots for Engel data.

Outliers and heteroscedasticity are apparent in the residual and normal probability plots in Figure 4.3.

4.4 Multiple Linear Regression

Rank-based regression offers a complete inference for the multiple linear regression. In this section we illustrate how to use `Rfit` to obtain estimates, standard errors, inference, and diagnostics based on an R fit of the model

$$Y_i = \alpha + \beta_1 x_{i1} + \ldots \beta_p x_{ip} + e_i \text{ for } i = 1, \ldots n \tag{4.11}$$

where $\beta_1, \ldots, \beta_p$ are regression coefficients, Y_i is a continuous response variable, $x_{i1}, \ldots, x_{ip}$ are a set of explanatory variables, e_i is the error term, and α is an intercept parameter. Interest is on inference for the set of parameters $\beta_1, \ldots \beta_p$. As in the simple linear model case, for inference, the errors are assumed to be iid with pdf f. Closed-form solutions exist for the least squares (LS); however, the R estimates (in most cases) must be solved iteratively.

4.4.1 Multiple Regression

In this subsection we discuss a dataset from Morrison (1983: p.64) (c.f. Hettmansperger and McKean 2011). The dataset is included in `Rfit`. The response variable is the level of free fatty acid (`ffa`) in a sample of prepubescent boys. The explanatory variables are `age` (in months), `weight` (in pounds), and `skin` fold thickness. In this subsection we illustrate the Wilcoxon analysis and in Exercise 4.7.3 the reader is asked to redo the analysis using bent scores. The model we wish to fit is

$$\text{ffa} = \alpha + \beta_1 \text{age} + \beta_2 \text{weight} + \beta_3 \text{skin} + \text{error}. \tag{4.12}$$

We use `Rfit` as follows to obtain the R fit of (4.12)

```
> fit<-rfit(ffa~age+weight+skin,data=ffa)

> summary(fit)

Call:
rfit.default(formula = ffa ~ age + weight + skin, data = ffa)

Coefficients:
              Estimate Std. Error t.value p.value
(Intercept)  1.49059    0.26761    5.57 2.4e-06 ***
age         -0.00113    0.00262   -0.43 0.66748
weight      -0.01535    0.00382   -4.02 0.00028 ***
skin         0.27480    0.13335    2.06 0.04641 *
---
Signif. codes:
0 '***' 0.001 '**' 0.01 '*' 0.05 '.' 0.1 ' ' 1
```

```
Multiple R-squared (Robust): 0.37731
Reduction in Dispersion Test: 7.4733 p-value: 0.00049
```

Displayed are estimates, standard errors, and Wald (t-ratio) tests for each of the individual parameters. The variable `age` is not statistically significant, while `weight` and `skin` are. In addition, there is a robust R^2 value which can be utilized in a manner similar to the usual R^2 value of LS analysis; here $R^2 = 0.38$. Finally, a test of all the regression coefficients excluding the intercept parameter is provided in the form of a **reduction in dispersion test**. In this example, we would reject the null hypothesis and conclude that at least one of the nonintercept coefficients is a significant predictor of free fatty acid.

The reduction in dispersion test is analogous to the LS's F-test based on the reduction in sums of squares. This test is based on the reduction (drop) in dispersion as we move from the reduced model (full model constrained by the null hypothesis) to the full model. As an example, for the free fatty acid data, suppose that we want to test the hypothesis:

$$H_0 : \beta_{\text{age}} = \beta_{\text{weight}} = 0 \text{ versus } H_A : \beta_{\text{age}} \neq 0 \text{ or } \beta_{\text{weight}} \neq 0. \tag{4.13}$$

Here, the reduced model contains only the regression coefficient of the predictor skin, while the full model contains all three predictors. The following code segment computes the reduction in dispersion test, returning the F test statistic (see expression (7.11)) and the corresponding p-value:

```
> fitF<-rfit(ffa~age+weight+skin,data=ffa)
> fitR<-rfit(ffa~skin,data=ffa)
> drop.test(fitF,fitR)

Drop in Dispersion Test
F-Statistic     p-value
 1.0838e+01  1.9736e-04
```

The command `drop.test` was designed with the functionality of the command `anova`—used for traditional analyses—in mind. In this case, as the p-value is small, the null hypothesis would be rejected at all reasonable levels of α.

4.4.2 Polynomial Regression

In this section we present an example of a polynomial regression fit using `Rfit`. The data are from Exercise 5 of Chapter 10 of Higgins (2003). The scatterplot of the data in Figure 4.4 reveals a curvature relationship between MPG (`mpg`) and speed (`sp`). Higgins (2003) suggests a quadratic fit, and that is how we proceed. That is, we fit the model

$$\text{speed} = \alpha + \beta_1 \text{mpg} + \beta_2 \text{mpg}^2 + \text{error}.$$

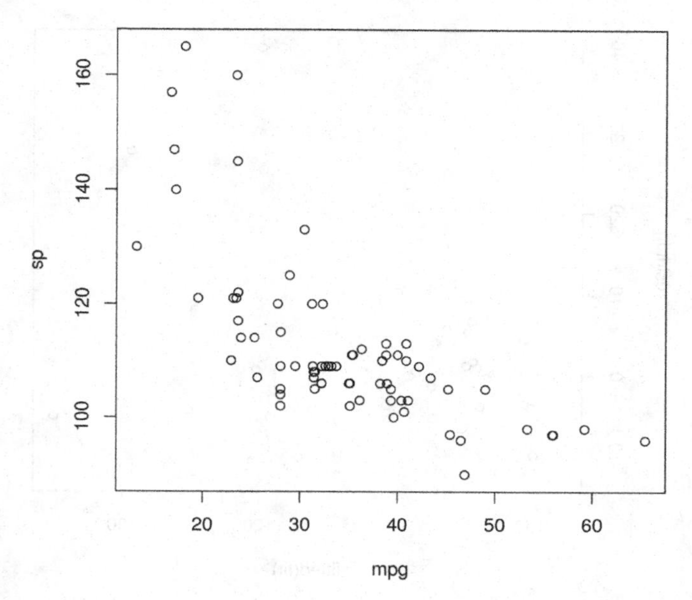

FIGURE 4.4
Scatterplot of miles per gallon vs. top speed.

To specify a squared term (or any function of the data to be interpreted arithmetically), use the `I` function:

```
> summary(fit<-rfit(sp~mpg+I(mpg^2),data=speed))

Call:
rfit.default(formula = sp ~ mpg + I(mpg^2), data = speed)

Coefficients:
             Estimate Std. Error t.value p.value
(Intercept) 160.61555    6.42727   24.99 < 2e-16 ***
mpg          -2.19068    0.35650   -6.14 3.1e-08 ***
I(mpg^2)      0.01909    0.00475    4.01 0.00013 ***
---
Signif. codes:
0 '***' 0.001 '**' 0.01 '*' 0.05 '.' 0.1 ' ' 1

Multiple R-squared (Robust): 0.54362
Reduction in Dispersion Test: 47.05 p-value: 0
```

Note that the quadratic term is highly significant. The residual plot, Figure 4.5, suggests that there is the possibility of heteroscedasticity and/or outliers; however, there is no apparent lack of fit.

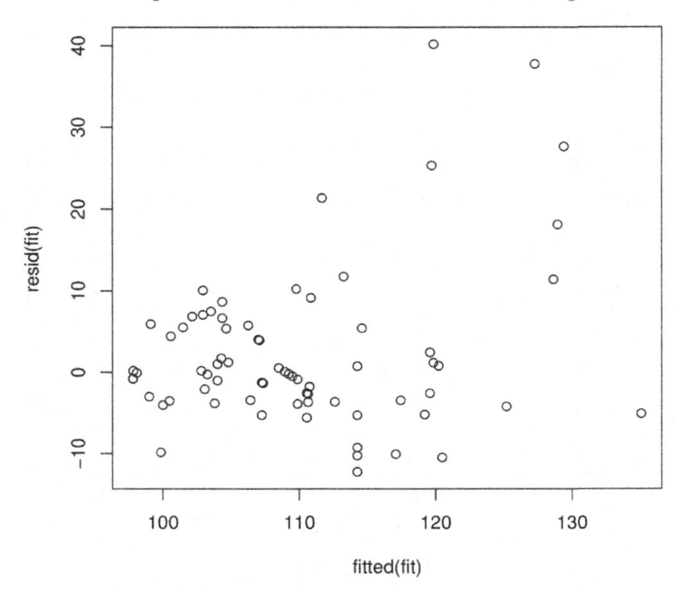

FIGURE 4.5
Residual plot based on the quadratic fit of the Higgins' data.

Remark 4.4.1 (Model Selection). For rank-based regression, model selection can be performed using forward, backwards, or step-wise procedures in the same way as in ordinary least squares. Procedures for penalized rank-based regression have been developed by Johnson and Peng (2008). In future versions of `Rfit`, we plan to include penalized model selection procedures. ■

4.5 Bootstrap

In this section we illustrate the use of the bootstrap for rank-based (R) regression. Bootstrap approaches for M estimation are discussed in Fox and Weisberg (2011), and we take a similar approach. While it is not difficult to write bootstrap functions in R, we make use of the `boot` library. Our goal is not a comprehensive treatment of the bootstrap, but rather we present an example to illustrate its use.

In this section we utilize the baseball data, which is a sample of 59 professional baseball players. For this particular example, we regress the weight of a baseball player on his height.

To use the `boot` function, we first define a function from which `boot` can calculate bootstrap estimates of the regression coefficients:

```
> boot.rfit<-function(data,indices) {
+     data<-data[indices,]
+     fit<-rfit(weight~height,data=data,tau='N')
+     coefficients(fit)[2]
+ }
```

Next use `boot` to obtain the bootstrap estimates, etc.

```
> bb.boot<-boot(data=baseball,statistic=boot.rfit,R=1000)
> bb.boot

ORDINARY NONPARAMETRIC BOOTSTRAP


Call:
boot(data = baseball, statistic = boot.rfit, R = 1000)


Bootstrap Statistics :
    original      bias    std. error
t1* 5.714278 -0.1575826   0.7761589
```

Our analysis is based on 1000 bootstrap replicates.

Figure 4.6 shows a histogram of the bootstrap estimates and a normal probability plot.

```
> plot(bb.boot)
```

Bootstrap confidence intervals are obtained using the `boot.ci` command. In this segment we obtain the bootstrap confidence interval for the slope parameter.

```
> boot.ci(bb.boot,type='perc',index=1)

BOOTSTRAP CONFIDENCE INTERVAL CALCULATIONS
Based on 1000 bootstrap replicates

CALL :
boot.ci(boot.out = bb.boot, type = "perc", index = 1)

Intervals :
Level     Percentile
95%   ( 3.75,  7.00 )
Calculations and Intervals on Original Scale
```

Exercise 4.7.4 asks the reader to compare the results with those obtained using large sample inference.

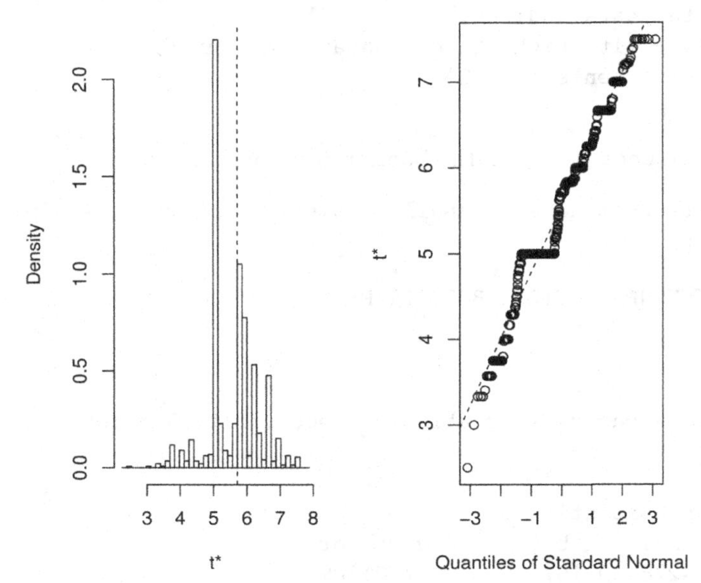

FIGURE 4.6
Bootstrap plots for R regression analysis for modeling weight versus height of the baseball data.

4.6 Nonparametric Regression

In this chapter we have been discussing linear models. Letting Y_i and $\boldsymbol{x}_i = [x_{i1}, \ldots, x_{ip}]^T$ denote the *ith* response and its associated vector of explanatory variables, respectively, these models are written as

$$Y_i = \alpha + \beta_1 x_{i1} + \beta_2 x_{i2} + \cdots + \beta_p x_{ip} + e_i, \quad i = 1, \ldots, n. \tag{4.14}$$

Note that these models are linear in the regression parameters β_j $j = 1, \ldots p$; hence, the name linear models. Next consider the model

$$Y_i = \beta_1 \exp\{\alpha + \beta_2 x_{i1} + \beta_3 x_{i3}\} + e_i, \quad i = 1, \ldots, n. \tag{4.15}$$

This model is not linear in the parameters and is an example of a nonlinear model. We explore such models in Section 8.6. The form of Model (4.15) is still known. What if, however, we do not know the functional form? For example, consider the model

$$Y_i = g(x_i) + e_i, \quad i = 1, \ldots, n, \tag{4.16}$$

where the function g is unknown. This model is often called a **nonparametric regression model**. There can be more than one explanatory variable x_i, but in this text we only consider one predictor. As with linear models, the goal is to fit the model. The fit usually shows local trends in the data, finding peaks and valleys which may have practical consequences. Further, based on the fit, residuals are formed to investigate the quality of fit. This is the main topic of this section. Before turning our attention to nonparametric regression models, we briefly consider polynomial models. We consider the case of unknown degree, so, although they are parametric models, they are not completely specified.

4.6.1 Polynomial Models

Suppose we are willing to assume that $g(x)$ is a sufficiently smooth function. Then by Taylor's Theorem, a polynomial may result in a good fit. Hence, consider polynomial models of the form

$$Y_i = \alpha + \beta_1(x_i - \overline{x}) + \beta_2(x_i - \overline{x})^2 + \cdots + \beta_p(x_i - \overline{x})^p + e_i, \quad i = 1, \ldots, n, \quad (4.17)$$

Here $\boldsymbol{x}$ is centered as shown in the model. A disadvantage of this model is that generally the degree of the polynomial is not known. One way of dealing with this unknown degree is to use residual plots based upon iteratively fitting polynomials of different degrees to determine a best fit; see Exercise 4.7.7 for such an example.

To determine the degree of a polynomial model, Graybill (1976) suggested an algorithm based on testing for the degree. Select a large (super) degree P which provides a satisfactory fit of the model. Then set $p = P$, fit the model, and test $\beta_p = 0$. If the hypothesis is rejected, stop and declare p to be the degree. If not, replace p with $p - 1$ and reiterate the test. Terpstra and McKean (2005) discuss the results of a small simulation study which confirmed the robustness of the Wilcoxon version of this algorithm. The `npsm` package contains the R function `polydeg` which performs this Wilcoxon version. We illustrate its use in the following example based on simulated data. In this section, we often use simulated data to check how well the procedures fit the model.

Example 4.6.1 (Simulated Polynomial Model)**.** In this example we simulated data from the polynomial $g(x) = 10 - 3x - 3x^2 + x^3$. One-hundred x values were generated from a $N(0, 3)$-distribution, while the added random noise, e_i, was generated from a t-distribution with 2 degrees of freedom and with a multiplicative scale factor of 15. The data are in the set `poly`. The next code segment shows the call to `polydeg` and the resulting output summary of each of its steps, consisting of the degree tested, the drop in dispersion test statistic, and the p-value of the test. Note that we set the super degree of the polynomial to 5.

```
> poly <- as.data.frame(poly)
> deg<-with(poly,polydeg(y,x,5,.05))
> deg$coll

    Deg  Robust F p-value
```

```
[1,]    5   2.33220 0.13008
[2,]    4   0.47418 0.49275
[3,]    3 228.60134 0.00000

> summary(deg$fitf)

Call:
rfit.default(formula = y ~ xr)

Coefficients:
            Estimate Std. Error t.value p.value
(Intercept)   9.4365     3.4535    2.73  0.0075 **
xr1          -7.1168     1.2537   -5.68 1.5e-07 ***
xr2          -1.5782     0.2140   -7.38 5.8e-11 ***
xr3           1.0789     0.0462   23.36 < 2e-16 ***
---
Signif. codes:
0 '***' 0.001 '**' 0.01 '*' 0.05 '.' 0.1 ' ' 1

Multiple R-squared (Robust): 0.77434
Reduction in Dispersion Test: 109.8 p-value: 0
```

Note that the routine determined the correct degree; i.e., a cubic. Based on the summary of the cubic fit, the 95% confidence interval for the leading coefficient β_3 traps its true value of 1. To check the linear and quadratic coefficients, the centered polynomial must be expanded. Figure 4.7 displays the scatterplot of the data overlaid by the rank-based fit of the cubic polynomial and the associated Studentized residual plot. The fit appears to be good which is confirmed by the random scatter of the Studentized residual plot. This plot also identifies several large outliers in the data, as expected, because the random errors follow a t-distribution with 2 degrees of freedom. ■

4.6.2 Nonparametric Regression

There are situations where polynomial fits will not suffice; for example, a dataset with many peaks and valleys. In this section, we turn our attention to the nonparametric regression model (4.16) and consider several nonparametric procedures which fit this model. There are many references for nonparametric regression models. An informative introduction, using R, is Chapter 11 of Faraway (2006); for a more authoritative account see, for example, Wood (2006). Wahba (1990) offers a technical introduction to smoothing splines using reproducing kernel Hilbert spaces.

Nonparametric regression procedures fit local trends producing, hopefully, a smooth fit. Sometimes they are called **smoothers**. A simple example is provided by a running average of size 3. In this case, the fit at x_i is $(Y_{i-1} + Y_i + Y_{i+1})/3$. Due to the non-robustness of this fit, often the mean is replaced by the median.

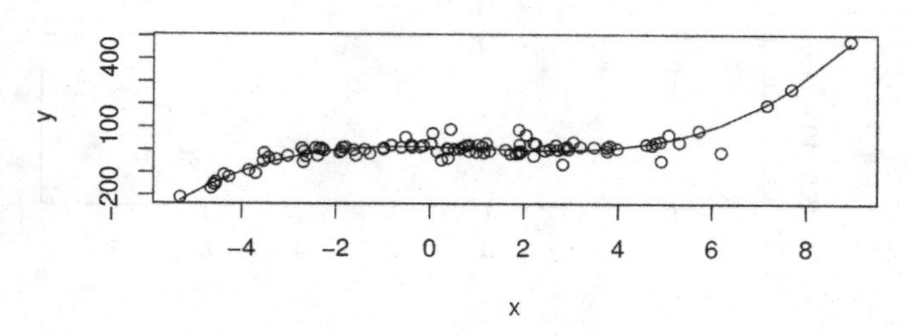

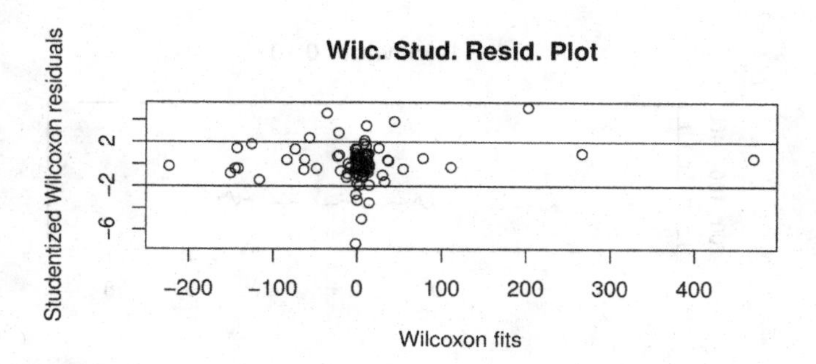

FIGURE 4.7
For the polynomial data of Example 4.6.1: The top panel displays the scatter-plot overlaid with the Wilcoxon fit, and the lower panel shows the Studentized residual plot.

The moving average can be thought of as a weighted average using the discrete distribution with mass sizes of 1/3 for the weights. This has been generalized to using continuous pdfs for the weighting. The density function used is called a **kernel** and the resulting fit is called a **kernel** nonparametric regression estimator. One such kernel estimator, available in base R, is the **Nadaraya–Watson** estimator which is defined at x by

$$\hat{f}_h(x) = \frac{1}{nh}\sum_{i=1}^{n} \frac{w_i}{\sum_{j=1}^{n} w_j} Y_i, \tag{4.18}$$

where the weights are given by

$$w_i = \frac{1}{h} K\left(\frac{x - x_i}{h}\right) \tag{4.19}$$

and the kernel $K(x)$ is a continuous pdf. The parameter h is called the bandwidth. Notice that h controls the amount of smoothing. Values of h too large often lead to overly smoothed fits, while values too small lead to overfitting (a jagged fit). Thus, the estimator (4.18) is quite sensitive to the bandwidth. On the other hand, it is generally not as sensitive to the choice of the kernel function. Often, the normal kernel is used. An R function that obtains the Nadaraya–Watson smoother is the function `ksmooth`.

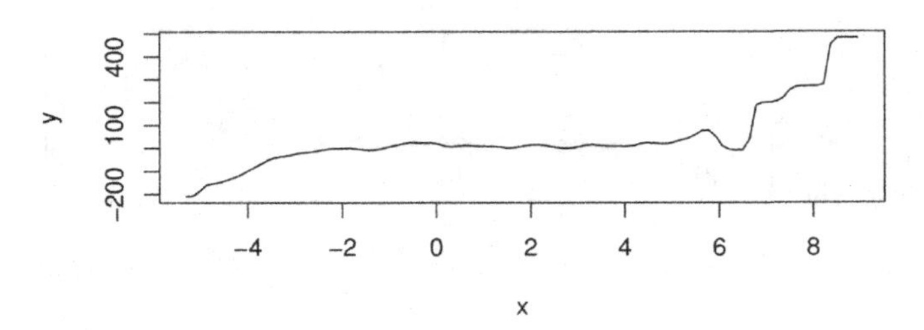

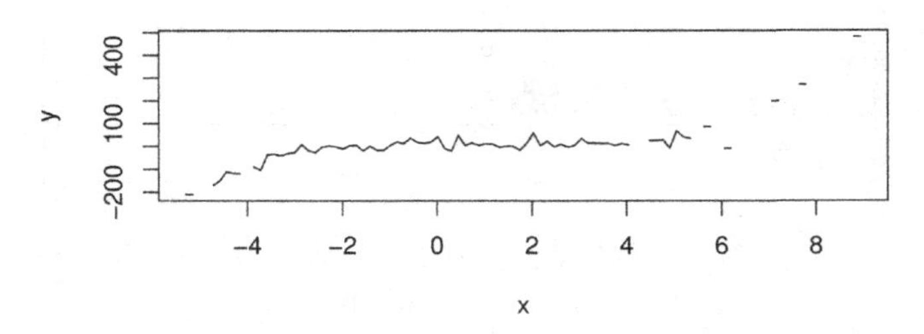

FIGURE 4.8
For the polynomial data of Example 4.6.1: The top panel displays the scatterplot overlaid with the `ksmooth` fit using the bandwidth set at 0.5 (the default value), while the bottom panel shows the fit using the bandwidth set at 0.10.

The following code segment obtains the `ksmooth` fit for the polynomial dataset of Example 4.6.1. In the first fit (top panel of Figure 4.8) we used the default bandwidth of $h = 0.5$, while in the second fit (lower panel of the figure) we set h at 0.10. We omitted the scatter of data to show clearly the sensitivity of the estimator (smooth) to the bandwidth setting. For both fits, in the call, we requested the normal kernel.

```
> par(mfrow=c(2,1))
> plot(y~x,xlab=expression(x),ylab=expression(y),pch=" ")
> title("Bandwidth 0.5")
> lines(ksmooth(x,y,"normal",0.5))
> plot(y~x,xlab=expression(x),ylab=expression(y),pch=" ")
> lines(ksmooth(x,y,"normal",0.10))
> title("Bandwidth 0.10")
```

The fit with the smaller bandwidth, 0.10, is much more jagged. The fit with the default bandwidth shows the trend, but notice the "artificial" valley it detected at about $x = 6.3$. In comparing this fit with the Wilcoxon cubic polynomial fit, this valley is due to the largest outlier in the data (see the Wilcoxon residual plot in Figure 4.7). The Wilcoxon fit was not impaired by this outlier.

The sensitivity of the fit to the bandwidth setting has generated a substantial amount of research on data-driven bandwidths. The R package `sm` (Bowman and Azzalini 2014) of nonparametric regression and density fits developed by Bowman and Azzalini contain such data-driven routines; see also Bowman and Azzalini (1997) for details. These are also kernel-type smoothers. We illustrate its computation with the following example.

Example 4.6.2 (Sine Cosine Model). For this example, we generated $n = 197$ observations from the model

$$y_i = 5\sin(3x_i) + 6\cos(x_i/4) + e_i, \tag{4.20}$$

where e_i are $N(0, 100)$ variates and x_i goes from 1 to 50 in increments of 0.25. The data are in the set `sincos`. The appropriate `sm` function is `sm.regression`. It has an argument `h` for bandwidth; but if this is omitted, a data-driven bandwidth is used. The fit is obtained as shown in the following code segment (the vectors `x` and `y` contain the data). The default kernel is the normal pdf and the option `display="none"` turns off the automatic plot.

```
> library(sm)
> fit <- sm.regression(x,y,display="none")
> fit$h              ##   Data driven bandwidth

[1] 4.211251
```

Figure 4.9 displays the data and the fit. Note that the procedure estimated the bandwidth to be 4.211. ∎

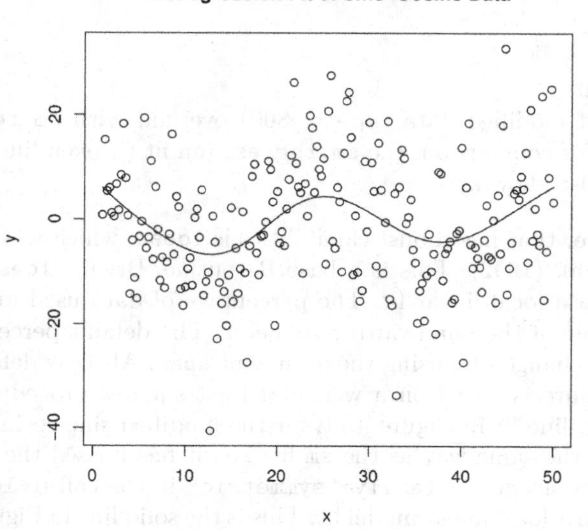

FIGURE 4.9
Scatterplot overlaid with `sm.regression` for the data of Example 4.6.2.

The smoother `sm.regression` is not robust. As an illustration, consider the sine-cosine data of Example 4.6.2. We modified the data by replacing y_{187} with the outlying value of 800. As shown in Figure 4.10, the `sm.regression` fit (solid line) of the modified data is severely impaired in the neighborhood of the outlier. The valley at $x = 36$ has essentially been missed by the fit. For ease of comparison, we have also displayed the `sm.regression` fit on the original data (broken line), which finds this valley.

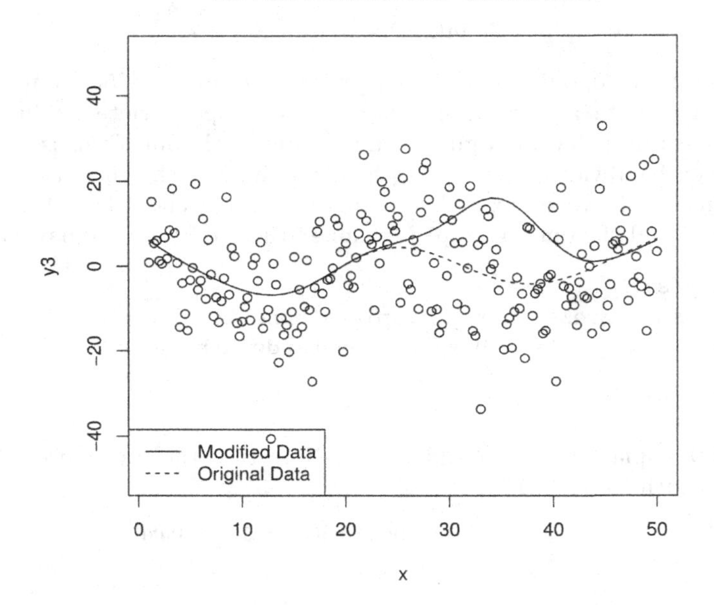

FIGURE 4.10
Scatterplot of modified data ($y_{137} = 800$) overlaid with `sm.regression` fit (solid line). For comparison, the `sm.regression` fit (broken line) on the original data is also shown.

A smoother that has robust capabilities is `loess`, which was developed by Cleveland et al. (1992). This is a base R routine. Briefly, `loess` smooths at a point x_i via a local linear fit. The percentage of data used for the local fit is the analogue of the bandwidth parameter. The default percentage is 75%, but it can be changed by using the argument `span`. Also, by default, the local fitting procedure is based on a weighted least squares procedure. As shown by the broken line fit in Figure 4.11, for the modified sine-cosine data, `loess` is affected in the same way as the `sm` fit; i.e., it has missed the valley at $x =$ 36. Setting the argument `family="symmetric"` in the call to `loess`, though, invokes a robust local linear model fit. This is the solid line in Figure 4.11 which is very close to the `sm` fit based on the original data. Exercise 4.7.15 asks the reader to create a similar graphic. The figure demonstrates the importance

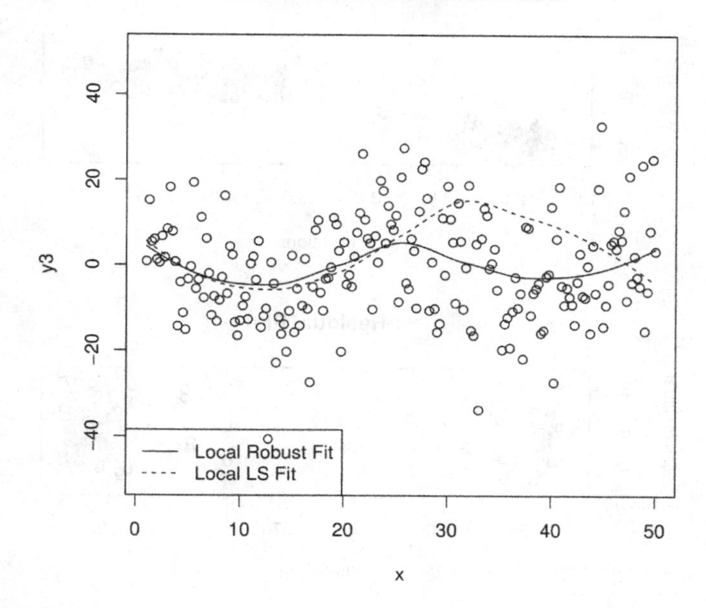

FIGURE 4.11
Scatterplot of modified data ($y_{137} = 800$) overlaid with `loess` fit (solid line). For comparison, the `sm.regression` fit (broken line) on the original data is also shown. Note that for clarity, the outlier is not shown.

in obtaining a robust fit in addition to a traditional fit. We close this section with two examples using real datasets.

Example 4.6.3 (Old Faithful). This dataset concerns the eruptions of Old Faithful, which is a geyser in Yellowstone National Park, Wyoming. The dataset is `faithful` in base R. As independent and dependent variables, we chose respectively the duration of the eruption and the waiting time between eruptions. In the top panel of Figure 4.12 we display the data overlaid with the `loess` fit. There is an increasing trend, i.e., longer eruption times lead to longer waiting times between eruptions. There appear to be two groups in the plot based on lower or higher duration of eruption times. As the residual plot shows, the loess fit has detrended the data. ∎

Example 4.6.4 (Maximum January Temperatures in Kalamazoo). The dataset `weather` contains weather data for the month of January for Kalamazoo, Michigan, from 1900 to 1995. For this example, our response variable is `avemax` which is the average maximum temperature in January. The top panel of Figure 4.13 shows the response variable over time. There seems to be little trend in the data. The lower panel shows the `loess` fits, local LS (solid line) and

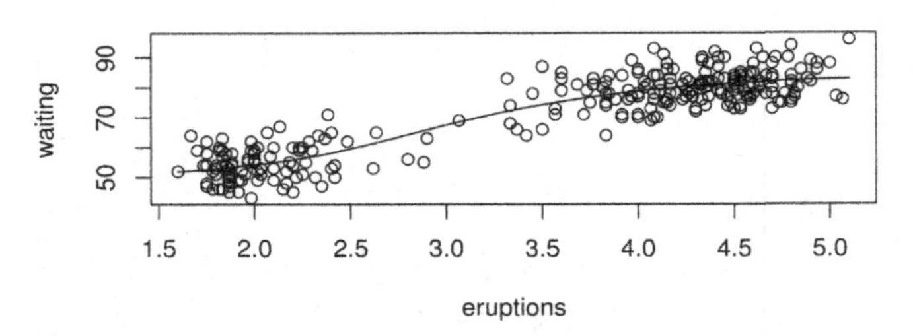

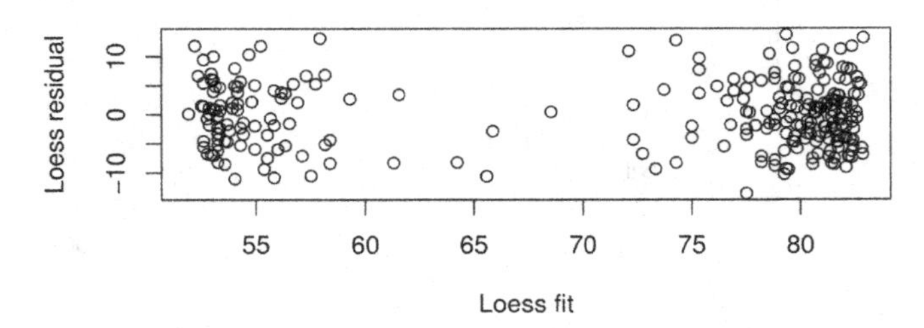

FIGURE 4.12
Top panel is the scatterplot of Old Faithful data, waiting time until the next eruption versus duration of eruption, overlaid with the loess fit. The bottom panel is the residual plot based on the loess fit.

local robust (broken line). The fits are about the same. They do show a slight pattern of a warming trend between 1930 and 1950. ∎

4.7 Exercises

4.7.1. Obtain a scatterplot of the `telephone` data. Overlay the least squares and R fits. Include a legend that indicates which line represents which fit.

4.7.2. Write an R function which given the results of a call to `rfit` returns the diagnostic plots: Studentized residuals versus fitted values, with ± 2 horizontal lines for outlier identification; normal $q-q$ plot of the Studentized residuals, with ± 2 horizontal lines outliers for outlier identification; histogram of residuals; and a boxplot of the residuals.

4.7.3. Consider the free fatty acid data discussed in Section 4.4.1 and available in the `Rfit` dataset `ffa`.

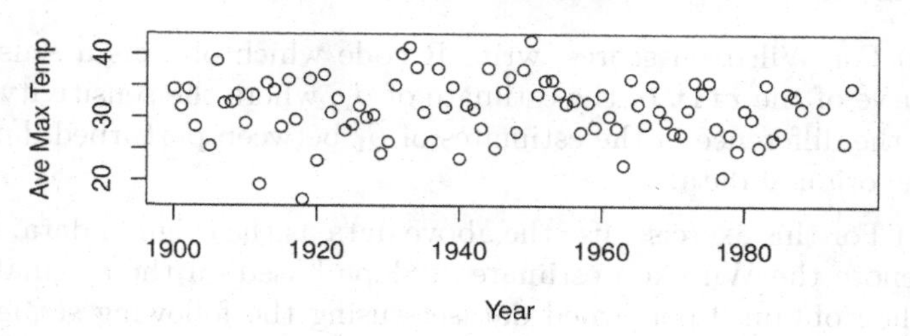

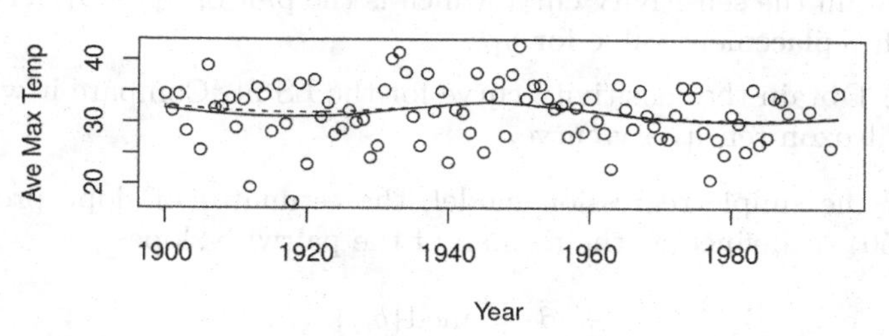

FIGURE 4.13
Top panel shows scatterplot of average maximum January temperature in Kalamazoo, Michigan, from 1900 to 1995. Lower panel displays the local LS (solid line) and the local robust (broken line) loess fits.

(a) For the Wilcoxon fit, obtain the Studentized residual plot and $q-q$ plot of the Studentized residuals. Comment on the skewness of the errors.

(b) Redo the analysis of the free fatty acid data using the bent scores (`bentscores1`). Compare the summary of the regression coefficients with those from the Wilcoxon fit. Why is the bent score fit more precise (smaller standard errors) than the Wilcoxon fit?

4.7.4. Using the `bb2010` data available in `npsm`, calculate a large sample confidence interval for the slope parameter when regressing weight on height. Compare the results to those obtained using the bootstrap discussed in Section 4.5.

4.7.5. Consider the following data:

x	1	2	3	4	5	6	7	8	9	10	11	12	13	14	15
y	−7	0	5	9	−3	−6	18	8	−9	−20	−11	4	−1	7	5

Consider the simple regression model: $Y = \beta_0 + \beta_1 x + e$.

(a) For Wilcoxon scores, write R code which obtains a sensitivity curve of the `rfit` of the estimate of β_1, where the sensitivity curve is the difference in the estimates of β_1 between perturbed data and the original data.

(b) For this exercise, use the above data as the original data. Let $\hat{\beta}_1$ denote the Wilcoxon estimate of slope based on the original data. Then obtain 9 perturbed datasets using the following sequence of replacements to y_{15}: $-995, -95, -25, -5, 5, 10, 30, 100, 1000$. Let $\hat{\beta}_{1j}$ be the Wilcoxon fit of the jth perturbed dataset for $j = 1, 2, \ldots, 9$. Obtain the sensitivity curve which is the plot of $\hat{\beta}_{1j} - \hat{\beta}_1$ versus the jth replacement value for y_{15}.

(c) Obtain the sensitivity curve for the LS fit. Compare it with the Wilcoxon sensitivity curve.

4.7.6. For the simple regression model, the estimator of slope proposed by Theil (1950) is defined as the median of the pairwise slopes:

$$\hat{\beta}_T = \text{med}\{b_{ij}\}$$

where $b_{ij} = (y_j - y_i)/(x_j - x_i)$ for $i < j$.

(a) Write an R function which takes as input a vector of response variables and a vector of explanatory variables and returns the Theil estimate.

(b) For a simple regression model where the predictor is a continuous variable, write an R function which computes the bootstrap percentile confidence interval for the slope parameter based on Theil's estimate.

(c) Show that Theil's estimate reduces to the the Hodges–Lehmann estimator for the two-sample location problem.

4.7.7. Hettmansperger and McKean (2011) discuss a dataset in which the dependent variable is the cloud point of a liquid, a measure of degree of crystallization in a stock, and the independent variable is the percentage of I-8 in the base stock. For the readers' convenience, the data can be found in the dataset `cloud` in the package `npsm`.

(a) Scatterplot the data. Based on the plot, is a simple linear regression model appropriate?

(b) Show by residual plots of the fits that the linear and quadratic polynomials are not appropriate but that the cubic model is.

(c) Use the R function `polydeg`, with a super degree set at 5, to determine the degree of the polynomial. Compare with Part (b).

4.7.8. Devore (2012) discusses a dataset on energy. The response variable is the energy output in watts, while the independent variable is the temperature difference in degrees K. A polynomial fit is suggested. The data are in the dataset `energy`.

(a) Scatterplot the data. What degree of polynomial seems suitable?

(b) Use the R function `polydeg`, with a super degree set at 6, to determine the degree of the polynomial.

(c) Based on a residual analysis, does the polynomial fit of Part (b) provide a good fit?

4.7.9. Consider the weather dataset, `weather`, discussed in Example 4.6.4. One of the variables is mean average temperature for the month of January (`meantmp`).

(a) Obtain a scatterplot of the mean average temperature versus the year. Determine the warmest and coldest years.

(b) Obtain the `loess` fit of the data. Discuss the fit in terms of years, (were there warm trends, cold trends?).

4.7.10. As in the last problem, consider the weather dataset, `weather`. One of the variables is total snowfall (in inches), `totalsnow`, for the month of January.

(a) Scatterplot total snowfall versus year. Determine the years of maximal and minimal snowfalls.

(b) Obtain the local LS and robust `loess` fits of the data. Compare the fits.

(c) Perform a residual analysis on the robust fit.

(d) Obtain a boxplot of the residuals found in Part (c). Identify the outliers by year.

4.7.11. In the discussion of Figure 4.8, the nonparametric regression fit by `ksmooth` detects an artificial valley. Obtain the locally robust loess fit of this dataset (`poly`) and compare it with the `ksmooth` fit.

4.7.12. Using the `bbsalaries` data, obtain the scatterplot between the variables `logSalary` and `era`. Then compute the Pearson's, Spearman's, and Kendall's analyses for these variables. Comment on the plot and analyses.

4.7.13. Write an R function which computes the t-test version of Spearman's procedure and returns it along with the corresponding p-value and the estimate of ρ_S.

4.7.14. Repeat Exercise 4.7.13 for Kendall's procedure.

4.7.15. Create a graphic similar to Figure 4.11.

4.7.16. Recall that, in general, the three measures of association estimate different parameters. Consider bivariate data (X_i, Y_i) generated as follows:

$$Y_i = X_i + e_i, \quad i = 1, 2, \ldots, n,$$

where X_i has a standard Laplace (double exponential) distribution, e_i has a standard $N(0, 1)$ distribution, and X_i and e_i are independent.

(a) Write an R script which generates this bivariate data. The supplied R function `rlaplace(n)` generates n iid Laplace variates. For $n = 30$, compute such a bivariate sample. Then obtain the scatterplot and the association analyses based on the Pearson's, Spearman's, and Kendall's procedures.

(b) Next write an R script which simulates the generation of these bivariate samples and collects the three estimates of association. Run this script for 10,000 simulations and obtain the sample averages of these estimates, their corresponding standard errors, and approximate 95% confidence intervals. Comment on the results.

4.7.17. The electronic memory game Simon was first introduced in the late 1970s. In the game there are four colored buttons which light up and produce a musical note. The device plays a sequence of light/note combinations and the goal is to play the sequence back by pressing the buttons. The game starts with one light/note and progressively adds one each time the player correctly recalls the sequence.[6]

Suppose the game were played by a set of statistics students in two classes (time slots). Each student played the game twice and recorded their longest sequence. The results are in the dataset `simon`.

Regression toward the mean is the phenomenon that if an observation is extreme on the first trial, it will be closer to the average on the second trial. In other words, students that scored higher than average on the first trial would tend to score lower on the second trial, and students who scored low on the first trial would tend to score higher on the second.

(a) Obtain a scatterplot of the data.

(b) Overlay an R fit of the data. Use Wilcoxon scores. Also overlay the line $y = x$.

(c) Obtain an R estimate of the slope of the regression line as well as an associated confidence interval.

(d) Do these data suggest a regression toward the mean effect?

---

[6]The game is implemented on the web. The reader is encouraged to use his or her favorite search engine and try it out.

5

ANOVA-Type Rank-Based Procedures

5.1 Introduction

In this chapter, the R functions in the packages `Rfit` and `npsm` for the computation of fits and inference for standard rank-based analyses of variance (ANOVA)[1] and analysis of covariance (ANCOVA) type designs are discussed. These include one-way, two-way, and k-way crossed designs that are covered in Sections 5.2–5.6. Both tests of general linear hypotheses and estimation of effects with standard errors and confidence intervals are emphasized. We also briefly present multiple comparison procedures (MCPs), in particular a robust Tukey–Kramer procedure, illustrating their computation via the package `Rfit`. We also consider the R computation of several traditional nonparametric methods for these designs including the Kruskal–Wallis (Section 5.2.2) and the Jonckheere–Terpstra tests for ordered alternatives (Section 5.7). In the last section, a generalization of the Fligner–Killeen procedure introduced in Chapter 3 to the k-sample scale problem is presented. The rank-based analyses covered in this chapter are for fixed effect models. Rank-based methods and their computation for mixed (fixed and random) models form the topic of Chapter 9

As a cursory reading, we suggest Section 5.2 and the two-way design material of Section 5.3, and the ordered alternative methods of Section 5.7. As usual, our emphasis is on how to easily compute these rank-based procedures using `Rfit`. Details of the robust rank-based inference for these fixed effect models are discussed in Chapter 4 of Hettmansperger and McKean (2011).

5.2 One-Way ANOVA

Suppose we want to determine the effect that a single factor A has on a response of interest over a specified population. Assume that A consists of k levels or treatments. In a completely randomized design (CRD), n subjects are randomly selected from the reference population and n_i of them are

[1]Though could be named ANODI for rank-based analysis.

randomly assigned to level i, $i = 1, \ldots k$. Let the jth response in the ith level be denoted by Y_{ij}, $j = 1, \ldots, n_i$, $i = 1, \ldots, k$. We assume that the responses are independent of one another and that the distributions among levels differ by at most shifts in location.

Under these assumptions, the **full model** can be written as

$$Y_{ij} = \mu_i + e_{ij} \quad j = 1, \ldots, n_i \ , \ i = 1, \ldots, k \ , \tag{5.1}$$

where the e_{ij}s are iid random variables with density $f(x)$ and distribution function $F(x)$ and the parameter μ_i is a convenient location parameter for the ith level, (for example, the mean or median of the ith level). This model is often referred to as a one-way design and its analysis as a one-way analysis of variance (ANOVA). Generally, the parameters of interest are the effects (pairwise contrasts),

$$\Delta_{ii'} = \mu_{i'} - \mu_i, \quad i \neq i', 1, \ldots, k. \tag{5.2}$$

We can express the model in terms of these simple contrasts. As in the R `lm` command, we reference the first level. Then the Model (5.1) can be expressed as

$$Y_{ij} = \begin{cases} \mu_1 + e_{1j} & j = 1, \ldots, n_1 \\ \mu_1 + \Delta_{i1} + e_{ij} & j = 1, \ldots, n_i, \ i = 2, \ldots, k. \end{cases} \tag{5.3}$$

Let $\boldsymbol{\Delta} = [\Delta_{21}, \Delta_{31}, \ldots, \Delta_{k1}]^T$. Upon fitting the model a residual analysis should be conducted to check these model assumptions. As the full model fit is based on a linear model, the diagnostic procedures discussed in Chapter 4 are implemented for ANOVA and ANCOVA models as well.

Observational studies can also be modeled this way. Suppose k independent samples are drawn, one from each of k populations. If we assume further that the distributions of the populations differ by at most a shift in locations, then Model (5.1) is appropriate. Usually, in the case of observational studies, it is necessary to adjust for covariates. These analyses are referred to as the analysis of covariance and are discussed in Section 5.6.

The analysis for the one-way design is usually a test of the hypothesis that all the effects, Δ_i's, are 0, followed by individual comparisons of levels. The hypothesis can be written as

$$H_0 : \mu_1 = \cdots = \mu_k \text{ versus } H_A : \mu_i \neq \mu_{i'} \text{ for some } i \neq i'. \tag{5.4}$$

Confidence intervals for the simple contrasts $\Delta_{ii'}$ can be used for the pairwise comparisons. We next briefly describe the general analysis for the one-way model and discuss its computation by `Rfit`.

A test of the overall hypothesis (5.4) is based on a reduction in dispersion test, first introduced in (7.1.3). For `Rfit`, assume that a score function φ has been selected; otherwise, `Rfit` uses the default Wilcoxon score function. As discussed in Section 5.3, let $\widehat{\boldsymbol{\Delta}}_\varphi$ be the rank-based estimate of $\boldsymbol{\Delta}$ when the full model (5.1) is fit. Let $D_\varphi(\text{FULL}) = D(\widehat{\boldsymbol{\Delta}}_\varphi)$ denote the full model dispersion, i.e., the minimum value of the dispersion function when this full model is fit. The reduced model is the location model

$$Y_{ij} = \mu + e_{ij} \quad j = 1, \ldots, n_i \,,\; i = 1, \ldots, k. \tag{5.5}$$

Because the dispersion function is invariant to location, the minimum dispersion at the reduced model is the dispersion of the observations; i.e., $D(\mathbf{0})$ which we call $D_\varphi(\text{RED})$. The reduction in dispersion is then $RD_\varphi = D_\varphi(\text{RED}) - D_\varphi(\text{FULL})$ and, hence, the drop in dispersion test statistic is given by

$$F_\varphi = \frac{RD_\varphi/(k-1)}{\widehat{\tau}_\varphi/2}, \tag{5.6}$$

where $\widehat{\tau}_\varphi$ is the estimate of scale discussed in Section 3.1. The approximate level α test rejects H_0, if $F_\varphi \geq F_{\alpha,k-1,n-k}$. The traditional LS test is based on a reduction of sums of squares. Replacing this by a reduction in dispersion the test based on F_φ can be summarized in an ANOVA table much like that for the traditional F-test; see page 298 of Hettmansperger and McKean (2011). When the linear Wilcoxon scores are used, we often replace the subscript φ by the subscript W; that is, we write the Wilcoxon rank-based F-test statistic as

$$F_W = \frac{RD_W/(k-1)}{\widehat{\tau}_W/2}. \tag{5.7}$$

The Rfit function `oneway.rfit` computes the robust rank-based one-way analysis. Its arguments are the vector of responses and the corresponding vector of levels. It returns the value of the test statistic and the associated p-value. We illustrate its computation with an example.

Example 5.2.1 (LDL Cholesterol of Quail). Hettmansperger and McKean (2011), page 295, discuss a study which investigated the effect that four drug compounds had on the reduction of low density lipid (LDL) cholesterol in quail. The drug compounds are labeled as I, II, III, and IV. The sample size for each of the first three levels is 10 while 9 quail received compound IV. The boxplots shown in Figure 5.1 attest to a difference in the LDL levels over treatments.

Using Wilcoxon scores, the results of `oneway.rfit` are shown in the following code segment.

```
> robfit = with(quail,oneway.rfit(ldl,treat))
> robfit

Call:
oneway.rfit(y = ldl, g = treat)

Overall Test of All Locations Equal

Drop in Dispersion Test
```

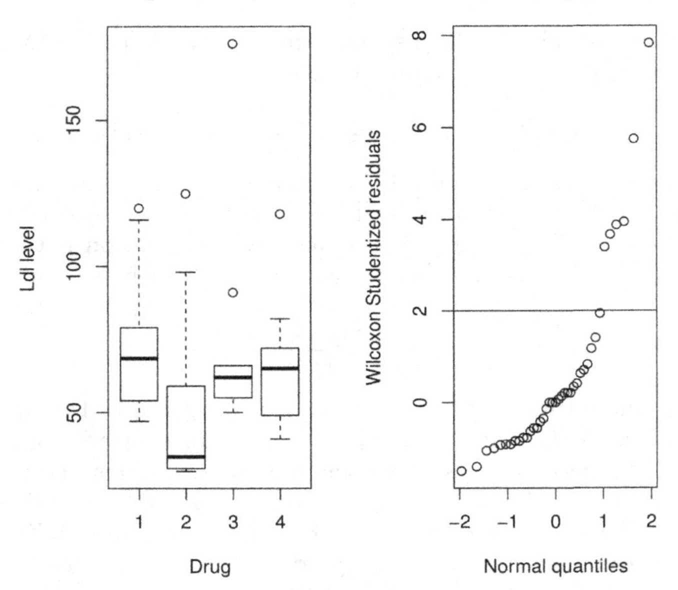

FIGURE 5.1
Plots for LDL cholesterol of quail example.

```
F-Statistic      p-value
   3.916944     0.016394


        Pairwise comparisons using Rfit

data:  ldl and treat

  1     2     3
2 0.005 -     -
3 0.632 0.016 -
4 0.560 0.024 0.907

P value adjustment method: none
```

The Wilcoxon test statistic has the value $F_W = 3.92$ with p-value 0.0164. Thus, the Wilcoxon test indicates that the drugs differ in their lowering of cholesterol effect. In contrast to the significant Wilcoxon test of the hypotheses (5.4), the LS-based F-test statistic has value 1.14 with the p-value 0.3451.

In practice, using the LS results, one would not proceed with comparisons of the drugs with such a large p-value. Thus, for this dataset, the robust and LS analyses would have different practical interpretations. Also, the coefficient of precision, (3.71), for this data between the Wilcoxon and LS analyses is $\widehat{\sigma}^2/\widehat{\tau}^2 = 2.72$. Hence, the Wilcoxon analysis is much more precise.

The resulting $q-q$ plot (see right panel of Figure 5.1) of the Studentized Wilcoxon residuals indicates that the random errors e_{ij} have a skewed distribution. R fits based on scores more appropriate than the Wilcoxon for skewed errors are discussed later.

5.2.1 Multiple Comparisons

The second stage of an analysis of a one-way design consists of pairwise comparisons of the treatments. The robust $(1-\alpha)100\%$ confidence interval to compare the *ith* and *ith′* treatments is given by

$$\widehat{\Delta}_{ii'} \pm t_{\alpha/2,n-1}\widehat{\tau}\sqrt{\frac{1}{n_i}+\frac{1}{n_{i'}}}. \tag{5.8}$$

Often there are many comparisons of interest. For example, in the case of all pairwise comparisons, $\binom{k}{2}$ confidence intervals are required. Hence, the overall family error rate is usually of concern. Multiple comparison procedures (MCPs) try to control the overall error rate to some degree. There are many robust versions of MCPs from which to choose. The `summary` function associated with `oneway.rfit` computes three of the most popular of such procedures. Assuming that the fit of the full model is in `robfit`, the syntax of the command is `summary(robfit,method="none")`. The argument of `method` produces the following MCPs:

`method="none"`	No adjustment made
`method="tukey"`	Tukey–Kramer
`method="bonferroni"`	Bonferroni

We give a brief description of these procedures, followed by an example using `Rfit`.

A protected least significant difference procedure (PLSD) consists of testing the hypothesis (5.4) at a specified level α. If H_0 is rejected, then the comparisons are based on the confidence intervals (5.8) with confidence coefficient $1-\alpha$. On the other hand, if H_0 is not rejected, then the analysis stops. Although this procedure does not control the overall family rate of error, the initial F-test offers protection, which has been confirmed (see, for example, McKean et al. (1989)) in simulation studies.

For the Tukey–Kramer procedure, Studentized range critical values replace the t-critical values in the intervals (5.8). When the traditional procedure is used, the random errors have a normal distribution, and the design is balanced then the Tukey–Kramer procedure has family error rate α. When either of these assumptions fail, then the Tukey–Kramer procedure has a family error rate approximately equal to α.

The Bonferroni procedure depends on the number of comparisons made. Suppose there are l comparisons of interest, then the Bonferroni procedure uses the intervals (5.8) with the critical value $t_{\alpha/(2l),n-k}$. The Bonferroni procedure has overall family rate $\leq \alpha$. If all pairwise comparisons are desired, then $l = \binom{k}{2}$.

Example 5.2.2 (LDL Cholesterol of Quail, Continued). For the quail data, we selected the Tukey–Kramer procedure for all six pairwise comparisons. The `Rfit` computation is:

```
> summary(robfit,method="tukey")

Multiple Comparisons
Method Used  tukey

  I J Estimate St Err Lower Bound CI Upper Bound CI
1 1 2      -25 8.2670       -47.2954        -2.7046
2 1 3       -4 8.2670       -26.2954        18.2954
3 1 4       -5 8.4936       -27.9064        17.9064
4 2 3       21 8.2670        -1.2954        43.2954
5 2 4       20 8.4936        -2.9064        42.9064
6 3 4       -1 8.4936       -23.9064        21.9064
```

The Tukey–Kramer procedure declares that the Drug Compounds I and II are statistically significantly different. ■

5.2.2 Kruskal–Wallis Test

Note that Model (5.1) generalizes the two-sample problem of Chapter 2 to k-samples. In this section, we discuss the Kruskal–Wallis test of the hypotheses (5.4) which is a generalization of the two-sample Mann-Whitney-Wilcoxon test.

Assume then that Model (5.1) holds for the responses Y_{ij}, $j = 1, \ldots n_i$ and $i = 1, \ldots, k$. As before let $n = \sum_{i=1}^{n} n_i$ denote the total sample size. Let R_{ij} denote the rank of the response Y_{ij} among all n observations; i.e., the ranking is done without knowledge of treatment. Let $\overline{R}_i$ denote the average of the ranks for sample i. The test statistic H is a standardized weighted average of the squared deviations of the $\overline{R}_i$ from the average of all ranks $(n+1)/2$. The test statistic is

$$H = \frac{12}{n(n+1)} \sum_{i=1}^{n} n_i \left(\overline{R}_i - \frac{n+1}{2} \right)^2 . \tag{5.9}$$

The statistic H is the **Kruskal–Wallis** test statistic; see Kruskal and Wallis (1952). Under H_0 H is distribution-free and there are some tables available for its exact distribution.[2] It also has an asymptotic χ^2-distribution with $k-1$

[2] See Chapter 6 of Hollander and Wolfe (1999).

degrees of freedom under H_0. The R command is `kruskal.test`. Assume that the responses are in vector `x`, and the group or treatment assignments are in vector `g`, then the call is `kruskal.test(x,g)`. In addition a `formula` can be used as in `kruskal.test(x~g)`. We illustrate the computation with the following example.

Example 5.2.3 (Mucociliary Efficiency). Hollander and Wolfe (1999), page 192, discuss a small study which assessed the mucociliary efficiency from the rate of dust in the three groups: normal subjects, subjects with obstructive airway disease, and subjects with asbestosis. The responses are the mucociliary clearance half-times of the subjects. The sample sizes are small: $n_1 = n_3 = 5$ and $n_2 = 4$. Hence, $n = 14$. The data are given in the R vectors `normal`, `obstruct`, and `asbestosis` in the following code segment which computes the Kruskal–Wallis test.

```
> normal <- c(2.9,3.0,2.5,2.6,3.2)
> obstruct <- c(3.8,2.7,4.0,2.4)
> asbestosis <- c(2.8,3.4,3.7,2.2,2.0)
> x <- c(normal,obstruct,asbestosis)
> g  <- c(rep(1,5),rep(2,4),rep(3,5))
> kruskal.test(x,g)

        Kruskal-Wallis rank sum test

data:  x and g
Kruskal-Wallis chi-squared = 0.7714, df = 2, p-value = 0.68
```

Based on this p-value, there do not appear to be differences among the groups for mucociliary efficiency. ∎

Corrections for ties for the Kruskal–Wallis test are discussed in Hollander and Wolfe (1999) and `kruskal.test` does make such adjustments in its calculation. As discussed in Hettmansperger and McKean (2011), the Kruskal–Wallis test is asymptotically equivalent to the drop in dispersion test (5.6) using Wilcoxon scores.[3]

5.3 Multi-Way Crossed Factorial Design

For a multi-way or k-way crossed factorial experimental design, the `Rfit` function `raov` computes the rank-based ANOVA for all $2^k - 1$ hypotheses, including

[3]This equivalence extends to local alternatives; hence, the Kruskal–Wallis and the drop in dispersion tests have the same asymptotic efficiency. In particular, for normal errors the relative efficiency of these tests relative to the traditional LS test is 0.955.

the main effects and interactions of all orders. These are the hypotheses in a standard ANOVA table for a k-way design. The design may be balanced or unbalanced. For simplicity, we briefly discuss the analysis in terms of a cell mean (median) model; see Hocking (1985) for details on the traditional LS analysis and Chapter 4 of Hettmansperger and McKean (2011) for the rank-based analysis. For clarity, we first discuss a two-way crossed factorial design before presenting the k-way design.

5.3.1 Two-Way

Let A and B denote the two factors in a two-way design with levels a and b, respectively. Let Y_{ijk} define the response for the kth replication at levels i and j of factors A and B, respectively. Then the full model can be expressed as

$$Y_{ijk} = \mu_{ij} + e_{ijk}, \quad k = 1, \ldots, n_{ij}, i = 1, \ldots, a; j = 1, \ldots, b, \tag{5.10}$$

where e_{ijk} are iid random variables with pdf $f(t)$. Since the effects of interest are contrasts in the μ_{ij}'s, these parameters can be either cell means or medians, (actually any location functional suffices). We refer to this model as the **cell mean** or **cell median** model. We assume that all $n_{ij} \geq 1$ and at least one n_{ij} exceeds 1. Let $n = \sum\sum n_{ij}$ denote the total sample size.

For the two-way model, the three hypotheses of interest are the main effects hypotheses and the interaction hypothesis.[4] For the two-way model (5.10) these hypotheses are:

$$\begin{aligned} H_{0A}: \overline{\mu}_{1\cdot} = \cdots = \overline{\mu}_{a\cdot} \ &\text{vs.}\ H_{1A}: \overline{\mu}_{i\cdot} \neq \overline{\mu}_{i'\cdot}, \text{ for some } i \neq i' \quad (5.11)\\ H_{0B}: \overline{\mu}_{\cdot 1} = \cdots = \overline{\mu}_{\cdot b} \ &\text{vs.}\ H_{1B}: \overline{\mu}_{\cdot j} \neq \overline{\mu}_{\cdot j'}, \text{for some } j \neq j' \quad (5.12)\\ H_{0AB}: \gamma_{ij} \equiv 0 \text{ for all } i,j \ &\text{vs.}\ H_{1AB}: \gamma_{ij} \neq 0, \text{ for some } (i,j), \quad (5.13) \end{aligned}$$

where γ_{ij} are the interaction parameters

$$\gamma_{ij} = \mu_{ij} - \overline{\mu}_{i\cdot} - \overline{\mu}_{\cdot j} + \overline{\mu}_{\cdot\cdot}.$$

$\overline{\mu}_{i\cdot} = \frac{1}{b}\sum_{j=1}^{b} \mu_{ij}$, $\overline{\mu}_{\cdot j} = \frac{1}{a}\sum_{j=1}^{a} \mu_{ij}$, and $\overline{\mu}_{\cdot j} = \frac{1}{ab}\sum_{i=1}^{b}\sum_{j=1}^{a} \mu_{ij}$. The function `raov` computes the drop in dispersion tests for each of these hypotheses as illustrated in the next example.

Example 5.3.1 (Serum LH Data). Hollander and Wolfe (1999) discuss a 2×5 factorial design for a study to determine the effect of light on the release of luteinizing hormone (LH). The factors in the design are: light regimes at two levels (constant light and 14 hours of light followed by 10 hours of darkness) and a luteinizing release factor (LRF) at 5 different dosage levels. The response is the level of LH, nanograms per milliliter of serum in blood samples. Sixty rats were put on test under these 10 treatment combinations, six rats per

[4] We have chosen Type III hypotheses which are easy to interpret even for severely unbalanced designs. Details of how `Rfit` performs this procedure are given in Section 7.4.

combination. The data are in the dataset serumLH. We first obtain the robust rank-based ANOVA for these data. The command is raov, and its syntax is shown in the output which follows.

```
> raov(serum~light.regime+LRF.dose+light.regime*LRF.dose,
+      data = serumLH)

Robust ANOVA Table
                      DF      RD Mean RD       F p-value
light.regime           1 1642.33 1642.33 58.2833 0.00000
LRF.dose               4 3027.67  756.92 26.8616 0.00000
light.regime:LRF.dose  4  451.46  112.86  4.0053 0.00678
```

We also obtained the LS-based ANOVA.

```
> summary(
+   aov(serum~light.regime+LRF.dose+light.regime*LRF.dose,
+       data = serumLH)
+   )

                      Df Sum Sq Mean Sq F value  Pr(>F)
light.regime           1 242189  242189   40.22 6.4e-08 ***
LRF.dose               4 545549  136387   22.65 1.0e-10 ***
light.regime:LRF.dose  4  55099   13775    2.29   0.073 .
Residuals             50 301055    6021
---
Signif. codes:
0 '***' 0.001 '**' 0.01 '*' 0.05 '.' 0.1 ' ' 1
```

Note that the analyses differ critically. The robust ANOVA clearly detects the interaction with p-value less than 0.01; while, interaction is not significant at the 5% level for the LS analysis. For both analyses, the average main effects are highly significant. If the level of significance was set at 0.05, then the two analyses would lead to different practical interpretations.

Next we examine the residuals by plotting them in a normal $q-q$ plot; these are based on the R fit of the full model using Wilcoxon scores. The normal $q-q$ plot of this fit is shown in the left panel of Figure 5.2. It is clear from this plot that the errors have much heavier tails than those of a normal distribution.

The empirical measure of precision (3.71) in this case is $\widehat{\sigma}^2/\widehat{\tau}_W^2 = 1.9$. Based on the $q-q$ plot, this large value is not surprising. The right panel in Figure 5.2 is an interaction plot of profiles based on a Wilcoxon fit. That is, in this case, we have plotted the profiles for the light regimes over the LRF levels, using the cell estimates based on the Wilcoxon fit. It is clear from this plot that interaction of the factors is present, which agrees with the robust test for interaction. The corresponding profile plot (not shown) based on cell sample means is similar, although the outliers caused some distortion. ■

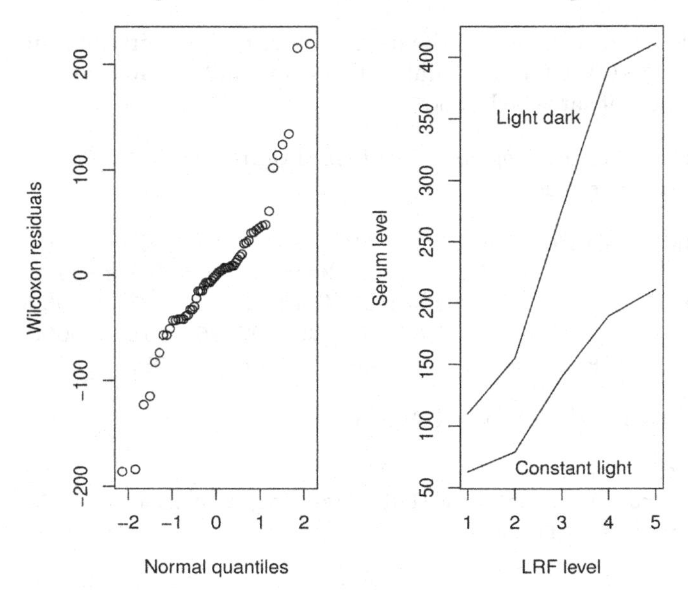

FIGURE 5.2
Plots for analysis of serum data.

5.3.2 k-Way

Consider a general k-way factorial design with factors $1, 2, \ldots, k$ having the levels $l_1, l_2, \ldots, l_k$, respectively. Let $Y_{i_1 i_2 \cdots i_k, j}$ be the jth response at levels $i_1\ i_2\ \cdots\ i_k$ of factors $1, 2, \ldots, k$ where $1 \leq i_1 \leq l_1$, $1 \leq i_2 \leq l_2$, $\ldots$, and $1 \leq i_k \leq l_k$ and $1 \leq j \leq n_{i_1 i_2 \cdots i_k}$. Let $n = \sum n_{i_1 i_2 \cdots i_k}$ denote the total sample size. Choose a location functional μ (can be the mean or median). Then the full model is

$$y_{i_1 i_2 \cdots i_k, j} = \mu_{i_1 i_2 \cdots i_k} + e_{i_1 i_2 \cdots i_k, j}, \tag{5.14}$$

where the errors $e_{i_1 i_2 \cdots i_k, j}$ are iid with cdf $F(t)$ and pdf $f(t)$.

For this model there are $2^k - 1$ hypotheses of interest. These include the k main effect hypotheses and the interaction hypotheses of all orders. Using the algorithms found in Hocking (1985), the corresponding hypotheses matrices are obtained and, hence, the appropriate reduced model design matrices can be computed. An illustrative example follows.

Example 5.3.2 (Plank Data). Abebe et al. (2001) discuss a dataset resulting from a three-way layout for a neurological experiment in which the time required for a mouse to exit a narrow elevated wooden plank is measured. The response is the log of time to exit. Interest lies in assessing the effects of three factors: the Mouse Strain (Tg+, Tg-), the mouse's Gender (female, male), and the mouse's Age (Aged, Middle, Young). The design is a $2 \times 2 \times 3$ factorial design. The data are in the `npsm` dataset `plank`.

Using raov, we computed the tests for main effects, as well as the second order and third order interactions. For comparison, we also obtained the corresponding LS-based tests.

```
> raov(response~strain:gender:age, data = plank)

Robust ANOVA Table
                   DF      RD Mean RD        F p-value
strain              1 4.42673 4.42673 13.83275 0.00049
gender              1 1.98920 1.98920  6.21589 0.01588
age                 2 0.83428 0.41714  1.30348 0.28031
strain:gender       1 0.50658 0.50658  1.58299 0.21395
strain:age          2 1.31007 0.65503  2.04686 0.13942
gender:age          2 2.38952 1.19476  3.73341 0.03055
strain:gender:age   2 0.36539 0.18269  0.57089 0.56852

> fi <- lm(response~.^3, data = plank)
> anova(fi)

Analysis of Variance Table

Response: response
                  Df Sum Sq Mean Sq F value Pr(>F)
strain             1    9.7    9.67    8.02 0.0066 **
gender             1    3.5    3.51    2.91 0.0941 .
age                2    0.8    0.39    0.32 0.7270
strain:gender      1    3.7    3.68    3.05 0.0868 .
strain:age         2    3.4    1.71    1.41 0.2525
gender:age         2    7.0    3.49    2.89 0.0646 .
strain:gender:age  2    0.4    0.20    0.17 0.8445
Residuals         52   62.8    1.21
---
Signif. codes:
0 '***' 0.001 '**' 0.01 '*' 0.05 '.' 0.1 ' ' 1
```

If the nominal level is $\alpha = 0.05$, then, based on the rank-based ANOVA, the factors strain and gender are significant, but, so is the factor age because of its significant two-way interaction with gender. In contrast, based on the LS ANOVA, only the factor strain is significant. Thus practical interpretations would differ for the two analyses.

For diagnostic checks of the model, the Rfit function raov returns the Rfit of the full model in the value fit. Using this value, we computed the Wilcoxon Studentized residuals. Figure 5.3 displays the normal $q-q$ plot and the residual plot based on these residuals. The $q-q$ plot suggests an error distribution with very thick tails. Other than the outliers, the residual plot reveals no serious lack of fit in the model.

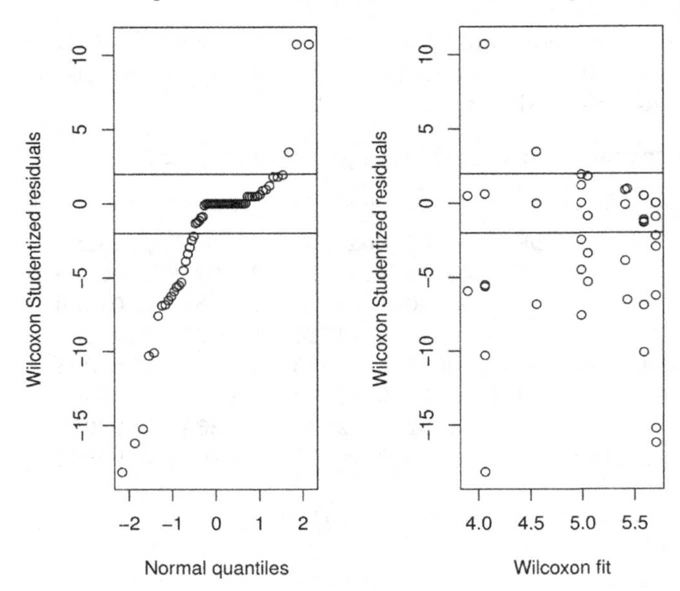

FIGURE 5.3
Diagnostic plots for plank data analysis.

The difference in the Wilcoxon and LS analyses are due to these outliers. Our empirical measure of efficiency is $\widehat{\sigma}^2/\widehat{\tau}_W^2 = 2.93$, which indicates that the Wilcoxon analysis is almost 3 times more efficient than the LS analysis on this dataset. ■

5.4 Contrasts

In practice, experimenters are often interested in estimating certain contrasts of interest along with their associated standard errors from which confidence intervals may be formed. For discussion, we consider the two-way interaction model of the previous section, but it is straightforward to apply the methods described to other situations.

Suppose that interaction has been detected in a two-way design. Because interaction is present, the full model is a one-way cell median model with ab levels as given in expression (5.10). Let μ_{ij} denote the median of cell (i, j), $i = 1, \ldots, a$ and $j = 1, \ldots, b$ and let $\hat{\mu}_{ij}$ be the rank-based estimator based on the full model fit. The contrasts of interest are contrasts in the cell medians μ_{ij}. Suppose then that we have q such contrasts of interest. For the computation of the rank-based estimates of the contrasts, along with their standard errors, let

H be the $q \times ab$ matrix whose rows are the vectors for these contrasts. Suppose that fittwc has captured the rank-based fit of the cell median model (5.10). Then R code for the rank-based estimates of the contrasts of interests and their associated standard errors is given by the following example code

```
> est <- H%*%fittwc$coef
> se <- sqrt(diagonal(H%*%vcov(fittwc)%*%t(H)))
```

Example 5.4.1 below illustrates this discussion for the serum LH data used in the last section.

One item that the user needs to check is that the μ_{ij}'s correctly identify with the intended columns of the one-way design matrix. One way to ensure this is to have the data sorted by level, first by the level of the first factor, then within the first factor by the level of the second factor. We say that such data are in **standard form**. This is easily accomplished in R as follows. Suppose that the three R vectors `response`, `fA`, and `fB` contain, respectively, the responses, levels of factor A, and levels of factor B. Suppose we combined these vectors in data frame `mat <- cbind(response,fA,fB)`. Then the data frame `mat2` computed by `mat2 <- mat[order(fA,fB),]` is in standard form. Note that the `Rfit` function `raov` presorts the data into standard form prior to computation of the rank-based fit.

Example 5.4.1 (Serum LH Data, continued). As discussed in Example 5.3.1, the rank-based analysis, using Wilcoxon scores, of this dataset detected interaction that was also evident in the profile plot of cell medians found in Figure 5.2. So the full model is the interaction (cell median) model, (5.10), with $10 = 2 \times 5$ (light regimes $\times$ LRF) cells. The columns of the R data frame serumLH contain the responses, `serum`, and the row factor, `light.regime`, and column factor, `LRF.dose`. This data frame is not in standard form. The R command `ds2 <- serumLH[order(light.regime, LRF.dose),]` results in `ds2` in standard form. The levels in the data frame are treated as categorical, so in its standard form the order of the levels of the factor `LRF.dose` are: "0", "10", "1250", "250", and "50". Suppose that we are interested in the contrasts that are the differences between the two light regimes (constant − intermittent) at each of the five dosage levels of the factor LRF. In the μ_{ij} notation, these are the five contrasts: $\mu_{11} - \mu_{21}, \mu_{12} - \mu_{22}, \mu_{13} - \mu_{23}, \mu_{14} - \mu_{24}$ and $\mu_{15} - \mu_{25}$. If there was no interaction, these differences of population cell medians would be the same. The next code segment obtains the rank-based estimates of these contrasts and their associated standard errors.

```
> library(npsm)
> attach(serumLH)
> I <- diag(rep(1,5)) ; hmat <- cbind(I,-I)
> hmat

     [,1] [,2] [,3] [,4] [,5] [,6] [,7] [,8] [,9] [,10]
[1,]    1    0    0    0    0   -1    0    0    0     0
```

```
[2,]    0    1    0    0    0    0   -1    0    0    0
[3,]    0    0    1    0    0    0    0   -1    0    0
[4,]    0    0    0    1    0    0    0    0   -1    0
[5,]    0    0    0    0    1    0    0    0    0   -1
```

```
> fit<-raov(serum~light.regime+LRF.dose+light.regime*LRF.dose,
+           data=serumLH)$fit  # rank-based fit
> cons <- hmat%*%fit$coef; cons<-as.vector(cons)
> se <- sqrt(diag(hmat%*%vcov(fit)%*%t(hmat)))
> tab <- rbind(cons,se)
> colnames(tab)<-c("0","10","1250","250","50")
> rownames(tab) <- c("Contr.","SE")
> tab

              0      10     1250      250       50
Contr. -47.000 -76.002 -200.365 -201.285 -136.000
SE      32.538  32.538   32.538   32.538   32.538
```

These differences are easily interpreted by considering the interaction plot, Figure 5.2. The first difference in column 1 of the table `tab` is the smallest gap between the profiles, and it is not significant. The second difference is the next smallest gap, but note that it is significantly different from 0. The other differences are also significantly different from 0. If interaction was not present, then these estimated differences would be about the same. The fact that four of them are significantly different from 0 confirms the significance of the test for interaction. Another set of four contrasts of interest would be the difference of these differences; i.e., contrast one versus contrast two, contrast two versus contrast three, etc.; see Exercise 5.9.9. ■

In the example, if we had computed the contrasts without first checking to see if the data frame was in standard form, then we might have mislabeled the column headings in the printed table of contrasts. One way of avoiding this dilemma is to treat the interaction model as a cell median model. Then for cell numbers, use $1, 2, \ldots, b$ for the cells in the first row, $b+1, b+2, \ldots, 2b$ for the cells in the second row, and so on. For an illustration, consider the last example using the original data frame serum. Since there are 6 replicates in each cell, the cell indicator for the observations followed by the design matrix are:

```
> ind <- rep(1:10,each=6)
> x <- model.matrix(~as.factor(ind)-1)
```

Next we obtain the fit and the estimates of the contrasts:

```
> hmat<-c(); zed <- rep(0,10)
> for(j in 1:5) {
+       ctm <- zed; ctm[j]<-1; ctm[j+5]<- -1
```

```
+      hmat <- rbind(hmat,ctm)}                          # matrix of contrasts
> fit<-rfit(serum~x-1,data=serumLH)                      # rank-based fit
> cons <- hmat%*%fit$coef; cons<-as.vector(cons)
> se <- sqrt(diag(hmat%*%vcov(fit)%*%t(hmat)))
> tab <- rbind(cons,se); colnames(tab)<-c("0","10","50","250","1250")
> rownames(tab) <- c("Contr.","SE")
> tab

              0       10       50      250     1250
Contr. -47.000 -76.002 -136.000 -201.285 -200.365
SE      32.538  32.538   32.538   32.538   32.538
```

These agree with the estimates that we obtained in Example 5.4.1.

If the contrasts are simple contrasts of the form $\mu_{ij} - \mu_{i'j'}$ then, since the full model is a one-way design, we can use the multiple comparison procedure to compute all $\binom{ab}{2}$ simple comparisons. For example, continuing with the R code in the last segment, we can compute PLSD comparisons for the serum data by the R command `summary(fit,method="none")`. This of course returns $\binom{10}{2} = 45$ comparisons, but we can peruse the output for the desired comparisons.

5.5 Additive Model

Consider a two-way design under the assumption that there is no interaction. Let A and B denote the factors with respective levels a and b. In this case, we can write the cell median μ_{ij} of Model (5.10) as

$$\mu_{ij} = \alpha^* + \beta^*_{Ai} + \beta^*_{Bj}, \quad i = 1, \ldots, a; j = 1, \ldots, b.$$

Hence, this model is called an additive model. This parametrization, though, is overparameterized, and we prefer to fit a full rank design. As in base R, we choose the full rank model that references the first level of the factors. Under this reparametrization, the cell medians are given by

$$\mu_{ij} = \alpha + \beta_{Ai} + \beta_{Bj}, \quad i = 2, \ldots, a; j = 2, \ldots, b, \tag{5.15}$$

where α is the intercept parameter; β_{Ai} is the expected shift in the response from level 1 of factor A to level i of factor A, $i \geq 2$; and β_{Bj} is the expected shift in the response from level 1 of factor B to level j of factor B, $j \geq 2$. For this parameterization, the main effect hypotheses are:

$$H_{0A}: \beta_{A2} = \cdots = \beta_{Aa} = 0 \text{ vs. } H_{1A}: \beta_{Ai} \neq 0, \ 2 \leq i \leq a \tag{5.16}$$

$$H_{0B}: \beta_{B2} = \cdots = \beta_{Bb} = 0 \text{ vs. } H_{1B}: \beta_{Bj} \neq 0, \ 2 \leq j \leq b. \tag{5.17}$$

These hypotheses can be tested using the drop in dispersion test. For identification of the parameters, as in the last section, we recommend that the data be sorted in standard form with respect to the factors A and B.

In terms of R code, suppose that the R vectors `y`, `f1`, and `f2`, contain the responses, the level of factor A, and the level of factor B, respectively. Assume that the data frame `mat<-data.frame(cbind(y,f1,f2))` is in standard form. Then the full rank design matrix and the Rfit of the full model are

```
> xadd <- model.matrix(~as.factor(f1)+as.factor(f2))
> xadd <- xadd[,-1]
> fitadd <- rfit(y~xadd)
```

Contrasts of interest, other than those as parameterized in the model, can be expressed in terms of the model parameters; for example, the expected shift in the response from level 2 of factor A to level 3 is $\beta_{A3} - \beta_{A2}$. Code for estimates of contrasts as well as tests of the main effect hypotheses are illustrated in the next example.

Example 5.5.1 (Additive Model). Consider the following data from a two-way design with factors A and B having respective levels 2 and 3 with several replicates per cell that are separated by commas.

	B1	B2	B3
A1	2, 15 20	14, 26 42, 47	35, 32 28, 29
A2	31, 20	37, 33 29, 34	24, 23 33, 30

We assume an additive model; see Exercise 5.9.14 for the test of interaction on this dataset. The R vectors containing the response and the associated factors are:

```
> yadd

 [1]  2 15 20 14 26 42 47 35 32 28 29 31 20 37 33 29 34 24
[19] 23 33 40

> FA

 [1] 1 1 1 1 1 1 1 1 1 1 1 2 2 2 2 2 2 2 2 2 2

> FB

 [1] 1 1 1 2 2 2 2 3 3 3 3 1 1 2 2 2 2 3 3 3 3
```

Note that the data are in standard form. In the next R segment, we obtain the rank-based (Wilcoxon scores) fit of the additive model and the tests of the main effects, H_{0A} and H_{0B}.

```
> xadd<-model.matrix(~as.factor(FA)+as.factor(FB))
> xadd<-xadd[,-1]
```

```
> fitadd <- rfit(yadd~xadd) # Fit of additive model
> xredA <- xadd[,-1]
> fitredA <- rfit(yadd~xredA)
> drop.test(fitadd,fitredA) # Test of main effect A

Drop in Dispersion Test
F-Statistic     p-value
    0.35238     0.56059

> xredB <- xadd[,-c(2,3)]
> fitredB <- rfit(yadd~xredB)
> drop.test(fitadd,fitredB) # Test of main effect B

Drop in Dispersion Test
F-Statistic     p-value
    4.03473     0.03682
```

With p-value 0.5606, there is little evidence that the levels of factor A impact the response. On the other hand, with p-value 0.0368, there is evidence at nominal $\alpha = 0.05$ that the different levels of factor B do have an effect on the response. To investigate factor B further, the next R segment obtains simple contrasts between the levels of factor B.

```
> c12 <- c(0,0,1,0); c13 <- c(0,0,0,1); c23 <- c(0,-1,0,1)
> hmat <- rbind(c12,c13,c23)
> est <- hmat%*%fitadd$coef                   # Estimates of contrasts
> SE <- sqrt(diag(hmat%*%vcov(fitadd)%*%t(hmat)))        # SEs
> tab <- cbind(est,SE)
> rownames(tab) <- c("1 to 2","1 to 3","2 to 3")
> colnames(tab) <- c("Est","SE"); tab

       Est     SE
1 to 2  14 5.4854
1 to 3  12 5.4854
2 to 3   9 7.1645
```

So the expected changes in the response from level 1 of factor B to levels 2 and 3 are the respective estimated increases of 14 and 12 units; furthermore, both estimated changes are significant at $\alpha = 0.05$. The change in shift from level 2 to level 3, however, is insignificant. We next display the summary table of the full model fit. As discussed above, the last two estimates are the estimated shifts from level 1 to respective levels 2 and 3 of factor B. These agree with the results in the above R segment.

```
> summary(fitadd)$coef

                  Estimate Std. Error t.value   p.value
```

```
(Intercept)             17     4.7248 3.59804 0.0022183
xaddas.factor(FA)2       3     4.2071 0.71308 0.4854730
xaddas.factor(FB)2      14     5.4854 2.55222 0.0206155
xaddas.factor(FB)3      12     5.4854 2.18762 0.0429575
```

■

Often experimenters conduct two-way designs with only one replicate in each cell. For these designs, the usual least squares and rank-based tests for interaction cannot be carried out. Often an additive model is assumed. The main effect rank-based tests discussed above can be conducted in the same way as well as estimates, along with their standard errors, of interesting contrasts. The R coding remains the same. Exercise 5.9.13 provides an example.

Common examples of one replicate, two-way designs are the randomized block designs where one of the factors is a blocking factor that is often thought of as a random factor. For a simple example, consider the effect that fertilizers have on the yield of a specified crop. The experiment concerns four fertilizers. Different plots are selected and each plot is divided into four subplots. Within a plot, the fertilizers are randomly assigned to the subplots. Here there are four fertilizers of interest. Fertilizer is the fixed effect of interest. On the other hand, the plots are essentially randomly selected, and it is usually not an effect of interest. This is an example of a randomized block design with plot as the random factor. The statistical analysis, though, must take into account the random factor. Appropriate rank-based analyses of randomized block designs are discussed in Chapter 8.

5.6 ANCOVA

It is common to collect additional information (e.g., demographics) in an experiment or study. When this information is included in the model to test for a treatment or group effect, it can be considered an **analysis of covariance** (ANCOVA). The goal of these covariates is to account for variation in the response variables. These analyses are often called **adjusted analyses**. Absent *a priori* information of which covariates are important, a model selection procedure can be implemented with the response on the covariates; i.e., treatment is included in the model *after* the covariates are selected. In this section, we discuss this adjusted analysis for rank-based procedures. Covariate variables are sometimes referred to as concomitant variables. The monograph by Huitema (2011) offers an informative discussion of the traditional ANCOVA, while Chapter 4 of Hettmansperger and McKean (2011) presents the rank-based ANCOVA which we discuss.

For notation, consider a one-way experimental design with k different treatment combinations with cell sample sizes n_i, $i = 1, \ldots k$. The discussion easily generalizes to other ANOVA designs. Suppose there are m covariate variables. Let Y_{ij} denote the response for the jth subject in the ith cell and let $x_{ijk}, k = 1, \ldots m$ denote the corresponding value of the kth covriate, **covariate**. The full model is:

$$Y_{ij} = \mu_i + \beta_1 x_{ij1} + \cdots + \beta_m x_{ijm} + e_{ij}. \tag{5.18}$$

Inference for the experimental effects (linear functions of $\boldsymbol{\mu}$) proceeds similar to the inference discussed in Sections 5.2–5.3, except that the inference is adjusted by the covariates. We can also test to see if the covariates *make a difference*; i.e., test that $\beta_1 = \cdots = \beta_m = 0$.

A major assumption, though, behind the adjusted model (5.18) is that the relationship between the covariates and the response variable is the same at each treatment combination. Thus, before considering adjusted inference, we usually test for no interaction between the experimental design and the covariates.

5.6.1 Computation of Rank-Based ANCOVA

For the general one-way or k-way ANCOVA models, we have written R functions which compute the rank-based ANCOVA analyses which are included in the R package `npsm`. We first discuss the computation of rank-based ANCOVA when the design is a one-way layout with k groups.

Computation of Rank-Based ANCOVA for a One-Way Layout

For one-way ANCOVA models, we have written two R functions which make use of `Rfit` to compute a rank-based analysis of covariance. The function `onecovaheter` computes the test of homogeneous slopes. It also computes a test of no treatment (group) effect, but this is based on the full model of heterogeneous slopes. It should definitely not be used if the hypothesis of homogeneous slopes is rejected. The second function `onecovahomog` assumes that the slopes are homogeneous and tests the hypothesis of no treatment effect; this is the adjusted analysis of covariance. It also tests the hypotheses that the covariates have a nonsignificant effect. The arguments to these functions are: the number of groups (the number of cells of the one-way design); a $n \times 2$ matrix whose first column contains the response variable and whose second column contains the responses' group identification; and the $n \times m$ matrix of covariates. The functions compute the analysis of covariances, summarizing the tests in ANCOVA tables. These tables are also returned in the value `tab` along with the full model fit in `fit`. For the function `onecovahomog` the full model is (5.18), while for the function `onecovaheter` the full model is Model (5.18) with the additional m columns of interaction effects between the group and covariates. We illustrate these functions with the following example.

Example 5.6.1 (Chateau Latour Wine Data). Sheather (2009) presents a dataset drawn from the Chateau Latour wine estate. The response variable is the quality of a vintage based on a scale of 1 to 5 over the years 1961 to 2004. The predictor is end of harvest, days between August 31st and the end of harvest for that year, and the factor of interest is whether or not it rained at harvest time. The data are in `latour` in the package `npsm`. We first compute the test for homogeneous slopes.

```
> data = latour[,c('quality','rain')]
> xcov = cbind(latour[,'end.of.harvest'])
> analysis = onecovaheter(2,data,xcov,print.table=TRUE)

Robust ANCOVA (Assuming Heterogeneous Slopes) Table
              df       RD      MRD      F   p-value
Groups         1 0.88293 0.88293 2.3955 0.1295608
Homog Slopes   1 2.80125 2.80125 7.6002 0.0087503
```

Based on the robust ANCOVA table, since the p-value is less than 0.01, there is definitely an interaction between the groups and the predictor. Hence, the test in the first row for treatment effect should be ignored.

To investigate this interaction, as shown in the left panel of Figure 5.4, we overlaid the fits of the two linear models over the scatterplot of the data.

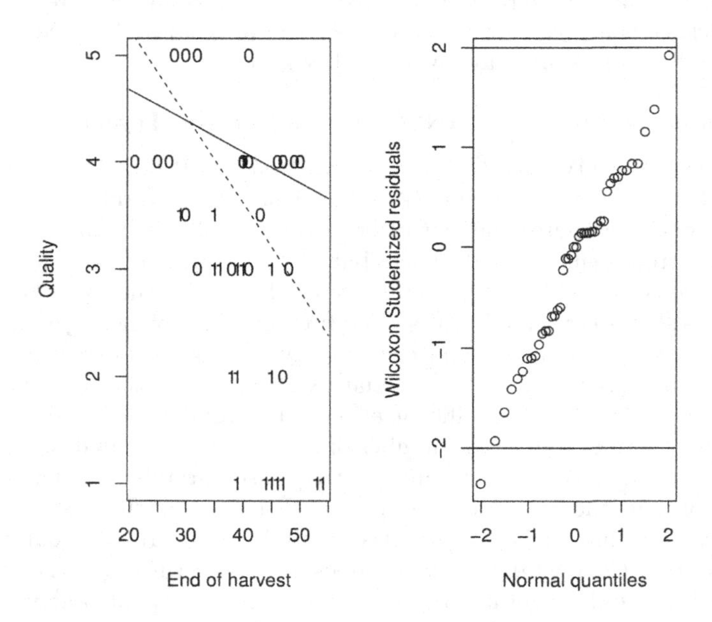

FIGURE 5.4
The left panel is the scatterplot of the data of Example 5.6.1. The dashed line is the fit for the group "rain at harvest time" and the solid line is the fit for the group "no rain at harvest time." The right panel is the normal probability plot of the Wilcoxon Studentized residuals.

The dashed line is the fit for the group "rain at harvest time," while the solid line is the fit for the group "no rain at harvest time." For both groups, the quality of the wine decreases as the harvest time increases, but the decrease is much worse if it rains. Because of this interaction, the tests in the first two rows of the ANCOVA table are not of much interest. Based on the plot, interpretations from confidence intervals on the difference between the groups at days 25 and 50 would seem to differ. ∎

If we believe the assumption of equal slopes is reasonable, then we may use the function `onecovahomog` which takes the same arguments as `onecovaheter`.

Computation of Rank-Based ANCOVA for a k-Way Layout

For the k-way layout, we have written the `Rfit` function `kancova` which computes the ANCOVA. Recall that the full model for the design is the cell mean (median) model. Under heterogeneous linear models, each cell in the model has a distinct linear model. This is the full model, Model 7.24, for testing homogeneous slopes. This function also computes the adjusted analysis, assuming that the slopes are homogeneous; that is, for these hypotheses the full model is Model 7.23. So these adjusted tests should be disregarded if there is reason to believe the slopes are different, certainly if the hypothesis of homogeneous slopes is rejected. For the adjusted tests, the standard hypotheses are those of the main effects and interactions of all orders as described in Section 5.3. A test for all covariate effects that are null is also computed. We illustrate this function in the following two examples.

Example 5.6.2. For this illustration we have simulated data from a 2×3 layout with one covariate. Denote the factors by A and B, respectively. It is a null model; i.e., for the data simulated, all effects were set at 0. The data are displayed by cell in the following table. The first column in each cell contains the response realizations, while the second column contains the corresponding covariate values.

Factor A	Factor B					
	B(1)		B(2)		B(3)	
A(1)	4.35	4.04	0.69	2.88	4.97	3.4
	5.19	5.19	4.41	5.3	6.63	2.91
	4.31	5.16	7.03	2.96	5.71	3.79
	5.9	1.43	4.14	5.33	4.43	3.53
			4.49	3.51	3.73	4.93
					5.29	5.22
					5.75	3.1
					5.65	3.89
A(2)	4.93	3.22	6.15	3.15	6.02	5.69
	5.1	4.73	4.94	2.01	4.27	4.2
	4.52	2.79	6.1	3.01	4.3	2.57
	5.53	5.63	4.93	3.87	4.47	3.75
	4.21	3.88	5.3	4.47	6.07	2.62
	5.65	3.85				

We use this example to describe the input to the function `kancova`. The design is a two-way with the first factor at 2 levels and the second factor at 3 levels. The first argument to the function is the vector of levels `c(2,3)`. Since there are two factors, the second argument is a matrix of the three columns: vector of responses, Y_{ikj}; level of first factor i; and level of second factor j. The third argument is the matrix of corresponding covariates. The data are in dataset `acov231`. The following code segment computes the rank-based ANCOVA.

```
> levs = c(2,3)
> data = acov231[,1:3]
> xcov = matrix(acov231[,4],ncol=1)
> temp = kancova(levs,data,xcov)

Robust ANCOVA Table
All tests except last row is with homogeneous slopes
as the full model.   For the last row the full model is
with heteroscedastic slopes.
            df       RD      MRD        F p-value
1 , 0        1 0.218132 0.218132 0.452753 0.50697
0 , 1        2 0.094502 0.047251 0.098074 0.90692
1 , 1        2 1.212041 0.606021 1.257855 0.30099
Covariate    1 0.136554 0.136554 0.283431 0.59898
Hetrog regr  5 2.338562 0.467712 1.008491 0.43723
```

The rank-based tests are all nonsignificant, which agrees with the null model used to generate the data. ■

Example 5.6.3 (2 × 2 with Covariate)**.** Huitema (2011), page 496, presents an example of a 2 × 2 layout with a covariate. The dependent variable is the number of novel responses under controlled conditions. The factors are type of reinforcement (Factor A at 2 levels) and type of program (Factor B at 2 levels); hence there are four cells. The covariate is a measure of verbal fluency. There are only 4 observations per cell for a total of $n = 16$ observations. Since there are 8 parameters in the heterogeneous slope model, there are only 2 observations per parameter. Hence, the results are tentative. The data are in the dataset `huitema496`. Using the function `kancova` with the default Wilcoxon scores, the following robust ANCOVA table is computed.

```
> levels = c(2,2);
> y.group = huitema496[,c('y','i','j')]
> xcov = huitema496[,'x']
> temp = kancova(levels,y.group,xcov)

Robust ANCOVA Table
All tests except last row is with homogeneous slopes
as the full model.   For the last row the full model is
with heteroscedastic slopes.
```

```
                df         RD        MRD        F  p-value
1 , 0            1    5.67402    5.67402  5.73779 0.035522
0 , 1            1    0.49396    0.49396  0.49951 0.494416
1 , 1            1    0.10701    0.10701  0.10822 0.748361
Covariate        1   12.27043   12.27043 12.40832 0.004777
Hetrog regr      3    8.51232    2.83744  5.24070 0.027182
```

The robust ANCOVA table indicates heterogeneous slopes, so we plotted the four regression models next as shown in Figure 5.5. The rank-based test of

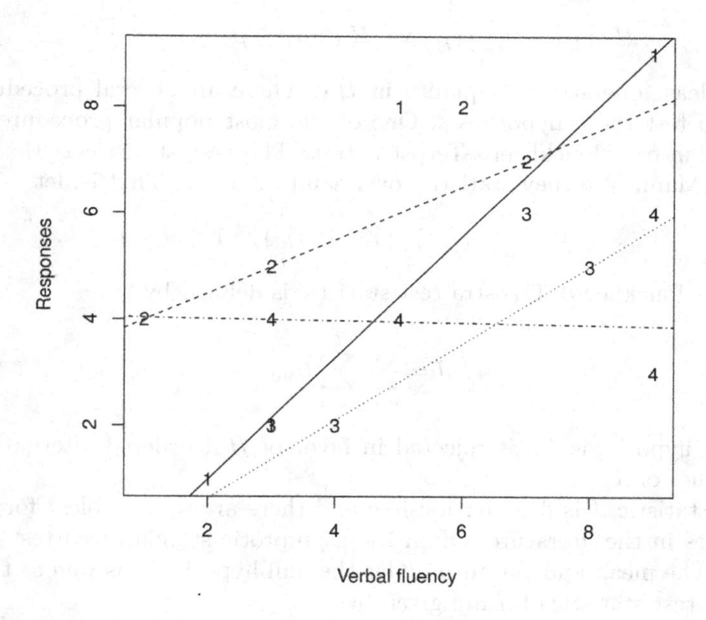

FIGURE 5.5
Scatterplot of the data of Example 5.6.3. The number 1 represents observations from the cell with $A = 1, B = 1$, 2 represents observations from the cell with $A = 1, B = 2$, 3 represents observations from the cell with $A = 2, B = 1$, and 4 represents observations from the cell with $A = 2, B = 2$.

homogeneity of the sample slopes agrees with the plot. In particular, the slope for the cell with $A = 2, B = 2$ differs from the others. Again, these results are based on small numbers and thus should be interpreted with caution. As a pilot study, these results may serve in the conduction of a power analysis for a larger study. ∎

5.7 Ordered Alternatives

Consider again the one-way ANOVA model, Section 5.2 with k levels of a factor or k treatments. The full model is given in expression (5.1). As in Section 5.2, let Y_{ij} denote the jth response in sample i for $j = 1, \ldots, n_i$ and $i = 1, \ldots, k$, and let μ_i denote the mean or median of Y_{ij}. In this section, suppose that a certain ordering of the means (centers) is a reasonable alternative to consider; that is, assume that the hypotheses are

$$H_0 : \mu_1 = \cdots = \mu_k \text{ vs. } H_A : \mu_1 \leq \mu_2 \leq \cdots \leq \mu_k, \tag{5.19}$$

with at least one strict inequality in H_A. There are several procedures designed to test these hypotheses. One of the most popular procedures is the distribution-free Jonckheere–Terpstra test. The test statistic is the sum of pairwise Mann–Whitney statistics over samples $u < v$. That is, let

$$U_{uv} = \#_{1 \leq j \leq n_u, 1 \leq l \leq n_v} \{Y_{ju} < Y_{lv}\}, \quad 1 \leq u < v \leq k. \tag{5.20}$$

Then the Jonckheere–Terpstra test statistic is defined by

$$J = \sum_{u=1}^{k-1} \sum_{v=2}^{k} U_{uv}. \tag{5.21}$$

The null hypothesis H_0 is rejected in favor of H_A, ordered alternatives, for large values of J.

The statistic J is distribution-free and there are some tables[5] for its critical values in the literature. Often the asymptotic standardized test statistic is used. The mean and variance under the null hypothesis as well as the standardized test statistic of J are given by

$$\begin{aligned} E_{H_0}(J) &= \frac{n^2 - \sum_{i=1}^k n_i^2}{4} \\ V_{H_0}(J) &= \frac{n^2(2n+3) - \sum_{i=1}^k n_i^2(2n_i+3)}{72} \\ z_J &= \frac{J - E_{H_0}(J)}{\sqrt{V_{H_0}(J)}}. \end{aligned} \tag{5.22}$$

An asymptotic level $0 < \alpha < 1$ test is to reject H_0 if $z_J \geq z_\alpha$.

In the case of tied observations, the contribution to U_{uv}, (5.20), is one-half. That is, if $Y_{ju} = Y_{lv}$ then instead of the contribution of 0 to U_{uv} as expression (5.20) dictates for this pair, the contribution is one-half. If there are many ties, then the tie correction to the variance[6] should be used. Included in **npsm**

---

[5] See Section 6.2 of Hollander and Wolfe (1999).

[6] See page 204 of Hollander and Wolfe (1999).

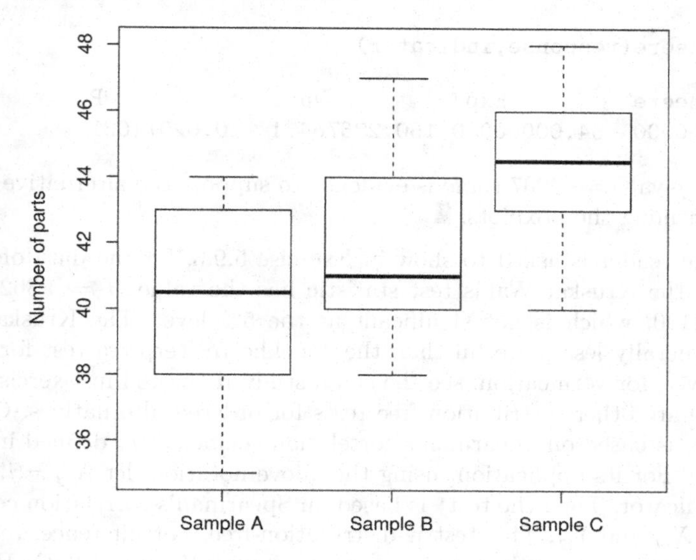

FIGURE 5.6
Comparison boxplots of three samples of Example 5.7.1.

is the function `jonckheere`. This software adjusts the null variance for ties, returning the test statistic and its asymptotic p-value. We illustrate its use in the following example.

Example 5.7.1 (Knowledge of Performance). Hollander and Wolfe (1999) discuss a study on workers whose job was a repetitive task. It was thought that knowledge of performance would improve their output. So 18 workers were randomly divided into three groups A, B, and C. Group A served as a control. These workers were given no information about their performance. The workers in Group B received some information about their performance, while those in Group C were given detailed information on their performance. The response was the number of parts each worker produced in the specified amount of time. The data appear in the next table. Figure 5.6 shows comparison boxplots of the samples.

	Parts Produced					
Group A	40	35	38	43	44	41
Group B	38	40	47	44	40	42
Group C	48	40	45	43	46	44

The alternative hypothesis is $H_A: \ \mu_A \leq \mu_b \leq \mu_c$ with at least one strict inequality, where μ_i is the mean output for a worker in Group i. In the next code segment, the arguments in the call `jonckheere(response,indicator)` are the vector of responses and the vector of group indicators.

```
> jonckheere(response,indicator)

  Jonckheere          ExpJ         VarJ            P
 79.00000000  54.00000000 150.28676471   0.02071039
```

With the p-value=0.0207 there is evidence to support the alternative hypothesis, confirming the boxplots. ■

As the reader is asked to show in Exercise 5.9.3, for the data of the last example, the Kruskal–Wallis test statistic has the value $H = 4.362$ with p-value 0.1130, which is not significant at the 5% level. The Kruskal–Wallis test is generally less powerful than the Jonckheere–Terpstra test for ordered alternatives; for verification, see the small study requested in Exercise 5.9.16.

There are other distribution-free tests for ordered alternatives. One such procedure is based on Spearman's correlation coefficient r_s defined in expression (4.8). For its application, using the above notation, let $X_{ij} = i$, i.e., the group indicator. Then the test[7] is based on Spearman's correlation coefficient between X_{ij} and Y_{ij}. This test is distribution-free. For inference, recall that under the null distribution, $z_s = \sqrt{n} r_s$ is asymptotically $N(0,1)$. Hence, z_s can be used as a standardized test statistic. A degree of freedom correction[8], though, makes use of the standardized test statistic t_s which is defined by

$$t_s = \frac{r_s}{\sqrt{(1 - r_s^2)/(n-2)}}. \tag{5.23}$$

We use the test statistics t_s in the next code segment to compute this Spearman procedure for the data in the last example.

```
> cor.test(indicator,response,method="spearman",
+     continuity=FALSE,exact=FALSE,alternative='greater')

        Spearman's rank correlation rho

data:  indicator and response
S = 488.4562, p-value = 0.01817
alternative hypothesis: true rho is greater than 0
sample estimates:
      rho
0.4959172
```

Thus the tests based on Spearman's ρ_S and the Jonckheere–Terpstra are essentially in agreement for this example.

A bootstrap confidence interval for ρ_S was discussed in Chapter 4; see Example 4.2.2 for the discussion of the R function `cor.boot.ci` which computes this bootstrap confidence interval. Note that r_s is an estimate of association and the confidence interval is a measure of the strength of this association.

[7] See Tryon and Hettmansperger (1973) and McKean et al. (2001) for details.
[8] See page 347 of Huitema (2011).

5.8 Multi-Sample Scale Problem

A general assumption in fixed-effects ANOVA-ANCOVA is the homogeneity of scale of the random errors from level (group) to level (group). For two levels (samples), we discussed the Fligner–Killeen test for homogeneity of scale in Section 3.5. This test generalizes immediately to the k-level (sample) problem.

For discussion, we use the notation of the k-cell (sample) model in Section 5.2 for the one-way ANOVA design. For this section, our full model is Model (5.1) except that the pdf of random errors for the jth level is of the form $f_j(x) = f[(x - \theta_j)/\sigma_j]/\sigma_j$, where θ_j is the median and $\sigma_j > 0$ is a scale parameter for the jth level, $j = 1, \ldots, k$. A general hypothesis of interest is that the scale parameters are the same for each level, i.e.,

$$H_0 : \sigma_1 = \cdots = \sigma_k \text{ versus } H_A : \sigma_j \neq \sigma_{j'} \text{ for some } j \neq j'. \tag{5.24}$$

Either this could be the hypothesis of interest or the hypothesis for a pre-test on homogeneous scales in the one-way ANOVA location model.

The Fligner–Killeen test of scale for two-samples, (3.47), easily generalizes to this situation. As in Section 3.5, define the folded-aligned sample as

$$Y_{ij}^* = Y_{ij} - \text{med}_{j'}\{Y_{ij'}\}, \quad j = 1, \ldots, n_i : i = 1, \ldots, k. \tag{5.25}$$

Let $n = \sum_{i=1}^n n_i$ denote the total sample size and let $R_{ij} = R|Y_{ij}^*|$ denote the rank of the absolute values of these items, from 1 to n. Define the scores $a^*(i)$ as

$$\begin{aligned} a(l) &= \left[\Phi^{-1}\left(\frac{l}{2(n+1)} + \frac{1}{2}\right)\right]^2 \\ \overline{a} &= \frac{1}{n}\sum_{l=1}^{n} a(l) \\ a^*(l) &= a(l) - \overline{a}. \end{aligned} \tag{5.26}$$

Then the Fligner–Killeen test statistic is

$$Q_{FK} = \frac{n-1}{\sum_{l=1}^n (a^*(l))^2} \sum_{i=1}^{k} \left\{ \sum_{j=1}^{n_i} a^*(R_{ij}) \right\}^2 . \tag{5.27}$$

An approximate level α-test[9] is based on rejecting H_0 if $Q_{FK} \geq \chi_\alpha^2(k-1)$.

As in Section 3.5, we can obtain rank-based estimates of the difference in scale. Let $Z_{ij} = \log(|Y_{i1}^*|)$, and let $\Delta_{1i} = \log(\sigma_i/\sigma_1)$, for $i = 2, \ldots, k$, where,

[9] See page 105 of Hájek and Šidák (1967).

without loss of generality, we have referenced the first level. Using $\Delta_{11} = 0$, we can write the log-linear model for the aligned, folded sample as

$$Z_{ij} = \Delta_{1i}^* + e_{ij}, \quad j = 1, \ldots n_i, i = 1, 2, \ldots, k. \tag{5.28}$$

As discussed in Section 3.5, the scores defined in expression (3.46) are appropriate for the rank-based fit of this model. Recall that they are optimal, when the random errors of the original samples are normally distributed. As discussed in Section 3.5, exponentiation of the regression estimates leads to estimation (and confidence intervals) for the ratio of scales $\eta_{1i} = \sigma_i/\sigma_1$.

The function `fkk.test` is a wrapper which obtains the R fit and analysis for this section. We demonstrate it in the following example.

Example 5.8.1 (Three Generated Samples). For this example, we generated three samples (rounded) from Laplace distributions. The samples have location and scale $(5, 1)$, $(10, 4)$, and $(10, 8)$ respectively. Hence in the above notation, $\eta_{21} = 4$ and $\eta_{31} = 8$. A comparison boxplot of the three samples is shown in Figure 5.7.

The following code segment computes the Fligner–Killeen test for these three samples. Note that the response variables are in the vector `response`, and that the vector `indicator` is a vector of group membership.

```
> fkk.test(response,indicator)

Table of estimates and  95  percent confidence intervals:
```

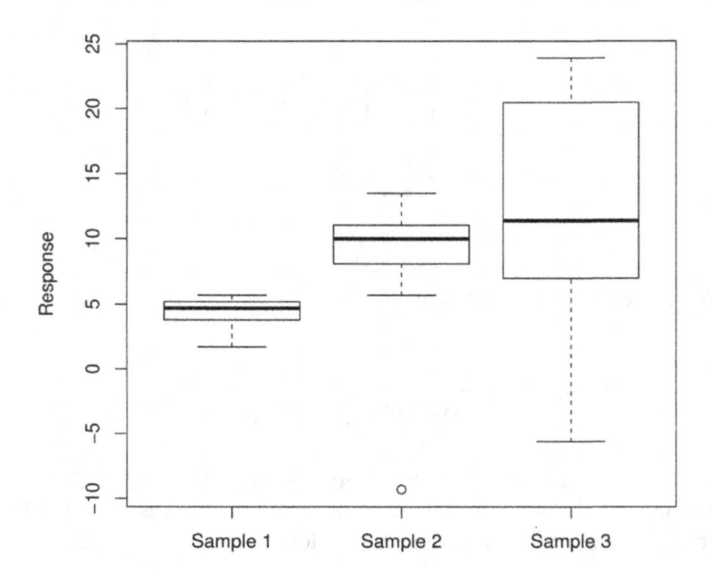

FIGURE 5.7
Comparison boxplots of three samples.

```
                      estimate  ci.lower ci.upper
xmatas.factor(iu)2 3.595758 0.9836632 13.14421
xmatas.factor(iu)3 8.445785 2.5236985 28.26458

 Test statistic =  8.518047  p-value =  0.0141361
```

Hence, based on the results of the test, there is evidence to reject H_0. The estimates of η_{21} and η_{31} are close to their true values. The respective confidence intervals are (0.98, 13.14) and (2.52, 28.26). ■

5.9 Exercises

5.9.1. Hollander and Wolfe (1999) report on a study of the length of YOY gizzard shad fish at four different sites of Kokosing Lake in the summer of 1984. The data are:

Site 1	Site 2	Site 3	Site 4
46	42	38	31
28	60	33	30
46	32	26	27
37	42	25	29
32	45	28	30
41	58	28	25
42	27	26	25
45	51	27	24
38	42	27	27
44	52	27	30

Let μ_i be the true median length of YOY gizzard shad at site i.

(a) Use the rank-based Wilcoxon procedure, (5.7), to test the hypothesis of equal medians.

(b) Based on Part (a), use Fisher's least significance difference to perform a multiple comparison on the differences in the medians. As discussed in Hollander and Wolfe, YOY gizzard shad are eaten by game fish, and for this purpose smaller fish are better. In this regard, based on the MCP analysis, which sites, if any, are preferred?

5.9.2. For the study discussed in Example 5.2.3, obtain the analysis based on the Wilcoxon test using F_W, (5.7). Then obtain the MCP analysis using Tukey's method. Compare this analysis with the Kruskal–Wallis analysis presented in the example.

5.9.3. For the data of Example 5.7.1 determine the value of the Kruskal–Wallis test and its p-value.

5.9.4. For a one-way design, instead of using the `oneway.rfit` function, the rank-based test based on F_W, expression (5.7), can be computed by the `Rfit` functions `rfit` and `drop.test`.

(a) Suppose the R vector `y` contains the combined samples, and the R vector `ind` contains the associated levels. Discuss why the following two code segments are equivalent:
`oneway.rfit(y,ind)` and
`fit<-rfit(y~factor(ind)); drop.test(fit)`

(b) Verify this equivalence for the data in Example 5.2.1.

5.9.5. Write an R script which compares empirically the power between the rank-based test based on F_W and the corresponding LS test for the following situation: 4 samples of size 10; location centers 10, 11, 12, and 15; and the random errors $3 * e_{ij}$ where the e_{ij} are iid with the common t-distribution having 3 degrees of freedom. Use a simulation size of 1000 and the level $\alpha = 0.05$. Which procedure has higher empirical power?

5.9.6. In Exercise 5.9.5, we simulated the empirical power of the rank-based and LS tests for a specific situation. For this exercise, check the validity of the rank-based and LS tests; i.e., set the location centers to be the same.

5.9.7. Suppose that we want a descriptive plot for a one-way design. Comparison boxplots are one such plot; however, if the level sample sizes n_i are small, then these plots can be misleading (quartiles and, hence, lengths of boxplots, can be adversely affected by a few outliers). Hence, for $n_i \leq 10$, we recommend comparison dotplots instead of boxplots. Consider the following data from a one-way design.

Level 1	66	45	42	53	71		
Level 2	38	53	47	23	42	50	
Level 3	82	26	95	70	80	82	75

(a) Obtain the comparison dotplots for the above data.

(b) Compute the Fligner–Killeen test of equal scales for these data.

5.9.8. Miliken and Johnson (1984) discuss a study pertaining to an unbalanced 2×3 crossed factorial design. For convenience, we present the data below. For their LS analysis, Miliken and Johnson recommend Type III hypotheses. As discussed in Section 7.4, the rank-based analysis based on the `Rfit` function `raov` obtains tests based on Type III hypotheses. The R function `lm`, however, does not. In this exercise, we show how to easily obtain LS analyses for Type III hypotheses using the functions `redmod` and `cellx` from `Rfit`. For the data, the factors are labeled T and B, and the responses are tabled as:

	B_1	B_2	B_3
T_1	19	24	22
	20	26	25
	21		25
T_2	25	21	31
	27	24	32
		24	33

The code below assumes that the response and indicator vectors are:

```
resp = c(19, 20, 21, 24, 26, 22, 25, 25, 25, 27, 21, 24, 24,
         31, 32, 33)
a = c(rep(1,8),rep(2,8))
b = c(1, 1, 1, 2, 2, 3, 3, 3, 1, 1, 2, 2, 2, 3, 3, 3)
```

(a) First obtain the analysis for interaction and main effects using the `Rfit` function `raov`. The hypotheses of this analysis are of Type III.

(b) The following script will obtain the LS Type III analysis for Factor T:

```
fitls <- lm(resp ~factor(a):factor(b))
cell <- rep(1:6,each=3)
cellmean <- cellx(cell)
ha <- c(1,1,1,-1,-1,-1)
xa <- redmod(cellmean,ha)
lmred <- lm(resp ~ xa)
anova(lmred,fitls)
```

Run this code and show that the LS test statistic computes to $F = 30.857$. Notice that this differs from the LS ANOVA based on `fitls`.

(c) Write code and run it for the LS Type III analysis of Factor B. *Hint:* the hypothesis matrix `hb` has two rows.

(d) Write code and run it for the LS Type III analysis of interaction. *Hint:* the hypothesis matrix `hint` has two rows. Notice that it agrees with the LS ANOVA based on `fitls`.

5.9.9. Obtain estimates and standard errors for the contrasts discussed at the end of Example 5.4.1. Discuss the results in terms of the example.

5.9.10. The base R dataset `ToothGrowth` contains the data of an experiment on the effect of vitamin C on the growth of teeth in guinea pigs; see also Bliss (1952). The factors are: `supp`, method of delivery of vitamin C (ascorbic acid (`VC`) or orange juice (`OJ`)) and `dose` of vitamin C (0.5, 1, or 2 mg/day). So the design is a 2×3 design. Sixty guinea pigs were selected for the experiment and 10 were assigned to each combination of the factors. The response is `len`, tooth length.

(a) Using Wilcoxon scores, obtain the analysis of this 2×3 design with interaction. Using $\alpha = 0.05$, interpret the results of the hypotheses tested in terms of the problem.

(b) Capture the fit in Part (a), using the code
`capf<-raov(len~supp+dose+supp*dose)`.
Next, obtain the Studentized residuals (`sr<-rstudent(capf$fit)`), and compute the $q-q$ plot of these residuals. Comment on outliers (identify any outlying points) and normality of the true error distribution.

(c) Since interaction was included in the model, the estimates of the cell medians are the fitted values. Table these in a 2×3 table. In terms of estimation, which combination of factors lead to the greatest tooth growth? Are they significantly different from other combinations (use `summary(capf$fit)` to get standard errors).

5.9.11. The base R dataset `PlantGrowth` contains the results of a one-way experiment on plant growth. The responses are the dried weights, `weight`, of plants grown under one of three levels, `group`; see also Dobson (1983). The levels are: Control (`ctrl`), Treatment 1 (`trt1`), and Treatment 2 (`trt2`). Assume large plant weights are desirable.

(a) Obtain a boxplot of weight versus group. Comment on the plot (which treatments appear to be better).

(b) Using Wilcoxon scores, test the hypotheses that the group medians are the same. Conclude using $\alpha = 0.05$.

(c) Select a multiple comparison procedure (MCP) giving the reasons for the method that you chose. Discuss the results of the MCP. Is there a "best" treatment? Why?

(d) Obtain a normal $q-q$ of the Wilcoxon Studentized residuals. Identify any outliers. Comment on the normality of the error distribution.

5.9.12. Recall that the `Rfit` dataset `baseball` contains data on 59 professional baseball players. As a one-way observational study, consider the batting averages of the fielders according to how they bat (`R` for right-handed hitters, `L` for left-handed hitters, and `S` for switch hitters). For the fielder hitting averages and grouping indicator, use the dataset: `aves<-average[field==1]` and `bats<-bat[field==1]`.

(a) Obtain a boxplot of the batting averages by group. Comment on the plot (which batting averages appear to be higher).

(b) Using Wilcoxon scores, test the hypotheses that the group medians are the same. Conclude using $\alpha = 0.05$.

(c) Select a multiple comparison procedure (MCP) giving the reasons for the method that you chose. Discuss the results of the MCP. Is there a "best" group? Why?

(d) Obtain a normal $q-q$ of the Wilcoxon Studentized residuals. Identify any outliers. Comment on the normality of the error distribution.

5.9.13. A two-way crossed factorial experiment involving two factors, A at 3 levels and B at 5 levels, resulted in the following data:

	B				
	22	19	42	39	11
A	13	23	33	30	29
	24	10	25	14	16

Fit the additive model using Wilcoxon scores.

(a) Test at level 0.05 the main effect hypotheses (5.16) and (5.17). Summarize the results in terms of the problem.

(b) Obtain all 10 simple contrasts between the levels of factor B along with standard errors. Form confidence intervals using the Bonferroni critical value. Summarize the results in terms of the problem.

(c) Perform part (b) for the 3 simple contrasts of factor A.

(d) Perform a residual analysis based on residual plot and normal $q-q$ plot using the Studentized residuals. Discuss.

5.9.14. Test for interaction between factors A and B at level 0.05 for the data in Example 5.5.1. Use Wilcoxon scores.

5.9.15. Page 436 of Hollander and Wolfe (1999), presents part of a study on the effects of cloud seeding on cyclones; see Wells and Wells (1967) for the original reference. For the reader's convenience, the data are contained in the dataset `SCUD`. The first column is an indicator for Control (2) or Seeded (1); column 2 is the predictor M, the geostrophic meridional circulation index; and column 3 is the response RI which is a measure of precipitation.

(a) Obtain a scatterplot of RI versus M. Use different plotting symbols for the Control and Seeded. Add the rank-based fits of the linear models for each. Comment on the plot.

(b) Using a rank-based analysis, test for homogeneous slopes for the two groups.

(c) If homogeneous slopes is "accepted" in (b), use a rank-based analysis to test for homogeneous groups.

(d) The test in Part (c) is adjusted for M. Is this adjustment necessary? Test at level $\alpha = 0.05$.

5.9.16. Using a simulation study, investigate the powers of the Jonckheere–Terpstra test and the Kruskal–Wallis test for the following situation: Samples of size 10 from four normal populations each having variance 1 and with the respective means of $\mu_1 = 0$, $\mu_2 = 0.45$, $\mu_3 = .90$, and $\mu_4 = 1.0$. Use the level of $\alpha = 0.05$ and a simulation size of 10,000.

5.9.17. In reading through Section 5.7 on ordered alternatives, the reader may have noticed the simplicity of the test based on Spearman's ρ_S over the test using the Jonckheere–Terpstra test statistic. Is it as powerful? As a partial answer, this exercise provides some empirical evidence. One may use `cor.test` to obtain a test based on Spearman's ρ. See, for example, the following code.

```
group <- c(rep(1,ni),rep(2,ni),rep(3,ni),rep(4,ni))
y1 <- rnorm(ni,0,1);y2 <- rnorm(ni,.15,1);
y3 <- rnorm(ni,.35,1); y4 <- rnorm(ni,.55,1)
y <- c(y1,y2,y3,y4)
cor.test(group,y,method='spearman',
    continuity=FALSE,exact=FALSE,alternative='less')
```

(a) Determine the situation (distributions, alternative, etc.) which the above code simulates.

(b) Based on the above situation, run a simulation to compare the empirical powers of the Jonckheere–Terpstra test and Spearman's ρ.

(c) Run a simulation where the error distribution is a t-distribution with 3 degrees of freedom.

(d) Run a simulation where the error distribution is a χ^2-distribution with 1.5 degrees of freedom.

5.9.18. Besides simplicity, another advantage of the test based Spearman's ρ_S over the Jonckheere–Terpstra is that the estimate of ρ_S is an easily understood correlational measure. In this setting, it offers a measure of the "strength" of the relationship. Use the function `cor.boot.ci` to obtain a bootstrap confidence interval for the data in Example 5.7.1.

5.9.19. Consider the malignant melanoma data in Example 3.1.1. See if the association found there still holds after adjusting for latitude and longitude.

5.9.20. In Example 7.3.1 the function `redmod` was used to obtain the reduced model design matrix for testing the hypothesis of whether or not the three lipid covariates (VLDL, LDL, and HDL) are significant predictors. Alternatively, the reduced model design matrix could have been formed by removing these predictors in the reduced model fit as illustrated in the following code segment.

```
> xred1.1 <- xfull[,-c(6,7,8)]
```

Verify that the p-value obtained from the two methods is the same value.

6

Categorical Data

In this chapter we examine the case of a categorical response variable. Categorical responses are types of random variables that are not measured on a numeric scale. Categorical variables may be ordered or unordered. We consider the case of unordered (except in the case of binary — where success is considered favorable to failure).

Categorical data occur commonly in practice. Demographic variables are often categorical. Outcomes in confirmatory clinical trials are often clinical outcomes; e.g., treatment response; i.e., clinically significant improvement. For example, in an oncology trial, remission may be considered the primary outcome with remission being considered a clinically significant improvement. A subject who is reported to have achieved treatment response is considered a **responder**. A primary statistical analysis of such a trial may compare the responder rates at a particular point in time.

6.1 Comparison of Two Probabilities

In this section we discuss inference for the comparison of two probabilities or rates. Parameters commonly of interest include the risk difference, risk ratio, and the odds ratio. Consider the example of two treatments T_1 and T_2 where the probability (or *risk*) of the event is p_1 and p_2 for the two treatments, respectively. Assume the event of interest is labeled E, so that $p_i = P(E|T_i)$ for $i = 1, 2$. The event not occurring is labeled E^c. Notice these are Bernoulli random variables. Assume random samples of size n_1 and n_2 are taken under the two experimental conditions. Let x_1 denote the number of success out of n_1 for treatment T_1 and x_2 denote the number of success out of n_2 for treatment T_2.

The **risk difference** is defined as $p_1 - p_2$ or vice versa. Also of interest in practice are the risk ratio and the odds ratio. The **risk ratio** is defined as

$$\frac{p_1}{p_2}$$

and the **odds ratio** is defined as

$$\frac{p_1/(1-p_1)}{p_2/(1-p_2)}.$$

A true or population risk difference of zero implies no treatment effect, and equivalently a true or population risk ratio or odds ratio of one implies no treatment effect.

If a study is conducted with n_1 subjects assigned to treatment T_1 and n_2 subjects assigned to treatment T_2, the results may be represented as in Table 6.1.

TABLE 6.1
Notation for results of a study with two treatments, T_1 and T_2.

	E	E^c
T_1	x_1	$n_1 - x_1$
T_2	x_2	$n_2 - x_2$

6.1.1 Risk Difference

Consider two Bernoulli random variables with probabilities of success p_1 and p_2. A parameter of interest is the difference in proportions $p_1 - p_2$. Inference concerns estimates of this difference, confidence intervals for it, and tests of hypotheses of the form

$$H_0 : p_1 = p_2 \text{ versus } H_A : p_1 \neq p_2. \tag{6.1}$$

Assume that a random sample is drawn from each of the distributions. The estimate of the difference in proportions or **risk difference** is the difference in sample proportions, i.e., $\hat{p}_1 - \hat{p}_2$.

It follows that the estimator $\hat{p}_1 - \hat{p}_2$ has an asymptotic normal distribution. Based on this, a $(1-\alpha)100\%$ asymptotic confidence interval for $p_1 - p_2$ is

$$\hat{p}_1 - \hat{p}_2 \pm z_{\alpha/2}\sqrt{\frac{\hat{p}_1(1-\hat{p}_1)}{n_1} + \frac{\hat{p}_2(1-\hat{p}_2)}{n_2}}. \tag{6.2}$$

For the hypothesis (6.1), there are two test statistics which are used in practice. The Wald-type test is the standardization of $\hat{p}_1 - \hat{p}_2$ based on its standard error, (the square-root term in expression (6.2)). The more commonly used test statistic is the scores test which standardizes under H_0. Under H_0 the population proportions are the same; hence, the following average

$$\hat{p} = \frac{n_1\hat{p}_1 + n_2\hat{p}_2}{n_1 + n_2} \tag{6.3}$$

is an estimate of the common proportion. The scores test statistic is given by

$$z = \frac{\hat{p}_1 - \hat{p}_2}{\sqrt{\hat{p}(1-\hat{p})}\sqrt{\frac{1}{n_1} + \frac{1}{n_2}}}. \tag{6.4}$$

This test statistic is compared with z-critical values. As in the one-sample problem, the χ^2-formulation, $\chi^2 = z^2$, is often used. We illustrate the R computation of this analysis with the next example.

Example 6.1.1 (Polio Vaccine). Rasmussen (1992), page 355, discusses one of the original clinical studies for the efficacy of the Salk polio vaccine which took place in 1954. The effectiveness of the vaccine was not known and there were fears that it could even cause polio since the vaccine contained live virus. Children with parental written consent were randomly divided into two groups. Children in the treatment group (1) were injected with the vaccine, while those in the control or placebo group (2) were injected with a biologically inert solution. Let p_1 and p_2 denote the true proportions of children who get polio in the treatment and control groups, respectively. The hypothesis of interest is the two-sided hypothesis (6.1). The following data are taken from Rasmussen (1992):

Group	No. Children	No. Polio Cases	Sample Proportion
Treatment	200,745	57	0.000284
Control	201,229	199	0.000706

The R function for the analysis is the same function used in the one-sample proportion problem, namely, `prop.test`. The first and second arguments are, respectively, the vectors `c(x1,x2)` and `c(n1,n2)`, where `x1` and `x2` are the number of successes for the two samples. The default hypotheses are the two-sided hypotheses (6.1). The following R segment provides the analysis for the polio vaccine data.

```
> prop.test(c(57,199),c(200745,201229),correct=FALSE)

        2-sample test for equality of proportions without
        continuity correction

data:  c(57, 199) out of c(200745, 201229)
X-squared = 78.5, df = 1, p-value <2e-16
alternative hypothesis: two.sided
95 percent confidence interval:
 -0.00086084 -0.00054912
sample estimates:
    prop 1     prop 2
0.00028394 0.00098892
```

The χ^2 test statistic is 78.474 with corresponding p-value zero to the level of machine precision. ■

6.1.2 Comparison of Two Rates

In Section 6.1.1 we reviewed large sample inference for the difference in two population proportions or risk difference. Also of interest in practice are the risk ratio and the odds ratio. A point estimate of the relative risk is

$$\frac{\hat{p}_1}{\hat{p}_2} = \frac{x_1/n_1}{x_2/n_2} \tag{6.5}$$

and a point estimate of the odds ratio is

$$\frac{\hat{p}_1/(1-\hat{p}_1)}{\hat{p}_2/(1-\hat{p}_2)} = \frac{x_1/(n_1-x_1)}{x_2/(n_2-x_2)}. \tag{6.6}$$

Confidence intervals for these two parameters are discussed in the next two subsections.

As our working example for this section, consider the results of an RCT comparing two treatments aspirin (A) and placebo (P) where the outcome E represents occurrence of myocardial infarction during follow-up of approximately 3 years (see c.f. Agresti (2002)). The estimate probabilities of the event are $\hat{p}_A = 18/676 = 0.0266$ and $\hat{p}_P = 28/684 = 0.0409$.

TABLE 6.2
Results of a RCT comparing aspirin (A) with placebo (P) on the risk of myocardial infarction (E) during approximately three years of follow-up.

	E	E^c
A	18	658
P	28	656

6.1.3 Confidence Interval for the Risk Ratio

Define $\theta_{RR} = p_1/p_2$ with point estimate as defined in Equation 6.5. To obtain a confidence interval, a common method is to first obtain the interval on a log scale and then back-transform to the ratio scale. A common method for determining a CI for the RR is to use a Taylor expansion for $\log\hat{\theta}_{RR}$.[1] So that an estimate of standard error is

$$\text{se}(\log\hat{\theta}_{RR}) = \sqrt{\frac{1-\hat{p}_1}{n_1\hat{p}_1} + \frac{1-\hat{p}_2}{n_2\hat{p}_2}}$$

and a Wald interval for θ_{RR} is the anti-log of

$$\log\hat{\theta}_{RR} \pm z_{\alpha/2}\sqrt{\frac{1-\hat{p}_1}{n_1\hat{p}_1} + \frac{1-\hat{p}_2}{n_2\hat{p}_2}}.$$

The estimate probabilities of the event are $\hat{p}_A = 18/676 = 0.0266$ and $\hat{p}_P = 28/684 = 0.0409$. A 95% confidence interval for $\log\theta_{RR}$ is

$$-0.4301 \pm 1.96 \cdot 0.2972.$$

[1]See, for example, Agresti (2002).

So a 95% CI for θ_{RR} is obtained by exponentiating and is given by

$$0.6505 \cdot \exp\{\pm\, 1.96 \cdot 0.2972\}$$

or (0.3633, 1.1647). Notice the 95% confidence interval contains the value 1. While there appears to be a reduction in the risk of MI by taking Aspirin relative to Placebo, the result would not be considered statistically significant at the 5% level.

Remark 6.1.1. Agresti (2002) recommends the score CI (c.f. Miettinen and Nurminen (1985)).

6.1.4 Confidence Interval for the Odds Ratio

Define $\theta_{OR} = (p_1/(1-p_1))/(p_2/(1-p_2))$ with point estimate as defined in Equation 6.6. Note the estimate is undefined if any of the components is zero; so, an estimate where 0.5 is added to each of the components is sometimes used.[2] For the working example, we have as a point estimate of the odds ratio

$$\hat{\theta}_{OR} = \frac{18}{658}/\frac{28}{656} = 0.6409.$$

As with the risk ratio, the standard error is often estimated based on $\log \hat{\theta}_{OR}$. So that an estimate of standard error is

$$\text{se}(\log \hat{\theta}_{OR}) = \sqrt{\frac{1}{x_1} + \frac{1}{n_1 - x_1} + \frac{1}{x_2} + \frac{1}{n_2 - x_2}}$$

and a Wald interval for θ_{OR} is the anti-log of

$$\log \hat{\theta}_{OR} \pm z_{\alpha/2}\sqrt{\frac{1}{x_1} + \frac{1}{n_1 - x_1} + \frac{1}{x_2} + \frac{1}{n_2 - x_2}}.$$

For our working example, to obtain a confidence interval, we first calculate an estimate of the standard error of $\log \hat{\theta}_{OR}$

$$\text{se}(\log \hat{\theta}_{OR}) = \sqrt{\frac{1}{18} + \frac{1}{658} + \frac{1}{28} + \frac{1}{656}} = 0.3071.$$

A 95% confidence interval for θ_{OR} is

$$0.6409 \cdot \exp\{\pm 1.96 \cdot 0.3071\}$$

or (0.3511, 1.17). Notice the 95% confidence interval contains the value 1. While there appears to be a reduction in the odds of MI by taking Aspirin relative to Placebo, the result would not be considered statistically significant at the 5% level.

[2] See, for example, Agresti (2002).

6.2 χ^2 Tests

In Section 6.1, we discussed inference for two Bernoulli (binomial) random variables; i.e., discrete random variables with a range consisting of two categories. In this section, we extend this discussion to discrete random variables whose range consists of a general number of categories. Generally, the χ^2-test statistic is essentially the sum over the categories of the squared and standardized differences between the observed and expected frequencies, where the expected frequencies are formulated under the assumption that the null hypothesis is true. In general, under the null hypothesis, this test statistic has an asymptotic χ^2-distribution with degrees of freedom equal to the number of free categories (cells) minus the number of parameters, if any, that need to be estimated to form the expected frequencies. As we note later, at times the exact null distribution can be used instead of the asymptotic distribution. For now, we present three general applications and their computation using R.

6.2.1 Goodness-of-Fit Tests for a Single Discrete Random Variable

Consider a discrete random variable X with range (categories) $\{1, 2, \ldots, c\}$. Let $p(j) = P[X = j]$ denote the probability mass function (pmf) of the distribution of X. Suppose the hypotheses of interest are:

$$H_0 : p(j) = p_0(j), j = 1, \ldots, c \text{ versus } H_A : p(j) \neq p_0(j), \text{ for some } j. \quad (6.7)$$

Suppose $X_1, \ldots, X_n$ is a random sample on X. Let $O_j = \#\{X_i = j\}$. The statistics $O_1, \ldots O_c$ are called the observed frequencies of the categories of X. The observed frequencies are constrained as $\sum_{j=1}^{c} O_j = n$; so, there are essentially $c - 1$ free cells. The expected frequencies of these categories under H_0 are given by $E_j = E_{H_0}[O_j]$, where E_{H_0} denotes expectation under the null hypothesis. There are two cases.

In the first case, the null distribution probabilities, $p_0(j)$, are completely specified. In this case, $E_j = np_0(j)$ and the test statistic is given by

$$\chi^2 = \sum_{j=1}^{c} \frac{(O_j - E_j)^2}{E_j}. \quad (6.8)$$

The hypothesis H_0 is rejected in favor of H_A for large values of χ^2. Note that the vector of observed frequencies, $(O_1, \ldots, O_c)^T$ has a multinomial distribution, so the exact distribution of χ^2 can be obtained. It is also asymptotically equivalent to the likelihood ratio test statistic,[3] and, hence, has an asymptotically χ^2-distribution with $c-1$ degrees of freedom under H_0. In practice, this asymptotic result is generally used. Let χ_0^2 be the realized value of the statistic

[3]See, for example, Exercise 6.5.11 of Hogg et al. (2019).

χ^2 when the sample is drawn. Then the p-value of the goodness-of-fit test is $1 - F_{\chi^2}(\chi_0^2; c-1)$, where $F_{\chi^2}(\cdot; c-1)$ denotes the cdf of a χ^2-distribution with $c-1$ degrees of freedom.

The R function `chisq.test` computes the test statistic (6.8). The input consists of the vectors of observed frequencies and the pmf $(p_0(1), \ldots, p_0(c))^T$. The uniform distribution ($p(j) \equiv 1/c$) is the default null distribution. The output includes the value of the χ^2-test statistic and the p-value. The return list includes values for the observed frequencies (`observed`), the expected frequencies (`expected`), and the residuals (`residuals`). These residuals are $(O_j - E_j)/\sqrt{E_j}$, $j = 1, \ldots, c$ and are often called the Pearson residuals. The squares of the residuals are the categories' contributions to the test statistic and offer valuable post-test information on which categories had large discrepancies from those expected under H_0.

Here is a simple example. Suppose we roll a die $n = 370$ times and we observe the frequencies $(58, 55, 62, 68, 66, 61)^T$. Suppose we are interested in testing to see if the die is fair; i.e., $p(j) \equiv 1/6$. Computation in R yields

```
> x <- c(58,55,62,68,66,61)
> chifit <- chisq.test(x)
> chifit

        Chi-squared test for given probabilities

data:  x
X-squared = 1.9027, df = 5, p-value = 0.8624

> round(chifit$expected,digits=4)

[1] 61.6667 61.6667 61.6667 61.6667 61.6667 61.6667

> round((chifit$residuals)^2,digits=4)

[1] 0.2180 0.7207 0.0018 0.6505 0.3045 0.0072
```

Thus there is no evidence to support the die being unfair.

In the second case for the goodness-of-fit tests, only the form of the null pmf is known. Unknown parameters must be estimated.[4] Then the expected values are the estimates of E_j based on the estimated pmf. The degrees of freedom, though, decrease by the number of parameters that are estimated.[5] The following example illustrates this case.

Example 6.2.1 (Birth Rate of Males to Swedish Ministers). This data is discussed on page 266 of Daniel (1978). It concerns the number of males in the first seven children for $n = 1334$ Swedish ministers of religion. The data are

[4] For this situation, generally we estimate the unknown parameters of the pmf by their maximum likelihood estimators. See Hogg et al. (2019).

[5] See Section 4.7 of Hogg et al. (2019).

No. of Males	0	1	2	3	4	5	6	7
No. of Ministers	6	57	206	362	365	256	69	13

For example, 206 of these ministers had two sons in their first seven children. The null hypothesis is that the number of sons is binomial with probability of success p, where success is a son. The maximum likelihood estimator of p is the number of successes over the total number of trials which is

$$\widehat{p} = \frac{\sum_{j=0}^{7} j \times O_j}{7 \times 1334} = 0.5140.$$

The expected frequencies are computed as

$$E_j = n\binom{7}{j}\widehat{p}^j(1-\widehat{p})^{7-j}.$$

The values of the pmf can be computed in R. The following code segment shows R computations of them along with the corresponding χ^2 test. As we have estimated p, the number of degrees of freedom of the test is $8-1-1=6$.

```
> oc<-c(6,57,206,362,365,256,69,13)
> n<-sum(oc)
> range<-0:7
> phat<-sum(range*oc)/(n*7)
> pmf<-dbinom(range,7,phat)
```

The estimated probability mass function is given in the following code segment.

```
> rbind(range,round(pmf,3))

       [,1]  [,2]  [,3]  [,4]  [,5]  [,6]  [,7]  [,8]
range 0.000 1.000 2.00 3.000 4.00 5.000 6.000 7.000
      0.006 0.047 0.15 0.265 0.28 0.178 0.063 0.009
```

The p-value is calculated using `pchisq` with the correct degrees of freedom (reduced by one due to the estimation of p).

```
> test.result<-chisq.test(oc,p=pmf)
> pchisq(test.result$statistic,df=6,lower.tail=FALSE)

X-squared
0.4257546
```

With a p-value $= 0.426$ we would not reject H_0. There is no evidence to refute a binomial probability model for the number of sons in the first seven children of a Swedish minister. The following provides the expected frequencies which can be compared with the observed.

```
> round(test.result$expected,1)

[1]   8.5  63.2 200.6 353.7 374.1 237.4  83.7  12.6
```

■

Confidence Intervals

In this section, we have been discussing tests for a discrete random variable with a range consisting of c categories, say, $\{1, 2, \ldots, c\}$. Write the distribution of X as $p_j = p(j) = P(X = j)$, $j = 1, 2, \ldots, c$. Using the notation at the beginning of this section, for a given j, the estimate of p_j is the proportion of sample items in category j; i.e., $\hat{p}_j = O_j/n$. Note that this is a binomial situation where category j is success and all other categories are failures. Hence from expression (2.4), an asymptotic $(1-\alpha)100\%$ confidence interval for p_j is

$$\hat{p}_j \pm z_{\alpha/2}\sqrt{\frac{\hat{p}_j(1-\hat{p}_j)}{n}}. \tag{6.9}$$

Another confidence interval of interest in this situation is for a difference in proportions, say, $p_j - p_k$, $j \neq k$. This parameter is the difference in two proportions in a multinomial setting; hence, the standard error[6] of this estimate is

$$\mathrm{SE}(\hat{p}_j - \hat{p}_k) = \sqrt{\frac{\hat{p}_j + \hat{p}_k - (\hat{p}_j - \hat{p}_k)^2}{n}}. \tag{6.10}$$

Thus, an asymptotic $(1-\alpha)100\%$ confidence interval for $p_j - p_k$ is

$$\hat{p}_j - \hat{p}_k \pm z_{\alpha/2}\sqrt{\frac{\hat{p}_j + \hat{p}_k - (\hat{p}_j - \hat{p}_k)^2}{n}}. \tag{6.11}$$

Example 6.2.2 (Birth Rate of Males to Swedish Ministers, continued)**.** Consider Example 6.2.1 concerning the number of sons in the first seven children of Swedish ministers. Suppose we are interested in the difference in the probabilities of all females or all sons. The following R segment estimates this difference along with a 95% confidence interval, (6.11), for it. The counts for these categories are, respectively, 6 and 13 with $n = 1334$.

```
> n <- 1334; p0 <- 6/n; p7 <- 13/n
> se <- sqrt((p0+p7-(p0-p7)^2)/n)
> lb <- p0-p7 - 1.96*se; ub <- p0-p7 + 1.96*se
> res<- c(p0-p7,lb,ub)
> res

[1] -0.005247376 -0.011645562  0.001150809
```

[6] See Section 6.5 of Hogg et al. (2019).

Since 0 is in the confidence interval, there is no discernible difference in the proportions at level 0.05. ■

A cautionary note is needed here. In general, many confidence intervals can be formulated for a given situation. For example, if there are c categories, then there are $\binom{c}{2}$ possible pairwise comparison confidence intervals. In such cases, the overall confidence may slip. This is a multiple comparison problem (MCP). There are several procedures to use, such, as the Bonferroni. With the Bonferroni procedure, suppose there are m confidence intervals of interest. Then if each confidence interval is obtained with confidence coefficient $(1 - (\alpha/m))$, the simultaneous confidence of all of the intervals is at least $1 - \alpha$. See Exercise 2.11.24.

6.2.2 Several Discrete Random Variables

A frequent application of goodness-of-fit tests concerns several discrete random variables, say $X_1, \ldots, X_r$, which have the same range $\{1, 2, \ldots, c\}$. The hypotheses of interest are

$$\begin{array}{ll} H_0: & X_1, \ldots, X_r \text{ have the same distribution} \\ H_A: & \text{Distributions of } X_i \text{ and } X_j \text{ differ for some } i \neq j. \end{array} \quad (6.12)$$

Note that the null hypothesis does not specify the common distribution. Information consists of independent random samples on each random variable. Suppose the random sample on X_i is of size n_i. Let $n = \sum_{i=1}^{r} n_i$ denote the total sample size. The observed frequencies are

$$O_{ij} = \#\{\text{sample items in sample drawn on } X_i \text{ such that } X_i = j\},$$

for $i = 1, \ldots, r$ and $j = 1, \ldots, c$. The set of $\{O_{ij}\}$s form a $r \times c$ matrix of observed frequencies. These matrices are often referred to as **contingency tables**. We want to compare these observed frequencies to the expected frequencies under H_0. To obtain these, we need to estimate the common distribution $(p_1, \ldots, p_c)^T$, where p_j is the probability that category j occurs. The nonparametric estimate of p_j is

$$\hat{p}_j = \frac{\sum_{i=1}^{r} O_{ij}}{n}, \quad j = 1, \ldots, c.$$

Hence, the estimated expected frequencies are $\hat{E}_{ij} = n_i \hat{p}_j$. This formula is easy to remember since it is the row total times the column total over the total number. The test statistic is the χ^2-test statistic, (6.8); that is,

$$\chi^2 = \sum_{i=1}^{r} \sum_{j=1}^{c} \frac{(O_{ij} - \hat{E}_{ij})^2}{\hat{E}_{ij}}. \quad (6.13)$$

For degrees of freedom, note that each row has $c - 1$ free cells because the sample sizes n_i are known. Further $c - 1$ estimates had to be made. So there

TABLE 6.3
Contingency Table for Type of Crime and Alcoholic Status Data.

Crime	Alcoholic	Nonalcoholic
Arson	50	43
Rape	88	62
Violence	155	110
Theft	379	300
Coining	18	14
Fraud	63	144

are $r(c-1)-(c-1) = (r-1)(c-1)$ degrees of freedom. Thus, an asymptotic level α test is to reject H_0 if $\chi^2 \geq \chi^2_{\alpha,(r-1)(c-1)}$, where $\chi^2_{\alpha,(r-1)(c-1)}$ is the α critical value of a χ^2-distribution with $(r-1)(c-1)$ degrees of freedom. This test is often referred to as the χ^2-test of homogeneity (same distributions). We illustrate it with the following example.

Example 6.2.3 (Type of Crime and Alcoholic Status). The contingency table, Table 6.3, contains the frequencies of criminals who committed certain crimes and whether or not they are alcoholics. We are interested in seeing whether or not the distribution of alcoholic status is the same for each type of crime. The data were obtained from Kendall and Stuart (1979).

To compute the test for homogeneity for this data in R, assume the contingency table is in the matrix `ct`. Then the command is `chisq.test(ct)`, as the following R session shows:

```
> c1 <- c(50,88,155,379,18,63)
> c2 <- c(43,62,110,300,14,144)
> ct <- cbind(c1,c2)
> chifit <- chisq.test(ct)
> chifit

        Pearson's Chi-squared test

data:  ct
X-squared = 49.7306, df = 5, p-value = 1.573e-09

> (chifit$residuals)^2

             c1          c2
[1,]  0.01617684  0.01809979
[2,]  0.97600214  1.09202023
[3,]  1.62222220  1.81505693
[4,]  1.16680759  1.30550686
[5,]  0.07191850  0.08046750
[6,] 19.61720859 21.94912045
```

The result is highly significant, but note that most of the contribution to the test statistic comes from the crime fraud. Next, we eliminate fraud and retest.

```
> ct2 <- ct[-6,]
> chisq.test(ct2)

        Pearson's Chi-squared test

data:  ct2
X-squared = 1.1219, df = 4, p-value = 0.8908
```

These results suggest that conditional on the criminal not committing fraud, there is no association between alcoholic status and type of crime. ■

Confidence Intervals

For a given category, say, j, it may be of interest to obtain confidence intervals for differences such as $P(X_i = j) - P(X_{i'} = j)$. In the notation of this section, the estimate of this difference is $(O_{ij}/n_i) - (O_{i'j}/n_{i'})$, where n_i and $n_{i'}$ are the respective sums of rows i and i' of the contingency table. Since the samples on these random variables are independent, the two-sample proportion confidence interval given in expression (6.2) can be used. The cautionary note regarding simultaneous confidence of the last section holds here, also.

6.2.3 Independence of Two Discrete Random Variables

The χ^2 goodness-of-fit test can be used to test the independence of two discrete random variables. Suppose X and Y are discrete random variables with respective ranges $\{1, 2, \ldots, r\}$ and $\{1, 2, \ldots, c\}$. Then we can write the hypothesis of independence between X and Y as

$$\begin{aligned} H_0 : P[X = i, Y = j] &= P[X = i]P[Y = j] \text{ for all } i \text{ and } j \text{ versus} \\ H_A : P[x = i, Y = j] &\neq P[X = i]P[Y = j] \text{ for some } i \text{ and } j. \end{aligned} \quad (6.14)$$

To test this hypothesis, suppose we have the observed the random sample $(X_1, Y_1), \ldots, (X_n, Y_n)$ on (X, Y). We categorize these data into the $r \times c$ contingency table with frequencies O_{ij} where

$$O_{ij} = \#_{1 \le l \le n}\{(X_l, Y_l) = (i, j)\}.$$

So the O_{ij}s are our observed frequencies, and there are initially $rc - 1$ free cells. The expected frequencies are formulated under H_0. Note that the maximum likelihood estimates (mles) of the marginal distributions of $P[X = i]$ and $P[Y = j]$ are the respective statistics $O_{i\cdot}/n$ and $O_{\cdot j}/n$. Hence, under H_0, the mle of $P[X = i, Y = j]$ is the product of these marginal distributions. So the expected frequencies are

$$\hat{E}_{ij} = n\frac{O_{i\cdot}}{n}\frac{O_{\cdot j}}{n} = \frac{i\text{th row total} \times j\text{th col. total}}{\text{total number}}, \quad (6.15)$$

which is the same formula as for the expected frequencies in the test for homogeneity. Further, the degrees of freedom are also the same. To see this, there are $rc-1$ free cells, and to formulate the expected frequencies we had to estimate $r-1+c-1$ parameters. Hence, the degrees of freedom are: $rc-1-r-c+2=(r-1)(c-1)$. Thus the R code for the χ^2-test of independence is the same as for the test of homogeneity. Several examples are given in the Exercises.

Confidence Intervals

Notice that the sampling scheme in this section consists of one-sample over $r \times c$ categories. Hence, it is the same scheme as in the beginning of this section, Section 6.2.1. The estimate of each probability $p_{ij} = P[X=i, Y=j]$ is O_{ij}/n and a confidence interval for p_{ij} is given by expression (6.9). Likewise, confidence intervals for differences of the form $p_{ij} - p_{i'j'}$ can be obtained by using expression (6.11).

6.2.4 McNemar's Test

McNemar's test for significant change is used in many applications. The data are generally placed in a contingency table, but the analysis is not the χ^2-goodness-of-fit tests discussed earlier. A simple example motivates the test. Suppose A and B are two candidates for a political office who are having a debate. Before and after the debate, the preference, A or B, of each member of the audience is recorded. Given a change in preference of candidate, we are interested in the difference in the change from B to A minus the change from A to B. If the estimate of this difference is significantly greater than 0, we might conclude that A won the debate.

For notation, assume we are observing a pair of discrete random variables X and Y. In most applications, the ranges of X and Y have two values, say, $\{0,1\}$.[7] In our simple debate example, the common range can be written as $\{\text{For A}, \text{For B}\}$. Note that there are four categories $(0,0),(0,1),(1,0),(1,1)$. Let p_{ij}, $i,j=0,1$, denote the respective probabilities of these categories. Consider the hypothesis

$$H_0 : p_{01} - p_{10} = 0 \text{ versus } H_A : p_{01} \neq p_{10}. \quad (6.16)$$

One-sided tests are of interest, also; for example, in the debate situation, the alternative $H_A : p_{01} > p_{10}$ expresses the claim that B wins the debate. Let $(X_1,Y_1),\ldots,(X_n,Y_n)$ denote a random sample on (X,Y). Let N_{ij}, $i,j=0,1$, denote the respective frequencies of the categories $(0,0),(0,1),(1,0),(1,1)$. For convenience, the data can be written in the contingency table

[7] See Hettmansperger and McKean (1973) for generalizations to more than two categories.

	0	1
0	N_{00}	N_{01}
1	N_{10}	N_{11}

The estimate of $p_{01} - p_{10}$ is $\hat{p}_{01} - \hat{p}_{10} = (N_{01}/n) - (N_{10}/n)$. This is the difference in two proportions in a multinomial setting; hence, the standard error of this estimate is given in expression (6.11). For convenience, we repeat it with the current notation.

$$\text{SE}(\hat{p}_{01} - \hat{p}_{10}) = \sqrt{\frac{\hat{p}_{01} + \hat{p}_{10} - (\hat{p}_{01} - \hat{p}_{10})^2}{n}}. \tag{6.17}$$

The Wald test statistic is the z-statistic which is the ratio of $\hat{p}_{01} - \hat{p}_{10}$ over its standard error. Usually, though, a scores test is used. In this case, the squared difference in the numerator of the standard error is replaced by 0, its parametric value under the null hypothesis. Then the square of the z-scores test statistic reduces to

$$\chi^2 = \frac{(N_{01} - N_{10})^2}{N_{01} + N_{10}}. \tag{6.18}$$

Under H_0, this test statistic has an asymptotic χ^2-distribution with 1 degree of freedom. Letting χ_0^2 be the realized values of the test statistic once the sample is drawn, the p-value of this test is $1 - F_{\chi^2}(\chi_0^2; 1)$. For a one-sided test, simply divide this p-value by 2.

Actually, an exact test is easily formulated. Note that this test is conditioned on the categories $(0, 1)$ and $(1, 0)$. Furthermore, the null hypothesis says that these two categories are equilikely. Hence under the null hypothesis, the statistic N_{01} has a binomial distribution with probability of success $1/2$ and $N_{01} + N_{10}$ trials. So the exact p-value can be determined from this binomial distribution. While either the exact or the asymptotic p-value is easily calculated by R, we recommend the exact p-value.

Example 6.2.4 (Hodgkin's Disease and Tonsillectomy)**.** Hollander and Wolfe (1999) report on a study concerning Hodgkin's disease and tonsillectomy. A theory purports that tonsils offer protection against Hodgkin's disease. The data in the study consist of 85 paired observations of siblings. For each pair, one member of the pair has Hodgkin's disease and the other does not. Whether or not each had a tonsillectomy was also reported. The data are:

		Sibling	
		Tonsillectomy (0)	No Tonsillectomy (1)
Hodgkin's Patients	Tonsillectomy (0)	26	15
	No Tonsillectomy (1)	7	37

If the medical theory is correct, then $p_{01} > p_{10}$. So we are interested in a one-sided test. The following R calculations show how easily the test statistic and p-value (including the exact) are calculated:

```
> teststat <- (15-7)^2/(15+7)
> pvalue <- (1 - pchisq(teststat,1))/2
> pexact <- 1 - pbinom(14,(15+7),.5)
> c(teststat,pvalue,pexact)

[1] 2.90909091 0.04404076 0.06690025
```

If the level of significance is set at $\alpha = 0.05$, then different conclusions may be drawn depending on whether or not the exact p-value is used. ∎

Remark 6.2.1. In practice, the p-values for the χ^2-tests discussed in this section are often the asymptotic p-values based on the χ^2-distribution. For McNemar's test, we have the option of an exact p-value based on a binomial distribution. There are other situations where an exact p-value is an option. One such case concerns contingency tables where both margins are fixed. For such cases, **Fisher's exact test** can be used; see, for example, Chapter 2 of Agresti (1996) for discussion. The R function for the analysis is `fisher.test`. One nonparametric example of this test concerns Mood's two-sample median test; see the discussion above Example 3.3.1. In this case, Fisher's exact test is based on a hypergeometric distribution. ∎

6.3 Logistic Regression

Logistic regression is used to model a binary response versus one or more explanatory variables. For example, the response could be the occurrence of the event in an outcomes trial (recovery from some condition or disease), and explanatory variables might consist of treatment indicators and demographic variables (covariates). We present a short introduction to logistic regression and suggest readers interested in a more in-depth treatment refer to texts such as Agresti (2002). The base R function `glm`[8] can be used to fit logistic regression models.

Generally, we model a binary response versus one or more explanatory variables as follows. Let y denote the response variable so that y takes on values of $\{0, 1\}$. At times we use the term *outcome* to indicate the event $y = 1$, in which case we say the outcome has occurred. We begin with the simple case of one explanatory variable and discuss the general case later in this section. Assume a numeric covariate x is collected and the probability of the outcome is $p(x)$; i.e., the random variable Y has the following pmdf

$$Y = \begin{cases} 1 \text{ w.p. } p(x) \\ 0 \text{ w.p. } 1 - p(x). \end{cases}$$

[8]Also common exponential family models of *generalized linear models* type.

Logistic regression models the log odds as a linear model:

$$\log\left(\frac{p(x)}{1-p(x)}\right) = \alpha + \beta x. \tag{6.19}$$

The left-hand side of (6.19) is commonly referred to as the **logit**. Solving for $p(x)$ gives the probability of the outcome for a given value of x:

$$p(x) = \frac{\exp\{\alpha + \beta x\}}{1 + \exp\{\alpha + \beta x\}}.$$

The function $p(x)$ is plotted in Figure 6.1 for the case of $\beta > 0$. We can see that there is an increasing relationship between x and the probability of the outcome. Note the function is monotonic. In general, with $\beta > 0$ the function

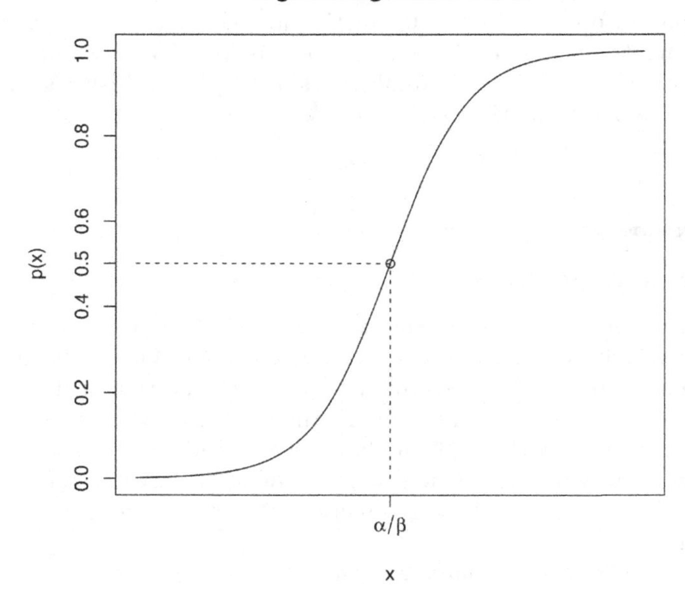

FIGURE 6.1
Logistic regression curve with $\beta > 0$. The rate of change of the curve is a function of β; i.e., larger values of β have a steeper tangent line. The value of α corresponds to $p(\alpha/\beta) = 0.5$.

is increasing, with $\beta < 0$ the function is decreasing, and with $\beta = 0$ the explanatory variable x has no predictive power; i.e., is the same as a fair coin flip.

Example 6.3.1. Consider the case where x is an indicator for active treatment versus control in an RCT; i.e., $x = 1$ for active (A) treatment and $x = 0$ for placebo (P). So that the odds ratio of the event under treatment relative to control is

$$\frac{O(x+1)}{O(x)} = \frac{O(\mathrm{A})}{O(\mathrm{P})} = \exp\{\beta\}.$$

■

Example 6.3.2 (1982 Milwaukee Brewers). Brief game summaries for the 1982 Milwaukee Brewers season are provided in the dataset `brewers1982`. The first few rows of the dataset are provided in the following code segment.

```
> head(brewers1982)

               Date Opp  R RA Time Attendance  home   win
1      Friday Apr 9 TOR 15  4 3:05      30216 FALSE  TRUE
2  Saturday Apr 10 TOR  2  3 2:35      11141 FALSE FALSE
3    Sunday Apr 11 TOR 14  5 2:37      10128 FALSE  TRUE
4   Tuesday Apr 13 CLE  9  8 3:27       6258 FALSE  TRUE
5 Wednesday Apr 14 CLE  2  6 2:50       7017 FALSE FALSE
6  Thursday Apr 15 CLE  1  8 2:32       9072 FALSE FALSE
```

As can be seen from the `home` column, the Brewers started the season on the road with series in Toronto (`TOR`) and Cleveland (`CLE`). The team won 3 out of their first 6 games.

Our goal is to explain the relationship between runs scored by the Brewers and the probability of a win. First, we look at the frequency distribution of runs in the next code segment.

```
> table(brewers1982$R)

 0  1  2  3  4  5  6  7  8  9 10 11 12 13 14 15
 1 10 28 21 18 13 12 16 13 10  7  6  1  1  4  2
```

In the next code segment, we find the proportion of times that the Brewers won for each value of `R`.

```
> pwin <- with(brewers1982,tapply(win,R,mean))
> pwin

    0     1     2     3     4     5     6     7     8     9
0.000 0.000 0.179 0.333 0.500 0.692 0.750 0.875 0.846 1.000
   10    11    12    13    14    15
1.000 1.000 1.000 1.000 1.000 1.000
```

Of course the team did not win the time that they were shutout (R=0), nor did they win any of the 10 times they scored only 1 run. As expected, the percentage of wins generally increases as the number of runs increases; see Figure 6.2.

FIGURE 6.2
Scatterplot of proportion of wins versus runs scores for the 1982 Milwaukee Brewers.

A logistic regression can be fit with the function `glm` by specifying `family=binomial` as illustrated in the next code segment.

```
> fit <- glm(win ~ R, data=brewers1982, family=binomial)
> coef(fit)

(Intercept)            R
   -2.96223      0.70606
```

So the fitted model is

$$\text{logit}(p(R)) = -2.962 + 0.706 * R.$$

Given an estimated β of 0.706, we see that the odds of a win approximately doubles for every additional run scored, $\exp\{\hat{\beta}\} = \exp\{0.706\} = 2.026$.

Fitting more than one explanatory variable is straightforward. For example, to fit a model with runs and an indicator for when the game was played at home, one could use the following code.

```
> fit2 <- glm(win~R+home,data=brewers1982,family=binomial)
> coef(fit2)
```

```
(Intercept)           R    homeTRUE
   -3.26739     0.72220     0.48172
```

Notice the coefficient for runs has changed slightly. The interpretation of the coefficient for home is, for a given number of runs (holding other variables constant), the team's odds of winning increased by $\exp\{0.482\} = 1.619$ when playing at home versus on the road. ■

6.4 Trees for Classification

We gave an overview of classification in Section 3.10. In this section, we expand a bit on that discussion.

In this section we introduce trees for classification: basic classification trees, as well as random forests. The basic tree produces a split on a single explanatory variable which creates the maximum similarity of the outcomes in two regions. The resulting regions of the tree are called nodes and the goal is to produce nodes with maximum purity. Bagging or bootstrap aggregating is the process of building many trees based on bootstrap samples with the resulting prediction being based on the majority vote of the bootstrap trees. A random forest attempts to decorrelate the trees, by, in addition to using a bootstrap sample, samples a subset of the variables to be used in each split of the tree. We provide a basic introduction to trees; readers interested in learning more are referred to the texts James et al. (2013), Hastie et al. (2017), as well as Kuhn and Johnson (2013).

In this section we consider a dataset on red wine quality for wines from the north of Portugal (Cortez et al. 2009). The goal is to predict the quality of wine from other measurements. The data may be downloaded from the UCI machine learning data repository[9] (Dua and Graff 2017).

The next code segment reads in the data.

```
> red_wine <- read.csv('data/winequality-red.csv',sep=';')
```

The dimension of the dataset is given in the following.

```
> dim(red_wine)

[1] 1599   12
```

We can see from the following code segment that most of the wines have a quality of 5 or 6.

```
> table(red_wine$quality)
```

[9] http://archive.ics.uci.edu/ml

```
  3   4   5   6   7   8
 10  53 681 638 199  18
```

For illustration, we create a new outcome variable based on if the quality was >= 6, representing higher quality wine.

```
> my_red_wine <- red_wine
> my_red_wine$quality <- as.factor(red_wine$quality >= 6)
```

We will first fit a classification tree with the variables `alcohol` and `sulphates`. The data are plotted in a scatterplot in Figure 6.3.

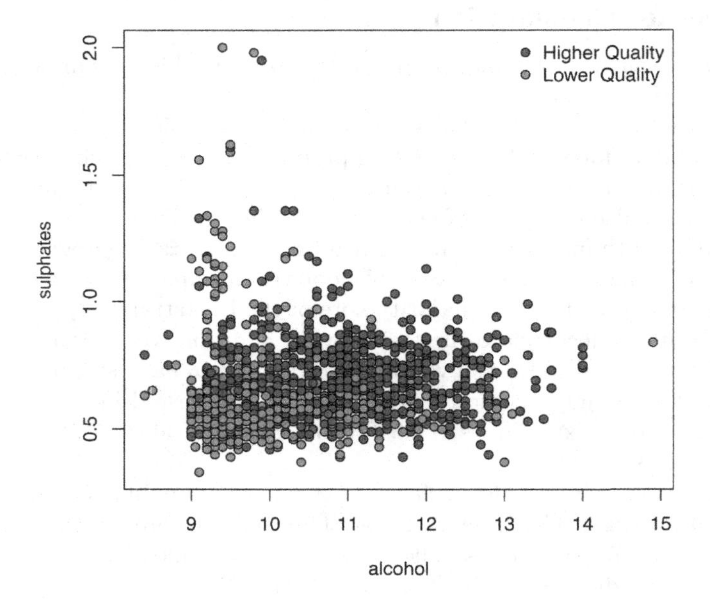

FIGURE 6.3
Scatterplot of the variables `alcohol` and `sulphates`.

We will use the `tree` package (Ripley 2023) to fit the basic classification tree. Note: the outcome variable for `tree` must be a factor or the function will attempt to fit a regression tree.[10]

A classification tree is fit via recursive binary partitioning with the goal of producing splits (of the design space) with the highest degree of *node purity*. That is, at each step, partitions or splits on each of the predictors are considered with the chosen split resulting in the maximum node similarity or minimum impurity. There are several measures of node impurity; `tree`

[10]Regression trees are similar in concept to classification trees in the case of a continuous outcome variable. Regression trees are discussed in Chapter 8.

implements deviance and Gini index, with deviance being the default. In the case of 2 classes, deviance is given as

$$-p \log p - (1-p)\log(1-p)$$

and Gini index is given as

$$2p(1-p).$$

As illustrated in Figure 6.4, the first split is based on the variable `alcohol` at the value 10.525, so that a split on alcohol produced higher purity than a split on sulphates. Wines with alcohol > 10.525 tend to have a higher quality rating, and those with alcohol < 10.525 tend to have lower quality rating. The tree, displayed in the right panel, which is grown from the root down, provides the predictions at the nodes. In this case, TRUE or FALSE, representing if the wine is higher quality or not.

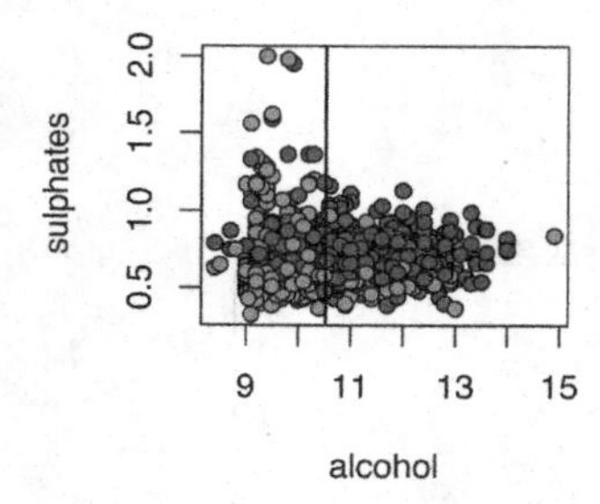

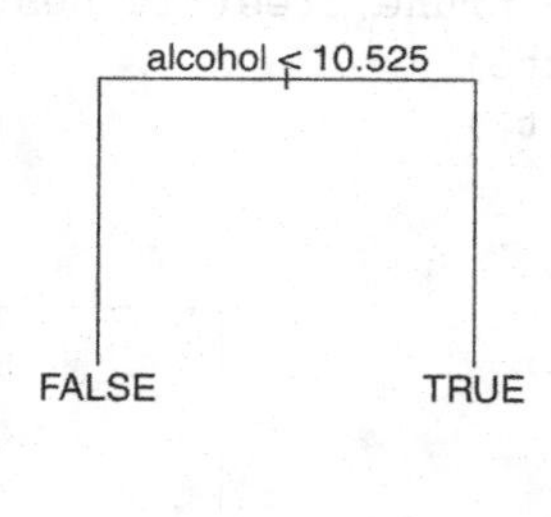

FIGURE 6.4
Illustration of the first binary split. *Left*: scatterplot with binary split represented by a vertical line. *Right*: fitted tree after first split.

The second binary partition then may split either of the two regions on either of the predictor variables. The process is repeated a prespecified number of times or until no more splits are possible. Trees have a tendency to overfit, and so pruning is performed to remove nodes which, after initially fitting the entire tree, do not end up increasing node purity.

We next train the classifier using cross-validation using misclassification to prune the tree. The function `cv.tree` in the package `class` (Ripley 2023) can be used to determine optimal model complexity. A tree with 7 nodes is displayed in Figure 6.5.

```
> set.seed(42)
> library(tree)
> fit <- tree(quality~.,data=my_red_wine)
> cvfit <- cv.tree(fit,FUN=prune.misclass)
> cvfit

$size
```

```
[1] 10  9  7  5  2  1

$dev
[1] 450 450 450 450 504 744

$k
[1]    -Inf   0.000   2.500   5.000  24.667 249.000

$method
[1] "misclass"

attr(,"class")
[1] "prune"         "tree.sequence"

> fit3 <- prune.tree(fit,best=7)
> plot(fit3)
> text(fit3)
```

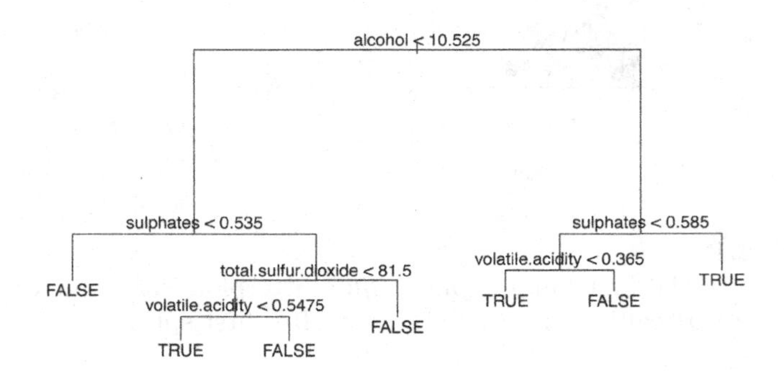

FIGURE 6.5
Fitted classification tree with 7 nodes.

6.4.1 Random Forest

A random forest builds many trees based on bootstrap samples. In determining each split of the tree, a subset of the variables is chosen to base the split on; this results in trees that are less correlated.[11] As a rule-of-thumb, in practice

[11]Bagged (bootstrap aggregated) trees are similar to random forests, except all of the variables are included as candidates at each split rather than a subset. In most cases, the decorrelating of the trees by selecting a subset of the variables in the case of a random forest fit results in a better fit. Thus random forests are recommended over bagged trees for most cases in practice.

the number of explanatory variables chosen at random for each split is

$$\lfloor\sqrt{p}\rfloor$$

where p is the number of explanatory variables. As we illustrate in the following code segment the package `randomForest` (Liaw and Wiener 2002) is used to develop a classifier by random forest.

We continue with the red wine example of the last subsection. The following code obtains a `randomForest` fit with the default options; namely 500 trees are fit based on 500 bootstrap samples of the data and, since the outcome variable is categorical (i.e. a `factor`) the number of variables sampled for each split is 3 (`floor(sqrt(ncol(x)))`).

Plots based on the fit are displayed in Figure 6.6.

```
> library(randomForest)
> fit <- randomForest(quality~.,data=my_red_wine)
> par(mfrow=c(1,2))
> plot(fit,col=c('black','grey70','grey30'),lty=c(1,2,2))
> legend('topright',legend=colnames(fit$err.rate),
+   col=c('black','grey70','grey30'),lty=c(1,2,2),bty='n')
> varImpPlot(fit)
```

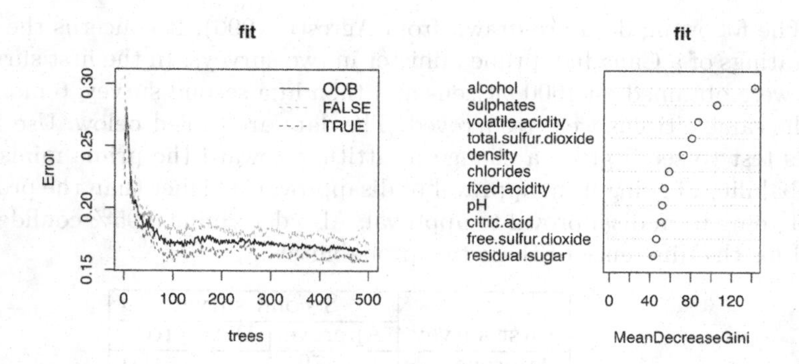

FIGURE 6.6
`randomForest` plots. Left : plot the error rates. Right : dotchart of variable importance.

The plot on the left of Figure 6.6 displays the error rate versus number of trees based on the out-of-bag (OOB) samples; i.e., those that were not in the bootstrap sample for a particular iteration. FALSE and TRUE represent the errors for those two particular labels while OOB is the overall error. The plot on the right of Figure 6.6 displays the ordered variable importance. The definition of the variable importance measure used in `varImpPlot` is the total decrease in node impurities, as measured by the Gini index, from splitting on the particular variable, averaged over all trees.[12]. From the plot we can see that, by far, `alcohol` is the most important factor in predicting if a wine is considered *high quality*.

6.5 Exercises

6.5.1. Rasmussen (1992) presents the following data on a survey of workers in a large factory on two variables: their feelings concerning a smoking ban (Approve, Do not approve, Not sure) and Smoking status (Never smoked, Ex-smoker, Current smoker). Use the χ^2-test to test the independence of these two variables. Using a post-test analysis, determine what categories contributed heavily to the dependence.

	Approval of the smoking ban		
Smoking status	Approve	Do not approve	Not sure
Never smoked	237	3	10
Ex-smoker	106	4	7
Current smoker	24	32	11

6.5.2. The following data are drawn from Agresti (1996). It concerns the approval ratings of a Canadian prime minister in two surveys. In the first survey, ratings were obtained on 1600 citizens and then in a second survey, 6 months later, the same citizens were resurveyed. The data are tabled below. Use McNemar's test to see if given a change in attitude toward the prime minister, the probability of going from approval to disapproval is higher than the probability of going from disapproval to approval. Also determine a 95% confidence interval for the difference of these two probabilities.

	Second survey	
First survey	Approve	Disapprove
Approve	794	150
Disapprove	86	570

6.5.3. Even though the χ^2-tests of homogeneity and independence are the same, they are based on different sampling schemes. The scheme for the test of independence is one-sample of bivariate data, while the scheme for the test of homogeneity consists of one-sample from each population. Let C be a contingency table with r rows and c columns. Assume for the test of homogeneity that the rows contain the samples from the r populations. Determine the (large sample) confidence intervals for each of the following parameters under both

[12] c.f., `help(importance,randomForest)`

schemes, where p_{ij} is the probability of cell (i, j) occurring. Write R code to obtain these confidence intervals assuming the input is a contingency table.

1. p_{11}.
2. $p_{11} - p_{12}$.

6.5.4. Mendel's early work on heredity in peas is well known. Briefly, he conducted experiments and the peas could be either round or wrinkled; yellow or green. So there are four possible combinations: RY, RG, WY, WG. If his theory is correct, the peas would be observed in a 9:3:3:1 ratio. Suppose the outcome of the experiment yielded the following observed data

RY	RG	WY	WG
315	108	101	32

Calculate a p-value and comment on the results.

6.5.5. Guimarães et al. (2021) describe an RCT in which adults hospitalized with COVID-19 pneumonia were randomized 1:1 to recieve Tofacitinib or Placebo. The primary outcome for the study was death or respiratory failure through Day 28. Results for the primary outcome are provided in the following table.

	E	E^c
T	26	118
P	42	103

a. Calculate a point estimate and confidence interval for the risk ratio of death or respiratory failure through Day 28.

b. Calculate a point estimate and confidence interval for the odds ratio of death or respiratory failure through Day 28.

6.5.6. Consider the `npsm` dataset `sim_class_4500` which contains 6 predictor variables and 1 binary class variable. Use the random forest method to train a classifier. Obtain a dotplot of variable importance. Comment.

6.5.7. In Section 6.4 we presented a plot of the error rates where we overrode the defaults for color (`col`) and line type (`lty`). Recreate the plot using the default colors and line types of `plot.randomForest`. Include a legend.

6.5.8. Produce graphics similar to the one in Figure 6.1. For each of the following sets of parameters, use a range for the x-axis of `xlim=c(-10,10)` and `ylim=c(0,1)` for the y-axis.

a. Using $\alpha = 0$, $\beta = 1$.

b. Using $\alpha = 1$, $\beta = 1$.

c. Using $\alpha = 0$, $\beta = 2$.

d. Using $\alpha = -1$, $\beta = -1$.

e. Using $\alpha = 0$, $\beta = 0$.

f. Using $\alpha = -1$, $\beta = -2$.

6.5.9. Create a graphic similar to Figure 6.2 using separate plotting symbols for home games and road games. Include a legend.

7

Linear Models

This chapter on linear models serves as background for the later chapters in the text which rely more heavily on this material.

7.1 Linear Models

In this section we provide a brief overview of rank-based methods for linear models. Our presentation is by no means comprehensive and is included simply as a convenient reference. We refer the reader interested in a thorough treatment to Chapters 3–5 of Hettmansperger and McKean (2011). This section uses a matrix formulation of the linear model; readers interested in application can skip this and the next section.

7.1.1 Estimation

In this section we discuss rank-based (R) estimation for linear regression models. As is the case with most of the modeling we discuss in this book, the geometry is similar to least squares. Throughout, we are interested in estimation and inference on the slope parameters in the following linear model

$$Y_i = \alpha + \beta_1 x_{i1} + \ldots + \beta_p x_{ip} + e_i \text{ for } i = 1, \ldots, n. \tag{7.1}$$

For convenience, we rewrite (7.1) as

$$Y_i = \alpha + \boldsymbol{x}_i^T \boldsymbol{\beta} + e_i \text{ for } i = 1, \ldots, n, \tag{7.2}$$

where Y_i is a continuous response variable, $\boldsymbol{x}_i$ is the vector of explanatory variables, α is the intercept parameter, $\boldsymbol{\beta}$ is the vector of regression coefficients, and e_i is the error term. For formal inference, the errors are assumed to be iid with continuous pdf $f(t)$ and finite Fisher information. Additionally, there are design assumptions that are the same as those for the least squares analysis.

Rewrite (7.2) in matrix notation as follows

$$\boldsymbol{Y} = \alpha \boldsymbol{1} + \boldsymbol{X}\boldsymbol{\beta} + \boldsymbol{e} \tag{7.3}$$

where $\boldsymbol{Y} = [Y_1, \ldots, Y_n]^T$ is a $n \times 1$ vector of response variable, $\boldsymbol{X} = [\boldsymbol{x}_1, \ldots, \boldsymbol{x}_n]^T$ is an $n \times p$ design matrix, and $\boldsymbol{e} = [e_1, \ldots, e_n]^T$ is an $n \times 1$ vector of error terms. Recall that the least squares estimator is the minimizer of Euclidean distance between $\boldsymbol{Y}$ and $\hat{\boldsymbol{Y}}_{LS} = \boldsymbol{X}\hat{\boldsymbol{\beta}}_{LS}$. To obtain the R estimator, we use a different measure of distance, Jaeckel's (1972) dispersion function, which is given by:

$$D(\boldsymbol{\beta}) = \|\boldsymbol{Y} - \boldsymbol{X}\boldsymbol{\beta}\|_{\varphi}, \tag{7.4}$$

where $\|\cdot\|_{\varphi}$ is a pseudo-norm defined as

$$\|\boldsymbol{u}\|_{\varphi} = \sum_{i=1}^{n} a(R(u_i))u_i, \tag{7.5}$$

the scores are generated as $a(i) = \varphi\left(\frac{i}{n+1}\right)$, and φ is a nondecreasing **score** function defined on the interval $(0, 1)$. Any of the score functions discussed in the previous chapter for the two-sample location problem can be used in a linear model setting and, therefore, in any of the models used in the remainder of this book. An adaptive procedure for the regression problem is discussed in Section 8.5.

It follows that $D(\boldsymbol{\beta})$, (7.4), is a convex function of $\boldsymbol{\beta}$ and provides a robust measure of distance between $\boldsymbol{Y}$ and $\boldsymbol{X}\boldsymbol{\beta}$. The R estimator of $\boldsymbol{\beta}$ is defined as

$$\hat{\boldsymbol{\beta}}_{\varphi} = \text{Argmin}\|\boldsymbol{Y} - \boldsymbol{X}\boldsymbol{\beta}\|_{\varphi}. \tag{7.6}$$

Note that closed-form solutions exist for least squares; however, this is not the case for rank estimation. The R estimates are obtained by minimizing a convex optimization problem. In `Rfit`, the R function `optim` is used to obtain the estimate of $\boldsymbol{\beta}$.

It can be shown, see for example Hettmansperger and McKean (2011), that the solution to (7.6) is consistent with the asymptotically normal distribution given by

$$\hat{\boldsymbol{\beta}}_{\varphi} \dot{\sim} N\left(\boldsymbol{\beta}, \tau_{\varphi}^2(\boldsymbol{X}^T\boldsymbol{X})^{-1}\right), \tag{7.7}$$

where τ_{φ} is the scale parameter which is defined in expression (3.19). Note that τ_{φ} depends on the pdf $f(t)$ and the score function $\varphi(u)$. In `Rfit`, the Koul, Sievers, and McKean (1987) consistent estimator of τ_{φ} is computed.

The intercept parameter, α, is estimated separately using a rank-based estimate of location based on the residuals $y_i - \boldsymbol{x}_i^T\hat{\beta}_{\varphi}$. Generally, the median is used, which is the default in `Rfit`, and which we denote by $\hat{\alpha}$. It follows that $\hat{\alpha}$ and $\hat{\boldsymbol{\beta}}_{\varphi}$ are jointly asymptotically normal with the variance-covariance matrix

$$\boldsymbol{V}_{\hat{\alpha},\hat{\boldsymbol{\beta}}_{\varphi}} = \begin{bmatrix} \kappa_n & -\tau_{\varphi}^2\overline{\boldsymbol{x}}'(\boldsymbol{X}'\boldsymbol{X})^{-1} \\ -\tau_{\varphi}^2(\boldsymbol{X}'\boldsymbol{X})^{-1}\overline{\boldsymbol{x}} & \tau_{\varphi}^2(\boldsymbol{X}'\boldsymbol{X})^{-1} \end{bmatrix}, \tag{7.8}$$

where $\kappa_n = n^{-1}\tau_S^2 + \tau_\varphi^2 \overline{\boldsymbol{x}}'(\boldsymbol{X}'\boldsymbol{X})^{-1}\overline{\boldsymbol{x}}$. The vector $\overline{\boldsymbol{x}}$ is the vector of column averages of $\boldsymbol{X}$ and τ_S is the scale parameter[1] $1/[2f(0)]$. The consistent estimator of τ_S, discussed in Section 1.5 of Hettmansperger and McKean (2011), is implemented in `Rfit`.

7.1.2 Diagnostics

Regression diagnostics are an essential part of the statistical analysis of any data analysis problem. In this section, we discuss Studentized residuals. Denote the residuals from the full model fit as

$$\hat{e}_i = Y_i - \hat{\alpha} - \boldsymbol{x}_i\hat{\boldsymbol{\beta}}. \tag{7.9}$$

Then the Studentized residuals are defined as

$$\frac{\hat{e}_i}{s(\hat{e}_i)}$$

where $s(\hat{e}_i)$ is the estimated standard error of $\hat{e}_i$ discussed in Chapter 3 of Hettmansperger and McKean (2011). In `Rfit`, the command `rstudent` is used to obtain Studentized residuals from an R fit of a linear model.

7.1.3 Inference

Based on the asymptotic distribution of $\hat{\boldsymbol{\beta}}_\varphi$, (7.7), we present inference for the vector of parameters $\boldsymbol{\beta}$. We discuss Wald type confidence intervals and tests of hypothesis. In addition to these procedures, R-analyses offer the drop in dispersion test which is an analog of the traditional LS test based on the reduction in sums of squares. An estimate of the scale parameter τ_φ is needed for inference and the Koul et al. (1987) estimator is implemented in `Rfit`.

From (7.7), Wald tests and confidence regions/intervals can easily be obtained. Let $\text{se}(\hat{\beta}_j)$ denote the standard error of $\hat{\beta}_j$. That is $\text{se}(\hat{\beta}_j) = \hat{\tau}_\varphi \left(\boldsymbol{X}^T\boldsymbol{X}\right)_{jj}^{-1/2}$ where $\left(\boldsymbol{X}^T\boldsymbol{X}\right)_{jj}^{-1}$ is the jth diagonal element of $\left(\boldsymbol{X}^T\boldsymbol{X}\right)^{-1}$. An approximate $(1-\alpha)*100\%$ confidence interval for β_j is

$$\hat{\beta}_j \pm t_{1-\alpha/2,n-p-1}\text{se}(\hat{\beta}_j).$$

Let bM be a full row rank $q \times p$ matrix. A Wald test of the hypothesis

$$H_0 : \boldsymbol{M\beta} = \boldsymbol{0} \text{ versus } H_A : \boldsymbol{M\beta} \neq \boldsymbol{0}, \tag{7.10}$$

is to reject H_0 if

$$\frac{(\boldsymbol{M}\hat{\boldsymbol{\beta}}_\varphi)^T[\boldsymbol{M}(\boldsymbol{X}^T\boldsymbol{X})^{-1}\boldsymbol{M}^T]^{-1}(\boldsymbol{M}\hat{\boldsymbol{\beta}})/q}{\hat{\tau}_\varphi^2} > F_{1-\alpha,q,n-p-1}.$$

---

[1]See Section 3.5 of Hettmansperger and McKean (2011).

Similar to the reduction in the sums of squares test of classical regression, rank-based regression offers a **drop in dispersion** test. Let Ω_F denote the full model space; i.e., the range (column space) of the design matrix $\boldsymbol{X}$ for the full Model (7.3). Let $D(\text{FULL})$ denote the minimized value of the dispersion function when the full model (7.3) is fit. That is, $D(\text{FULL}) = D(\widehat{\boldsymbol{\beta}}_\varphi)$. Geometrically, $D(\text{FULL})$ is the distance between the response vector $\boldsymbol{Y}$ and the space Ω_F. Let Ω_R denote the reduced model subspace of Ω_F; that is, $\Omega_R = \{\boldsymbol{v} \in \Omega_F : \boldsymbol{v} = \boldsymbol{X\beta} \text{ and } \boldsymbol{M\beta} = \boldsymbol{0}\}$. Let $D(\text{RED})$ denote the minimum value of the dispersion function when the reduced model is fit; i.e., $D(\text{RED})$ is the distance between the response vector $\boldsymbol{Y}$ and the space Ω_R. The reduction in dispersion is $RD = D(RED) - D(FULL)$. The drop in dispersion test is a standardization of the reduction in dispersion which is given by

$$F_\varphi = \frac{\text{RD}/q}{\hat{\tau}/2}. \tag{7.11}$$

An approximate level α test is to reject H_0 if

$$F_\varphi = \frac{\text{RD}/q}{\hat{\tau}/2} > F_{1-\alpha,q,n-p-1}. \tag{7.12}$$

The default ANOVA and ANCOVA `Rfit` analyses described in Chapter 5 use Type III general linear hypotheses (effect being tested is adjusted for all other effects). The Wald type test satisfies this by its formulation. This is true also of the drop in dispersion test defined above; i.e., the reduction in dispersion is between the reduced and full models. Of course, the reduced model design matrix must be computed. This is easily done, however, by using a QR-decomposition of the row space of the hypothesis matrix $\boldsymbol{M}$; see page 210 of Hettmansperger and McKean (2011). For default tests, `Rfit` uses this development to compute a reduced model design matrix for a specified matrix $\boldsymbol{M}$. In general, the `Rfit` function `redmod(xmat,amat)` computes a reduced model design matrix for the full model design matrix `xmat` and the hypothesis matrix `amat`. For traditional LS tests, the corresponding reduction in sums-of-squares is often referred to as a Type III sums-of-squares.

7.1.4 Confidence Interval for a Mean Response

Consider the general linear model (7.3). Let $\boldsymbol{x}_0$ be a specified vector of the independent variables. Although we need not assume finite expectation of the random errors, we do in this section, which allows us to use familiar notation. In practice, a problem of interest is to estimate $\eta_0 = E(Y|\boldsymbol{x}_0)$ for a specified vector of predictors $\boldsymbol{x}_0$. We next consider solutions for this problem based on a rank-based fit of Model (7.3). Denote the rank-based estimates of α and $\boldsymbol{\beta}$ by $\hat{\alpha}$ and $\hat{\boldsymbol{\beta}}$.

The estimator of η_0 is of course

$$\hat{\eta}_0 = \hat{\alpha} + \boldsymbol{x}_0^T\hat{\boldsymbol{\beta}}. \tag{7.13}$$

It follows that $\hat{\eta}_0$ is asymptotically normal with mean η_0 and variance, using expression (7.8),

$$v_0 = \text{Var}(\hat{\eta}_0) = [1\ \boldsymbol{x}_0^T]\hat{\boldsymbol{V}}_{\hat{\alpha},\hat{\boldsymbol{\beta}}_\varphi}\begin{bmatrix} 1 \\ \boldsymbol{x}_0 \end{bmatrix}. \tag{7.14}$$

Hence, an approximate $(1-\alpha)100\%$ confidence interval for η_0 is

$$\hat{\eta}_0 \pm t_{\alpha/2,n-p-1}\sqrt{\hat{v}_0}, \tag{7.15}$$

where $\hat{v}_0$ is the estimate of v_0, (τ_S and τ_φ are replaced, respectively, by $\hat{\tau}_S$ and $\hat{\tau}_\varphi$).

Using `Rfit` this confidence interval is easily computed. Suppose `x0` contain the explanatory variables for which we want to estimate the mean response. The following illustrates how to obtain the estimate and standard error, from which a confidence interval can be computed. Assume the full model fit has been obtained and is in `fit` (e.g., `fit<-rfit(y~X)`).

```
x10 <- c(1,x0)
st.err <- sqrt( t(x10)%*%vcov(fit)%*%x10 )
eta0 <- t(x10)%*%fit$coef
```

We illustrate this code with the following example.

Example 7.1.1. The following responses were collected consecutively over time. For convenience, take the vector of time to be `t <- 1:15`.

t	1	2	3	4	5	6	7	8
y	0.67	0.75	0.74	-5.57	0.76	2.42	0.16	1.52
t	9	10	11	12	13	14	15	
y	2.91	2.74	12.65	5.45	4.81	4.17	3.88	

Interest centered on the model for linear trend, $y = \alpha + \beta t + e$, and in estimating the expected value of the response for the next time period $t = 16$. Using Wilcoxon scores and the above code, it is easy to show that predicted value at time 16 is 5.43 with the 95% confidence interval $(3.38, 7.48)$; see Exercise 7.6.1. Note, in practice this might be considered an extrapolation and consideration must be made as to whether the time trend will continue. ∎

There is a related problem consisting of a predictive interval for Y_0 a new (independent of $\boldsymbol{Y}$) response variable which follows the linear model at $\boldsymbol{x}_0$. Note that Y_0 has mean η_0. Assume finite variance, σ^2, of the random errors. Then $Y_0 - \hat{\eta}$ has mean 0 and (asymptotic) variance $\sigma^2 + v_0$, where v_0 is given in expression (7.14). If in addition we assume that the random errors have a normal distribution, then we could assume (asymptotically) that the difference $Y_0 - \hat{\eta}$ is normally distributed. Based on these results, a predictive interval is easily formed. There are two difficult problems here. One is the assumption of normality and the second is the robust estimation of σ^2. Preliminary Monte

Carlo results show that estimation of σ by MAD of the residuals leads to liberal predictive intervals. Also, we certainly would not recommend using the sample variance of robust residuals. Currently, we are investigating a bootstrap procedure.

7.2 Weighted Regression

For a linear model, weighted regression can be used, for example, when variances are heteroscedastic or for problems that can be solved by iterated reweighted procedures (least squares or rank-based). We begin with a general discussion that includes a simple R function which computes weighted rank-based estimates using `Rfit`.

Consider the linear model with $\boldsymbol{Y}$ as an $n \times 1$ vector of responses and $\boldsymbol{X}$ as the $n \times p$ design matrix. Write the model as

$$\boldsymbol{Y} = \alpha\mathbf{1} + \boldsymbol{X}\boldsymbol{\beta} + \boldsymbol{e} = \boldsymbol{X}_1\boldsymbol{b} + \boldsymbol{e}, \tag{7.16}$$

where $\boldsymbol{X}_1 = [\mathbf{1}\ \boldsymbol{X}]$ and $\boldsymbol{b} = (\alpha, \boldsymbol{\beta}')'$. We have placed the subscript 1 on the matrix $\boldsymbol{X}_1$ because it contains a column of ones (intercept). Let $\boldsymbol{W}$ be an $n \times n$ matrix of weights. For example, if the covariance matrix of $\boldsymbol{e}$ is $\boldsymbol{\Sigma}$, then the weight matrix that yields homogeneous variances is $\boldsymbol{W} = \boldsymbol{\Sigma}^{-1/2}$, where $\boldsymbol{\Sigma}^{-1/2}$ is the usual square root of a positive definite matrix,[2] (of course, in practice $\boldsymbol{W}$ needs to be estimated). The weighted model is:

$$\boldsymbol{W}\boldsymbol{Y} = \boldsymbol{W}\boldsymbol{X}_1\boldsymbol{b} + \boldsymbol{W}\boldsymbol{e}.$$

For easier notation, let $\boldsymbol{Y}^* = \boldsymbol{W}\boldsymbol{Y}$, $\boldsymbol{X}^* = \boldsymbol{W}\boldsymbol{X}_1$, and $\boldsymbol{e}^* = \boldsymbol{W}\boldsymbol{e}$. The matrix $\boldsymbol{X}^*$ does not have a subscript 1 because generally the weights eliminate the intercept. Then we can write the model as

$$\boldsymbol{Y}^* = \boldsymbol{X}^*\boldsymbol{b} + \boldsymbol{e}^*. \tag{7.17}$$

Let Ω^* be the column space of $\boldsymbol{X}^*$. This is our subspace of interest.

As mentioned above, usually in Model (7.17) there is no longer an intercept parameter; i.e., we have a case of **regression through the origin**. This is an explicit assumption for the following discussion; that is, we assume that the column space of $\boldsymbol{X}^*$ does not contain a column of ones. Because the rank-based

[2] See Remark 7.2.1.

estimators minimize a pseudo-norm we have

$$\begin{aligned}\|\boldsymbol{Y}^* - \boldsymbol{X}^*\boldsymbol{\beta}\|_\varphi &= \sum_{i=1}^n a(R(y_i^* - \boldsymbol{x}_i^{*\prime}\boldsymbol{\beta}))(y_i^* - \boldsymbol{x}_i^{*\prime}\boldsymbol{\beta}) \\ &= \sum_{i=1}^n a(R(y_i^* - (\boldsymbol{x}_i^* - \overline{\boldsymbol{x}}^*)'\boldsymbol{\beta}))(y_i^* - (\boldsymbol{x}_i^* - \overline{\boldsymbol{x}}^*)'\boldsymbol{\beta}) \\ &= \sum_{i=1}^n a(R(y_i^* - \alpha - (\boldsymbol{x}_i^* - \overline{\boldsymbol{x}}^*)'\boldsymbol{\beta}))(y_i^* - \alpha - (\boldsymbol{x}_i^* - \overline{\boldsymbol{x}}^*)'\boldsymbol{\beta}),\end{aligned} \quad (7.18)$$

where $\boldsymbol{x}_i^*$ is the ith row of $\boldsymbol{X}^*$ and $\overline{\boldsymbol{x}}^*$ is the vector of column averages of $\boldsymbol{X}^*$. Hence, based on this result, the estimator of the regression coefficients based on the R fit of Model (7.17) is estimating the regression coefficients of the centered model, i.e., the model with the design matrix $\boldsymbol{X}_c = \boldsymbol{X}^* - \overline{\boldsymbol{X}^*}$. Thus the rank-based fit of the vector of regression coefficients is in the column space of $\boldsymbol{X}_c$, not the column space of $\boldsymbol{X}^*$.

Dixon and McKean (1996) proposed the following[3] solution. Assume that (7.17) is the true model, but first obtain the rank-based fit of the model:

$$\boldsymbol{Y}^* = \mathbf{1}\alpha_1 + \boldsymbol{X}^*\boldsymbol{b} + \boldsymbol{e}^* = [\mathbf{1}\ \boldsymbol{X}^*]\begin{bmatrix}\alpha_1 \\ \boldsymbol{b}\end{bmatrix} + \boldsymbol{e}^*, \quad (7.19)$$

where the true α_1 is 0. Let $\boldsymbol{U}_1 = [\mathbf{1}\ \boldsymbol{X}^*]$ and let Ω_{U_1} denote the column space of $\boldsymbol{U}_1$. Let $\widehat{\boldsymbol{Y}}_{U_1} = \mathbf{1}\widehat{\alpha}_1 + \boldsymbol{X}^*\widehat{\boldsymbol{b}}$ denote the R fitted value based on the fit of Model (7.19), where, as usual, the intercept parameter α_1 is estimated by the median of the residuals of the rank-based fit. Note that the subspace of interest Ω^* is a subspace of Ω_1, i.e., $\Omega^* \subset \Omega_{U_1}$. Secondly, project this fitted value onto the desired space Ω^*; i.e., let $\widehat{\boldsymbol{Y}}^* = \boldsymbol{H}_{\Omega^*}\widehat{\boldsymbol{Y}}_{U_1}$, where $\boldsymbol{H}_{\Omega^*} = \boldsymbol{X}^*(\boldsymbol{X}^{*\prime}\boldsymbol{X}^*)^{-1}\boldsymbol{X}^{*\prime}$. Thirdly, and finally, estimate $\boldsymbol{b}$ by solving the equation

$$\boldsymbol{X}^*\widehat{\boldsymbol{b}} = \widehat{\boldsymbol{Y}}^*, \quad (7.20)$$

that is, $\widehat{\boldsymbol{b}}^* = (\boldsymbol{X}^{*\prime}\boldsymbol{X}^*)^{-1}\boldsymbol{X}^{*\prime}\widehat{\boldsymbol{Y}}_{U_1}$. This is the rank-based estimate for Model (7.20).

The asymptotic variance of $\widehat{\boldsymbol{b}}^*$ is given by

$$\begin{aligned}\text{AsyVar}(\widehat{\boldsymbol{b}}^*) &= \tau_S^2(\boldsymbol{X}^{*\prime}\boldsymbol{X}^*)^{-1}\boldsymbol{X}^{*\prime}\boldsymbol{H}_1\boldsymbol{X}^*(\boldsymbol{X}^{*\prime}\boldsymbol{X}^*)^{-1} \\ &\quad + \tau_\varphi^2(\boldsymbol{X}^{*\prime}\boldsymbol{X}^*)^{-1}\boldsymbol{X}^{*\prime}\boldsymbol{H}_{\boldsymbol{X}_c}\boldsymbol{X}^*(\boldsymbol{X}^{*\prime}\boldsymbol{X}^*)^{-1},\end{aligned} \quad (7.21)$$

where $\boldsymbol{H}_1$ and $\boldsymbol{H}_{\boldsymbol{X}_c}$ are the projection matrices onto a column of ones and the column space of $\boldsymbol{X}_c$, respectively.

We have written an R function which obtains the weighted rank-based fit. The function has the following arguments

[3] See also, Page 287 of Hettmansperger and McKean (2011).

```
> library(npsmReg2)
> args(wtedrb)

function (x, y, wts = diag(rep(1, length(y))), scores = wscores)
NULL
```

where `y` and `x` are the R vector of responses and design matrix, respectively. These responses and design matrix are the non-weighted values as in Model (7.16). The weights are in the matrix `wts` with the default as the identity matrix. The `scores` option allows for different scores to be used with the Wilcoxon scores as the default. Values returned are: fitted values (`$yhatst`), residuals (`$ehatst`), estimate of regression coefficients (`$bstar`), standard errors of estimates (`$se`), and estimated covariance matrix of estimates (`$vc`).

The two main applications for the function `wtedrb` are:

1. Models of the form (7.16) for which a weighted regression is desired. Let $\boldsymbol{W}$ denote the weight matrix. Then the model to be fitted is given in expression (7.17). In this case, the matrix $\boldsymbol{X}^*$ is the design matrix argument for the function `wtedrb` and the matrix $\boldsymbol{W}$ is the weight matrix. It is assumed that a vector of ones is not in the column space of $\boldsymbol{X}^*$.
2. A **regression through the origin** is desired. In this case, the model is
$$\boldsymbol{Y} = \boldsymbol{X}\boldsymbol{\beta} + \boldsymbol{e} = \boldsymbol{U}_1\boldsymbol{b} + \boldsymbol{e}, \tag{7.22}$$
where the column space of $\boldsymbol{X}$ does not contain a vector of ones while that of $\boldsymbol{U}_1$ does. For this model, the design matrix argument is $\boldsymbol{X}$ and there is no weight matrix; i.e., the default identity matrix is used. Some cautionary notes on using this model are discussed in Exercise 7.6.7.

Exercise 7.6.4 discusses a weighted regression. The next example serves as an illustration of regression through the origin.

Example 7.2.1 (Crystal Data). Hettmansperger and McKean (2011) discuss a dataset[4] where regression through the origin was deemed important. The response variable y is the weight of a crystalline form of a certain chemical compound and the independent variable x is the length of time that the crystal was allowed to grow. For convenience, we display the data in Table 7.1.

The following code segment computes the rank-based fit of this model (the responses are in the vector `y` and the vector for the independent variable is in `x`). Note that `x` is the designed matrix used and that there is no weight argument (the default identity matrix is used). Figure 7.1 contains the scatterplot of the data overlaid by the rank-based fit.

[4] See Graybill and Iyer (1994) for initial reference for this dataset.

TABLE 7.1
Crystal Data.

Time (hours)	2	4	6	8	10	12	14
Weight (grams)	0.08	1.12	4.43	4.98	4.92	7.18	5.57
Time (hours)	16	18	20	22	24	26	28
Weight (grams)	8.40	8.881	10.81	11.16	10.12	13.12	15.04

```
> wtedfit <- wtedrb(x,y)
> wtedfit$bstar

        [,1]
[1,] 0.50652

> wtedfit$se

[1] 0.02977
```

The rank-based estimate of slope is 0.507 with a standard error of 0.030; hence, the result is significantly different from 0. ■

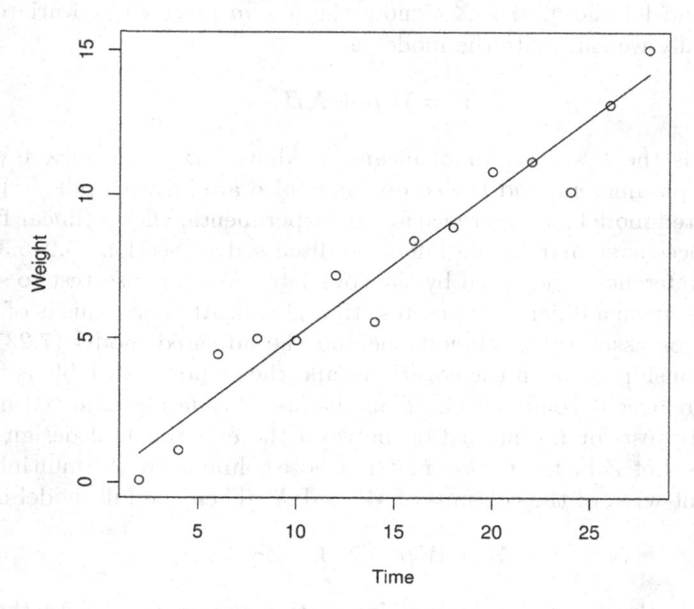

FIGURE 7.1
Scatterplot of the crystal data overlaid by the rank-based fit.

Remark 7.2.1. Weighted rank-based estimates can be used to fit general linear models where the variance of the random errors is an $n \times n$ matrix $\mathbf{\Sigma} > 0$. In this case, the weight matrix is $\mathbf{\Sigma}^{-1/2}$, where $\mathbf{\Sigma}^{-1/2} = \mathbf{\Gamma}\mathbf{\Lambda}^{-1/2}\mathbf{\Gamma}'$ and $\mathbf{\Lambda}$ and $\mathbf{\Gamma}$ are, respectively, the diagonal matrix of eigenvalues and eigenvectors of $\mathbf{\Sigma}$. Generally in practice, $\mathbf{\Sigma}$ has to be estimated; see Bilgic (2012) for such an iterated reweighted procedure for hierarchical linear models. Dixon and McKean (1996) and Carroll and Ruppert (1982) developed robust (R and M, respectively) for iteratively estimating the regression coefficients and the weights when $\mathbf{\Sigma}$ is diagonal and the form of the heteroscedasticity is known; for example, when scale varies directly with response. ■

7.3 ANCOVA

ANCOVA was first introduced in Section 5.6. In this section we extend the discussion and introduce the matrix form of the model. As before, consider a one-way experimental design with k different treatment combinations with cell sample sizes n_i, $i = 1, \ldots k$. Suppose there are m covariates. Let Y_{ij} denote the response for the jth subject in the ith cell and let $\boldsymbol{x}_{ij}$ denote the associated $m \times 1$ vector of covariates for this subject. Let $\boldsymbol{Y}$ denote the $n \times 1$ vector of the responses Y_{ij}. Let the matrix $\boldsymbol{W}$ denote the $n \times k$ cell-mean model design matrix, and let the matrix $\boldsymbol{X}$ denote the $n \times m$ matrix of covariates. Then symbolically we can write the model as

$$\boldsymbol{Y} = \boldsymbol{W}\boldsymbol{\mu} + \boldsymbol{X}\boldsymbol{\beta} + \boldsymbol{e}, \tag{7.23}$$

where $\boldsymbol{\mu}$ is the $k \times 1$ vector of means (medians), $\boldsymbol{\beta}$ is the $m \times 1$ vector of covariate parameters, and the components of $\boldsymbol{e}$ are iid with pdf $f(t)$. This is the **adjusted model** and inference for the experimental effects (linear functions of $\boldsymbol{\mu}$) proceeds similar to the inference discussed in Sections 5.2–5.3, except that the inference is adjusted by the covariates. We can also test to see if the covariates make a difference; i.e., test that $\boldsymbol{\beta} = \boldsymbol{0}$. At times, this is of interest.

A major assumption, though, behind the adjusted model (7.23) is that the relationship between the covariates and the response variable is the same at each treatment combination. Thus, before considering adjusted inference, we usually test for no interaction between the experimental design and the covariates. Let $\boldsymbol{Z}$ be the $n \times km$ matrix whose columns are the multiplications, component-wise, of the columns of $\boldsymbol{W}$ and $\boldsymbol{X}$. Then the full model is

$$\boldsymbol{Y} = \boldsymbol{W}\boldsymbol{\mu} + \boldsymbol{X}\boldsymbol{\beta} + \boldsymbol{Z}\boldsymbol{\gamma} + \boldsymbol{e}, \tag{7.24}$$

where $\boldsymbol{\gamma}$ is the $km \times 1$ vector of interaction parameters. Note that under Model (7.24) each treatment combination has its own linear model with the covariates.

The hypothesis of homogeneous slopes (the linear models of all treatments combinations are the same) is

$$H_{OI} : \boldsymbol{\gamma} = \mathbf{0} \text{ versus } H_{AI} : \boldsymbol{\gamma} \neq \mathbf{0} \tag{7.25}$$

Note that studies are often not powered to detect a difference in slopes, and a failure to reject the hypothesis (7.25) does not mean the *true* slopes are necessarily the same. If hypothesis (7.25) is rejected, then inference on the effects will often be quite misleading; see, for example, the scatterplot of two groups found in Example 5.6.1. In such cases, confidence intervals for simple contrasts between groups at specified factor values can be carried out. These are often called **pick-a-point** analyses.[5]

7.3.1 Computation of Rank-Based ANCOVA

For the general one-way or k-way ANCOVA models, we have written R functions which compute the rank-based ANCOVA analyses which are included in the R package `npsm`.

Computation of Rank-Based ANCOVA for a k-Way Layout

For the k-way layout, we have written the `Rfit` function `kancova` which computes the ANCOVA. Recall that the full model for the design is the cell mean (median) model. Under heterogeneous linear models, each cell in the model has a distinct linear model. This is the full model, Model 7.24, for testing homogeneous slopes. This function also computes the adjusted analysis, assuming that the slopes are homogeneous; that is, for these hypotheses the full model is Model 7.23. So these adjusted tests should be disregarded if there is reason to believe the slopes are different, certainly if the hypothesis of homogeneous slopes is rejected. For the adjusted tests, the standard hypotheses are those of the main effects and interactions of all orders as described in Section 5.3. A test for all covariate effects that are null is also computed. We illustrate this function in the following two examples.

The rank-based tests computed by these functions are based on reductions of dispersion as we move from reduced to full models. Hence, as an alternative to these functions, the computations can also be obtained using the `Rfit` functions `rfit` and `drop.test` with only a minor amount of code. In this way, specific hypotheses of interest can easily be computed, as we show in the following example.

Example 7.3.1 (Triglyceride and Blood Plasma Levels). The data for this example are drawn from a clinical study discussed in Hollander and Wolfe (1999). The data consist of triglyceride levels on 13 patients. Two factors, each at two levels, were recorded: Sex and Obesity. The concomitant variables are chylomicrons, age, and three lipid variables (very low-density lipoproteins (VLDL),

[5]See Huitema (2011) and Watcharotone et al. (2017).

low-density lipoproteins (LDL), and high-density lipoproteins (HDL)). The data are in the npsm dataset `blood.plasma`. The next code segment displays a subset of it. Also, notice the dataset is a `matrix` rather than a `data.frame`.

```
> is.matrix(blood.plasma)

[1] TRUE

> head(blood.plasma)

     Total Sex Obese Chylo VLDL  LDL  HDL Age
[1,] 20.19   1     1  3.11 4.51 2.05 0.67  53
[2,] 27.00   0     1  4.90 6.03 0.67 0.65  51
[3,] 51.75   0     0  5.72 7.98 0.96 0.60  54
[4,] 51.36   0     1  7.82 9.58 1.06 0.42  56
[5,] 28.98   1     1  2.62 7.54 1.42 0.36  66
[6,] 21.70   0     1  1.48 3.96 1.09 0.23  37
```

The design matrix for the full model to test the hypotheses of no interaction between the factors and the covariates would have 24 columns, which, with only 13 observations, is impossible. Instead, we discuss the code to compute several tests of hypotheses of interest. The full model design matrix consists of the four dummy columns for the cell means and the 5 covariates. This design matrix is created in the following code segment.

```
> xcell = cellx(blood.plasma[,2:3])
> colnames(xcell)<-c('00','01','10','11')
> xcov = blood.plasma[,4:8]
> xfull = cbind(xcell,xcov)
```

In the following code segment, this design matrix is in the R matrix `xfull` while the response, total triglyceride, is in the column `Total` of `blood.plasma`. The resulting full model fit and its summary are given by:

```
> fitfull = rfit(blood.plasma[,'Total'] ~ xfull-1)
> summary(fitfull)

Call:
rfit.default(formula = blood.plasma[, "Total"] ~ xfull - 1)

Coefficients:
           Estimate Std. Error t.value p.value
xfull00       8.590      9.300    0.92  0.4079
xfull01      -3.004      8.320   -0.36  0.7363
xfull10     -12.616     10.113   -1.25  0.2802
xfull11     -11.588     10.327   -1.12  0.3246
xfullChylo    1.741      0.532    3.27  0.0307 *
```

```
xfullVLDL      2.878       0.417     6.91  0.0023 **
xfullLDL       3.797       2.771     1.37  0.2424
xfullHDL     -11.470       4.611    -2.49  0.0677 .
xfullAge       0.249       0.134     1.87  0.1353
---
Signif. codes:
0 '***' 0.001 '**' 0.01 '*' 0.05 '.' 0.1 ' ' 1

Multiple R-squared (Robust): 0.95802
Reduction in Dispersion Test: 11.409 p-value: 0.01619

> disp(fitfull$betahat, xfull, fitfull$y, fitfull$scores)

       [,1]
[1,] 26.444
```

The last line is the minimum value of the dispersion function based on the fit of the full model. The one hypothesis of interest discussed in Hollander and Wolfe is whether or not the three lipid covariates (VLDL, LDL, and HDL) are significant predictors; these are columns 6, 7, 8 in `xfull`. The answer appears to be "yes" based on the above summary of the `Rfit` of the full model. The following code performs a formal test using the hypothesis matrix `hmat`:

```
> hmat = rbind(c(rep(0,5),1,rep(0,3)),
+              c(rep(0,6),1,rep(0,2)),
+              c(rep(0,7),1,rep(0,1)))
> hmat

     [,1] [,2] [,3] [,4] [,5] [,6] [,7] [,8] [,9]
[1,]    0    0    0    0    0    1    0    0    0
[2,]    0    0    0    0    0    0    1    0    0
[3,]    0    0    0    0    0    0    0    1    0

> xred1 <- redmod(xfull,hmat)
> fitr1 <- rfit(blood.plasma[,'Total']~xred1-1)
> drop.test(fitfull,fitr1)

Drop in Dispersion Test
F-Statistic      p-value
  10.407738     0.023245
```

Hence, based on this p-value (0.0232), it seems that the lipid variables are related to triglyceride levels. The next few lines of code test to see if the factor sex has an effect on triglycerides.

```
> hmat=rbind(c(1,1,-1,-1,rep(0,5)))
> xred3 = redmod(xfull,hmat)
```

```
> fitr3 = rfit(blood.plasma[,'Total']~xred3-1)
> drop.test(fitfull,fitr3)

Drop in Dispersion Test
F-Statistic     p-value
 21.7920622   0.0095311
```

Based on the p-value of 0.0095, it appears that the factor sex also has an effect on triglyceride levels. Finally, consider the effect of obesity on triglyceride level.

```
> hmat=rbind(c(1,-1,1,-1,rep(0,5)))
> xred2 = redmod(xfull,hmat)
> fitr2 = rfit(blood.plasma[,'Total']~xred2-1)
> drop.test(fitfull,fitr2)

Drop in Dispersion Test
F-Statistic     p-value
  11.543597    0.027338
```

Thus, the rank-based test for obesity results in the test statistic 11.54 with p-value 0.0273. ■

7.4 Methodology for Type III Hypotheses Testing

In this section, we briefly describe how Rfit obtains Type III hypotheses for the two-way and k-way designs. Consider first the hypotheses for the two-way design which are given in expressions (5.11)–(5.13). For our discussion, assume that the data are stacked as the $n \times 1$ vector $\boldsymbol{Y}$ by cell and row-by-row in row order; i.e., in terms of the subscripts ijk, k runs the fastest and i runs the slowest. Let $\boldsymbol{\mu}$ denote the corresponding $ab \times 1$ vector of parameters, and let $\boldsymbol{W}$ denote the $n \times ab$ incidence matrix. Then the full model can be written as $\boldsymbol{Y} = \boldsymbol{W}\boldsymbol{\mu} + \boldsymbol{e}$, where $\boldsymbol{e}$ denotes the vector of random errors.

For the two-way model, the three hypotheses of interest are the main effects hypotheses and the interaction hypothesis given respectively by (5.11)–(5.13). Following Hocking (1985), the hypotheses matrices $\boldsymbol{M}$ can easily be computed in terms of Kronecker products. For a positive integer s, define the augmented matrix

$$\boldsymbol{\Delta}_s = [\boldsymbol{I}_{s-1} \;\; -\boldsymbol{1}_{s-1}], \tag{7.26}$$

where $\boldsymbol{I}_{s-1}$ is the identity matrix of order $s-1$ and $\boldsymbol{1}_{s-1}$ denotes a vector of $(s-1)$ ones. For our two-way design with A at a levels and B at b levels, the hypothesis matrices of average main effects and interaction are given by

$$\begin{aligned}
\text{For Hypothesis (5.11): } \boldsymbol{M}_A &= \boldsymbol{\Delta}_a \otimes \frac{1}{b}\mathbf{1}_b^T \\
\text{For Hypothesis (5.12): } \boldsymbol{M}_B &= \frac{1}{a}\mathbf{1}_a^T \otimes \boldsymbol{\Delta}_b \\
\text{For Hypothesis (5.13): } \boldsymbol{M}_{A\times B} &= \boldsymbol{\Delta}_a \otimes \boldsymbol{\Delta}_b,
\end{aligned}$$

where $\otimes$ denotes the Kronecker product. Based on these hypothesis matrices, `Rfit` computes[6] reduced model design matrices.

Hypothesis matrices for higher order designs can be computed[7] similarly. For example, suppose we have the four factors A, B, C and D with respective levels a, b, c and d. Then the hypotheses matrices to test the interaction between B and D and the 4-way interaction are respectively given by

$$\begin{aligned}
\boldsymbol{M}_{B\times D} &= \frac{1}{a}\mathbf{1}_a^T \otimes \boldsymbol{\Delta}_b \otimes \frac{1}{c}\mathbf{1}c^T \otimes \boldsymbol{\Delta}_d \\
\boldsymbol{M}_{A\times B\times C`B\times D} &= \boldsymbol{\Delta}_a \otimes \boldsymbol{\Delta}_b \otimes \boldsymbol{\Delta}_c \otimes \boldsymbol{\Delta}_d.
\end{aligned}$$

The corresponding reduced model design matrices can be easily computed to obtain the tests of the hypotheses.

7.5 Aligned Rank Tests

An aligned rank test is a nonparametric method which allows for adjustment of covariates in tests of hypotheses. In the context of a randomized experiment to assess the effect of some intervention, one might want to adjust for baseline covariates in the test for the intervention. In perhaps the simplest context of a two-sample problem, the test is based on the Wilcoxon rank sum from the residuals of a robust fit of a model on the covariates. Aligned rank tests were first developed by Hodges and Lehmann (1962) for use in randomized block designs. They were developed for the linear model by Adichie (1978); see also Puri and Sen (1985) and Chiang and Puri (1984). Kloke and Cook (2014) discuss aligned rank tests and consider an adaptive scheme in the context of a clinical trial.

For simplicity, suppose that we are testing a treatment effect, and each subject is randomized to one of k treatments. For this section consider the model

$$Y_i = \alpha + \boldsymbol{w}_i^T\boldsymbol{\Delta} + \boldsymbol{x}_i^T\boldsymbol{\beta} + e_i \qquad (7.27)$$

[6] See page 209 of Hettmansperger and McKean (2011).

[7] See Hocking (1985).

where $\boldsymbol{w}_i$ is a $(k-1)\times 1$ incidence vector denoting the treatment assignment for the ith subject, $\boldsymbol{\Delta} = [\Delta_2, \ldots, \Delta_K]^T$ is a vector of unknown treatment effects, $\boldsymbol{x}_i$ is a $p\times 1$ vector of (baseline) covariates, $\boldsymbol{\beta}$ is a vector of unknown regression coefficients, and e_i denotes the error term. The goal of the experiment is to test

$$H_0 : \boldsymbol{\Delta} = \mathbf{0}. \tag{7.28}$$

In this section we focus on developing an aligned rank tests for Model (7.27). We write the model as

$$\boldsymbol{Y} = \alpha\mathbf{1} + \boldsymbol{W}\boldsymbol{\Delta} + \boldsymbol{X}\boldsymbol{\beta} + \boldsymbol{e} = \alpha\mathbf{1} + \boldsymbol{Z}\boldsymbol{b} + \boldsymbol{e}. \tag{7.29}$$

Then the full model gradient is

$$\boldsymbol{S}(\boldsymbol{b}) = \boldsymbol{Z}^T\boldsymbol{a}(R(\boldsymbol{Y} - \boldsymbol{Z}\boldsymbol{b})).$$

First, fit the reduced model

$$\boldsymbol{Y} = \alpha\mathbf{1} + \boldsymbol{X}\boldsymbol{\beta} + \boldsymbol{e}.$$

Then, plug the reduced model estimate $\hat{\boldsymbol{b}}_r = [\mathbf{0}^T\hat{\boldsymbol{\beta}}_r^T]^T$ into the full model:

$$\boldsymbol{S}(\hat{\boldsymbol{b}}_r) \doteq \begin{bmatrix} \boldsymbol{W}^T\boldsymbol{a}(R(\boldsymbol{Y} - \boldsymbol{X}\hat{\boldsymbol{\beta}}_r)) \\ \mathbf{0} \end{bmatrix}.$$

Define $\hat{\boldsymbol{S}}_1 = \boldsymbol{W}^T\boldsymbol{a}(R(\boldsymbol{Y} - \boldsymbol{X}\hat{\boldsymbol{\beta}}_r))$ as the first $k-1$ elements of $\boldsymbol{S}(\hat{\boldsymbol{b}}_r)$. Then the aligned rank test for (7.28) is based on the test statistic

$$\hat{\boldsymbol{S}}_1^T[\boldsymbol{W}^T\boldsymbol{W} - \boldsymbol{W}^T\boldsymbol{H}_{\boldsymbol{X}}\boldsymbol{W}]^{-1}\hat{\boldsymbol{S}}_1 = \hat{\boldsymbol{S}}_1^T[\boldsymbol{W}^T\boldsymbol{H}_{\boldsymbol{X}^\perp}\boldsymbol{W}]^{-1}\hat{\boldsymbol{S}}_1$$

where $\boldsymbol{H}_{\boldsymbol{X}}$ is the projection matrix onto the space spanned by the columns of $\boldsymbol{X}$. For inference, this test statistic should compared to χ^2_{K-1} critical values.

In the package npsm, we have included the function aligned.test which performs the aligned rank test. A simple simulated example illustrates the use of the code.

```
> set.seed(43329)
> k<-3   # number of treatments
> p<-2   # number of covariates
> n<-10  # number of subjects per treatment
> N<-n*k # total sample size
> y<-rnorm(N)
> x<-matrix(rnorm(N*p),ncol=p)
> g<-rep(1:k,each=n)
> aligned.test(x,y,g)

statistic =  2.3949 , p-value =  0.30196
```

Since the data were simulated to have no treatment effect, it is not that surprising that the p-value is non-significant.

7.6 Exercises

7.6.1. Consider the data of Example 7.1.1.

(a) Obtain a scatterplot of the data.

(b) Obtain the Wilcoxon fit of the linear trend model. Overlay the fit on the scatterplot. Obtain the Studentized residual plot and normal $q-q$ plots. Identify any outliers and comment on the quality of fit.

(c) Obtain a 95% confidence interval for the slope parameter and use it to test the hypothesis of 0 slope.

(d) Estimate the mean of the response when time has the value 16 and find the 95% confidence interval for it which was discussed in Section 7.1.4.

7.6.2. Bowerman et al. (2005) present a dataset concerning the value of a home (x) and the upkeep expenditure (y). The data are in `qhic`. The variable x is in the thousands of dollars while the y variable is in the tens of dollars.

(a) Obtain a scatterplot of the data.

(b) Use Wilcoxon Studentized residual plots, values of $\hat{\tau}$, and values of the robust R^2 to decide whether a linear or a quadratic model fits the data better.

(c) Based on your model, estimate the expected expenditures (with a 95% confidence interval) for a house that is worth $155,000.

(d) Repeat (c) for a house worth $250,000.

7.6.3. Rewrite the `aligned.test` function to take an additional design matrix as its third argument instead of group/treatment membership. That is, for the model $\boldsymbol{Y} = \alpha\boldsymbol{1} + \boldsymbol{X}_1\boldsymbol{\beta}_1 + \boldsymbol{X}_2\boldsymbol{\beta}_2 + \boldsymbol{e}$, test the hypothesis $H_0 : \boldsymbol{\beta}_2 = \boldsymbol{0}$.

7.6.4. One form of heteroscedasticity that occurs in regression models is when the observations are collected over time and the response varies with time. Good statistical practice dictates plotting the responses and other variables versus time, if time may be a factor. Consider the following data:

	1	2	3	4	5	6	7	8	9	10
x	46	37	34	30	33	24	30	49	54	47
y	132	105	94	71	84	135	10	132	148	132
x	24	33	42	50	47	55	25	44	38	53
y	67	105	104	132	142	130	60	133	204	163

(a) Scatterplot the data. Obtain the Wilcoxon fit and the residual plot. Do the data appear to be heteroscedastic?

(b) These data were collected over time. For such cases, it is best to plot the residuals versus time. Obtain this plot for our dataset (Wilcoxon residuals versus time, `t<-1:20`). Comment on the heteroscedasticity.

(c) Assume the response varies directly with time; i.e., $\text{Var}(e_i) = i\sigma^2$, where e_i denotes the random error for the ith case. Appropriate weights in this case are: `diag(1/i)`, where `i<-1:20`. Use the function `wtedrb` to fit the data with these weights. Remember to use the design matrix `xmat<-cbind(rep(1,20),x)`, where `x` is the vector containing the x's.

(d) Compare the precision of fits in parts (a) and (c).

7.6.5. Consider a model where the responses vary directly with time order of the observations:

$$Y_i = 5 - 3x_i + e_i \text{ for } i = 1, \ldots 20$$

where the explanatory variables $(x_1, \ldots, x_{20})$ are generated from a standard normal distribution and the errors $(e_1, \ldots, e_{20})$ are generated from a contaminated normal distribution with $\epsilon = 0.2$ and $\sigma = 9$. Run a simulation comparing the usual Wilcoxon fit with the weighted Wilcoxon fit using the weights `diag(1/t)`. Remember to use the design matrix `xmat<-cbind(rep(1,20),x)` for the weighted Wilcoxon fit. Comment on the simulation, in particular address the following.

(a) Empirical mean square errors for the estimates of the slope parameter $\beta_1 = -3$.

(b) Validity of 95% confidence intervals for the slope parameter $\beta_1 = -3$.

7.6.6. Hamilton (1992) presents a dataset concerning the number of accidental oil spills (x) at sea and the amount of oil loss (y) in millions of metric tons for the years 1973–1975. The data are:

	1	2	3	4	5	6	7
x	36.00	48.00	45.00	29.00	49.00	35.00	65.00
y	84.50	67.10	188.00	204.20	213.10	260.50	723.50
	8	9	10	11	12	13	
x	32.00	33.00	9.00	17.00	15.00	8.00	
y	135.60	45.30	1.70	387.80	24.20	15.00	

Hamilton suggests a regression through the origin model for this data.

(a) Obtain the scatterplot y versus x and overlay the the rank-based and LS fits of the regression through the origin models.

(b) Obtain the residual plots and $q-q$ plots of the fits in Part (a). Comment on which fit is better.

(c) Obtain a 95% confidence interval for the slope parameter based on the rank-based fit.

7.6.7. In using a regression through the origin model, one assumes that the intercept parameter is 0. This is a strong assumption. If there is little or no data with x relatively near 0, this is a form of extrapolation and there is little evidence for which to verify the assumption. One simple diagnostic check is to fit an intercept model, also, and then to compare the two fits.

(a) For the data in Exercise 7.6.6, besides the rank-based fit of the regression through the origin model, obtain the rank-based fit of the intercept model. Overlay these fits on the scatterplot of the data. Which, if any, fit is better? Why?

(b) One check for the assumption of a 0 intercept is to use a confidence interval for the intercept as a diagnostic confirmation. Obtain the confidence interval for the rank-based fit.

(c) The fallacy in Part (b), of course, is that it may be a form of extrapolation as discussed at the beginning of this exercise. Is this extrapolation a concern for this problem? Why?

8

Topics in Regression

In Chapter 4 we introduced rank-based fitting of linear models using `Rfit`. In this chapter, we discuss further topics for rank-based regression. These include high breakdown fits, diagnostic procedures, weighted regression, nonlinear models, and autoregressive time series models. We also discuss optimal scores for a family of skew-normal distributions and present an adaptive procedure for regression estimation based on a family of Winsorized Wilcoxon scores.

8.1 Introduction

Let $\boldsymbol{Y} = [Y_1, \ldots, Y_n]^T$ denote an $n \times 1$ vector of responses. Then the matrix version of the linear model (4.11) is

$$\boldsymbol{Y} = \alpha \boldsymbol{1} + \boldsymbol{X}\boldsymbol{\beta} + \boldsymbol{e} \tag{8.1}$$

where $\boldsymbol{X} = [\boldsymbol{x}_1, \ldots, \boldsymbol{x}_n]^T$ is an $n \times p$ design matrix, and $\boldsymbol{e} = [e_1, \ldots, e_n]^T$ is an $n \times 1$ vector of error terms. Assume for discussion that $f(t)$ and $F(t)$ are the pdf and cdf of e_i, respectively. Assumptions differ for the various sectional topics.

Recall from expression (7.6) that the rank-based estimator $\hat{\boldsymbol{\beta}}_\varphi$ is the vector that minimizes the rank-based distance between $\boldsymbol{Y}$ and $\boldsymbol{X}\boldsymbol{\beta}$; i.e., $\hat{\boldsymbol{\beta}}_\varphi$ is defined as

$$\hat{\boldsymbol{\beta}}_\varphi = \text{Argmin}\|\boldsymbol{Y} - \boldsymbol{X}\boldsymbol{\beta}\|_\varphi, \tag{8.2}$$

where the norm is defined by

$$\|\boldsymbol{v}\|_\varphi = \sum_{i=1}^{n} a[R(Y_i - \boldsymbol{x}_i^T\boldsymbol{\beta})](y_i - \boldsymbol{x}_i^T\boldsymbol{\beta}), \quad \boldsymbol{v} \in R^n, \tag{8.3}$$

and the scores $a(i) = \varphi[i/(n+1)]$ for a specified score function $\varphi(u)$ defined on the interval $(0, 1)$ and satisfying the standardizing conditions given in (3.12).

Note that the norm is invariant to the intercept parameter; but, once $\boldsymbol{\beta}$ is estimated, the intercept α is estimated by the median of the residuals. That is,

$$\hat{\alpha} = \text{med}_i\{Y_i - \boldsymbol{x}_i^T\hat{\boldsymbol{\beta}}_\varphi\}. \tag{8.4}$$

The rank-based residuals are defined by

$$\hat{e}_i = Y_i - \hat{\alpha} - \boldsymbol{x}_i^T \hat{\boldsymbol{\beta}}_\varphi, \quad i = 1, 2, \ldots, n. \tag{8.5}$$

Recall that the joint asymptotic distribution of the rank-based estimates is multivariate normal with the covariance structure as given in (7.8).

As discussed in Chapter 3, the rank-based estimates are generally highly efficient estimates. Further, as long as the score function is bounded, the influence function of $\hat{\boldsymbol{\beta}}_\varphi$ is bounded in the $\boldsymbol{Y}$-space (response space). As with LS estimates, though, the influence function is unbounded in the $\boldsymbol{x}$-space (factor space). In the next section, we present a rank-based estimate which has bounded influence in both spaces and which can attain the maximal 50% breakdown point.

8.2 High Breakdown Rank-Based Fits

High breakdown rank-based (HBR) estimates were developed by Chang et al. (1999) and are fully discussed in Section 3.12 of Hettmansperger and McKean (2011). To obtain HBR fits of linear models, a suite of R functions (`ww`) was developed by Terpstra and McKean (2005). We use a modified version, **`hbrfit`**, of `ww` to compute HBR fits.[1]

The objective function for HBR estimation is a weighted Wilcoxon dispersion function given by

$$\|\boldsymbol{v}\|_{HBR} = \sum_{i<j} b_{ij} |v_i - v_j| \tag{8.6}$$

where $b_{ij} \geq 0$ and $b_{ij} = b_{ji}$. The HBR estimator of $\boldsymbol{\beta}$ minimizes this objective function, which we denote by

$$\hat{\boldsymbol{\beta}}_{HBR} = \text{Argmin}\|\boldsymbol{Y} - \boldsymbol{X}\boldsymbol{\beta}\|_{HBR}. \tag{8.7}$$

As with the rank-based estimates of Chapter 4, the intercept α is estimated as the median of the residuals; that is,

$$\hat{\alpha} = \text{med}_i\{Y_i - \boldsymbol{x}^T \hat{\boldsymbol{\beta}}_{HBR}\}. \tag{8.8}$$

As shown in Chapter 3 of Hettmansperger and McKean (2011), if all the weights are one (i.e., $b_{ij} \equiv 1$), then $\| \cdot \|_{HBR}$ is the Wilcoxon norm. Thus the question is, what weights should be chosen to yield estimates which are robust to outliers in both the $\boldsymbol{x}$- and $\boldsymbol{Y}$-spaces? In Section 8.2.1, we discuss the HBR weights implemented in **`hbrfit`** which achieve 50% breakdown. For now, though, we illustrate their use and computation with several examples.

[1] See **`https://github.com/kloke/book`** for more information.

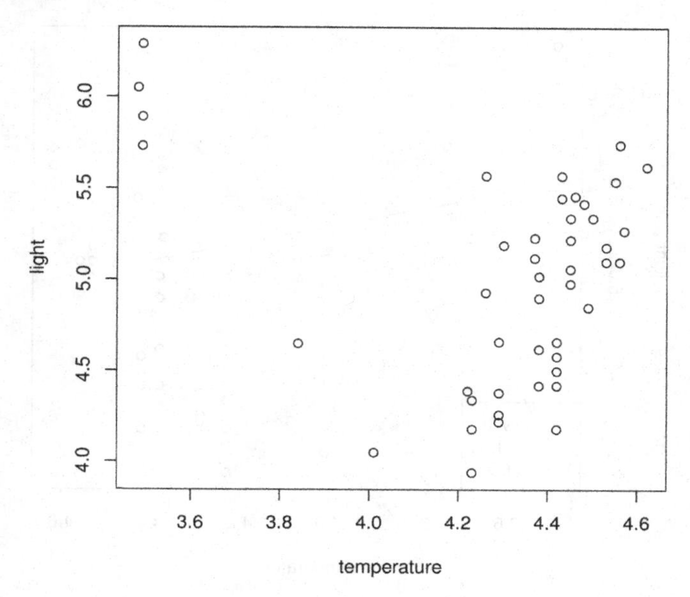

FIGURE 8.1
Scatterplot of stars data.

Stars Data

In this subsection we present an example to illustrate the usage of the weighted Wilcoxon code `hbrfit` to compute HBR estimates. This example uses the `stars` dataset which is from Rousseeuw et al. (1987). The data are from an astronomy study on the star cluster CYG OB1. The cluster contains 47 stars. Measurements were taken on light intensity and temperature. The response variable is log light intensity, and the explanatory variable is log temperature. As is apparent in the scatterplot displayed in Figure 8.1 there are several outliers: there are four stars with lower temperature and higher light intensity than the other members of the cluster. These four stars are labeled giant stars in this dataset. The others are labeled main sequence stars, except for the two with log temperatures 3.84 and 4.01 which are between the giant and main sequence stars.

In Figure 8.2, the Wilcoxon (WIL), high breakdown (HBR), and least squares (LS) fits are overlaid on the scatterplot. As seen in Figure 8.2, both the least squares and Wilcoxon fit are affected substantially by the outliers; the HBR fit, however, is robust.

The HBR fit is computed as

```
> fitHBR<-hbrfit(light ~ temperature,data=stars)
```

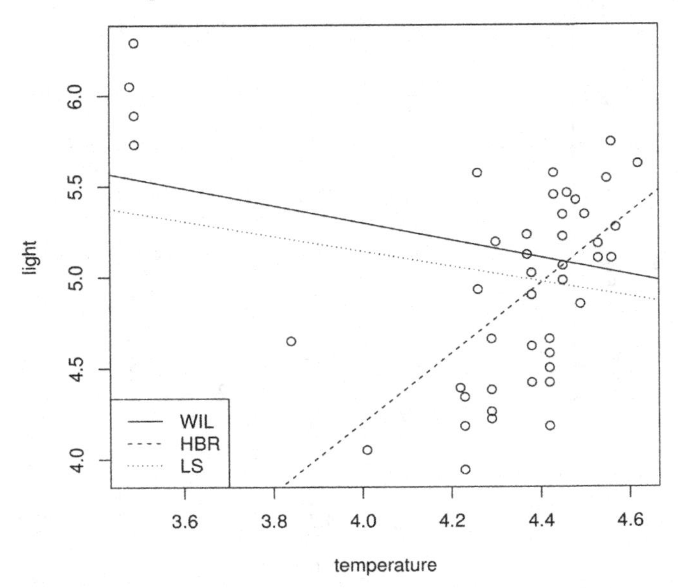

FIGURE 8.2
Scatterplot of stars data with fitted regression lines overlaid.

As we have emphasized throughout the book, the use of residuals, in particular Studentized residuals, are essential to the model building process. Studentized residuals are available through the command `rstudent`. In addition, a set of diagnostic plots can be obtained using `diagplot`. For HBR fit of the stars data, Figure 8.3 displays these diagnostic plots, which resulted from the code:

```
> diagplot(fitHBR)
```

Note from these plots in Figure 8.3 that the Studentized residuals of the HBR fit clearly identify the 4 giant stars. They also identify the two stars between the giant and main sequence stars.

Finally, we may examine the estimated regression coefficients and their standard errors in the table of regression coefficients with the command `summary`; i.e.,

```
> summary(fitHBR)

Call:
hbrfit(formula = light ~ temperature, data = stars)

Coefficients:
            Estimate Std. Error t.value p.value
(Intercept)   -3.469      1.647   -2.11   0.041 *
temperature    1.917      0.381    5.02 8.5e-06 ***
```

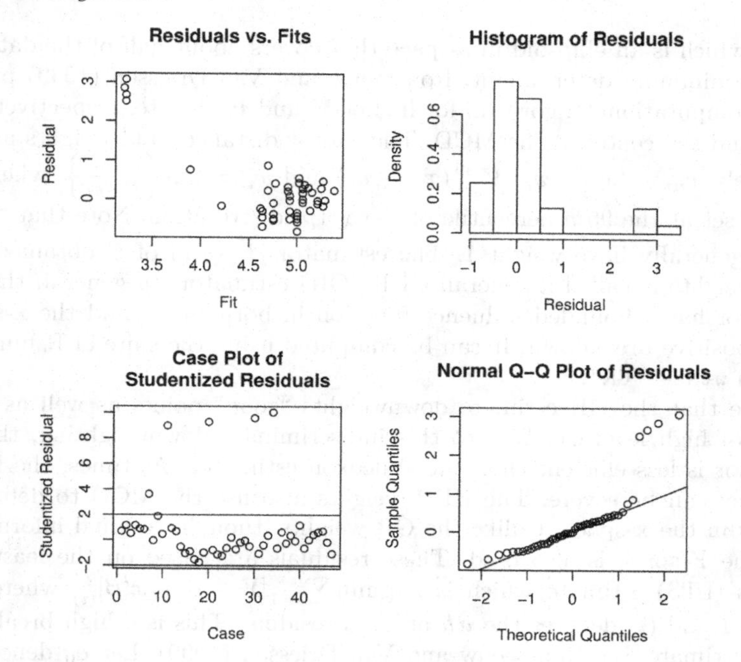

FIGURE 8.3
Diagnostic plots based on the HBR fit of the stars data.

```
---
Signif. codes:
0 '***' 0.001 '**' 0.01 '*' 0.05 '.' 0.1 ' ' 1

Wald Test: 25.249 p-value: 1e-05
```

The estimate of intercept is -3.47 (se $= 1.65$). The estimate of slope is 1.92 (se $= 0.38$). Critical values based on a t-distribution with $n - p - 1$ degrees of freedom are recommended for inference; for example, p-values in the coefficients table for the stars data are based on a t_{45} distribution. Also displayed is a Wald test of $H_0 : \boldsymbol{\beta} = \mathbf{0}$.

8.2.1 Weights for the HBR Fit

Let $\boldsymbol{X}_c$ be the centered design matrix. For weights, it seems reasonable to downweight points far from the center of the data. The traditional distances are the leverage values $h_i = n^{-1} + \boldsymbol{x}_{ci}'(\boldsymbol{X}_c'\boldsymbol{X}_c)^{-1}\boldsymbol{x}_{ci}$, where $\boldsymbol{x}_{ci}$ is the vector of the ith row of $\boldsymbol{X}_c$. Because the leverage values are based on the LS variance-covariance scatter matrix, they are not robust. The weights for the HBR estimate make use of the high breakdown minimum covariance determinant,

MCD, which is an ellipsoid in p-space that covers about half of the data and yet has minimum determinant. Rousseeuw and Van Driessen (1999) present a fast computational algorithm for it. Let $\boldsymbol{V}$ and $\boldsymbol{v}_c$ denote, respectively, the MCD and the center of the MCD. The robust distances and weights are, respectively, $v_{ni} = (\boldsymbol{x}_{ci} - \boldsymbol{v}_c)'\boldsymbol{V}^{-1}(\boldsymbol{x}_{ci} - \boldsymbol{v}_c)$ and $w_i = \min\left\{1, \frac{c}{v_{ni}}\right\}$, where c is usually set at the $95th$ percentile of the $\chi^2(p)$ distribution. Note that "good" points generally have weight 1. The estimator $\widehat{\boldsymbol{\beta}}^*$ (8.7) of $\boldsymbol{\beta}$ obtained with these weights is called a generalized R (GR) estimator. In general, this GR estimator has a bounded influence function in both the Y and the **x**-spaces and a positive breakdown. It can be computed using the suite of R functions `ww` with `wts = "GR"`.

Note that the GR estimate downweights "good" points as well as "bad" points of high leverage. Due to this indiscriminate downweighting, the GR estimator is less efficient than the Wilcoxon estimator. At times, the loss in efficiency can be severe. The HBR weights also use the MCD to determine weights in the **x**-space. Unlike the GR weights, though, residual information from the Y-space is also used. These residuals are based on the **least trim squares** (LTS) estimate which is $\text{Argmin} \sum_{i=1}^{h} [Y - \alpha - \boldsymbol{x}'\boldsymbol{\beta}]^2_{(i)}$ where $h = [n/2] + 1$ and (i) denotes the ith ordered residual. This is a high breakdown initial estimate; see Rousseeuw and Van Driessen (1999). Let $\widehat{\mathbf{e}}_0$ denote the residuals from this initial fit.

Define the function $\psi(t)$ by $\psi(t) = 1,\ t,\ \text{ or } \ -1$ according as $t \geq 1$, $-1 < t < 1$, or $t \leq -1$. Let σ be estimated by the initial scaling estimate $\text{MAD} = 1.483\ \text{med}_i |\widehat{e}_i^{(0)} - \text{med}_j\{\widehat{e}_j^{(0)}\}|$. Letting $Q_i = (\boldsymbol{x}_i - \boldsymbol{v}_c)'\boldsymbol{V}^{-1}(\boldsymbol{x}_i - \boldsymbol{v}_c)$, define

$$m_i = \psi\left(\frac{b}{Q_i}\right) = \min\left\{1, \frac{b}{Q_i}\right\} .$$

Consider the weights

$$\widehat{b}_{ij} = \min\left\{1, \frac{c\hat{\sigma}}{|\hat{e}_i^{(0)}|}\frac{\hat{\sigma}}{|\hat{e}_j^{(0)}|} \min\left\{1, \frac{b}{\hat{Q}_i}\right\} \min\left\{1, \frac{b}{\hat{Q}_j}\right\}\right\} , \tag{8.9}$$

where b and c are tuning constants. Following Chang et al. (1999), b is set at the upper $\chi^2_{.05}(p)$ quantile and c is set as

$$c = [\text{med}\{a_i\} + 3MAD\{a_i\}]^2,$$

where $a_i = \hat{e}_i^{(0)}/(MAD \cdot Q_i)$. From this point of view, it is clear that these weights downweight both outlying points in factor space and outlying responses. Note that the initial residual information is a multiplicative factor in the weight function. Hence, a good leverage point will generally have a small (in absolute value) initial residual which will offset its distance in factor space. These are the weights used for the HBR fit computed by `hbrfit`.

In general, the HBR estimator has a 50% breakdown point, provided the initial estimates used in forming the weights also have a 50% breakdown point. Further, its influence function is a bounded function in both the Y and the $\mathbf{x}$-spaces, is continuous everywhere, and converges to zero as $(\mathbf{x}^*, Y^*)$ get large in any direction. The asymptotic distribution of $\widehat{\boldsymbol{\beta}}_{HBR}$ is asymptotically normal. As with all high breakdown estimates, $\widehat{\boldsymbol{\beta}}_{HBR}$ is less efficient than the Wilcoxon estimates, but it regains some of the efficiency loss of the GR estimate. See Section 3.12 of Hettmansperger and McKean (2011) for discussion.

8.3 Robust Diagnostics

Diagnostics are an essential part of any analysis. The assumption of a model is a very strong statement and should not be taken lightly. As we have stressed throughout the book, diagnostic checks should be made to confirm the adequacy of the model and check the quality of fit. In this section, we explore additional diagnostics based on both highly efficient and high breakdown robust fits. These diagnostics are primarily concerned with the determination of highly influential points on the fit.

For motivation, we consider a simple dataset with two predictors and $n = 30$ data points. The scatterplot of the columns of the two-dimensional ($p = 2$) design matrix, $\boldsymbol{X}$, is shown in Figure 8.4. The values of the x's are drawn from uniform distributions. The design matrix and observations are in the dataset `diagdata`. In the following R code segment, we create a design matrix and two vectors of responses.

```
> library(npsmReg2)
> data(diagdata)
> x <- diagdata[,c('x1','x2')]
> ybad <- diagdata[,c('ybad')]
> ygood <- diagdata[,c('ygood')]
```

Consider the four points in the upper-right corner of the plot, which are the 27th through 30th data points (see also the following code chunk).

```
> tail(diagdata)

      x1    x2   ybad  ygood
25 0.554 0.109  1.318  1.318
26 0.024 0.083 -0.154 -0.154
27 0.870 0.910 -7.001  7.001
28 0.860 0.920 -7.397  7.397
29 0.850 0.900 -9.191  9.191
30 0.900 0.850 -8.269  8.269
```

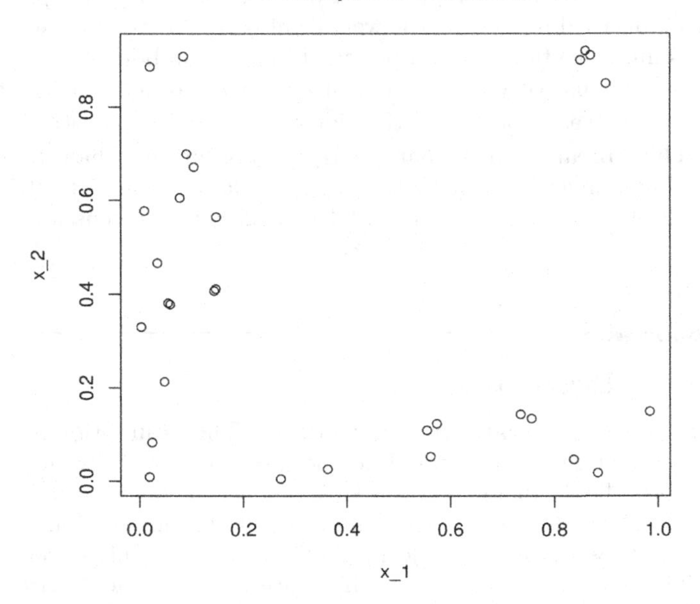

FIGURE 8.4
Scatterplot of the columns of the design matrix for the simple example.

As the following two sets of responses show, these points are potentially influential points on fits. The first set of responses is drawn from the model

$$Y_i = 5x_{i1} + 5x_{i2} + e_i, \tag{8.10}$$

where $e_1, \ldots, e_n$ were drawn independently from a $N(0,1)$ distribution. We label this the "good" dataset. For this set, the responses for cases 27 through 30 are, respectively, 7.001, 7.397, 9.191, and 8.269, which follow the model. To form the second set of responses, we negated these four responses; i.e., the observations for cases 27 through 30 are, respectively, $-7.001, -7.397, -9.191$, and -8.269, which of course do not follow the model. We label this second set, the "bad" dataset. We obtain the LS, Wilcoxon, and HBR fits of the two models, summarizing them in Table 8.1.

On the good dataset, all three fits agree. On the bad dataset, both the LS and Wilcoxon fits are impaired, while the HBR fit exhibits robustness. Thus, the 27th through 30th data points are potentially influential points.

Note first that traditional diagnostic procedures are generally not efficient at these tasks. Their geometry is based on the Euclidean norm which is sensitive to outlying points in the $\boldsymbol{x}$-space. For example, for this simple dataset, the vector of column means of the design matrix is $(0.369, 0.399)$, while the vector of column medians is $(0.148, 0.38)$. Hence, the outliers have influenced

TABLE 8.1
Estimates of the regression coefficients for the simple datasets ("good" data and "bad" data).

		Estimates		
Dataset	Method	Intercept	X_1	X_2
"Good" Data	LS	0.12	4.45	4.74
"Good" Data	W	0.10	4.42	4.64
"Good" Data	HBR	-0.11	4.62	4.91
"Bad" Data	LS	5.56	-4.49	-5.96
"Bad" Data	W	6.07	-4.59	-5.83
"Bad" Data	HBR	-0.11	4.55	4.79

the center of design based on the mean. The leverage values for the four outlying points are, respectively, 0.192, 0.193, 0.183, and 0.181. Note that these are less than the usual benchmark for points of high leverage which is given by $2(p+1)/n = 0.2$. Thus, if the leverage rule is followed strictly, these four points would not have been identified.

Traditional delete-one diagnostics are often used to check the quality of the fit. For example, consider the diagnostic DFFITS_i which is the standardized change in the LS fitted value for case i when case i is deleted; see Belsley et al. (1980). For datasets containing a cluster of outliers in the $\boldsymbol{X}$-space, though, when one of the cases in the cluster is deleted, there are still the remaining cases in the cluster which will impair the LS fit in the same way as in the LS fit based on all the data. Hence, DFFITS_i will generally be small and the points will not be detected. For the simple dataset, the values of DFFITS_{27} through DFFITS_{30} are, respectively, $-0.517, -0.583, -0.9$, and -0.742. Since the benchmark is $2\sqrt{(p+1)/n} = 0.632$, Cases 27 and 28 are not detected as influential points. The other two cases are detected, but they are borderline. For example, Case 7, which is not a point of concern, has the DFFITS value of 0.70. Hence, for this example, the diagnostic DFFITS_i has not been that successful.

Note that for this simple example, the HBR estimates remain essentially the same for both the "good" and "bad" datasets. We now present diagnostics based on the HBR estimates and the robust distances and LTS residuals that are used to form the HBR weights. These diagnostics are robust and are generally successful in detecting influential cases and in detecting differences between highly efficient and high breakdown robust fits.

8.3.1 Graphics

In general, consider a linear model of the form (8.1). In Section 8.2, we defined the robust distances $\sqrt{Q_i}$, where $Q_i = (\boldsymbol{x}_i - \boldsymbol{v}_c)'\boldsymbol{V}^{-1}(\boldsymbol{x}_i - \boldsymbol{v}_c)$, $i = 1, \ldots, n$, $\boldsymbol{V}$ is the minimum covariance determinant (MCD), and $\boldsymbol{v}_c$ is the center of the ellipsoid $\boldsymbol{V}$. Recall that these were used to obtain the weights in the HBR fit,

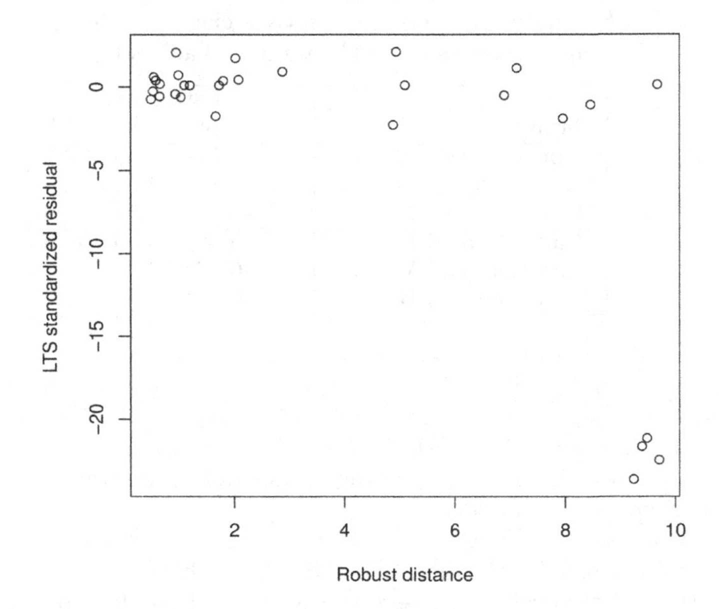

FIGURE 8.5
Standardized LTS residuals versus robust distances for the simple example with bad responses.

see expression (8.9). Another part of these weights utilizes the standardized residuals based on the LTS fit. Rousseeuw and van Zomeren (1990) proposed as a diagnostic, the plot of these standardized residuals versus the robust distances. For the simple example with the set of bad responses, this plot is found in Figure 8.5. Note that the 4 influential cases are clearly separated from the other cases. Hence, for this example, the diagnostic plot was successful. The next segment of R code obtains the robust distances, standardized LTS residuals, and the diagnostic plot shown in Figure 8.5. The observations for the second set of data (bad data) are in the R vector `ybad`, while the design matrix is in the R matrix `x`. Some caution is necessary here, because standardized residuals are not corrected for locations of the residuals in the $\boldsymbol{X}$-space as Studentized residuals are.

```
> rsdfitlts <- ltsreg(x,ybad)$resid
> srsd <- rsdfitlts /mad(rsdfitlts )
> rdis <- sqrt(robdistwts(x,ybad)$robdis2)
> plot(rdis,srsd,xlab="Robust distance",
+ ylab="LTS standardized residual")
> title(main="Standardized LTS Residuals vs. Robust Distances")
```

8.3.2 Procedures for Differentiating between Robust Fits

Recall that for the simple dataset with bad responses the differences between the Wilcoxon and HBR fits are readily apparent; see Table 8.1. We next discuss a set of formal diagnostics based on the difference between fits; see[2] McKean et al. (1996). Consider a general linear model, say Model (8.1). The difference between estimates includes the intercept, so, for this section, let $\boldsymbol{b}^T = (\alpha, \boldsymbol{\beta}^T)$ denote the combined parameters. Then the difference between the HBR and Wilcoxon regression estimates is the vector $\widehat{\boldsymbol{b}}_D = \widehat{\boldsymbol{b}}_W - \widehat{\boldsymbol{b}}_{HBR}$. An effective standardization is the estimate of the variance-covariance of $\widehat{\mathbf{b}}_W$. A statistic which measures the total difference between the fits is

$$\text{TDBETAS} = \widehat{\boldsymbol{b}}_D^T \widehat{\boldsymbol{A}}_W^{-1} \widehat{\boldsymbol{b}}_D, \tag{8.11}$$

where $\boldsymbol{A}_W$ is the asymptotic Wilcoxon covariance matrix for linear models. Large values of TDBETAS indicate a discrepancy between the fits. A useful cutoff value is $(4(p+1)^2)/n$.

If $TDBETAS_R$ exceeds its benchmark, then usually we want to determine the individual cases causing the discrepancy between the fits. Let $\widehat{y}_{W,i}$ and $\widehat{y}_{HBR,i}$ denote the respective Wilcoxon and HBR fits for the ith case. A Studentized statistic which detects the observations that differ in fit is

$$CFITS_i = (\widehat{y}_{R,i} - \widehat{y}_{HBR,i})/\sqrt{n^{-1}\widehat{\tau}_S^2 + h_{c,i}\widehat{\tau}^2}. \tag{8.12}$$

An effective benchmark for CFITS_i is $2\sqrt{(p+1)/n}$. Note that the standardization of $CFITS_i$ accounts for the location of the ith case in the $\boldsymbol{X}$-space.

The objective of the diagnostic CFITS is *not* outlier deletion; rather, the intent is to identify the *few critical* data points for closer study, because these points are causing discrepancies between the highly efficient and high breakdown fits of the data. In this regard, the proposed benchmarks are meant as a heuristic aid, not a boundary to some formal critical region.

In the same way, the difference between the LS fit and either the Wilcoxon or HBR fits can be investigated. In general, though, we are interested in the difference between a highly efficient robust fit and a high breakdown robust fit. In all comparison cases, the standardization of the diagnostics is with the variance-covariance matrix of the Wilcoxon fit. For computation, the function `fitdiag`, in the collection `hbrfit`, computes these diagnostics for the Wilcoxon, HBR, GR, LS, and LTS fits. Its argument `est` specifies the difference to compute; for example, if `est=c("WIL","HBR")`, then the diagnostics between the Wilcoxon and HBR fits are computed, while `est=c("LTS","WIL")` computes the diagnostics between the Wilcoxon and LTS fits. Besides the diagnostics, the associated benchmarks are returned.

From the computation, the value of TDBETAS is 43.93 which far exceeds the benchmark of 1.2; hence, the diagnostic has been successful. Even more

[2]See also, McKean and Sheather (2009).

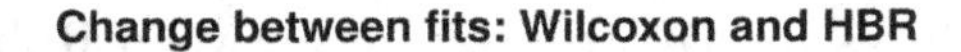

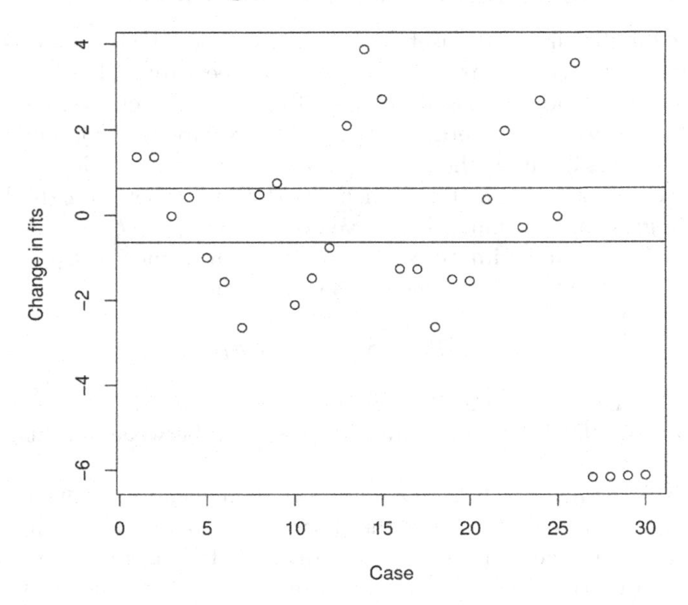

FIGURE 8.6
Plot of the changes in fits (CFITS) between the Wilcoxon and HBR fits for the simple dataset with the set of bad responses. The horizontal lines are set at the benchmark.

importantly, though, is that the diagnostic CFITS in Figure 8.6 clearly flags the four influential cases. These are the points at the bottom right corner of the plot. In reading plots, such as the CFITS plot, the large gaps are important. In this case, the four influential points clearly stand out and are the ones to investigate first. As the reader is asked to show, Exercise 8.11.5, for this dataset with the set of good responses, TDBETAS is less than its benchmark and none of the CFITS values, in absolute value, exceed their benchmark.

Example 8.3.1 (Fit Diagnostics for Stars Data). The next code segment computes the robust distances and the diagnostics for the difference between the Wilcoxon and HBR fits of the stars data, discussed in Section 8.2.

```
> dwilhbr = fitdiag(stars$temperature,stars$light,est=c("WIL","HBR"))
> tdbetas <- round(c(dwilhbr$tdbeta,dwilhbr$bmtd),digits=2)
> rdis <- sqrt(robdistwts(stars$temperature,stars$light)$robdis2)
```

The diagnostic TDBETAS has the value 67.92 which greatly exceeds its benchmark of 0.34. Thus numerically indicating that the HBR and Wilcoxon fits differ. The CFITS plot, right panel of Figure 8.7, clearly shows the four giant stars (Cases 11, 20, 30, and 34). It also finds the two stars between the giant stars and the main sequence stars, namely Cases 7 and 14. The

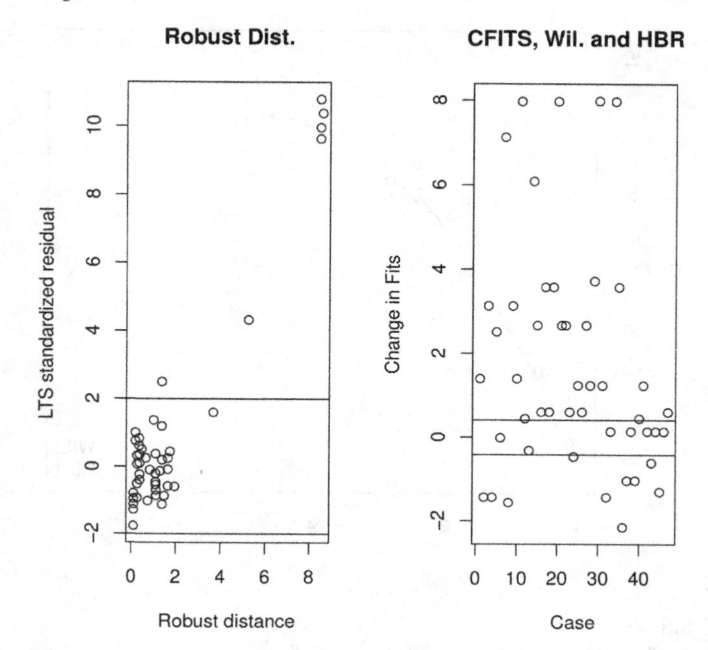

FIGURE 8.7
Robust distance plots and CFITS plot between HBR and Wilcoxon fits, stars data.

robust distance plot conveys similar information. For the record, the diagnostic TDBETAS for the difference in fits between the Wilcoxon and LTS fits is 265.39, which far exceeds the benchmark. ■

The following generated dataset illustrates the curvature problem for high breakdown fits.

Example 8.3.2 (Curvature Data). Hettmansperger and McKean (2011), page 267, consider a simulated quadratic model with $N(0,1)$ random errors and absolute contaminated normal xs. The model is $Y = 5.5|x| - 0.6x^2 + e$. The scatterplot of the data overlaid with the Wilcoxon and HBR fits are shown in Figure 8.8. The Wilcoxon and HBR estimates are:

```
> library(quantreg)
> library(hbrfit)
> curvature <- as.data.frame(curvature)
> summary(rfit(y~x+I(x^2),data=curvature))

Call:
rfit.default(formula = y ~ x + I(x^2), data = curvature)

Coefficients:
```

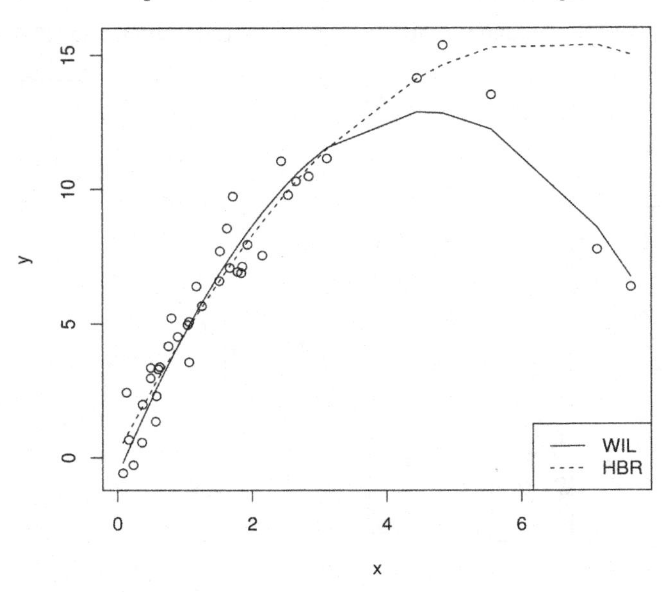

FIGURE 8.8
Scatterplot of quadratic data overlaid with fits.

```
              Estimate Std. Error t.value p.value
(Intercept)   -0.6650     0.4216   -1.58    0.12
x              5.9469     0.3265   18.21  <2e-16 ***
I(x^2)        -0.6525     0.0455  -14.35  <2e-16 ***
---
Signif. codes:
0 '***' 0.001 '**' 0.01 '*' 0.05 '.' 0.1 ' ' 1

Multiple R-squared (Robust): 0.84228
Reduction in Dispersion Test: 98.796 p-value: 0

> summary(hbrfit(y~x+I(x^2),data=curvature))

Call:
hbrfit(formula = y ~ x + I(x^2), data = curvature)

Coefficients:
              Estimate Std. Error t.value p.value
(Intercept)    0.0118     0.3999    0.03    0.98
x              4.9789     0.4702   10.59 9.4e-13 ***
I(x^2)        -0.4201     0.0922   -4.56 5.5e-05 ***
---
Signif. codes:
```

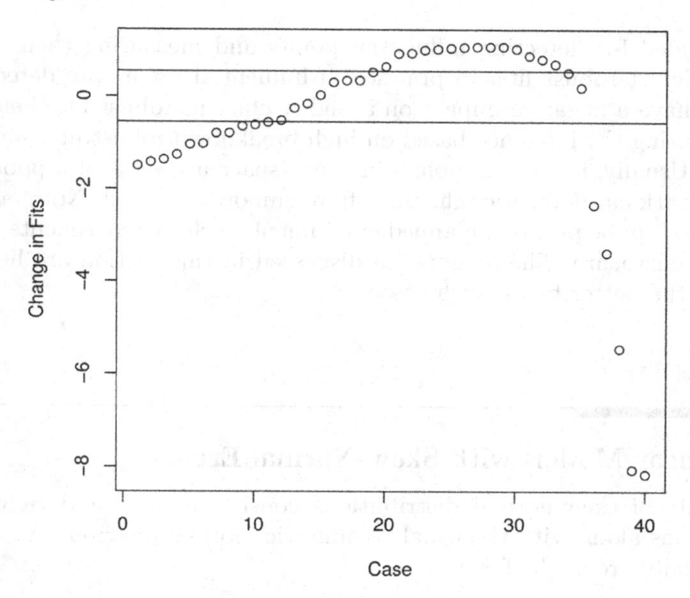

FIGURE 8.9
Plot of CFITS between HBR and Wilcoxon fits, quadratic data.

```
0 '***' 0.001 '**' 0.01 '*' 0.05 '.' 0.1 ' ' 1

Wald Test: 191.31 p-value: 0
```

As the summary of coefficients and the scatterplot show, the Wilcoxon and HBR fits differ. The Wilcoxon fit follows the model, while the HBR fit did not detect the curvature as well as the Wilcoxon fit. For the Wilcoxon and HBR fits, the value of TDBETAS is 67.637 with benchmark 0.9; so the diagnostics on the difference of the two fits agrees with this assessment. Figure 8.9 shows the corresponding plot of the diagnostic CFITS.

The points corresponding to the largest absolute change in CFITS are at the region of most curvature in the quadratic model. ■

As in the last example, high breakdown fits may be impaired, if curvature occurs "far" from the robust center of the data; see McKean et al. (1994) for a study on this concern for polynomial models. In the last example, the diagnostic CFITS did pinpoint this region.

8.3.3 Concluding Remarks

As the examples in this and the last sections show, in the case of messy datasets, influential points in the x-space can have a negative impact on highly efficient robust fits as well as LS fits. Hence, the use of diagnostics is recommended for detecting influential points and measuring their effect on highly efficient robust fits. In practice, if influential points are detected and found to have a negative impact on a highly efficient robust fit, then we recommend using the inference based on high breakdown robust fits, such as the HBR fit. Usually, influential points in the $\boldsymbol{x}$-space are signs of a poor design. For observational data, though, this often cannot be helped. Note, also, that such points can be part of a planned experiment, such as experiments designed to detect curvature. The diagnostics discussed in this section are helpful for assessing the better fit for such cases.

8.4 Linear Models with Skew-Normal Errors

The family of skew-normal distributions consists of left- and right-skewed distributions along with the usual, symmetric, normal distribution. The pdfs in this family are of the form

$$f(x;\alpha) = 2\phi(x)\Phi(\alpha x), \tag{8.13}$$

where the parameter α satisfies $-\infty < \alpha < \infty$ and, $\phi(x)$ and $\Phi(x)$ are the pdf and cdf of a standard normal distribution, respectively. If a random variable X has this pdf, we say that X has a standard skew-normal distribution with parameter α and write $X \sim SN(\alpha)$. If $\alpha = 0$, then X has a standard normal distribution. Further, X is distributed left-skewed if $\alpha < 0$ and right-skewed if $\alpha > 0$. This family of distributions was introduced by Azzalini (1985), who discussed many of its properties. We are interested in using this family of distributions for error distributions in the linear model,

$$Y_i = f(\boldsymbol{x}_i;\boldsymbol{\beta}) + e_i, \quad i = 1, 2, \ldots, n, \tag{8.14}$$

where $e = b\epsilon_i$, where $\epsilon_i \sim SN(\alpha)$, for some $-\infty < \alpha < \infty$, and the scale parameter $b > 0$. We next discuss rank-based fits and inference for such models; see McKean and Kloke (2014) for more information.

Since these fits are scale equivariant, there is no need to estimate the scale parameter b. Likewise, for inference on the vector of parameters $\boldsymbol{\beta}$, there is no need to estimate the shape parameter α. Wilcoxon scores could be used or the bent scores designed for skewed error distributions. However, to get an idea of what scores to consider, we next discuss the optimal scores for a specified α.

To obtain the optimal rank-based scores, because of equivariance, we need only the form (down to scale and location) of the pdf. So, for the derivation of the scores, assume that the random variable $X \sim SN(\alpha)$ with pdf (8.13). Then as Exercise 8.11.8 shows:

$$-\frac{f'(x;\alpha)}{f(x;\alpha)} = x - \frac{\alpha\phi(\alpha x)}{\Phi(\alpha x)}. \tag{8.15}$$

Denote the inverse of the cdf of X by $F^{-1}(u;\alpha)$. Then the optimal score function for X is

$$\varphi_\alpha(u) = F^{-1}(u;\alpha) - \frac{\alpha\phi(\alpha F^{-1}(u;\alpha))}{\Phi(\alpha F^{-1}(u;\alpha))}. \tag{8.16}$$

For all values of α, this score function is strictly increasing over the interval $(0,1)$; see Azzalini (1985). As expected, for $\alpha = 0$, the scores are the normal scores introduced in Chapter 3. Due to the first term on the right-side of expression (8.16), all the score functions in this family are unbounded, indicating that the skew-normal family of distributions is light-tailed. Thus, the influence functions of the rank-based estimators based on scores in this family are unbounded in the $\boldsymbol{Y}$-space and, hence, are not robust.

Similar to behavior of the rank-based estimator based on normal scores discussed in Chapter 3, the rank-based estimators based on skew-normal scores are less sensitive to contaminated distributions than the MLEs, as illustrated by the senstivity study in Section 8.4.1 and the simulation study in Section 8.4.2.

We also need the derivative of (8.16) to complete the installation of these scores in `scores`. Let $l(x)$ denote the function defined in expression (8.15). Then as Exercise 8.11.10 shows, the derivative of the optimal score function is

$$\varphi'_\alpha(u) = l'[F^{-1}(u;\alpha)]\frac{1}{2\phi[F^{-1}(u;\alpha)]\Phi[\alpha F^{-1}(u;\alpha)]}, \tag{8.17}$$

where

$$l'(x) = 1 + \frac{\alpha^2\phi(\alpha x)[\alpha x\Phi[\alpha F^{-1}(u;\alpha)] + \phi(\alpha x)]}{\Phi^2(\alpha x)}.$$

Hence, to install this class of scores in `Rfit`, we only need the computation of the quantiles $F^{-1}(u;\alpha)$. Azzalini (2014) developed the R package `sn` which computes the quantile function $F^{-1}(u;\alpha)$ and, also, the corresponding pdf and cdf. The command `qsn(u,alpha=alpha)` returns $F^{-1}(u;\alpha)$, for $0 < u < 1$. We have added the class `skewns` to the book package `npsm`. Figure 8.10 displays these skew-normal scores for $\alpha = -7, 1,$ and 5 in the right panels of the figure and the corresponding pdf in the left panels.

Note that the pdf for $\alpha = -7$ is left-skewed, while those for positive α values are right-skewed. Unsurprisingly, the pdf for $\alpha = 1$ is closer to symmetry than the pdfs of the others. The score function for the left-skewed pdf emphasizes relatively the right-tails over the left-tails, while the reverse is true for the right-skewed pdfs.

8.4.1 Sensitivity Analysis

For the sensitivity analysis, we generated $n = 50$ observations from a linear model of the form $y_i = x_i + e_i$, where x_i has a $N(0,1)$ distribution and e_i has a $N(0,10^2)$ distribution. The x_is and e_is are all independent. The generated

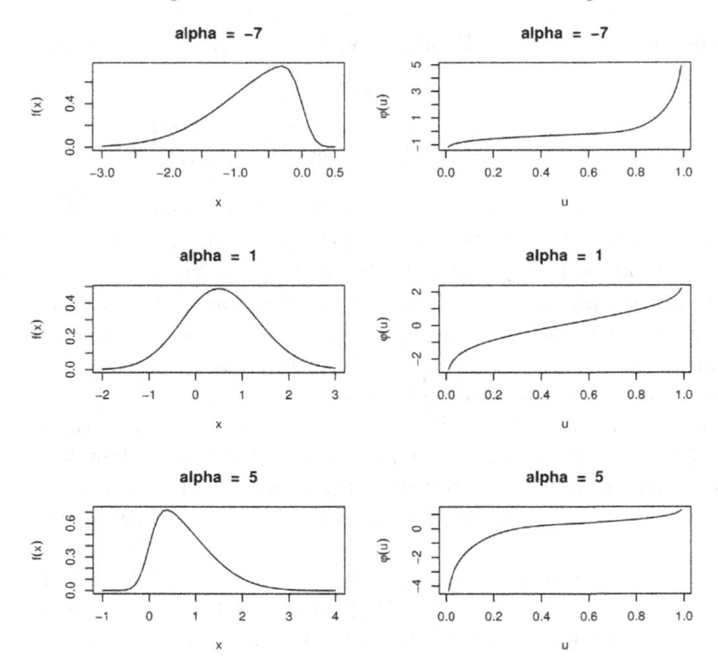

FIGURE 8.10
Plots displaying the pdfs of the three skew-normal distributions with shape parameter $\alpha = -7, 1$, and 5, along with the corresponding optimal scores.

data can be found in the data file **sensxy**. We added outliers of the form

$$y_{50} \leftarrow y_{50} + \Delta, \tag{8.18}$$

where Δ ranges through the values in the top row of Table 8.2. For an estimator $\hat{\beta}$, its sensitivity curve at Δ is

$$S(\Delta; \hat{\beta}) = \hat{\beta} - \hat{\beta}(\Delta), \tag{8.19}$$

where $\hat{\beta}$ and $\hat{\beta}(\Delta)$ denote the estimates of β on the original and modified data (8.18), respectively. We obtained the sensitivity curves for the estimators: Wilcoxon, normal scores, skew-normal ($\alpha = 3$), skew-normal ($\alpha = 5$), skew-normal ($\alpha = 7$), and maximum likelihood estimates (MLEs). The MLEs were computed by the package **sn**. The following code segment illustrates setting of the parameter in **skewns**:

```
> s5 <- skewns;   s5@param <- c(5)
```

That is, we first obtain a copy of the skew-normal scores object (**skewns**), and then we set the value of the α parameter to 5.

For all values of Δ, the changes in all of the rank-based estimates were less than 0.004; see McKean and Kloke (2014) for details. Thus the rank-based skew-normal estimators, including the normal scores estimator, appear to be less sensitive to contamined samples than the MLE. We show these changes in Table 8.2; hence, for this study, the MLE was not robust.

TABLE 8.2
Values of the sensitivity function for the MLE.

Δ	0	20	40	60	80	100	1000	2000
mle	0.00	−0.07	−0.07	−0.00	0.12	0.30	−5.80	−6.32

8.4.2 Simulation Study

We conclude this section with the results of a small simulation study concerning rank-based procedures based on skew-normal scores. The model simulated is

$$y_i = \beta_0 + \beta x_i + \theta c_i + e_i, \tag{8.20}$$

where x_i is distributed $N(0,1)$; e_i is distributed from a selected error distribution; $i = 1, \ldots, 100$; the x_is and e_is are all independent; and the variable c_i is a treatment indicator with values of either 0 or 1. We selected two error distributions for the study. One is a skew-normal distribution with shape parameter $\alpha = 5$, while the other is a contaminated version of a skew-normal. The contaminated errors are of the form

$$e_i = (1 - I_{\epsilon,i})W_i + I_{\epsilon,i}V_i, \tag{8.21}$$

where W_i has a skew-normal distribution with shape parameter $\alpha = 5$, V_i has $N(\mu_c = 10, \sigma_c^2 = 36)$ distribution, $I_{\epsilon,i}$ has a binomial $(1, \epsilon = 0.15)$ distribution, and W_i, V_i, and $I_{\epsilon,i}$ are all independent. This contaminated distribution is skewed with heavy right-tails. The design is slightly unbalanced with $n_1 = 45$ and $n_2 = 55$. Without loss of generality, β, θ, and β_0 were set to 0.

For procedures, we selected 7 rank-based procedures based on skew-normal scores with the respective values of α set at $2, 3, \ldots, 8$; the Wilcoxon procedure; and the MLE procedure. Hence, the asymptotically efficient rank-based procedure (score with $\alpha = 5$) is one of the selected procedures. The results presented are the empirical relative efficiencies (RE), which for each estimator is the ratio of the empirical mean-square error (MSE) of the MLE to the empirical MSE of the estimator; hence, values of this ratio less than 1 are favorable to the MLE, while values greater than 1 are favorable to the estimator. Secondly, we present the empirical confidence coefficients for nominal 95% confidence intervals. For all the procedures, we chose asymptotic confidence intervals of the form $\hat{\beta} \pm 1.96SE(\hat{\beta})$. We used a simulation size of 10,000.

The results are presented in Table 8.3. For the skew-normal errors, for both parameters β and θ, all the rank-based estimators except the Wilcoxon

TABLE 8.3
Summary of results of simulation study of rank-based (RB) procedures and the maximum likelihood procedure for the skew-normal distribution with shape $\alpha = 5$ and a skew-normal contaminated distribution.

	Skew-Normal Errors				Contaminated Errors			
	β		θ		β		θ	
Proced.	RE	Conf.	RE	Conf.	RE	Conf.	RE	Conf.
RB $\alpha = 2$	1.02	0.96	1.04	0.96	6.61	0.98	10.84	0.98
RB $\alpha = 3$	1.09	0.96	1.11	0.96	7.43	0.97	12.24	0.98
RB $\alpha = 4$	1.13	0.96	1.15	0.96	7.79	0.97	12.91	0.98
RB $\alpha = 5$	1.14	0.96	1.16	0.96	7.85	0.96	13.10	0.97
RB $\alpha = 6$	1.13	0.95	1.16	0.96	7.73	0.96	13.02	0.97
RB $\alpha = 7$	1.11	0.95	1.14	0.95	7.49	0.95	12.72	0.97
RB $\alpha = 8$	1.09	0.95	1.12	0.95	7.17	0.95	12.30	0.96
RB Wil.	0.78	0.95	0.79	0.95	4.70	0.96	7.56	0.97
MLE	1.00	0.93	1.00	0.93	1.00	0.96	1.00	0.99

estimator are more efficient than the MLE estimator. Note that the most efficient estimator for both β and θ is the rank-based estimator based on the score function with $\alpha = 5$; although, the result is not significantly different than the results for a few of the nearby (α close to 5) rank-based estimators. In terms of validity, the empirical confidences of all the rank-based estimators are close to the nominal confidence of 0.95. In this study the MLE is valid, also.

For the contaminated error distribution, the rank-based estimators are much more efficient than the MLE procedure. Further, the estimator with scores based on $\alpha = 5$ is still the most empirically powerful in the study. It has empirical efficiency of 785% relative to the MLE for β and 1310% for θ. Even the Wilcoxon procedure is over 400% more efficient than the MLE.

8.5 Hogg-Type Adaptive Procedure

Score selection was discussed in Section 3.7.1 for the two-sample problem. The two-sample location model, though, is a linear model (see expression (3.23)); hence, the discussion on scores in Chapter 3 pertains to regression models of this chapter also. Thus, for Model (8.1) if we assume that the pdf of the distribution of the errors is $f(t) = f_0[(t-a)/b]$, where f_0 is known and a and b are not, then the scores generated by the score function

$$\varphi_{f_0}(u) = -\frac{f_0'(F_0^{-1}(u))}{f_0(F_0^{-1}(u))}; \tag{8.22}$$

lead to fully efficient rank-based estimates (asymptotically equivalent to maximum likelihood estimates). For example, if we assume that the random errors of Model (8.1) are normally distributed, then rank-based fits based on normal scores are asymptotically equivalent to LS estimates. Other examples are discussed in Sections 8.4 and 3.7.1. Further, a family of score functions for log-linear models is discussed in Section 8.9.

Suppose, though, that we do not know the form. Based on the derivation of the optimal scores (8.22) given in Hettmansperger and McKean (2011), estimates based on scores "close" to the optimal scores tend to have high efficiency. A practical approach is to select a family of score functions which are optimal for a rich class of distributions and then use a data driven "selector" to choose a score from this family for which to obtain the rank-based fit of the linear model. We call such a procedure a data driven **adaptive scheme**. This is similar to Hogg's adaptive scheme for the two-sample problem (see Section 3.8), except our interest here is in obtaining a good fit of the linear model and not in tests concerned with location parameters.

Adaptive schemes should be designed for the problem at hand. For example, perhaps it is clear that only right-skewed distributions for the random errors need to be considered. In this case, the family of scores should include scores appropriate for right-skewed distributions.

For discussion, we consider a generic Hogg-type adaptive scheme designed for light- to heavy-tailed distributions, that can be symmetric or skewed (both left- and right-). The scheme utilizes the class of bent (Winsorized Wilcoxons) scores discussed in detail in Section 3.7.1. As discussed in Chapter 3, the four types of bent scores are appropriate for our family of distributions of interest. For example, consider the `bentscores4`. These are optimal for distribution with a "logistic" middle and "exponential" tails. Scores corresponding to heavier tailed distributions have larger intervals where the score function is flat. Such scores are optimal for symmetric distributions if the bends are at c and $1 - c$, for $0 < c < (1/2)$; else, they are optimal for skewed distributions. If the distribution has a longer right- than left-tail, then correspondingly the optimal score will have a longer flat interval on the right than on the left.

For our scheme, we have selected the nine bent scores which are depicted in Figure 8.11. The scores in the first column are for left-skewed distributions, those in the second column are for symmetric distributions, while those in the third column are for right-skewed distributions. The scores in the first row are for heavy-tailed distributions, those in the second row are for moderate-tailed distributions, and those in the third row are for light-tailed distributions.

Recall that our goal is to fit a linear model; hence, the selection of the score must be based on the residuals from an initial fit. For the initial fit, we have chosen to use the Wilcoxon fit. Wilcoxon scores are optimal for the logistic distribution which is symmetric and of moderate tail weight, slightly heavier tails than those of a normal distribution. Let $\hat{\mathbf{e}} = (\hat{e}_1, \ldots, \hat{e}_n)^T$, denote the vector of Wilcoxon residuals. As a selector, we have chosen the pair of statistics (Q_1, Q_2) proposed by Hogg. These are defined in expression (3.72)

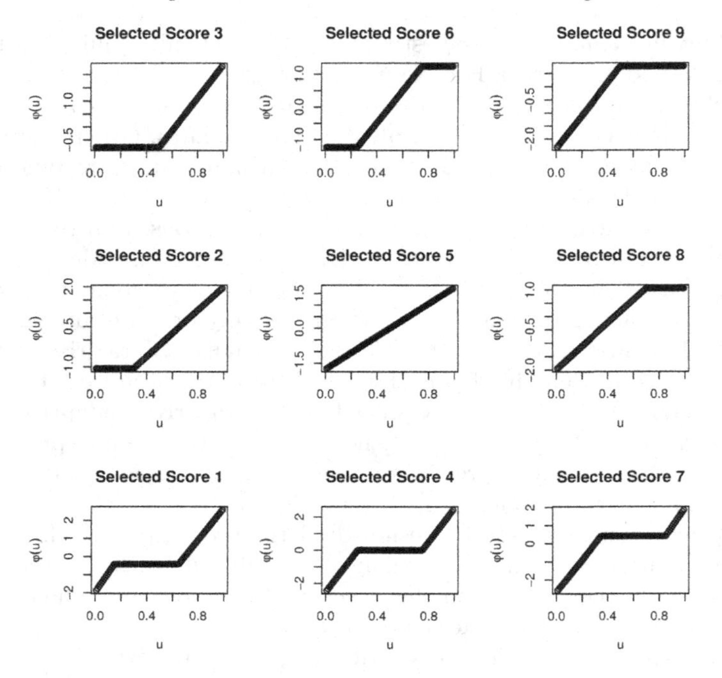

FIGURE 8.11
The nine bent scores for the generic Hogg-type adaptive scheme.

of Chapter 3. For the reader's convenience, we restate them here:

$$Q_1 = \frac{\bar{U}_{0.05} - \bar{M}_{0.5}}{\bar{M}_{0.5} - \bar{L}_{0.05}} \text{ and } Q_2 = \frac{\bar{U}_{0.05} - \bar{L}_{0.05}}{\bar{U}_{0.5} - \bar{L}_{0.5}}, \tag{8.23}$$

where $U_{0.05}$ is the mean of the Upper 5%, $M_{0.5}$ is the mean of the Middle 50%, and $L_{0.05}$ is the mean of the Lower 5% of the residuals $\hat{\mathbf{e}}$. Recall that Q_1 is a measure of skewness, while Q_2 is a measure of tail heaviness.

Cutoff values for the selection are required. In an investigation of this adaptive scheme for linear models, based on large simulation studies, Al-Shomrani (2003) developed the following cutoff values:

$$\begin{aligned} c_{lq1} &= 0.36 + (0.68/n) \\ c_{uq1} &= 2.73 - (3.72/n) \\ c_{lq2} &= \begin{cases} 2.17 - (3.01/n) & n < 25 \\ 2.24 - (4.68/n) & n \geq 25 \end{cases} \\ c_{uq2} &= \begin{cases} 2.63 - (3.94/n) & n < 25 \\ 2.95 - (9.37/n) & n \geq 25 \end{cases} \end{aligned}$$

Note that the scores are numbered 1 through 9 in Figure 8.11. Using these numbers, the scheme's selected score is:

$$
\begin{array}{rll}
Q_1 \le c_{lq1}, Q_2 \le c_{lq2} & \text{Select} & \text{Score \#1} \\
Q_1 \le c_{lq1}, c_{lq2} < Q_2 \le c_{uq2} & \text{Select} & \text{Score \#2} \\
Q_1 \le c_{lq1}, Q_2 > c_{uq2} & \text{Select} & \text{Score \#3} \\
c_{lq1} < Q_1 \le c_{uq1}, Q_2 \le c_{lq2} & \text{Select} & \text{Score \#4} \\
c_{lq1} < Q_1 \le c_{uq1}, c_{lq2} < Q_2 \le c_{uq2} & \text{Select} & \text{Score \#5} \\
c_{lq1} < Q_1 \le c_{uq1}, Q_2 > c_{uq2} & \text{Select} & \text{Score \#6} \\
Q_1 > c_{uq1}, Q_2 \le c_{lq2} & \text{Select} & \text{Score \#7} \\
Q_1 > c_{uq1}, c_{lq2} < Q_2 \le c_{uq2} & \text{Select} & \text{Score \#8} \\
Q_1 > c_{uq1}, Q_2 > c_{uq2} & \text{Select} & \text{Score \#9}
\end{array} \quad (8.24)
$$

We have written an auxiliary R function `adaptor` which computes this adaptive scheme. The response vector $\boldsymbol{Y}$ and design matrix $\boldsymbol{X}$ form the input, while the output includes both the initial (Wilcoxon) and selected fits, the scores selected and, for convenience, the number of the selected score. We illustrate its computation with the following examples.

Example 8.5.1 (Adaptive Scheme on Generated Exponential Errors). The first example consists of simulated data. We consider a regression model with two predictors each having a $N(0,1)$ distribution and with sample size $n = 40$. For an error distribution, we chose an exponential distribution. We set all regression coefficients to 0, so that the model generated is:

$$y_i = 0 + 0 \cdot x_{i1} + 0 \cdot x_{i2} + e_i.$$

The following code segment computes the adaptive scheme for this dataset. It shows the respective summaries of the selected score fit and the Wilcoxon fit.

```
> xmat <- adapteg[,1:2]
> y <- adapteg[,3]
> adapt <- adaptor(xmat,y)
> summary(adapt$fitsc)

Call:
rfit.default(formula = y ~ xmat, scores = sc, delta = delta,
    hparm = hparm)

Coefficients:
            Estimate Std. Error t.value p.value
(Intercept)   0.6076     0.2211    2.75  0.0092 **
xmatx1       -0.0839     0.0950   -0.88  0.3829
xmatx2        0.0276     0.0981    0.28  0.7798
---
```

```
Signif. codes:
0 '***' 0.001 '**' 0.01 '*' 0.05 '.' 0.1 ' ' 1

Multiple R-squared (Robust): 0.029845
Reduction in Dispersion Test: 0.56912 p-value: 0.5709

> summary(adapt$fitwil)

Call:
rfit.default(formula = y ~ xmat, delta = delta, hparm = hparm)

Coefficients:
            Estimate Std. Error t.value p.value
(Intercept)    0.636      0.229    2.78  0.0085 **
xmatx1        -0.130      0.141   -0.92  0.3630
xmatx2         0.109      0.146    0.74  0.4619
---
Signif. codes:
0 '***' 0.001 '**' 0.01 '*' 0.05 '.' 0.1 ' ' 1

Multiple R-squared (Robust): 0.046195
Reduction in Dispersion Test: 0.896 p-value: 0.41687

> adapt$iscore

[1] 9
```

In this case the adaptive scheme correctly chose score function #9; i.e., it selected the score for right-skewed error distributions with heavy tails.

```
> precision <- (adapt$fitsc$tauhat/adapt$fitwil$tauhat)^2
> precision

[1] 0.45174
```

The ratio of the squared $\hat{\tau}$'s (the selected score function to the Wilcoxon score function) is 0.452; hence, the selected fit is more precise in terms of standard errors than the Wilcoxon fit. ■

For our second example, we chose a real dataset.

Example 8.5.2 (Free Fatty Acid Data). In this dataset (`ffa`), the response is the free fatty acid level of 41 boys, while the predictors are age (in months), weight (lbs), and skin-fold thickness. It was initially discussed on page 64 of Morrison (1983) and more recently in Kloke and McKean (2012). The Wilcoxon Studentized residual, and $q-q$ plots are shown in the top panels of Figure 8.12. Note that the residual plot indicates right-skewness, which is definitely confirmed by the $q-q$ plot. For this dataset, our adaptive scheme selected score function #8, **bentscores2**, with bend at $c = 0.75$ (moderately heavy tailed and right-skewed), which confirms the residual plot. The bottom panels of Figure 8.12 display the Studentized residual and q–q plots based on the fit from the selected scores.

For the next code segment, the R matrix `xmat` contains the three predictors, while the R vector `ffalev` contains the response. The summaries of both the initial Wilcoxon fit and the selected score fit are displayed along with the value of the precision.

```
> xmat <- as.matrix(ffa[,1:3])
> y <- ffa[,4]
> adapt <- adaptor(xmat,y)
> summary(adapt$fitwil)

Call:
rfit.default(formula = y ~ xmat, delta = delta, hparm = hparm)

Coefficients:
            Estimate Std. Error t.value p.value
(Intercept)  1.49059    0.26761    5.57 2.4e-06 ***
```

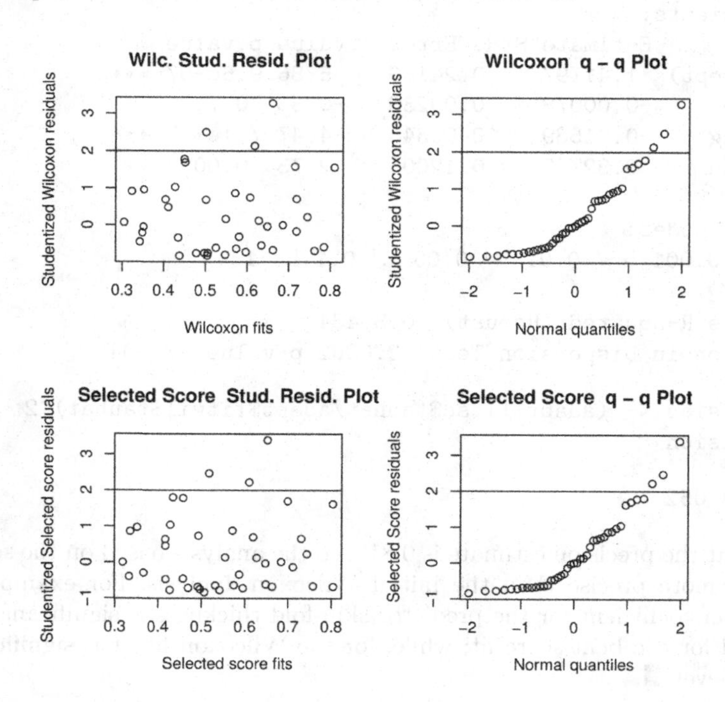

FIGURE 8.12
Wilcoxon (top pair) and the selected score (bottom pair) Studentized residual plots for Example 8.5.2.

```
xmatage     -0.00113   0.00262   -0.43 0.66748
xmatweight  -0.01535   0.00382   -4.02 0.00028 ***
xmatskin     0.27480   0.13335    2.06 0.04641 *
---
Signif. codes:
0 '***' 0.001 '**' 0.01 '*' 0.05 '.' 0.1 ' ' 1

Multiple R-squared (Robust): 0.37731
Reduction in Dispersion Test: 7.4733 p-value: 0.00049

> adapt$iscore

[1] 8

> summary(adapt$fitsc)

Call:
rfit.default(formula = y ~ xmat, scores = sc, delta = delta,
    hparm = hparm)

Coefficients:
             Estimate Std. Error t.value p.value
(Intercept)  1.41797   0.24186    5.86 9.6e-07 ***
xmatage     -0.00078   0.00236   -0.33  0.7425
xmatweight  -0.01539   0.00344   -4.47 7.1e-05 ***
xmatskin     0.32732   0.12005    2.73  0.0097 **
---
Signif. codes:
0 '***' 0.001 '**' 0.01 '*' 0.05 '.' 0.1 ' ' 1

Multiple R-squared (Robust): 0.39434
Reduction in Dispersion Test: 8.0302 p-value: 3e-04

> precision <- (adapt$fitsc$tauhat/adapt$fitwil$tauhat)^2
> precision

[1] 0.81052
```

Note that the precision estimate is 0.811, so the analysis based on the selected score is more precise than the initial Wilcoxon analysis. For example, the regression coefficient for the predictor skin-fold thickness is significant at the 1% level for the bent score fit; while, for the Wilcoxon fit, it is significant at the 5% level. ■

Remark 8.5.1. Kapenga and McKean (1989) and Naranjo and McKean (1997) developed estimators for the score function $\varphi(u)$ based on residuals from a fit of the linear model. These and several other adaptive schemes including the

Hogg-type scheme discussed above were compared in a large Monte Carlo study by Al-Shomrani (2003). These schemes are ultimately for fitting the linear model, and they are all based on residuals. So not surprisingly, their associated inference is somewhat liberal. In the Al-Shomrani's study, however, the Hogg-type scheme was less liberal than the other schemes in the study. In general, the Hogg-type scheme outperformed the others in terms of validity and empirical power. Okyere (2011) extended the Hogg-type scheme to mixed linear models. ■

8.6 Nonlinear

In this section we present example code to obtain Wilcoxon estimates for the general nonlinear regression problem. The model that we consider is

$$y_i = f(\boldsymbol{\theta}; \boldsymbol{x}_i) + e_i \tag{8.25}$$

where $\boldsymbol{\theta}$ is a $k \times 1$ vector of unknown parameters, y_i is a response variable, and $\boldsymbol{x}_i$ is a $p \times 1$ vector of explanatory variables.

As a simple working example consider the simple nonlinear model

$$f(\theta; x_i) = \exp\{\theta x_i\}, \quad i = 1, \ldots, n. \tag{8.26}$$

Suppose the errors are normally distributed. Setting $n = 25$, $\theta = 0.5$, and using uniformly distributed x's, a simulated example of this model is generated by the next code segment. Figure 8.13 displays a scatterplot of the data.

```
> n<-25
> theta<-0.5
> x<-runif(n,1,5)
> f<-function(x,theta) { exp(theta*x) }
> y<-f(x,theta)+rnorm(n,sd=0.5)
```

The rank-based fit of Model (8.25) is based on minimizing the same norm that was used for linear models. That is, for a specified score function $\varphi(u)$, the rank-based estimate of $\boldsymbol{\theta}$ is

$$\hat{\boldsymbol{\theta}}_\varphi = \text{Argmin}\|\boldsymbol{Y} - f(\boldsymbol{\theta}; \boldsymbol{x})\|_\varphi, \tag{8.27}$$

where $\boldsymbol{Y}$ is the $n \times 1$ vector and the components of $f(\boldsymbol{\theta}; \boldsymbol{x})$ are the $f(\boldsymbol{\theta}; \boldsymbol{x}_i)$s. For the traditional LS estimate, the squared-Euclidean norm is used instead of $\|\cdot\|_\varphi$. The properties of the rank-based nonlinear estimator were developed by Abebe and McKean (2007). More discussion of these rank-based estimates can be found in Section 3.14 of Hettmansperger and McKean (2011), including the estimator's influence function. Based on this influence function, the rank-based

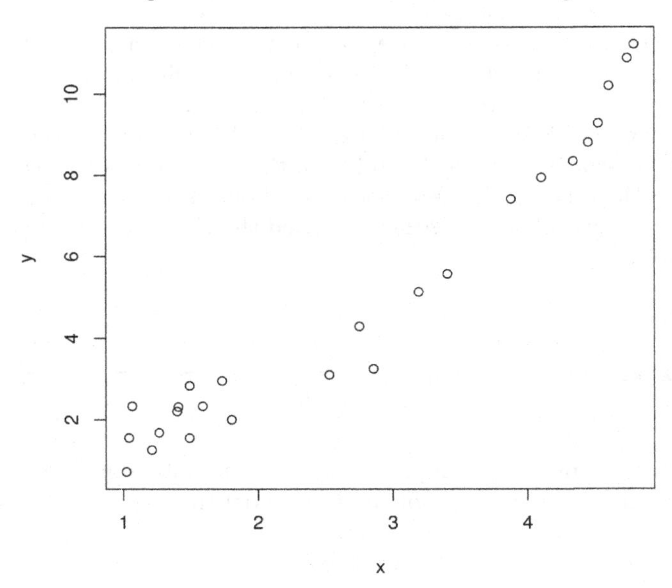

FIGURE 8.13
Scatterplot of the simulated data based on Model (8.26).

estimator is robust in the $\boldsymbol{Y}$-space, but not in the $\boldsymbol{x}$-space. The nonlinear HBR estimator developed by Abebe and McKean (2013) is robust with bounded influence in both the $\boldsymbol{Y}$-space and the $\boldsymbol{x}$-spaces. As in the case of linear models, the HBR estimator minimizes the weighted Wilcoxon norm; we recommend the high breakdown weights given in expression (8.9).

In this section, we discuss the Wilcoxon fit of a nonlinear model. The discussion for the HBR fit is in Section 8.6.4. We begin by discussing a simple Newton algorithm for the estimator and its subsequent computation using `Rfit`. For nonlinear fitting, the usual computational algorithm is a Gauss–Newton type procedure, which is based on a Taylor series expansion of $f(\boldsymbol{\theta}; \boldsymbol{x})$. Let $\hat{\boldsymbol{\theta}}^{(0)}$ be an initial estimate of $\boldsymbol{\theta}$. The Jacobian at $\boldsymbol{\theta}$ is the $n \times k$ matrix

$$\boldsymbol{J}(\boldsymbol{\theta}) = \left[\frac{\partial f_i(\boldsymbol{\theta})}{\partial \theta_j}\right]. \tag{8.28}$$

The expansion of $f(\boldsymbol{\theta})$ about $\hat{\boldsymbol{\theta}}^{(0)}$ is

$$f(\boldsymbol{\theta}) \approx f(\hat{\boldsymbol{\theta}}^{(0)}) + J(\hat{\boldsymbol{\theta}}^{(0)})\boldsymbol{\Delta}, \tag{8.29}$$

where $\boldsymbol{\Delta} = \boldsymbol{\theta} - \hat{\boldsymbol{\theta}}^{(0)}$. Hence, an approximation to the norm is

$$\|\boldsymbol{Y} - f(\boldsymbol{\theta})\|_\varphi \approx \|\{[\boldsymbol{Y} - f(\hat{\boldsymbol{\theta}}^{(0)})] - J(\hat{\boldsymbol{\theta}}^{(0)})\boldsymbol{\Delta}(\boldsymbol{\theta})\}\|_\varphi. \tag{8.30}$$

Note that the quantity within the braces on the right side defines a linear model with the quantity in brackets serving as the dependent variable, $J(\hat{\boldsymbol{\theta}}^{(0)})$ serving as the design matrix, and $\boldsymbol{\Delta}(\boldsymbol{\theta})$ serving as the vector of regression coefficients. For this linear model, let $\hat{\boldsymbol{\Delta}}$ be the rank-based estimate of $\boldsymbol{\Delta}$. Then, the first step estimate of $\boldsymbol{\theta}$ is

$$\hat{\boldsymbol{\theta}}^{(1)} = \hat{\boldsymbol{\theta}}^{(0)} + \hat{\boldsymbol{\Delta}}. \tag{8.31}$$

Usually at this point of the algorithm, convergence is tested based on the relative increments in the estimates and the dispersion function. If convergence has not been achieved, then $\hat{\boldsymbol{\theta}}^{(0)}$ is replaced by $\hat{\boldsymbol{\theta}}^{(1)}$, and the next step proceeds similar to the first step.

Often there is no intercept in the nonlinear model. In this case, the steps in the above algorithm consist of regressions through the origin. For rank-based estimation, this is handled by the adjustment described in Section 7.2. Our R code for the ranked-based nonlinear fit, discussed in Section 8.6.2, automatically makes this adjustment. First, we discuss implementation of the rank-based nonlinear procedure.

8.6.1 Implementation of the Wilcoxon Nonlinear Fit

Consider the rank-based fit of the general nonlinear model (8.25) using Wilcoxon scores. Let $\hat{\boldsymbol{\theta}}_W$ denote the estimator (8.27). As shown in Abebe and McKean (2007), under general conditions, the asymptotic variance-covariance matrix of $\hat{\boldsymbol{\theta}}_W$ is given by

$$\tau_W^2 \left(J(\boldsymbol{\theta}))^T J(\boldsymbol{\theta})\right)^{-1}, \tag{8.32}$$

where $J(\boldsymbol{\theta})$ is the Jacobian evaluated at the true vector of parameters $\boldsymbol{\theta}$ and τ_W is the scale parameter given in expression (3.19). The only difference for the asymptotic variance of the LS estimator is that the variance of the random errors, σ^2, replaces τ_W^2. Hence, the asymptotic relative efficiency (ARE) of the Wilcoxon estimator relative to the LS estimator is σ^2/τ_W^2; i.e., the same ARE as in linear models. In particular, at normal errors, this relative efficiency is 0.955.

Provided the Jacobian is a continuous function of $\boldsymbol{\theta}$, $J(\hat{\boldsymbol{\theta}}_W)^T J(\hat{\boldsymbol{\theta}}_W)$ is a consistent estimator of $J(\boldsymbol{\theta})^T J(\boldsymbol{\theta})$. Further, the same estimator of τ_W that we used in the linear model case (Koul et al. (1987)) but here based on the residuals $\hat{e}_W = \boldsymbol{Y} - f(\hat{\boldsymbol{\theta}}_W; \boldsymbol{x})$ is a consistent estimator of τ. Thus, the vector of standard errors of $\hat{\boldsymbol{\theta}}_W$ is

$$SE(\hat{\boldsymbol{\theta}}_W) = \text{Diagonal}\left\{\left[\hat{\tau}_W^2 J(\hat{\boldsymbol{\theta}}_W)^T J(\hat{\boldsymbol{\theta}}_W)^{-1}\right]^{1/2}\right\}. \tag{8.33}$$

Note, for future reference, that these standard errors are essentially the standard errors of the approximate linear model on the last step of the Gauss–Newton algorithm.

8.6.2 R Computation of Rank-Based Nonlinear Fits

We have written R software for the computation of the nonlinear rank-based estimates which utilizes the Gauss–Newton algorithm described above. Currently, it has options for the Wilcoxon and HBR fits. The function is `wilnl` and is included in the package `npsmReg2`.[3] Its defining R statement with default values is

```
wilnl = function(x,y,theta0,fmodel,jmodel,numstp=50,
            eps=.001,wts.type="WIL",
             intest="HL",intercept=FALSE)
```

Definitions of these arguments are:

- y is the $n \times 1$ vector of responses and `x` is the $n \times p$ matrix of predictors.
- `theta0` is the initial estimate (starting value) of $\boldsymbol{\theta}$. The routine assumes it to be a $k \times 1$ **matrix**.
- `fmodel` and `jmodel` are user supplied R functions, one for the model and the other for the Jacobian. The arguments to these functions are matrices. These are described most easily by the ensuing discussion of examples.
- `numstp` and `eps` are the total number of Newton steps and the tolerance for stopping, respectively.
- `wts.type="WIL"` or `wts.type="HBR"` obtain, respectively, in this case, the Wilcoxon or the HBR rank-based nonlinear fits.
- The rank-based algorithm uses an estimate of a (pseudo) intercept. It is either the Hodges–Lehmann estimate, `intest="HL"` or the median, `intest="MED"`. We recommend the default value `intest="HL"`, which generally leads to more efficient estimates, unless the data are highly skewed. In the later case, we recommend the median.
- Generally, nonlinear models do not have an intercept parameter. For such models, set the argument `int` at its default value, i.e., `int="NO"`. Occasionally, models do contain an intercept and, for these models, set `int` the value `int="YES"`. For models containing an intercept, one column of the Jacobian matrix consists of ones. In the user-supplied Jacobian function, make this the first column of the Jacobian.

The returned `list` file includes the following items of interest: the estimate of $\boldsymbol{\theta}$, `$coef`; the standard errors of the estimates, `$se`; the estimate of the scale parameter τ, `$tauhat`; the residuals, `$resid`; and the results of each step, `$coll`, (the step values of $\boldsymbol{\theta}$ and $\|\boldsymbol{\theta}\|_2^2$).

[3]See `https://github.com/kloke/book` for more information.

The user supplied functions are most easily described by discussing a few examples. Consider first the simple working model (8.26). Recall that the nonlinear function is $f(\theta, x) = \exp\{\theta x\}$. The arguments are matrices **x** and **theta**. In the example, **x** is 25×1 and **theta** is 1×1. The following model function, **expmod**, suffices:

```
expmod <- function(x,theta){ exp(x%*%theta) }
```

For the Jacobian, there is only one partial derivative given by $\partial f/\partial\theta = x\exp\{\theta x\}$. Hence our Jacobian function, **expjake**, is

```
expjake <- function(x,theta){ x*exp(x%*%theta) }
```

The analysis of a simple dataset follows in the next example.

8.6.3 Examples

Example 8.6.1 (Computation of Rank-Based Estimates for Model (8.26)). The following rounded data are a generated realization of the Model (8.26).

x	4.4	4.3	1.4	2.2	3.1	4.2	2.0	3.1	4.0	1.9	2.9	1.4	3.2
y	8.4	8.4	1.8	3.2	4.3	7.6	3.3	4.8	7.5	2.4	4.0	1.9	6.0
x	2.5	4.4	3.6	4.3	1.5	1.0	4.1	1.1	1.9	4.0	1.1	2.2	
y	3.6	8.5	5.3	8.8	1.4	1.8	7.1	1.4	2.9	7.3	2.6	2.3	

For the following code segment, the 25×1 matrix **x** contains the x values, while the vector y contains the y values. As a starting value, the true parameter $\theta = 0.5$ is used. For comparison, we computed the LS fit of this nonlinear model using the R function **nls**.

```
> expmod <- function(x,theta){ exp(x%*%theta) }
> expjake <- function(x,theta){ x*exp(x%*%theta) }
> fitwil <- wilnl(x,y,0.5,expmod,expjake)
> fitwil

Call:
wilnl(x = x, y = y, theta0 = 0.5, fmodel = expmod,
    jmodel = expjake)

Coefficients:
        [,1]
[1,] 0.49042

> summary(fitwil)

Call:
wilnl(x = x, y = y, theta0 = 0.5, fmodel = expmod,
    jmodel = expjake)
```

```
Coefficients:
      Estimate Std. Error t.value
[1,] 0.4904200  0.0050235  97.625

Number of iterations:  2

> fitls <- nls(y~exp(x*theta),start=list(theta=0.5))
> summary(fitls)

Formula: y ~ exp(x * theta)

Parameters:
      Estimate Std. Error t value Pr(>|t|)
theta 0.491728   0.004491   109.5   <2e-16 ***
---
Signif. codes:  0 '***' 0.001 '**' 0.01 '*' 0.05 '.' 0.1 ' ' 1

Residual standard error: 0.4636 on 24 degrees of freedom

Number of iterations to convergence: 2
Achieved convergence tolerance: 5.876e-06
```

The Wilcoxon and LS results are quite similar, which is to be expected since the error distribution selected is normal. The Wilcoxon algorithm converged in three steps. ■

For a second example, we altered this simple model to include an intercept.

Example 8.6.2 (Rank-Based Estimates for Model (8.26) with an Intercept). Consider the intercept version of Model (8.26); i.e., $f(\boldsymbol{\theta}; x) = \theta_1 + \exp\{\theta_2 x\}$. For this example, we set $\theta_1 = 2$. We use the same data as in the last example, except that 2 is added to all the components of the vector y. The code segment follows. Note that both the model and Jacobian functions have been altered to include the intercept. As starting values, we chose the true parameters $\theta_1 = 2$ and $\theta_2 = 0.5$.

```
> y <- y + 2
> expmod <- function(x,theta){ theta[1]+exp(x%*%theta[2]) }
> expjake <- function(x,theta){ cbind(rep(1,length(x[,1])),
+            x*exp(x%*%theta[2])) }
> fitwil = wilnl(x,y,as.matrix(c(2,.5),ncol=1), expmod,
+             expjake,intercept=TRUE)
> fitwil

Call:
wilnl(x = x, y = y, theta0 = as.matrix(c(2, 0.5), ncol = 1),
```

```
    fmodel = expmod, jmodel = expjake, intercept = TRUE)

Coefficients:
          [,1]
[1,] 2.0436573
[2,] 0.4887782

> summary(fitwil)

Call:
wilnl(x = x, y = y, theta0 = as.matrix(c(2, 0.5), ncol = 1),
    fmodel = expmod, jmodel = expjake, intercept = TRUE)

Coefficients:
      Estimate Std. Error t.value
[1,] 2.0436573  0.1636433  12.489
[2,] 0.4887782  0.0080226  60.925

Number of iterations:  2
```

Note that the Wilcoxon estimate of the intercept is close to the true value of 2. For both parameters, the asymptotic 95% confidence intervals (± 1.96SE) trap the true values. ■

The 4-Parameter Logistic Model

The 4-parameter logistic is a nonlinear model which is often used in pharmaceutical science for dose-response situations. The function is of the form

$$y = \frac{a-d}{1+(x/c)^b} + d, \tag{8.34}$$

where x is the dose of the drug and y is the response. The exponent b is assumed to be negative; hence, as $x \to 0$ (0 concentration) $y \to d$ and as $x \to \infty$ (full concentration) $y \to a$. So a and d are the expected values of the response under minimum and maximum concentration of the drug, respectively. It follows that the value of $(a+d)/2$ is the 50% response rate and that this occurs at $x = c$. In terms of biological assays, the value c is called the IC_{50}, the amount of concentration of the drug required to inhibit a biological process by 50%. See Crimin et al. (2012) for discussion of this model and the robust Wilcoxon fit of it.

Usually in pharmaceutical science, the dose of the drug is in log base 10 units. Also, this is an intercept model. Let $z = \log_{10} x$ and, to isolate the intercept, let $s = a - d$. Then the 4-parameter logistic model can be equivalently expressed as

$$Y_i = \frac{s}{1+\exp\{b[z_i \log(10) - \log(c)]\}} + d + e_i, \quad i = 1, 2, \ldots, n. \tag{8.35}$$

In this notation, as $z \to -\infty$, $E(Y_i) \to d$ and as $z \to \infty$, $E(Y_i) \to a$. Figure 8.14 shows the LS and Wilcoxon fits of this model for a realization of the model discussed in Example 8.6.3. From this scatterplot of the data, guesstimates of the asymptotes a and d and the IC_{50} c are readily obtained for starting values. For the Jacobian, the four partial derivatives of the model function f are given by

$$\begin{aligned}
\frac{\partial f}{\partial d} &= 1 \\
\frac{\partial f}{\partial s} &= \frac{1}{1+\exp\{b[z_i\log(10)-\log(c)]\}} \\
\frac{\partial f}{\partial c} &= s\left\{1+\exp\{b[z_i\log(10)-\log(c)]\}\right\}^{-2} \\
&\quad \times \left\{\frac{b}{c}\exp\{b[z\log(10)-\log(c)]\}\right\} \\
\frac{\partial f}{\partial b} &= -s\left\{1+\exp\{b[z_i\log(10)-\log(c)]\}\right\}^{-2} \\
&\quad \times \left\{[z\log(10)-\log(c)]\exp\{b[z\log(10)-\log(c)]\}\right\} \qquad (8.36)
\end{aligned}$$

In the next example, we obtain the robust fit of a realization of Model (8.35).

Example 8.6.3 (The 4-Parameter Logistic). We generated a realization of size $n = 24$ from Model (8.35) with doses ranging from 0.039 to 80 with two repetitions at each dose. The data are in the set `eg4parm`. For this situation, normal random errors with standard deviation one were generated. The parameters were set at $a = 10$, $b = -1.2$, $c = 3$, and $d = 110$; hence, $s = -100$. The scatterplot of the data is displayed in Figure 8.14. The functions for the model and the Jacobian are displayed in the next code segment. This is followed by the computation of the LS and Wilcoxon fits. These fits are overlaid on the scatterplot of Figure 8.14. The segment of code results in a comparison of the LS and Wilcoxon estimates of the coefficients and their associated standard errors. Note that the parameter s was fit, so a transformation is needed to obtain the estimates and standard errors of the original parameters.

```
> func <- function(z,theta){
+       d = theta[1]; s = theta[2]; c = theta[3]; b = theta[4]
+       func <- (s/(1 + exp(b*(z*log(10) - log(c))))) + d
+       func
+ }
> jake = function(z,theta){
+       d = theta[1]; s = theta[2]; c = theta[3]; b = theta[4]
+       xp = 1 + exp(b*(z*log(10) - log(c)))
+       fd = 1; fs = 1/xp
+       fc = s*(xp^(-2))*((b/c)*exp(b*(z*log(10) - log(c))))
+       fb = -s*(xp^(-2))*((z*log(10) - log(c))*exp(b*(z*log(10)
          - log(c))))
```

```
+       jake = cbind(fd,fs,fc,fb); jake
+ }

> fitwil = wilnl(z,y, theta0,func,jake,intercept=TRUE)
> summary(fitwil)

Call:
wilnl(x = z, y = y, theta0 = theta0, fmodel = func,
    jmodel = jake, intercept = TRUE)

Coefficients:
    Estimate Std. Error t.value
fd 105.81755    5.60053 18.8942
   -95.07300   10.31646 -9.2157
     3.55830    0.83385  4.2673
    -1.81164    0.68261 -2.6540

Number of iterations:  7

> fitls <- nls(y~(s/(1 + exp(b*(z*log(10) - log(c))))) + d,
+ start=list( b =-1.2, c = 3,d = 110,s = -100))
> summary(fitls)

Formula: y ~ (s/(1 + exp(b * (z * log(10) - log(c))))) + d

Parameters:
  Estimate Std. Error t value Pr(>|t|)
b -2.1853    0.8663  -2.523 0.020230 *
c  3.0413    0.6296   4.831 0.000102 ***
d 105.4357   5.4575  19.319 2.09e-14 ***
s -93.0179   9.2404 -10.066 2.83e-09 ***
---
Signif. codes:  0 '***' 0.001 '**' 0.01 '*' 0.05 '.' 0.1 ' ' 1

Residual standard error: 16.02 on 20 degrees of freedom

Number of iterations to convergence: 11
Achieved convergence tolerance: 4.751e-06

> resid = fitwil$residuals
> yhat = fitwil$fitted.values
> ehatls <- summary(fitls)$resid
> yhatls <- y -ehatls
```

Note that the LS fit has been impacted by the outlier with the $\log_{10}$ dose at approximately 0.7. The LS function `nls` interchanged the order of the coefficients. The standard errors of the LS estimates are slightly less.

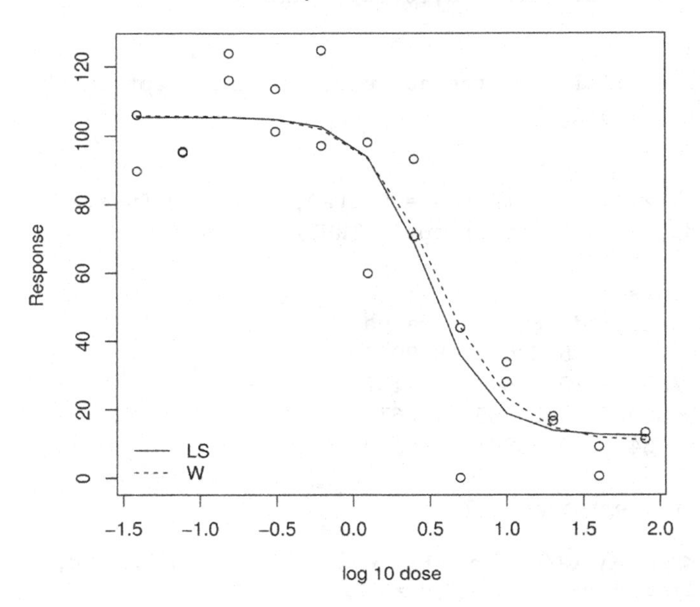

FIGURE 8.14
Scatterplot of the 4-parameter logistic data overlaid by Wilcoxon (W) and least squares (LS) nonlinear fits.

To demonstrate the robustness of the Wilcoxon fit, we changed the last response item from 11.33 to 70.0. The summary of the Wilcoxon and LS fits follows, while Figure 8.15 contains the scatterplot of the data and the overlaid fits.

```
> summary(fitwil)

Call:
wilnl(x = z, y = y, theta0 = theta0, fmodel = func,
    jmodel = jake, intercept = TRUE)

Coefficients:
    Estimate Std. Error t.value
fd 105.01857    6.08552 17.2571
   -89.58696   10.53863 -8.5008
     3.35834    0.81774  4.1068
    -2.20656    1.03166 -2.1388

Number of iterations:  3

> summary(fitls)
```

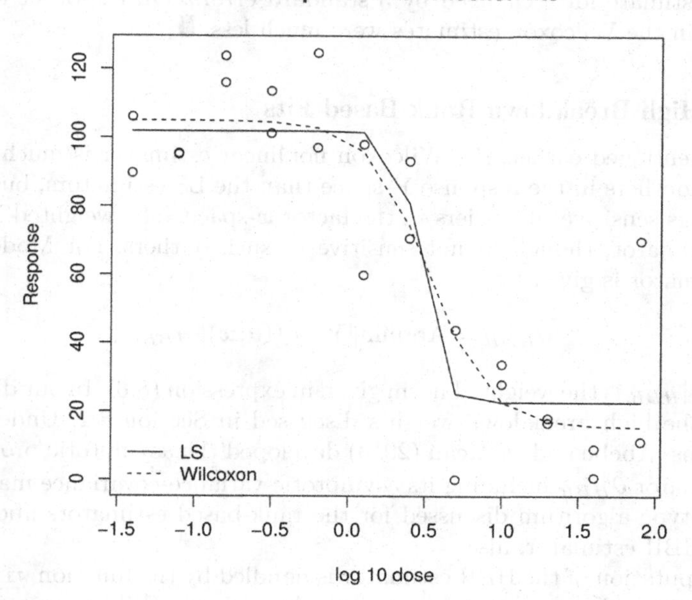

FIGURE 8.15
Scatterplot of the 4-parameter logistic for the changed data, overlaid by the Wilcoxon (W) and least squares (LS) nonlinear fits.

```
Formula: y ~ (s/(1 + exp(b * (z * log(10) - log(c))))) + d

Parameters:
  Estimate Std. Error t value Pr(>|t|)
b  -6.4036     9.2048  -0.696  0.49464
c   2.9331     0.7652   3.833  0.00104 **
d 101.9051     5.7583  17.697 1.10e-13 ***
s -79.1769     9.2046  -8.602 3.73e-08 ***
---
Signif. codes:  0 '***' 0.001 '**' 0.01 '*' 0.05 '.' 0.1 ' ' 1

Residual standard error: 19.8 on 20 degrees of freedom

Number of iterations to convergence: 32
Achieved convergence tolerance: 7.87e-06

>
```

As Figure 8.15 shows, the LS fit was impaired by the outlier. The LS estimates of the parameters a and d changed by over a standard error and the LS estimate for b changed by 3 standard errors. On the other hand, the changes in the Wilcoxon estimates were much less. ∎

8.6.4 High Breakdown Rank-Based Fits

As we mentioned earlier, the Wilcoxon nonlinear estimator is much less sensitive to outliers in the response Y-space than the LS estimators, but, similar to LS, it is sensitive to outliers in the factor $\boldsymbol{x}$-space. The weighted Wilcoxon HBR estimator, though, is not sensitive to such outliers. For Model (8.25), this estimator is given by

$$\hat{\boldsymbol{\theta}}_{HBR} = \text{Argmin}\|\boldsymbol{Y} - f(\boldsymbol{\theta};\boldsymbol{x})\|_{HBR}, \tag{8.37}$$

where $\|\cdot\|_{HBR}$ is the weighted norm given in expression (8.6). In our discussion, we use the high breakdown weights discussed in Section 8.2. Under general conditions, Abebe and McKean (2013) developed the asymptotic properties of the estimator $\hat{\boldsymbol{\theta}}_{HBR}$ including its asymptotic variance-covariance matrix. The Newton-type algorithm discussed for the rank-based estimators above works for the HBR estimator, also.

Computation of the HBR estimator is handled by the function `wilnl` using `wts="HBR"` as the argument for weights. As with the Wilcoxon nonlinear estimator, a consistent estimator of the variance-covariance matrix of the HBR estimator is obtained from the last linear step. The adjustment for the regression through the origin is handled similar to the Wilcoxon nonlinear estimator and, as with the Wilcoxon, the intercept estimator can be either the median of the residuals `int="MED"` or the Hodges–Lehmann `int="HL"`. We demonstrate the computation for our simple working example.

Example 8.6.4 (HBR Estimator for Example 8.6.1). The following code segment uses the same data as in Example 8.6.1. In particular the 25×1 R matrix x contains the independent variable and the 25×1 R vector y contains the dependent variable.

```
> expmod <- function(x,theta){ exp(x%*%theta) }
> expjake <- function(x,theta){ x*exp(x%*%theta) }
> fithbr = wilnl(x,y,0.5,expmod,expjake,wts.type="HBR")
> fithbr$coef

          [,1]
[1,] 0.4904202

>
```

For this data, there are no outliers in the x-space, so the HBR and Wilcoxon fits coincide. To demonstrate the robustness of the HBR estimator to outliers in factor space, we changed the last value of `x` from 2.3 to 8.0. The results are

```
> x[25,1] <- 8.0
> fitwil <- wilnl(x,y,0.5,expmod,expjake,wts.type="WIL")
> fitwil$coef

          [,1]
[1,] 0.2864679

> fithbr <- wilnl(x,y,0.5,expmod,expjake,wts.type="HBR")
> fithbr$coef

         [,1]
[1,] 0.488722

> fitls <- nls(y~exp(x*theta),start=list(theta=0.5))
> fitls$coef
NULL
```

Note that the Wilcoxon estimate of θ changed (absolutely) by 0.2, while the HBR estimate remained about the same. Also, the LS fit did not converge. ■

The diagnostics, TDBETAS and CFITS, which differentiate among the LS, Wilcoxon, and HBR fits that were discussed for linear models in Section 8.3 extend straightforwardly to nonlinear models, including their benchmarks; see Abebe and McKean (2013) for details.

8.7 Time Series

Let $\{X_t\}$ be a sequence of random variables observed over time, $t = 1, 2, \ldots, n$. A regression model frequently used in practice is the autoregressive model. This is a time series model where the observation at time t is a function of past observations plus some random noise. In this section, we discuss rank-based procedures for general order p autoregressive models. A related model consists of a linear model with random errors that follow a time series. Some discussion on robust procedures for these types of models is discussed in Section 6.6.3 of Hettmansperger and McKean (2011).

We say that X_t follows an autoregressive time series of order p, $X_t \sim \text{AR}(p)$, if

$$\begin{aligned} X_t &= \phi_0 + \phi_1 X_{t-1} + \phi_2 X_{t-2} + \cdots + \phi_p X_{t-p} + e_t \\ &= \phi_0 + \boldsymbol{Y}'_{t-1}\boldsymbol{\phi} + e_t, \quad t = 1, 2, \ldots, n \end{aligned} \tag{8.38}$$

where $p \geq 1$, $\boldsymbol{Y}_{t-1} = (X_{t-1}, X_{t-2}, \ldots, X_{t-p})'$, $\boldsymbol{\phi} = (\phi_1, \phi_2, \ldots, \phi_p)'$, and $\boldsymbol{Y}_0$ is an observable random vector independent of $\boldsymbol{e}$. The stationarity assumption requires that the solutions to the following equation,

$$x^p - \phi_1 x^{p-1} - \phi_2 x^{p-2} - \cdots - \phi_p = 0 \tag{8.39}$$

lie in the interval $(-1, 1)$; see, for example, Box et al. (2008). We further assume that the components of $\boldsymbol{e}$, e_t, are iid with cdf $F(x)$ and pdf $f(x)$, respectively.

Model (8.38) is a regression model with the tth response given by X_t and the *tth* row of the design matrix given by $(1, X_{t-1}, \ldots, X_{t-p})$, $t = p+1, \ldots, n$. Obviously the time series plot, X_t versus t, is an important first diagnostic. Also lag plots of the form X_t versus X_{t-j} are informative on the order p of the autoregressive series. We discuss determination of the order later.

As in Chapter 3, let $\varphi(u)$ denote a general score function which generates the score $a_\varphi(i) = \varphi[i/(n+1)]$, $i = 1, \ldots, n$. Then the rank-based estimate of $\boldsymbol{\phi}$ is given by

$$\begin{aligned} \widehat{\boldsymbol{\phi}} &= \text{Argmin} D_\varphi(\boldsymbol{\phi}) \\ &= \text{Argmin} \sum_{t=p+1}^{n} a_\varphi[R(X_t - \boldsymbol{Y}'_{t-1}\boldsymbol{\phi})](X_t - \boldsymbol{Y}'_{t-1}\boldsymbol{\phi}), \end{aligned} \tag{8.40}$$

where $R(X_t - \boldsymbol{Y}'_{t-1}\boldsymbol{\phi})$ denotes the rank of $X_t - \boldsymbol{Y}'_{t-1}\boldsymbol{\phi}$ among $X_1 - \boldsymbol{Y}'_0\boldsymbol{\phi}, \ldots$, $X_n - \boldsymbol{Y}'_{n-1}\boldsymbol{\phi}$. Koul and Saleh (1993) developed the asymptotic theory for these rank-based estimates. Outlying responses, though, also appear on the right side of the model; hence, error distributions with even moderately heavy tails produce outliers in factor space (points of high leverage). With this in mind, the HBR estimates of Section 8.2 should also be fitted. The asymptotic theory for the HBR estimators for Model (8.38) was developed by Terpstra et al. (2000) and Terpstra et al. (2001); see also, Chapter 5 of Hettmansperger and McKean (2011) for discussion.

The rank-based and HBR fits of the AR(p) are computed by `Rfit`. As simulations studies have confirmed, the usual standard errors from regression serve as good approximations to the asymptotic standard errors. Also, Terpstra et al. (2003) developed Studentized residuals for rank-based fits based on an AR(1) model. They showed, though, that Studentized residuals from the rank-based fit of the regression model were very close approximations to the AR(1) Studentized residuals. In this section, we use the regression Studentized residuals. We have written a simple R function, `lagmat`, which returns the design matrix given the vector of responses and the order of the autoregressive series. We illustrate this discussion with a simple example of generated data.

Example 8.7.1 (Generated AR(2) Data). In this example, we consider a dataset consisting of $n = 50$ variates generated form an AR(2) model with $\phi_1 = 0.6$, $\phi_2 = -0.3$, and random noise which followed a Laplace distribution with median 0 and scale parameter 10. The time series is plotted in the upper panel of Figure 8.16. The following R code segment obtains the Wilcoxon fit of the AR(2) model.

```
> library(npsmReg2)
> data <- lagmat(ar2,2)
> x <- data[,1]
> lag1 <- data[,2]
> lag2 <- data[,3]
> wil <- rfit(x ~ lag1 + lag2)
> summary(wil)

Call:
rfit.default(formula = x ~ lag1 + lag2)

Coefficients:
            Estimate Std. Error t.value p.value
(Intercept)   -2.442      2.972   -0.82  0.4157
lag1           0.701      0.121    5.79 6.5e-07 ***
lag2          -0.363      0.123   -2.96  0.0049 **
---
Signif. codes:
0 '***' 0.001 '**' 0.01 '*' 0.05 '.' 0.1 ' ' 1

Multiple R-squared (Robust): 0.39935
Reduction in Dispersion Test: 14.959 p-value: 1e-05
```

Note that the estimates of ϕ_1 and ϕ_2 are, respectively, 0.70 and -0.36 which are close to their true values of 0.6 and -0.3. With standard errors of 0.12, the corresponding 95% confidence intervals trap these true values. The lower panel of Figure 8.16 displays the plot of the Studentized Wilcoxon residuals. At a few time points, the residual series is outside of the ± 2 bounds, but these are mild discrepancies. The largest in absolute value has the residual value of about -4 at time 8. We next compute the HBR fit of the AR(2) model and obtain the diagnostic TDBETAS between the Wilcoxon and HBR fits.

```
> hbr <- hbrfit(x~lag1+lag2)
> summary(hbr)

Call:
hbrfit(formula = x ~lag1 + lag2)

Coefficients:
            Estimate Std. Error t.value p.value
(Intercept)   -2.376      2.926   -0.81  0.4211
lag1           0.704      0.139    5.07 7.2e-06 ***
lag2          -0.364      0.108   -3.37  0.0016 **
```

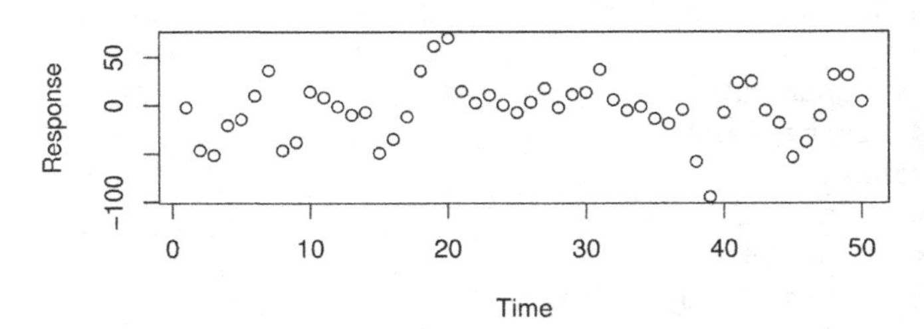

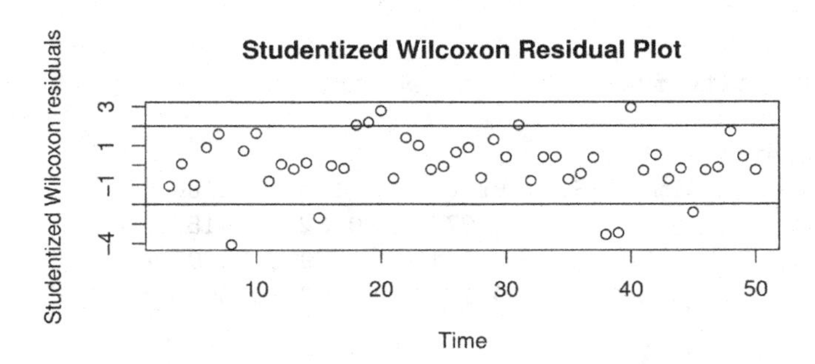

FIGURE 8.16
The upper panel shows the time series plot of the generated AR(2) data, while the lower panel is a plot of the Studentized Wilcoxon residuals from the fit of the AR(2) model.

```
---
Signif. codes:
0 '***' 0.001 '**' 0.01 '*' 0.05 '.' 0.1 ' ' 1

Wald Test: 13.5 p-value: 3e-05

> dnost <- fitdiag(cbind(lag1,lag2),x,est=c("WIL","HBR"))
> round(c(dnost$tdbeta,dnost$bmtd),3)

[1] 0.001 0.750
```

Notice that the HBR fit is quite close to the Wilcoxon fit. This is confirmed by TDBETAS which has the value 0.001 (benchmark of 0.75). ∎

There are essentially two types of outliers for time series. The first type consists of the **innovative outliers** (IO). These can occur when the error distribution has heavy-tails. If an outlier occurs at time t ($|e_t|$ is large), then this generally leads to a response outlier, X_t, at time t, i.e., an outlier on the left-side of the model. In subsequent times, it appears on the right-side and becomes incorporated into the model. These IO outliers generally lead to good points of high leverage. The dataset generated in Example 8.7.1 illustrates IO outliers. **Additive outliers** form a second type of outliers. These are patched into the series by a contaminating process, often leading to bad leverage points; see page 413 of Terpstra et al. (2001) for details. Both types of outliers occur in practice. For this reason, we recommend fitting both highly efficient and high breakdown rank-based estimates to the series and using the diagnostic TDBETAS to see if the fits differ. Terpstra et al. (2001) performed a large Monte Carlo study of rank-based and M estimates of time series involving both types of outliers. In all of the situations they simulated, the rank-based estimates either performed the best or nearly best over all procedures considered.

8.7.1 Order of the Autoregressive Series

In practice, before fitting the autoregressive time series model, (8.38), we must decide on its order p. One approach consists of subsequent fitting of models beginning with a model of large order. As discussed in Terpstra et al. (2001) for rank-based procedures, a more precise statement of the algorithm is: First, select a value P of maximal order; i.e., the residual analysis shows that the model fits well. Next, select a level α for the testing. Then the algorithm is given by

(0) Set $p = P$.

(1) While $p > 0$, fit Model (8.38) with order p.

(2) Let $\boldsymbol{\phi}_2 = (\phi_p, \ldots, \phi_P)'$. Then use the Wald test procedure to test $H_0 : \boldsymbol{\phi}_2 = \mathbf{0}$ versus $H_A : \boldsymbol{\phi}_2 \neq \mathbf{0}$.

(3) If H_0 is rejected, then stop and declare p to be the order; otherwise, set $p = p - 1$ and go to (1).

We return to the last example and use this algorithm to determine the order of the AR model based on the HBR fit.

Example 8.7.2 (Generated AR(2) Data, Continued). We have written a R function, `arorder`, to implement the above algorithm. The sequential testing is based on rank-based Wald's test of $\boldsymbol{A\phi} = \mathbf{0}$ where $\boldsymbol{A}$ is the constraint matrix as described in Step 2 of the algorithm. The user must select a maximum order of the AR model. For the example, we have selected `maxp=4`. Besides this, as shown in the code segment, the estimates and the covariance matrix of the estimates are inputted. It returns the stepwise results of the algorithm. We illustrate its use for the HBR fit.

```
> data <- lagmat(ar2,4)
> x <- data[,1]
> xmat <- data[,2:(4+1)]
> hbr <- hbrfit(x~xmat)
> varcov = vcov(hbr,details=T)
```

```
> est <- hbr$coef
> alg <- arorder(length(x),4,est,varcov)
> alg$results

 [,1]       [,2]        [,3]
    4 0.8331220 0.366707515
    3 0.4832491 0.620244664
    2 4.6797785 0.006679725
```

In this case, the algorithm correctly identifies the order of the autoregressive series, which is 2. ■

This algorithm performed well in a simulation study by Terpstra and McKean (2005). A similar algorithm for the order of a polynomial regression model was discussed in Section 4.6.1; see Graybill (1976) for discussion.

8.8 Cox Proportional Hazards Models

Let T denote the time until the event of an experimental unit. Let $\boldsymbol{x}$ denote the corresponding $p \times 1$ vector of covariates. Assume that T is a continuous random variable with respective pdf and cdf denoted by $f(t)$ and $F(t)$. Let $S(t) = 1 - F(t)$ denote the survival time of T. Let T_0 denote a baseline response; i.e., a response in the absence of all covariate effects.

The hazard function of T, which is often interpreted as the instantaneous chance of the event (death), is defined as

$$h(t) = \frac{f(t)}{S(t)};$$

see expression (8.43) for a formal definition. For a simple but much used example, assume that T_0 has the exponential distribution with pdf $f(t) = \lambda_0 \exp\{-\lambda_0 t)$, $t > 0$. Then, it is straighforward to show that the hazard function of T_0 has the constant value of λ_0. The proportional hazards model assumes that the hazard function of T is given by

$$\lambda(t; \boldsymbol{x}) = \lambda_0 e^{\boldsymbol{\beta}^T \boldsymbol{x}} \tag{8.41}$$

where $\boldsymbol{x}$ is a $p \times 1$ vector of covariates and $\boldsymbol{\beta}$ is a $p \times 1$ vector of parameters. Note that the hazard function of T is proportional to that of T_0.

To illustrate these ideas, assume that T_0 has constant hazard λ_0. Suppose the only covariate is an indicator variable w which is either 0 or 1 depending on whether a subject is not treated or treated. Assuming a proportional hazards model, the hazard function of T is given by

$$\lambda(t; w) = \lambda_0 e^{w\Delta}. \tag{8.42}$$

The hazard ratio of the experimental treatment relative to the control is then e^{Δ}. That is, Δ has the interpretation of log hazard; so $\Delta < 0$ indicates treatment is more favorable (results in lower hazard) than control, and a value of $\Delta > 0$ indicates that the treatment is less favorable than control (results in higher hazard). Further examples are given in Section 7.4.2 of Cook and DeMets (2008).

The proportional hazards model developed by Cox (1972) is a semiparametric model which does not necessarily specify the hazard function; only the relative effect of covariates is estimated. In the simple case under discussion, it can be used to estimate the parameter Δ as shown in the following example.

Example 8.8.1 (Hemorrhage data example, Continued). As a first example, we again consider Example 3.11.3 concerning the hemorrhage data. Using the function `coxph` from the `survival` package, we obtain an estimate Δ and corresponding inference.

```
> library(survival)
> fit<-coxph(Surv(time,recur)~genotype,data=hemorrhage)
> summary(fit)

Call:
coxph(formula = Surv(time, recur) ~ genotype, data = hemorrhage)

  n= 70, number of events= 18
   (1 observation deleted due to missingness)

         coef exp(coef) se(coef)     z Pr(>|z|)
genotype 1.33      3.79     0.57 2.34    0.019 *
---
Signif. codes:
0 '***' 0.001 '**' 0.01 '*' 0.05 '.' 0.1 ' ' 1

         exp(coef) exp(-coef) lower .95 upper .95
genotype      3.79      0.264      1.24      11.6

Concordance= 0.622  (se = 0.061 )
Likelihood ratio test= 6.61  on 1 df,   p=0.01
Wald test            = 5.46  on 1 df,   p=0.02
Score (logrank) test = 6.28  on 1 df,   p=0.01
```

From the output, we see $\hat{\Delta} = 1.33$ which indicates an increased risk for Group 2, those with heterozygous genotype. We also observe the estimated risk of hemorrhage for being heterozygous (Group 2) is 3.79 over being homozygous (Group 1). A 95% confidence interval is also given as (1.24, 11.57). Notice that the value of the score test statistics is the same as in Example 3.11.3. ∎

More generally, assume that the baseline hazard function is $\lambda_0(t)$. Assume that the hazard function of T is

$$\lambda(t; \boldsymbol{x}) = \lambda_0(t) e^{\boldsymbol{\beta}^T \boldsymbol{x}}.$$

Notice that the hazard ratio of two covariate patterns (e.g., for two subjects) is independent of baseline hazard

$$\frac{\lambda(t; \boldsymbol{x}_1)}{\lambda(t; \boldsymbol{x}_2)} = e^{\boldsymbol{\beta}(\boldsymbol{x}_1 - \boldsymbol{x}_2)}.$$

We close this section with the following example concerning an investigation with treatment at two levels and several covariates.

Example 8.8.2 (DES for treatment of prostate cancer)**.** The following example is taken from Collett (2003); data are available from the publisher's website. Under investigation in this clinical trial was the pharmaceutical agent diethylstilbestrol, (DES); subjects were assigned treatment to 1.0 mg DES (treatment = 2) or to placebo (treatment = 1). Covariates include age, serum hemoglobin level, size, and the Gleason Index.

In Exercise 8.11.21, the reader is asked to obtain the full model fit for the Cox proportional hazards model. Several of the explanatory variables are nonsignificant, though in practice one may want to include important risk factors such as age in the final model. For demonstration purposes, we have dropped `age` and `shb` from the model. As discussed in Collett (2003), the most important predictor variables are size and index.

```
> f2<-coxph(Surv(time,event=status)~as.factor(treatment)+size+index,
+ data=prostate)
> summary(f2)

Call:
coxph(formula = Surv(time, event = status) ~ as.factor(treatment) +
    size + index, data = prostate)

  n= 38, number of events= 6

                          coef exp(coef) se(coef)      z Pr(>|z|)
as.factor(treatment)2 -1.11272   0.32866  1.20313 -0.925   0.3550
size                   0.08257   1.08608  0.04746  1.740   0.0819 .
index                  0.71025   2.03450  0.33791  2.102   0.0356 *
---
Signif. codes:  0 '***' 0.001 '**' 0.01 '*' 0.05 '.' 0.1 ' ' 1

                      exp(coef) exp(-coef) lower .95 upper .95
as.factor(treatment)2    0.3287     3.0426   0.03109     3.474
size                     1.0861     0.9207   0.98961     1.192
index                    2.0345     0.4915   1.04913     3.945
```

```
Concordance= 0.873  (se = 0.132 )
Rsquare= 0.304   (max possible= 0.616 )
Likelihood ratio test= 13.78  on 3 df,    p=0.003226
Wald test            = 10.29  on 3 df,    p=0.01627
Score (logrank) test = 14.9   on 3 df,   p=0.001903
```

These data suggest that the Gleason Index is a significant risk factor of mortality (p-value=0.0356). Size of tumor is marginally significant (p-value=0.0819). Given that $\hat{\Delta} = -1.11272 < 1$, it appears that DES lowers risk of mortality; however, the p-value $= 0.3550$ is nonsignificant. ■

8.9 Accelerated Failure Time Models

In this section we consider analysis of survival data based on an accelerated failure time model. We assume that all survival times are observed. Rank-based analysis with censored survival times is considered in Jin et al. (2003).

Consider a study on experimental units (subjects) in which data are collected on the time until failure of the subjects. Hence, the setup for this section is the same as in the previous two sections of this chapter, with time until event replaced by time until failure. For such an experiment or study, let T be the time until failure of a subject, and let $\boldsymbol{x}$ be the vector of associated covariates. The components of $\boldsymbol{x}$ could be indicators of an underlying experimental design and/or concomitant variables collected to help explain random variability. Note that $T > 0$ with probability one. Generally, in practice, T has a skewed distribution. As in the last section, let the random variable T_0 denote the baseline time until failure. This is the response in the absence of all covariates.

In this section, let $g(t;\boldsymbol{x})$ and $G(t;\boldsymbol{x})$ denote the pdf and cdf of T, respectively. In the last section, we introduced the hazard function $h(t)$. A more formal definition of the hazard function is the limit of the rate of instantaneous failure at time t; i.e.,

$$\begin{aligned} h(t;\boldsymbol{x}) &= \lim_{\Delta t \downarrow 0} \frac{P[t < T \le t + \Delta t | T > t; \boldsymbol{x}]}{\Delta t} \\ &= \lim_{\Delta t \downarrow 0} \frac{g(t;\boldsymbol{x})\Delta t}{\Delta t(1 - G(t;\boldsymbol{x})} = \frac{g(t;\boldsymbol{x})}{1 - G(t;\boldsymbol{x})}. \end{aligned} \tag{8.43}$$

Models frequently used with failure time data are the log-linear models

$$Y = \alpha + \boldsymbol{x}^T\boldsymbol{\beta} + \epsilon, \tag{8.44}$$

where $Y = \log T$ and ϵ is random error with respective pdf and cdf $f(s)$ and $F(s)$. We assume that the random error ϵ is free of $\boldsymbol{x}$. Hence, the baseline response is given by $T_0 = \exp\{\epsilon\}$. Let $h_0(t)$ denote the hazard function of T_0. Because

$$T = \exp\{Y\} = \exp\{\alpha + \boldsymbol{x}^T\boldsymbol{\beta} + \epsilon\} = \exp\{\alpha + \boldsymbol{x}^T\boldsymbol{\beta}\}\exp\{\epsilon\} = \exp\{\alpha + \boldsymbol{x}^T\boldsymbol{\beta}\}T_0,$$

it follows that the hazard function of T is

$$h_T(t; \boldsymbol{x}) = \exp\{-(\alpha + \boldsymbol{x}^T\boldsymbol{\beta})\}h_0(\exp\{-(\alpha + \boldsymbol{x}^T\boldsymbol{\beta})\}t). \tag{8.45}$$

Notice that the effect of the covariate $\boldsymbol{x}$ either accelerates or decelerates the instantaneous failure time of T; hence, log-linear models of the form (8.44) are generally called **accelerated failure time models**.

If T_0 has an exponential distribution with mean $1/\lambda_0$, then the hazard function of T simplifies to:

$$h_T(t; \boldsymbol{x}) = \lambda_0 \exp\{-(\alpha + \boldsymbol{x}^T\boldsymbol{\beta})\}; \tag{8.46}$$

i.e., Cox's proportional hazard function given by expression (8.41) of the last section. In this case, it follows that the density function of ϵ is the extreme-valued pdf given by

$$f(s) = \lambda_0 e^s \exp\{-\lambda_0 e^s\}, \quad -\infty < s < \infty. \tag{8.47}$$

Accelerated failure time models are discussed in Kalbfleisch and Prentice (2002). As a family of possible error distributions for ϵ, they suggest the generalized log F family; that is, $\epsilon = \log T_0$, where down to a scale parameter, T_0 has an F-distribution with $2m_1$ and $2m_2$ degrees of freedom. In this case, we say that $\epsilon = \log T_0$ has a $GF(2m_1, 2m_2)$ distribution. Kalbfleisch and Prentice discuss this family for $m_1, m_2 \geq 1$; while McKean and Sievers (1989) extended it to $m_1, m_2 > 0$. This provides a rich family of distributions. The distributions are symmetric for $m_1 = m_2$; positively skewed for $m_1 > m_2$; negatively skewed for $m_1 < m_2$; moderate to light-tailed for $m_1, m_2 > 1$; and heavy-tailed for $m_1, m_2 \leq 1$. For $m_1 = m_2 = 1$, ϵ has a logistic distribution, while as $m_1 = m_2 \to \infty$ the limiting distribution of ϵ is normal. Also, if one of m_i is one, while the other approaches infinity, then the GF distribution approaches an extreme valued-distribution, with pdf of the form (8.47). So at least in the limit, the accelerated GF models encompass the proportional hazards models. See Kalbfleisch and Prentice (2002) and Section 3.10 of Hettmansperger and McKean (2011) for discussion.

The accelerated failure time models are linear models so the rank-based fit and associated inference using Wilcoxon scores can be used for analyses. By a prudent choice of a score function, though, this analysis can be optimized. We next discuss optimal score functions for these models and show how to compute analyses based on the them using `Rfit`. We begin with the proportional hazards model and then discuss the scores for the generalized log F-family.

Suppose a proportional hazard model is appropriate, where the baseline random variable T_0 has an exponential distribution with mean $1/\lambda_0$. Then ϵ

has the extreme valued pdf given by (8.47). Then as shown in Exercise 8.11.22 the optimal rank-based score function is $\varphi(u) = -1 - \log(1-u)$, for $0 < u < 1$. A rank-based analysis using this score function is asymptotically fully efficient. These scores are in the package `Rfit` under the name `logrank.scores`. The left panel of Figure 8.17 contains a plot of these scores, while the right panel shows a graph of the corresponding extreme valued pdf, (8.47). Note that the density has very light right-tails and much heavier left-tails. To guard against the influence of large (absolute) observations from the left-tails, the scores are bounded on the left, while their behavior on the right accommodates light-tailed error structure. The scores, though, are unbounded on the right and, hence, the resulting rank-based analysis is not robust in response space. In the sensitivity analysis discussed in McKean and Sievers (1989), however, the rank-based estimates based on log-rank scores were much less sensitive to outliers than their maximum likelihood estimates counterparts.

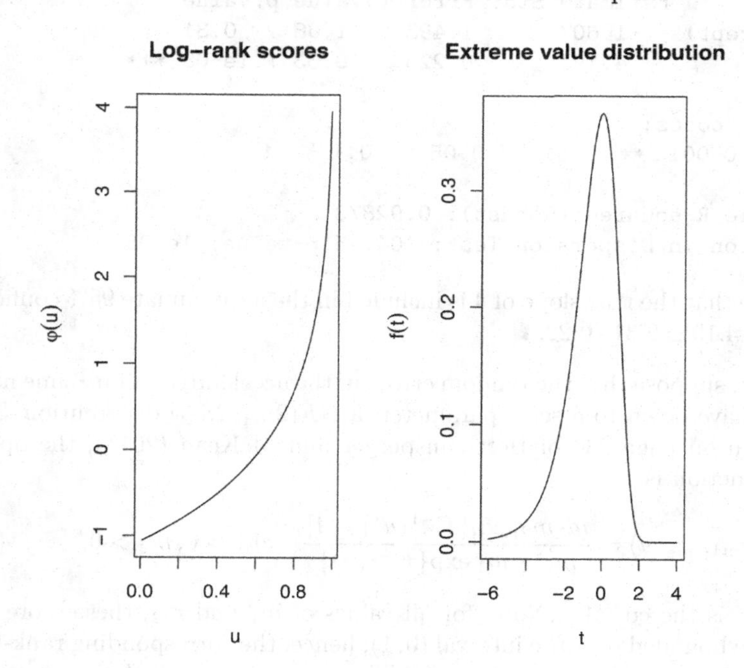

FIGURE 8.17
(*Left*) Log-rank score function. (*Right*) Probability density function for the corresponding extreme value distribution.

We illustrate the use of these scores in the next example.

Example 8.9.1 (Simulated Exponential Data). The data for this model are generated from a proportional hazards model with $\lambda = 1$ based on the code below.

```
eps <- log(rexp(10))
x <- 1:10
y <- round(4*x+eps,digits=2)
```

The actual data used are given in Exercise 8.11.28. Using `Rfit` with the log-rank score function, we obtain the fit of this dataset:

```
> fit <- rfit(y~x,scores=logrank.scores)
> summary(fit)

Call:
rfit.default(formula = y ~ x, scores = logrank.scores)

Coefficients:
            Estimate Std. Error t.value p.value
(Intercept)   -1.607      1.493   -1.08    0.31
x              4.191      0.225   18.63 7.1e-08 ***
---
Signif. codes:
0 '***' 0.001 '**' 0.01 '*' 0.05 '.' 0.1 ' ' 1

Multiple R-squared (Robust): 0.92873
Reduction in Dispersion Test: 104.25 p-value: 1e-05
```

Note that the true slope of 4 is included in the approximate 95% confidence interval $4.19 \pm 2.31 \cdot 0.22$. ■

Next, suppose that the random errors in the accelerated failure time model, (8.44), have down to a scale parameter, a $GF(2m_1, 2m_2)$ distribution. Then, as shown on page 234 of Hettmansperger and McKean (2011), the optimal score function is

$$\varphi_{m1_1,m_2}(u) = \frac{m_1 m_2[\exp\{F^{-1}(u)\} - 1]}{m_2 + m_1 \exp\{F^{-1}(u)\}}, \quad m_1 > 0, m_2 > 0, \tag{8.48}$$

where F is the cdf of ϵ. Note, for all values of m_1 and m_2, these score functions are bounded over the interval $(0, 1)$; hence, the corresponding rank-based analysis is robust in response space. These scores are called the generalized log-F scores (GLF). The library `Rfit` contains the `logGFscores` to be used for these cases. For this code, we have used the fact that the *pth* quantile of the $F_{2m_1,2m_2}$ cdf satisfies

$$q = \exp\{F_\epsilon^{-1}(p)\} \text{ where } q = F^{-1}_{2m_1,2m_2}(p).$$

The default values are set at $m_1 = m_2 = 1$, which gives the Wilcoxon scores.

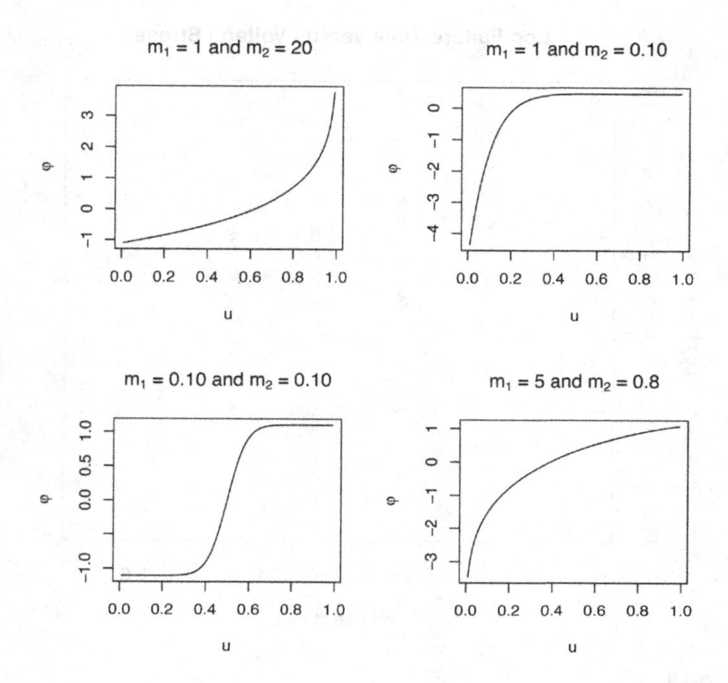

FIGURE 8.18
GLF scores for various settings of m_1 and m_2.

Figure 8.18 shows the diversity of these scores for different values of m_1 and m_2. It contains plots of four of the scores. The upper-left corner graph displays the scores for $m_1 = 1$ and $m_2 = 20$. These are suitable for error distributions which have moderately heavy (heaviness of a logistic distribution) left-tails and very light right-tails. In contrast, the scores for $m_1 = 1$ and $m_2 = 0.10$ are appropriate for moderately heavy left-tails and very heavy right-tails. The lower left panel of the figure is a score function designed for heavy-tailed and symmetric distributions. The final plot shows the score function for $m_1 = 5$ and $m_2 = 0.8$, which is appropriate for moderate left-tails and heavy right-tails. But note from the degree of downweighting that the right-tails for this last case are clearly not as heavy as for the two cases with $m_2 = 0.10$.

The next example serves as an application of the log F-scores.

Example 8.9.2 (Insulating Fluid Data). Hettmansperger and McKean (2011) present an example involving failure time (T) of an electrical insulating fluid subject to seven different levels of voltage stress (x). The data are in the dataset `insulation`. Figure 8.19 shows a scatterplot of the log of failure time ($Y = \log T$) versus the voltage stress. As voltage stress increases, time until failure of the insulating fluid decreases. It appears that a simple linear model suffices. In their discussion, Hettmansperger and McKean recommend a

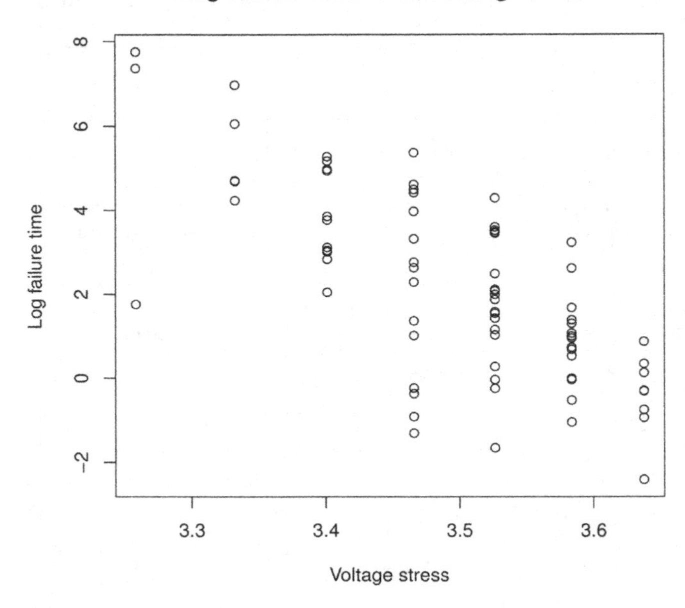

FIGURE 8.19
Log failure times of the insulation fluid versus the voltage stress.

rank-based fit based on generalized log F-scores with $m_1 = 1$ and $m_2 = 5$. This corresponds to a distribution with left-tails as heavy as a logistic distribution and right-tails lighter than a logistic distribution; i.e., moderately skewed left. The following code-segment illustrates computation of the rank-based fit of these data based on this log F-score.

```
> myscores <- logGFscores
> myscores@param=c(1,5)
> fit <- rfit(log.time~log.stress,scores=myscores,data=insulation)
> summary(fit)

Call:
rfit.default(formula = log.time ~ log.stress, data = insulation,
    scores = myscores)

Coefficients:
            Estimate Std. Error t.value p.value
(Intercept)    64.03       6.51    9.83 4.5e-15 ***
log.stress    -17.68       1.86   -9.50 1.9e-14 ***
---
```

```
Signif. codes:
0 '***' 0.001 '**' 0.01 '*' 0.05 '.' 0.1 ' ' 1

Multiple R-squared (Robust): 0.50992
Reduction in Dispersion Test: 76.997 p-value: 0
```

Not surprisingly, the estimate of the slope is highly significant. As a check on goodness-of-fit, Figure 8.20 presents the Studentized residual plot and the q–q plot of the Studentized residuals versus the quantiles of a log F-distribution with the appropriate degrees of freedom 2 and 10. This plot is fairly linear,

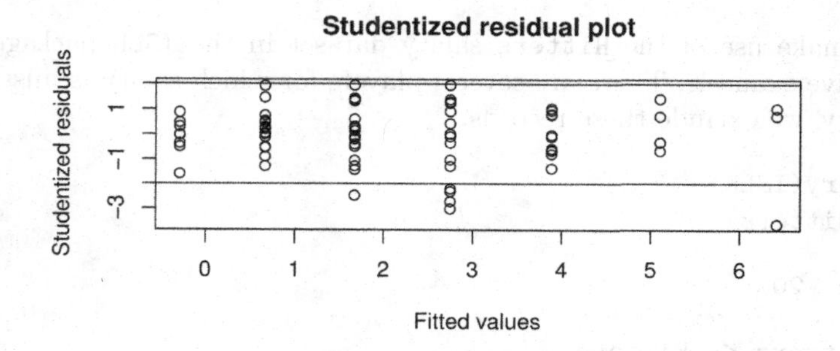

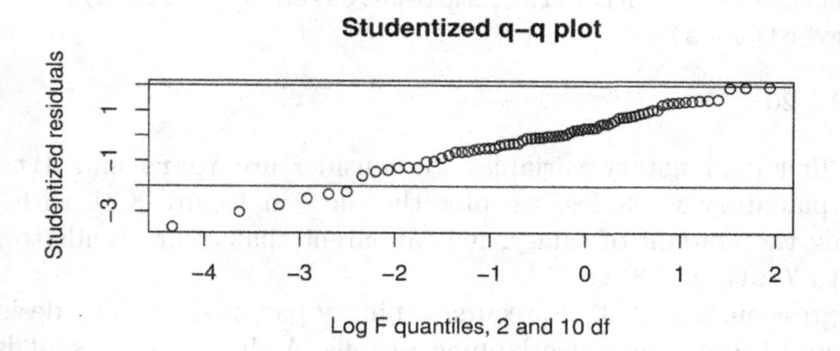

FIGURE 8.20
The top panel contains the Studentized residual plot of the rank-based fit using generalized log F-scores with 2 and 10 degrees of freedom. The bottom panel shows the $q-q$ plot of Studentized residuals versus log F-scores with 2 and 10 degrees of freedom.

indicating[4] that an appropriate choice of scores was made. The residual plot indicates a good fit. The outliers on the left are mild and, based on the $q-q$

[4]See the discussion in Section 3.10 of Hettmansperger and McKean (2011).

plot, follow the pattern of the log F-distribution with 2 and 10 degrees of freedom. ■

8.10 Trees for Regression

Trees for regression are similar in concept to those for classification discussed in Section 6.4 with the response variable being numeric instead of categorical. In addition to the basic regression tree, there are also bagged trees and random forests.

We make use of the `Hitters` salary dataset in the `ISLR` package as an illustrative example. There are several players for which salary is missing; for simplicity, we exclude these records.

```
> library(ISLR)
> dim(Hitters)

[1] 322  20

> my_hitters <- Hitters
> my_hitters <- my_hitters[complete.cases(my_hitters),]
> dim(my_hitters)

[1] 263  20
```

The first explanatory variables we consider are `Years` and `Hits`. Using these explanatory variables, we plot the data in Figure 8.21 with shading indicating the amount of salary; it is apparent that salary tends to increase with both `Years` and `Hits`.

A regression tree is fit via recursive binary partitioning. The design space is partitioned into K non-overlapping regions. A single binary split is chosen based on a single explanatory variable. The split is chosen to minimize the sum of squared errors (SSE) where the estimated value is the mean of data points falling into the region. Denote the K regions of the design space as $R_1, R_2, \ldots, R_K$. Let $I_1, I_2, \ldots, I_K$ denote the set of indices of design points falling into each of the regions; i.e., $I_j = \{i : \boldsymbol{x}_i \in R_j\}$. Let n_j denote the number of observations in set j. The first split chooses a variable which minimizes

$$\sum_{i \in I_1} (y_i - \hat{y}_1)^2 + \sum_{i \in I_2} (y_i - \hat{y}_2)^2$$

where $\hat{y}_j = \frac{1}{n_j} \sum_{i \in I_j} y_i$. Subsequent splits are determined by the binary split, which further reduces SSE the most.

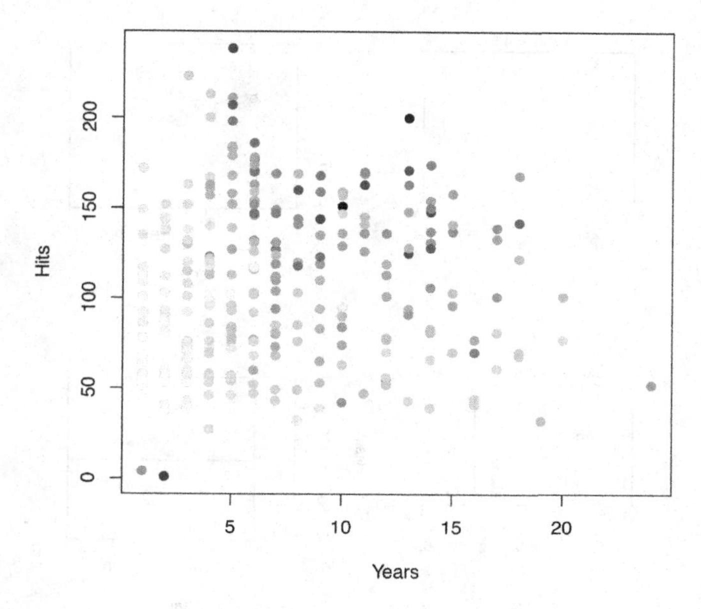

FIGURE 8.21
Scatterplot of `my_hitters` dataset. Gray scale indicates log(Salary); i.e. darker data points indicate higher salary.

Figure 8.22 illustrates the first two splits when growing a regression tree to predict log(Salary) based on the explanatory variables `Hits` and `Years`. Notice the salaries of players who have played less than 4.5 years tend to be less than those who have played greater than 4.5 years. In this case, the first split determines $R_1 = \{\texttt{years} < 4.5\}$ and $R_2 = \{\texttt{years} > 4.5\}$. The second split then subdivides R_2 so that $R_1 = \{\texttt{years} < 4.5\}$, $R_2 = \{\texttt{years} > 4.5, \texttt{Hits} < 117.5\}$, $R_3 = \{\texttt{years} > 4.5, \texttt{Hits} > 117.5\}$; $\hat{y}_1 = 5.107$, $\hat{y}_2 = 5.998$, $\hat{y}_3 = 6.740$.

Bagged Trees and Random Forests

Bootstrap aggregated trees (a.k.a. bagged trees) are trees built based on the average response over many trees based on bootstrap samples. The fitted trees tend to be correlated, so, to decorrelate the fitted trees, random forests are often fit, where the covariates to be considered at each split are sampled at random from the entire set of covariates in the model. In most cases, the predictions from random forest will dominate the predictions of bagged trees, so they are generally preferred when there are enough covariates. As a rule

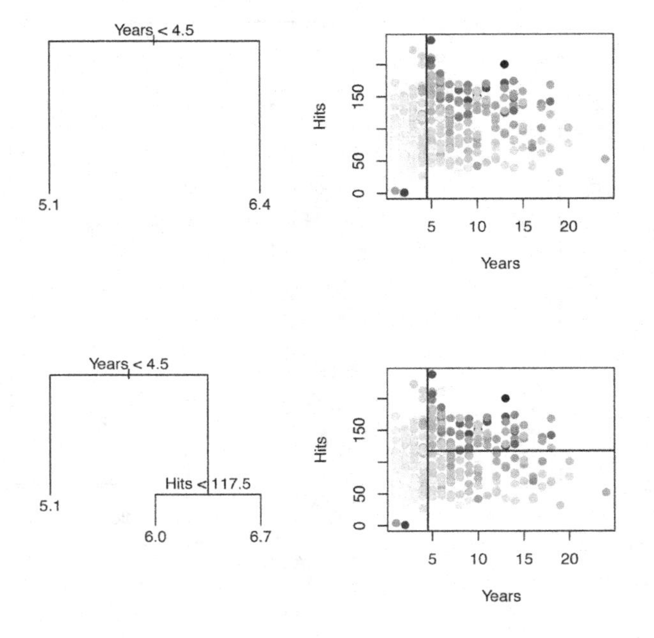

FIGURE 8.22
Display of fit two binary splits of a regression tree fit with log(Salary) as the outcome variable with Hits and Years as predictor variables.

of thumb, the number of covariates selected at random for each split for a regression tree is

$$\left\lfloor \frac{p}{3} \right\rfloor$$

where p is the number of explanatory variables.

For illustration, we begin with a bagged tree fit. Fitting a bagged tree can be done by using the **randomForest** function with the option **mtry** set to the number of covariates, 2 in this case (for **Years** and **Hits**).

```
> library(randomForest)
> fit <- randomForest(log(Salary)~ Years + Hits,data=my_hitters,mtry=2)
```

In Figure 8.23 we plot the predicted values based on our bagged fitted model.

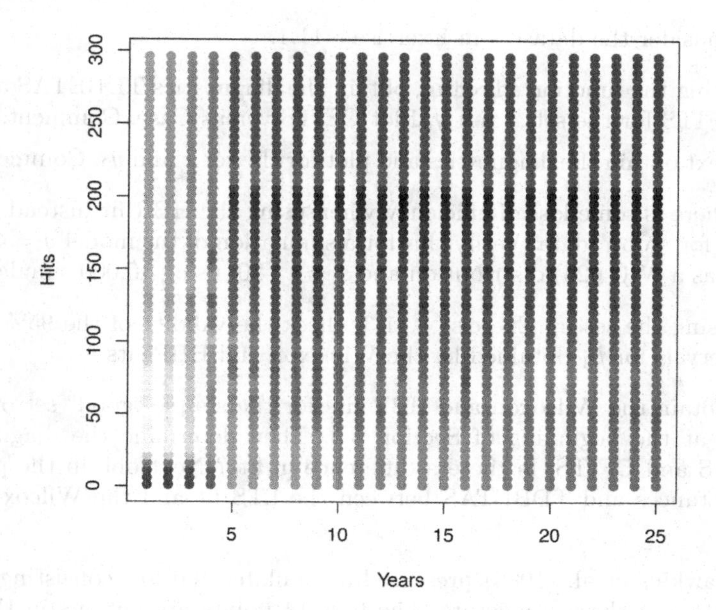

FIGURE 8.23
Grid of fitted values for `Salary` data based on bagged tree fit.

8.11 Exercises

8.11.1. To see the effect on fits that "good" and "bad" points of high leverage can have, consider the following dataset:

x	1	2	3	4	5	6	7	8	9	10	20
y	5	7	6	14	14	25	29	33	31	41	75
y_2	5	7	6	14	14	25	29	33	31	41	20

The point $x = 20$ is a point of high leverage. The data for $\boldsymbol{y}$ (rounded) are realizations from the model $y = 4x + e$, where e has a $N(0, 9)$ distribution. Hence, the value of y follows the model and is a "good" point of high leverage. Notice that $\boldsymbol{y}_2$ is the same as $\boldsymbol{y}$, except the last component of $\boldsymbol{y}_2$ has been changed to 20 and, thus, is a "bad" point of high leverage.

(a) Obtain the scatterplot for x and y, the Wilcoxon and HBR fits, and overlay these fits on the scatterplot.

(b) Obtain the scatterplot for x and y_2, the Wilcoxon and HBR fits, and overlay these fits on the scatterplot.

(c) Comment on the differences among the fits and plots.

8.11.2. Consider the datasets in Exercise 8.11.1.

(a) Using the function `fitdiag`, obtain the diagnostics TDBETAS and CFITS for the set x and y. Plot CFITS versus Case. Comment.

(b) Next obtain the diagnostics and plot for the set x and y_2. Comment.

8.11.3. There is some loss of efficiency when using the HBR fit instead of the Wilcoxon for "good" data. Verify this for a simulation of the model $y = 4x + e$, where e has a $N(0, 625)$ distribution and $x = 1 : 20$, using 10,000 simulations.

8.11.4. Using the set up Exercise 8.11.3, check the validity of the 95% confidence intervals for β_1 obtained by the Wilcoxon and HBR fits.

8.11.5. Obtain the Wilcoxon and HBR fits for the simple "good" set of data discussed at the beginning of Section 8.3. Then determine the diagnostics TDBETAS and CFITS. Is there a difference in fits? Next obtain the plot of robust distances and TDBETAS between the LTS fit and the Wilcoxon fit. Comment.

8.11.6. Hawkins et al. (1984) presented a simulated dataset consisting of 75 data points and three predictors. The first 14 points are outliers in the $\boldsymbol{X}$-space, while the remaining 61 points follow a linear model. Of the 14 outliers, the first 10 do not follow the model, while the next 4 do follow the model. The data is in the `hawkins` dataset in the package `npsmReg2`.

(a) Obtain the Wilcoxon fit and plot its Studentized residuals versus Case number. Comment.

(b) Obtain the HBR fit and plot its Studentized residuals versus Case number. Comment.

(c) Obtain the diagnostics TDBETAS and CFITS between the HBR and Wilcoxon fits. Plot CFITS versus Case. Did the diagnostics discern the embedded outlier structure for this dataset?

8.11.7. Use the `bonds` data to examine the relationship between the bid prices for US treasury bonds (`BidPrice`) and the size of the bond's periodic payment rate (`CouponRate`). Consider `BidPrice` as the response variable and `CouponRate` as the explanatory variable. Obtain the Wilcoxon and HBR fits of this dataset. Using Studentized residuals and the diagnostics TDBETAS and CFITS, show the heteroscedasticity and determine the outliers. See Sheather (2009) for discussion on the *bow tie* pattern in the residual plot. The data are in the dataset `bonds`.

8.11.8. Assuming $X \sim SN(\alpha)$ for some $-\infty < \alpha < \infty$, obtain the derivation of expression (8.15).

8.11.9. As in Figure 8.10, obtain the plots of the pdfs and their associated scores for $\alpha = 1, 2, \ldots, 10$. Comment on the trends in the plots as α increases.

8.11.10. Assuming $X \sim SN(\alpha)$ for some $-\infty < \alpha < \infty$, show that the derivative of the optimal scores satisfies expression (8.17).

8.11.11. Simulate 50 observations from the model

$$Y_i = 0.01 * x_{i1} + 0.15 * x_{i2} + 0 * x_{i3} + e_i,$$

where x_{i1}, x_{i2} are deviates from a standard normal distribution and $e_i \sim SN(-8)$.

(a) Obtain the rank-based fit of these data using skew-normal scores with $\alpha = -8$. Obtain confidence intervals for the 3 (nonintercept) regression parameters. Did the confidence intervals trap the true values?

(b) Obtain the Studentized residuals for the fit obtained in Part (a). Using these residuals, obtain the residual plot and the normal $q-q$ plot. Comment on the fit.

(c) Obtain the Wilcoxon fit of these data. What is the estimated precision of fit in Part (a) over the Wilcoxon fit?

8.11.12. On Page 204, Bowerman et al. (2005) present a dataset concerning sales prices of houses in a city in Ohio. The variables are: x_1 is the total square footage; x_2 is the number of rooms; x_3 is the number of bedrooms; and x_4 is the age of the house at the time data were collected; y is the sale price in \$10,000. The sample size is $n = 60$. For the reader's convenience, the data are in the dataset `homesales` in the package `npsmReg2`. Consider the linear model $y = \alpha + \sum_{i=1}^{4} x_i\beta_i + e$.

(a) Use the Hogg-type adaptive scheme discussed in Section 8.5 on these data; i.e., use the function `adaptor`. Which score function did it select?

(b) Comment on the estimated regression coefficients as to significance and what they mean in terms of the problem.

(c) Using the Studentized residuals from Part (b), perform a residual analysis which includes at least a residual plot and a $q-q$ plot. Identify all outliers.

(d) Based on your analysis in Part (c), what, if any, other models would you fit to these data?

8.11.13. Apply the Hogg-type adaptive scheme of Section 8.5 to data of Exercise 8.11.23. Compare the selected fit with that of the rank-based fits obtained in Exercise 8.11.23. Which of the three is best in terms of precision?

8.11.14. For the data of Exercise 8.11.24, run the adaptive analysis of Section 8.5. Which score function did it select?

8.11.15. Bowerman et al. (2005) discuss the daily viscosity measurements of a manufactured chemical product XB-77-5 for a series of 95 days. Using traditional methods, they determined that the best autoregressive model for this data had order 2. For the reader's convenience, we have placed this data in the dataset `viscosity`.

(a) Plot this time series data.

(b) As in Example 8.7.1, using the Wilcoxon scores, determine the order of the autoregressive series for this data. Use as the maximum order, $p = 4$. Do the results agree with the order determined by Bowerman et al. (2005)?

(c) Obtain the Wilcoxon fit for the autoregressive model using the order determined in the last part. Obtain confidence intervals for the autoregressive parameters.

(d) Write the model expression for the unknown observation at time $t = 96$. Using the fitted model in Part (c), predict the expected viscosity for Day 96.

(e) Determine a confidence interval for the $E(y_{96})$; see Section 7.1.4.

(f) Continue Parts (d) and (e) for the observations at times 97 and 98.

8.11.16. For the time series in Exercise 8.11.15, assume that the order of the autoregressive model is 2. Compute the diagnostic TDBETAS between the Wilcoxon and HBR fits. Obtain the plot of the corresponding diagnostic CFITS versus time, also. Comment on the diagnostics.

8.11.17. Seber and Wild (1989) present a dataset for the nonlinear model of the form $y = f(x; \boldsymbol{\theta}) + e$, where

$$f(x; \boldsymbol{\theta}) = \frac{\theta_1 x}{\theta_2 + x}.$$

The response y is the enzyme velocity in an enzyme-catalyzed chemical reaction and x denotes the concentration of the substrate. The 12 data points are:

x	2.00	2.00	0.67	0.67	0.40	0.40
y	0.0615	0.0527	0.0334	0.0258	0.0138	0.0258
x	0.29	0.29	0.22	0.22	0.20	0.20
y	0.0129	0.0183	0.0083	0.0169	0.0129	0.0087

(a) Obtain a scatterplot of the data.

(b) Write the R functions for the model and the Jacobian.

(c) Using $\boldsymbol{\theta}_0 = (.1, 1.8)$ as the initial estimate, obtain the Wilcoxon fit of the model. Obtain 95% confidence intervals for θ_1 and θ_2.

(d) Overlay the scatterplot with the fitted model.

(e) Obtain a residual plot and a $q-q$ plot of the residuals. Comment on the quality of the fit.

8.11.18. Obtain the HBR fit of the nonlinear model discussed in Exercise 8.11.17. Then change the first two response to $y_1 = y_2 = 0.04$. Obtain both the Wilcoxon and HBR fits for this changed data. Which fit changed less? Why?

8.11.19. The dataset `gamnl` contains simulated data from the nonlinear model

$$f(x;\boldsymbol{\theta}) = \theta_2^{\theta_1} x^{\theta_1 - 1} e^{-\theta_2 x}.$$

The first and second columns of the data contain the respective x's and y's.

(a) Obtain a scatterplot of the data.

(b) Write the R functions for the model and the Jacobian.

(c) For starting values, the model suggests taking the log of both sides of the model expression and then fit a linear model; but since some of the y values are negative, use $\log(y+2)$ instead.

(d) Obtain the Wilcoxon fit of the model and 95% confidence intervals for θ_1 and θ_2.

(e) Overlay the scatterplot with the fitted model.

(f) Obtain a residual plot and a $q-q$ plot of the residuals. Comment on the quality of the fit.

8.11.20. Devore (2012) presents a dataset on the wear life for solid film lubricant. We consider the model

$$Y_i = \frac{\theta_1}{x_{1i}^{\theta_2} x_{2i}^{\theta_3}} + e_i,$$

where Y_i the wear life in hours of a Mil-L-8937-type film, x_{i1} is load in psi, and x_{i2} is the speed of the film in rpm. The data are the result of a 3×3 crossed design with three replicates and are in dataset `wearlife`.

(a) Write the R functions for the model and the Jacobian.

(b) For starting values, as the model suggests, take logs of both sides. Show that $\theta_1 = 36,000$, $\theta_2 = 1.15$, and $\theta_3 = 1.24$ are reasonable starting values.

(c) Obtain the Wilcoxon fit and find 95% confidence intervals for each of the parameters.

(d) Obtain the residual versus fitted values plot and the normal $q-q$ plot of the residuals. Comment on the quality of the fit.

(e) Obtain a scatterplot of x_{i1} versus x_{i2}. There are, of course, 9 treatment combinations. At each combination, plot the predicted wear life based on the fitted model. Obtain 95% confidence intervals for these predictions. Discuss the plot in terms of the model.

8.11.21. Obtain full model fit of the prostate cancer data discussed in Example 8.8.2. Include age, serum haemoglobin level, size, and Gleason Index. Comment on the similarity or dissimilarity of the estimated regression coefficients to those obtained in Example 8.8.2.

8.11.22. Show that the optimal rank-based score function is $\varphi(u) = -1 - \log(1-u)$, for $0 < u < 1$ for random variables which have an extreme valued distribution (8.47). In this case, the generated scores are called the log-rank scores.

8.11.23. Consider the dataset `rs`. This is simulated data from a simple regression model with the true slope parameter at 0.5. The first column is the independent variable x, while the second column is the dependent variable y. Obtain the following three fits of the model: least squares, Wilcoxon rank-based, and rank-based using `logGFscores` with $m_1 = 1$ and $m_2 = 0.10$.

(a) Scatterplot the data and overlay the three fits.

(b) Obtain Studentized residual plots of all three fits.

(c) Based on Parts (a) and (b), which fit is worse?

(d) Compare the two rank-based fits in terms of precision (estimates of τ_φ). Which fit is better?

8.11.24. Generate data from a linear model with log-F errors with degrees of freedom 4 and 8 using the following code:

```
n <- 75; m1 <- 2; m2 <- 4; x<-rnorm(n,50,10)
errs1 <- log(rf(n,2*m1,2*m2)); y1 <- x + 30*errs1
```

(a) Using `logGFscores`, obtain the optimal scores for this dataset.

(b) Obtain side-by-side plots of the pdf of the random errors and the scores. Comment on the plot.

(c) Fit the simple linear model for this data using the optimal scores. Obtain a residual analysis including a Studentized residual plot and a normal q–q plot. Comment on the plots and the quality of the fit.

(d) Obtain a histogram of the residuals for the fit in Part (c). Overlay the histogram with an estimate of the density and compare it to the plot of the pdf in Part (b).

(e) Obtain a summary of the fit of the simple linear model for this data using the optimal scores. Obtain a 95% confidence interval for the slope parameter β. Did the interval trap the true parameter?

(f) Use the fit to obtain a confidence interval for the expected value of y when $x = 60$.

8.11.25. For the situation described in Exercise 8.11.24, obtain a simulation study comparing the mean squared errors of the estimates of slope using fits based on Wilcoxon scores and the optimal scores. Use 10,000 simulations.

8.11.26. Consider the failure time data discussed in Example 8.9.2. Recall that the generalized log F-scores with $2m_1 = 2$ and $2m_2 = 10$ degrees of freedom were used to compute the rank-based fit. The Studentized residuals from this fit were then used in a q–q plot to check goodness-of-fit based on the strength of linearity in the plot, where the population quantiles were obtained from a log F-distribution with 2 and 10 degrees of freedom. Obtain the rank-based fits based on the Wilcoxon scores, normal scores, and log F-scores with $2m_1 = 10$ and $2m_2 = 2$. For each, obtain the q–q plot of Studentized residuals using as population quantiles the normal distribution, the logistic distribution, and the log F-distribution with 10 and 2 degrees of freedom, respectively. Compare the plots. Which, if any, is most linear?

8.11.27. Suppose we are investigating the relationship between a response Y and an independent variable x. In a planned experiment, we record responses at r values of x, $x_1 < x_2 < \cdots < x_r$. Suppose n_i independent replicates are obtained at x_i. Let Y_{ij} denote the response for the j*th* replicate at x_i. Then the model for a linear relationship is

$$Y_{ij} = \alpha + x_{ij}\beta + e_{ij}, \quad i = 1, \ldots, r; j = 1, \ldots, n_i. \tag{8.49}$$

In this setting, we can obtain a **lack-of-fit** test. For this test, the null hypothesis is Model (8.49). For the alternative, we take the most general model which is a one-way design with r groups; i.e., the model

$$Y_{ij} = \mu_i + e_{ij}, \quad i = 1, \ldots, r; j = 1, \ldots, n_i, \tag{8.50}$$

where μ_i is the median (or mean) of the i*th* group (responses at x_i). The rank-based drop in dispersion is easily formulated to test these hypotheses. Select a score function φ. Let $D(\text{RED})$ denote the minimum value of the dispersion function when Model (8.49) is fit and let $D(\text{FULL})$ denote the minimum value of the dispersion function when Model (8.50) is fit. The F_φ test statistic is

$$F_\varphi = \frac{[D(\text{RED}) - D(\text{FULL})]/(r-2)}{\hat{\tau}_\varphi}.$$

This test statistic should be compared with F-critical values having $r - 2$ and $n - r$ degrees of freedom, where $n = \sum_i n_i$ is the total sample size. In general the drop in dispersion test is computed by the function `drop.test`. Carry out this test for the data in Example 8.9.2 using the log F-scores with $2m_1 = 2$ and $2m_2 = 10$ degrees of freedom.

8.11.28. The data for Example 8.9.1 are:

x	1	2	3	4	5	6	7	8	9	10
y	2.84	6.52	6.87	16.43	18.17	25.24	28.15	31.65	36.37	38.84

(a) Using `Rfit`, verify the analysis presented in Example 8.9.1.

(b) Obtain Studentized residuals from the fit. Comment on the residual plot.

(c) Obtain the $q-q$ plot of the sorted residuals of Part (b) versus the quantiles of the random variable ε which is distributed as the log of an exponential. Comment on linearity in the $q-q$ plot.

9

Cluster Correlated Data

9.1 Introduction

Often in practice, data are collected in clusters. Examples include block designs, repeated measure designs, and designs with random effects. Generally, the observations within a cluster are dependent. Thus the independence assumption of fixed effects linear models breaks down. These models generally include fixed effects, also. Inference (estimation, confidence intervals, and tests of linear hypotheses) for the fixed effects is often of primary importance.

Several rank-based approaches have been considered for analyzing cluster-correlated data. Kloke, McKean, and Rashid (2009) extended the rank-based analysis for linear models discussed in Chapters 4–5 to many cluster models which occur in practice. In their work, the authors, in addition to developing general theory for cluster-correlated data, develop the application of a simple mixed model with one random effect and an arbitrary number of fixed effects and covariates. Kloke and McKean (2011) discuss a rank-based analysis when the blocks have a compound symmetric variance-covariance structure.

In this chapter we illustrate extensions of the rank-based methods discussed in earlier chapters to data which have cluster-correlated responses. For our purpose, we consider an experiment done over a number of blocks (clusters) where the observations within a block are correlated. We begin (Section 9.2) by discussing Friedman's nonparametric test for a randomized block design. In Section 9.3, we present the rank-based analysis of Kloke, McKean, and Rashid (2009). Besides tests of general linear hypotheses, this analysis includes estimation with standard errors of fixed effects as well as diagnostic procedures to check the quality of fit. Section 9.4 offers a discussion of robust estimation of variance components. These estimates are also used in the estimation of standard errors and in the Studentized residuals. We end the chapter with a discussion of rank-based procedures for generalized estimating equation (GEE) models, which in terms of assumptions are the most general. Computation by R and R packages of these analyses is highlighted throughout the chapter.

For this chapter we use a common notation which we provide now. Suppose we have m blocks or clusters. Within the kth cluster, there are n_k

measurements. We may model the ith measurement within the kth cluster as

$$Y_{ki} = \alpha + \boldsymbol{x}_{ki}^T \boldsymbol{\beta} + e_{ki} \text{ for } k = 1, \ldots m, i = 1, \ldots, n_k, \quad (9.1)$$

where $\boldsymbol{x}_{ki}$ is a vector of covariates. The errors between clusters are assumed to be independent and the errors within a block are assumed to be correlated.

At times, dependent data fit into a multivariate framework; i.e., a multivariate multiple regression model. Rank-based analyses for such data are discussed in Chapter 10.

9.2 Friedman's Test

The first nonparametric test for cluster-correlated data was developed by Friedman (1937). The goal is to compare the effect of n treatments. Each treatment is applied to each of m experimental units or clusters. In this test a separate ranking is calculated for each of the clusters. The rankings are then averaged for each of the treatments and then compared. If there is a large difference between the average rankings, the null hypothesis of no treatment effect is rejected.

Suppose we have n treatments and m clusters each of size n. Suppose all the treatments are randomly assigned once within a cluster. Let Y_{kj} denote the measurement (response) for the jth treatment within cluster (experimental unit) k. Assume the model is

$$Y_{kj} = \alpha + \beta_j + b_k + \epsilon_{kj}, \quad k = 1, \ldots, m, j = 1, \ldots, n, \quad (9.2)$$

where α is an intercept parameter, β_j is the jth treatment effect, b_k is the random effect due to cluster k, and ϵ_{kj} is the $jkth$ random error. Assume that the random errors are iid and are independent of the random effects.

Let R_{kj} denote the rank of Y_{kj} among $Y_{k1}, \ldots, Y_{kn}$. Let

$$\bar{R}_{\cdot j} = \frac{\sum_{k=1}^m R_{kj}}{m}.$$

The test statistic is given by

$$T = \frac{12m}{n(n+1)} \sum_{j=1}^n \left(R_{\cdot j} - \frac{n+1}{2} \right)^2.$$

Under H_0, the test statistic T has an asymptotic χ^2_{n-1} distribution. We illustrate the R computation of Friedman's test with the following example.

Example 9.2.1 (Rounding First Base). This example is discussed in Hollander and Wolfe (1999). In the game of baseball, three methods were evaluated for rounding first base (for an illustration see Figure 7.1 of Hollander and Wolfe 1999). Label these methods as round out, narrow angle, and wide angle. Each method was evaluated twice for each of $m = 22$ baseball players. The average time of the two runs, for each of the three methods, for each player, are in the dataset `firstbase`. Hence, there are 22 blocks (clusters) and one fixed effect (method of base rounding) at three levels. The R function `friedman.test` can take either a numeric data matrix, separate arguments for the response vector, the group vector, and the block vector or a formula.

```
> friedman.test(as.matrix(firstbase))

        Friedman rank sum test

data:  as.matrix(firstbase)
Friedman chi-squared = 11.1429, df = 2, p-value = 0.003805
```

Hence, the difference between methods of rounding first base is significant. Friedman's test is for an overall difference in the methods. Note that it offers no estimate of the effect size between the different methods. Using the rank-based analysis discussed in the next section, we can both test the overall hypothesis and estimate the effect sizes, with standard errors. ■

9.3 Joint Rankings Estimator

Kloke et al. (2009) showed that rank-based analysis can be extended to cluster-correlated data. In this section we summarize these methods and present examples which illustrate the computation; as we demonstrate, the function to determine the fit is `jrfit`, and there are, in addition, several of the standard linear model helper functions.

Assume an experiment is done over m blocks or clusters. Note that we use the terms block and cluster interchangeably. Let n_k denote the number of measurements taken within the kth block. Let Y_{ki} denote the response variable for the ith experimental unit within the kth block; let $\boldsymbol{x}_{ki}$ denote the corresponding vector of covariates. Note that the design is general in that $\boldsymbol{x}_{ki}$ may contain, for example, covariates, baseline values, or treatment indicators. The response variable is then modeled as

$$Y_{ki} = \alpha + \boldsymbol{x}_{ki}^T\boldsymbol{\beta} + e_{ki} \text{ for } k = 1, \ldots, m, i = 1, \ldots, n_k, \tag{9.3}$$

where α is the intercept parameter, $\boldsymbol{\beta}$ is a $p \times 1$ vector of unknown parameters, and e_{ki} is a random error. We assume that the errors within a block are correlated (i.e., e_{ki} & $e_{ki'}$) but the errors between blocks are independent (i.e., e_{ki} & $e_{k'j}$). Further, we assume that e_{ki} has pdf and cdf $f(x)$ and $F(x)$, respectively. Now write model (9.3) in block vector notation as

$$\boldsymbol{Y}_k = \alpha \mathbf{1}_{n_k} + \boldsymbol{X}_k \boldsymbol{\beta} + \boldsymbol{e}_k. \tag{9.4}$$

where $\mathbf{1}_{n_k}$ is an $n_k \times 1$ vector of ones, $\boldsymbol{X}_k = [\boldsymbol{x}_{k1} \ldots \boldsymbol{x}_{kn_k}]^T$ is a $n_k \times p$ design matrix, and $\boldsymbol{e}_k = [e_{k1}, \ldots e_{kn_k}]^T$ is a $n_k \times 1$ vector of error terms. Let $N = \sum_{k=1}^{m} n_k$ denote the total sample size. Let $\boldsymbol{Y} = (\boldsymbol{Y}_1^T, \ldots, \boldsymbol{Y}_m^T)^T$ be the $N \times 1$ vector of all measurements (responses) and consider the matrix formulation of the model as

$$\boldsymbol{Y} = \alpha \mathbf{1}_N + \boldsymbol{X}\boldsymbol{\beta} + \boldsymbol{e} \tag{9.5}$$

where $\mathbf{1}_N$ is an $N \times 1$ vector of ones and $\boldsymbol{X} = [\boldsymbol{X}_1^T \ldots \boldsymbol{X}_m^T]^T$ is a $N \times p$ design matrix and $\boldsymbol{e} = [\boldsymbol{e}_1^T, \ldots \boldsymbol{e}_m^T]^T$ is a $N \times 1$ vector of error terms. Since there is an intercept in the model, we may assume, without loss of generality, that $\boldsymbol{X}$ is centered.

Select a set of rank scores $a(i) = \varphi[i/(N+1)]$ for a nondecreasing score function φ which is standardized as usual, ($\int \varphi(u)\,du = 0$ and $\int \varphi^2(u)\,du = 1$). As with `Rfit`, the default score function for `jrfit` is the Wilcoxon, i.e., $\varphi(u) = \sqrt{12}[u - (1/2)]$. Then the rank-based estimator of $\boldsymbol{\beta}$ is given by

$$\hat{\boldsymbol{\beta}}_\varphi = \text{Argmin}\|\boldsymbol{y} - \boldsymbol{X}\boldsymbol{\beta}\|_\varphi \text{ where } \|\boldsymbol{v}\|_\varphi = \sum_{t=1}^{N} a(R(v_t))v_t, \quad \boldsymbol{v} \in R^N. \tag{9.6}$$

The convex function $D(\boldsymbol{\beta}) = \|\boldsymbol{y} - \boldsymbol{X}\boldsymbol{\beta}\|_\varphi$ is the dispersion function of Jaeckel (1972).

For formal inference, Kloke et al. (2009) develop the asymptotic distribution of the $\hat{\boldsymbol{\beta}}_\varphi$ under the assumption that the marginal distribution functions of the random vector $\boldsymbol{e}_k$ are the same. This includes two commonly assumed error structures: exchangeable within-block errors as well as the components of $\boldsymbol{e}_k$ following a stationary time series, such as autoregressive of general order. This asymptotic distribution of $\hat{\boldsymbol{\beta}}$ is given by

$$\hat{\boldsymbol{\beta}}_\varphi \dot{\sim} N_p \left(\boldsymbol{\beta}, \tau_\varphi^2 (\boldsymbol{X}^T\boldsymbol{X})^{-1} \left(\sum_{k=1}^{m} \boldsymbol{X}_k^T \boldsymbol{\Sigma}_{\varphi_k} \boldsymbol{X}_k \right) (\boldsymbol{X}^T\boldsymbol{X})^{-1} \right)$$

where $\boldsymbol{\Sigma}_k = \text{var}(\varphi(F(\boldsymbol{e}_k)))$ and $F(\boldsymbol{e}_k) = [F(e_{k1}), \ldots, F(e_{kn_k})]^T$. To estimate τ_φ, `jrfit` uses the estimator purposed by Koul et al. (1987).

9.3.1 Estimates of Standard Error

In this section we discuss several approaches to estimating the standard error of the R estimator defined in (9.6). Kloke et al. (2009) develop the inference under the assumption of exchangeable within-block errors; Kloke and McKean (2013) considered two additional estimates and examined the small sample properties of each.

Let $\boldsymbol{V} = \left(\sum_{k=1}^{m} \boldsymbol{X}_k^T \boldsymbol{\Sigma}_{\varphi_k} \boldsymbol{X}_k \right)$. Let σ_{ij} be the (i,j)th element of $\boldsymbol{\Sigma}_{\varphi k}$. That is $\sigma_{ij} = \text{cov}(\varphi(F(e_{1i})), \varphi(F(e_{1j})))$.

Compound Symmetric

Kloke et al. (2009) discuss estimates of $\boldsymbol{\Sigma}_{\varphi k}$ when the within-block errors are exchangeable. Under the assumption of exchangeable errors, $\boldsymbol{\Sigma}_{\varphi k}$ reduces to a compound symmetric matrix; i.e., $\boldsymbol{\Sigma}_{\varphi k} = [\sigma_{ij}]$ where

$$\sigma_{ij} = \begin{cases} 1 \text{ if } i = j \\ \rho_\varphi \text{ if } i \neq j \end{cases}$$

and $\rho_\varphi = \text{cov}(\varphi(F(e_{11})), \varphi(F(e_{12})))$. An estimate of ρ_φ is

$$\hat{\rho}_\varphi = \frac{1}{M - p} \sum_{k=1}^{m} \sum_{i>j} a(R(\hat{e}_{ki}))a(R(\hat{e}_{kj}))$$

where $M = \sum_{k=1}^{m} \binom{n_k}{2}$.

One advantage of this estimate is that it requires estimation of only one additional parameter. A main disadvantage is that it requires the somewhat strong assumption of exchangeability. Furthermore, the simulation studies of Kloke and McKean (2013) show that for small values of m, $m < 50$, rank-based analyses using this estimator are liberal; i.e., empirical levels were often greater than nominal α. When $m \geq 50$, though, the empirical levels were close to nominal α.

Empirical

A natural estimate of $\boldsymbol{\Sigma}_\varphi$ is the unstructured variance-covariance matrix using the sample correlations. To simplify notation, let $a_{ki} = a(R(\hat{e}_{ki}))$. Estimate σ_{ij} with

$$\hat{\sigma}_{ij} = \sum_{k=1}^{m} (a_{ki} - \bar{a}_{\cdot i})(a_{kj} - \bar{a}_{\cdot j})$$

where $\bar{a}_{\cdot i} = \sum_{k=1}^{m} a_{ki}$.

The advantage of this estimator is that it is general and makes no additional simplifying assumptions. In simulation studies, Kloke and McKean (2013) demonstrate that the sandwich estimator discussed next works at least as well for large samples as this empirical estimate.

Sandwich Estimator

Another natural estimator of $\boldsymbol{V}$ is the sandwich estimator, which for the problem at hand is defined as

$$\frac{m}{m - p} \sum_{k=1}^{m} \boldsymbol{X}_k^T a(R(\hat{\boldsymbol{e}}_k)) a(R(\hat{\boldsymbol{e}}_k))^T \boldsymbol{X}_k.$$

Kloke and McKean (2013) demonstrate that this estimate works well for large samples and should be used when possible. The advantage of this estimator is that it does not require additional assumptions. Furthermore, the simulation studies of Kloke and McKean (2013) indicate that rank-based analyses using this estimator had empirical levels close to nominal α for $m \geq 50$, while for $m < 50$ empirical levels were often conservative, i.e., less than nominal α. Hence, the study showed that over the set of situations simulated, the rank-based analyses, using the sandwich estimator, were valid (i.e., generally maintained the correct type I error rate). See Kloke and McKean (2013) for more details. The sandwich estimator is the default estimator in `jrfit`.

9.3.2 Inference

Simulation studies suggest using a t-distribution for tests of hypothesis of the form

$$H_0 : \beta_j = 0 \text{ versus } H_A : \beta_j \neq 0.$$

Specifically, when the standard error (SE) is based on the estimate of the compound symmetric structure, we may reject the null hypothesis at level α provided

$$\left| \frac{\hat{\beta}_j}{\mathrm{SE}(\hat{\beta}_j)} \right| > t_{\alpha, N-p-1-1}.$$

On the other hand, if the sandwich estimator is used, (Section 9.3.1), we test the hypothesis using $df = m$. That is, we reject the null hypothesis at level α if

$$\left| \frac{\hat{\beta}_j}{\mathrm{SE}(\hat{\beta}_j)} \right| > t_{\alpha, m}.$$

These inferences are the default when utilizing the `summary` functions of `jrfit`. We illustrate this discussion with the following examples.

9.3.3 Examples

In this section we present several examples. The first is a simulated example for which we illustrate the package `jrfit`. Following that, we present several real examples.

Simulated Dataset

To fix ideas, we present an analysis of a simulated dataset utilizing both the compound symmetry and sandwich estimators discussed in the previous section.

The setup is as follows:

```
> m<-160  # blocks
> n<-4    # observations per block
> p<-1    # baseline covariate
> k<-2    # trtmnt groups
```

First, we set up the design and simulate a baseline covariate which is normally distributed.

```
> trt<-as.factor(rep(sample(1:k,m,replace=TRUE),each=n))
> block<-rep(1:m,each=n)
> x<-rep(rnorm(m),each=n)
```

Next, we set the overall treatment effect to be $\Delta = 0.5$. We simulate the responses as follows. We simulate the block effects from a t-distribution with 3 degrees of freedom and the random errors from a t-distribution with 5 degrees of freedom. Note that the assumption for exchangeable errors is met.

```
> delta<-0.5
> w<-trt==2
> Z<-model.matrix(~as.factor(block))
> e<-rt(m*n,df=5)
> b<-rt(m,df=3)
> y<-delta*w+Z%*%b+e
```

Note the regression coefficient for the covariate was set to 0.

First, we analyze the data with the compound symmetry assumption. The three required arguments to `jrfit` are the design matrix, the response vector, and the vector denoting block membership. In future releases, we plan to incorporate a model statement as we have done in `Rfit` similar to the one in `friedman.test`.

```
> library(jrfit)
> X<-cbind(w,x)
> fit<-jrfit(X,y,block,var.type='cs')
> summary(fit)

Coefficients:
  Estimate Std. Error t-value   p.value
  1.395707   0.165898  8.4130 2.636e-16 ***
w 0.256514   0.252201  1.0171    0.3095
x 0.083595   0.130023  0.6429    0.5205
---
Signif. codes:  0 '***' 0.001 '**' 0.01 '*' 0.05 '.' 0.1 ' ' 1
```

Notice, by default the intercept is displayed in the output. If the inference on the intercept is of interest, then set the option `int` to `TRUE` in the `jrfit` `summary` function. The cell medians model can also be fit as follows.

```
> library(jrfit)
> W<-model.matrix(~trt-1)
> X<-cbind(W,x)
```

```
> fit<-jrfit(X,y,block,var.type='cs')
> summary(fit)

Coefficients:
     Estimate Std. Error t-value   p.value
trt1 1.395707   0.165898  8.4130 2.636e-16 ***
trt2 1.652221   0.168449  9.8085 < 2.2e-16 ***
x    0.083595   0.130023  0.6429    0.5205
---
Signif. codes:  0 '***' 0.001 '**' 0.01 '*' 0.05 '.' 0.1 ' ' 1
```

Next we present the same analysis utilizing the sandwich estimator.

```
> X<-cbind(w,x)
> fit<-jrfit(X,y,block,var.type='sandwich')
> summary(fit)

Coefficients:
  Estimate Std. Error t-value   p.value
  1.395707   0.164288  8.4955 1.294e-14 ***
w 0.256514   0.247040  1.0383    0.3007
x 0.083595   0.113422  0.7370    0.4622
---
Signif. codes:  0 '***' 0.001 '**' 0.01 '*' 0.05 '.' 0.1 ' ' 1

> X<-cbind(W,x)
> fit<-jrfit(X,y,block,var.type='sandwich')
> summary(fit)

Coefficients:
     Estimate Std. Error t-value   p.value
trt1 1.395707   0.164288  8.4955 1.294e-14 ***
trt2 1.652221   0.166176  9.9426 < 2.2e-16 ***
x    0.083595   0.113422  0.7370    0.4622
---
Signif. codes:  0 '***' 0.001 '**' 0.01 '*' 0.05 '.' 0.1 ' ' 1
```

For this example, the results of the analysis based on the compound symmetry method and the analysis based on the sandwich method are quite similar.

Crabgrass Data

Cobb (1998) presented an example of a complete block design concerning the weight of crabgrass. The fixed factors in the experiment were the density of the crabgrass (four levels) and the levels (two) of the three nutrients nitrogen, phosphorus, and potassium. So $p = 6$. Two complete blocks of the experiment were carried out, so altogether there are $N = 64$ observations. In this experiment, block is a random factor. Under each set of experimental conditions, crabgrass was grown in a cup. The response is the dry weight of a unit (cup) of crabgrass, in milligrams. The R analysis of these data were first discussed in Kloke et al. (2009).

The model is a mixed model with one random effect

$$Y_{ki} = \alpha + \boldsymbol{x}_{ki}^T \boldsymbol{\beta} + b_k + \epsilon_{ki} \text{ for } k = 1, 2 \text{ and } j = 1, \ldots, 32.$$

The example below illustrates the rank-based analysis of these data using `jrfit`.

```
> library(jrfit)
> data(crabgrass)
> x<-crabgrass[,1:6]; y<-crabgrass[,7]; block<-crabgrass[,8]
> fit<-jrfit(x,y,block,var.type='sandwich')
> summary(fit)

Coefficients:
    Estimate Std. Error t-value p.value
      28.325      4.937    5.74 0.02906 *
N     39.899      4.430    9.01 0.01210 *
P     10.950      2.461    4.45 0.04698 *
K      1.599      0.645    2.48 0.13124
D1    24.050      0.379   63.53 0.00025 ***
D2     7.951      1.363    5.83 0.02815 *
D3     3.250      0.379    8.59 0.01330 *
---
Signif. codes:
0 '***' 0.001 '**' 0.01 '*' 0.05 '.' 0.1 ' ' 1
```

Based on the summary of the fit, the factors nitrogen and density are significant. The Studentized residual plot based on the Wilcoxon fit is given in Figure 9.1. Note the one large outlier in this plot. As discussed in Cobb (1998) and Kloke et al. (2009), this outlier occurred in the data. It impairs the traditional analysis of the data but has little effect on the robust analysis.

Electric Resistance Data

Presented in Stokes et al. (1995), these data are from an experiment to determine if five electrode types performed similarly. Each electrode type (etype) was applied to the arm of 16 subjects. Hence, there are 16 blocks and one fixed factor at 5 levels.

The classical nonparametric approach to addressing the question of a difference between the electrode types is to use Friedman's test (Friedman 1937), which is the analysis that Stokes et al. (1995) used. As discussed in Section 9.2, Friedman's test is available in base R via the function `friedman.test`. We illustrate its use with the electrode dataset available in `jrfit`.

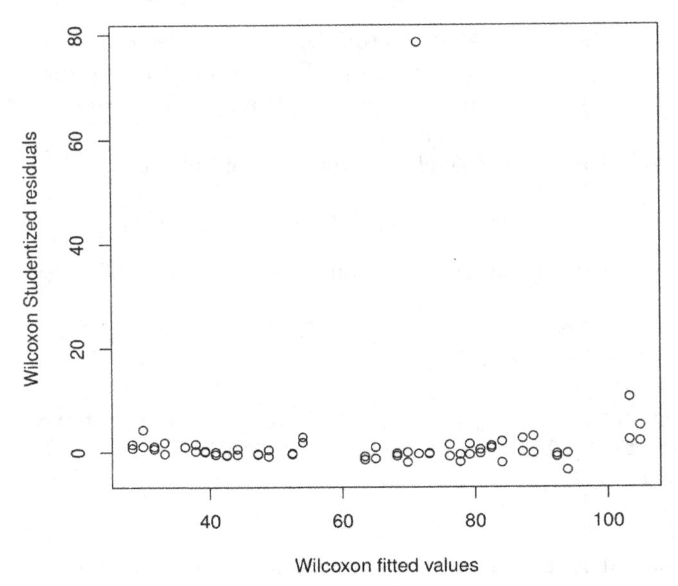

FIGURE 9.1
Plot of fitted values vs. Studentized residuals for crabgrass data.

```
> friedman.test(resistance~etype|subject,data=eResistance)

        Friedman rank sum test

data:  resistance and etype and subject
Friedman chi-squared = 5.4522, df = 4, p-value = 0.244
```

From the comparison boxplots presented in Figure 9.2 we see there are several outliers in the data.

First, we consider a cell medians model where we estimate the median resistance for each type of electrode. There the model is $y_{ki} = \mu_i + b_k + e_{ki}$ where μ_i represents the median resistance for the ith type of electrode, b_k is the kth subject (random) effect, and e_{ki} is the error term encompassing other variability. The variable **etype** is a factor from which we create the design matrix.

```
> x<-with(eResistance,model.matrix(~etype-1))
> fit<-jrfit(x,eResistance$resistance,eResistance$subject)
> summary(fit)

Coefficients:
       Estimate Std. Error t-value p.value
etype1    123.8       54.8    2.26  0.0381 *
```

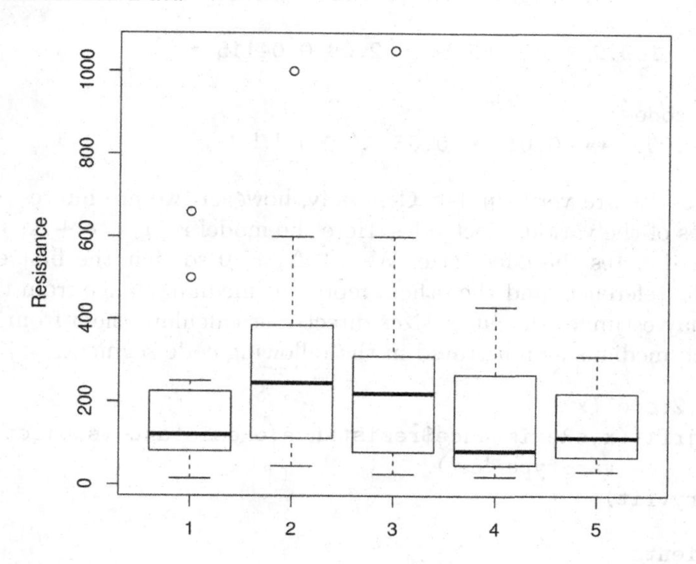

FIGURE 9.2
Comparison boxplots of resistance for five different electrode types.

```
etype2     211.2          53.7     3.94  0.0012 **
etype3     158.5          55.9     2.84  0.0119 *
etype4     106.2          55.3     1.92  0.0730 .
etype5     109.2          50.9     2.14  0.0477 *
---
Signif. codes:
0 '***' 0.001 '**' 0.01 '*' 0.05 '.' 0.1 ' ' 1
```

The above analysis was computed using the default sandwich estimator of the variance. For comparison, we next display the inference using compound symmetry. Note, in practice, this can be a strong assumption and may lead to incorrect inference. However, given the nature of this experiment, there is likely to be little carry-over effect; hence, we feel comfortable with this assumption.

```
> fit<-jrfit(x,eResistance$resistance,eResistance$subject,
+          var.type='cs')
> summary(fit)

Coefficients:
       Estimate Std. Error t-value p.value
etype1    123.8        52.5    2.36 0.02112 *
etype2    211.2        52.5    4.02 0.00014 ***
etype3    158.5        52.5    3.02 0.00350 **
etype4    106.2        52.5    2.02 0.04687 *
```

```
etype5     109.2          52.5     2.08 0.04115 *
---
Signif. codes:
0 '***' 0.001 '**' 0.01 '*' 0.05 '.' 0.1 ' ' 1
```

The results are very similar. Generally, however, we are interested in the effect sizes of the various electrodes. Here the model is $y_{ki} = \alpha + \Delta_i + b_k + e_{ki}$ where Δ_i denotes the effect size. We set $\Delta_1 = 0$ so that the first electrode type is the reference, and the others represent median change from the first.

We may estimate the effect sizes directly or calculate them from the estimated cell medians as illustrated in the following code segment.

```
> x<-x[,2:ncol(x)]
> fit<-jrfit(x,eResistance$resistance,eResistance$subject,
+            var.type='cs')
> summary(fit)

Coefficients:
       Estimate Std. Error t-value p.value
          123.8       52.5    2.36   0.021 *
etype2     87.4       33.2    2.63   0.010 *
etype3     34.7       33.2    1.05   0.299
etype4    -17.6       33.2   -0.53   0.598
etype5    -14.6       33.2   -0.44   0.662
---
Signif. codes:
0 '***' 0.001 '**' 0.01 '*' 0.05 '.' 0.1 ' ' 1
```

Next we illustrate a Wald test of the hypothesis that there is no effect due to the different electrodes; i.e.,

$$H_0 : \Delta_i = 0 \text{ for all } k = 2, \ldots, 5 \text{ versus } H_A : \Delta_i \neq 0 \text{ for some } k = 2, \ldots, 5.$$

```
> est<-fit$coef[2:5]
> vest<-fit$varhat[2:5,2:5]
> tstat<-t(est)%*%chol2inv(chol(vest))%*%est/4
> df2<-length(eResistance$resistance)-16-4-1
> pval<-pf(tstat,4,df2,lower.tail=FALSE)
> pval   #Wald test

         [,1]
[1,] 0.012229
```

Note that the overall test for effects is highly significant (p-value=0.0122). This is a much stronger result than that of Friedman's test which was nonsignificant with p-value=0.2440.

9.4 Robust Variance Component Estimators

Consider a cluster-correlated model with a compound symmetry (CS) variance-covariance structure; i.e., a simple mixed model. In many applications, we are interested in estimating the variance components and/or the random effects. For several of the fitting procedures discussed in this chapter, under CS structure, the iterative fitting of the fixed effects depends on estimates of the variance components. Even in the case of the JR fit of Section 9.3, variance components estimates are needed for standard errors and for the Studentization of the residuals. In this section, we discuss a general procedure for the estimation of the variance components and then focus on two procedures, one robust and the other highly efficient.

Consider then the cluster-correlated model (9.3) of the last section. Under CS structure, we can write the model as

$$Y_{ki} = \alpha + \boldsymbol{x}_{ki}^T\boldsymbol{\beta} + b_k + \epsilon_{ki}, \quad k = 1, \ldots, m, i = 1, \ldots, n_k, \tag{9.7}$$

where ϵ_{ki}'s are iid with pdf $f(t)$, b_k's are iid with pdf $g(t)$, and the ϵ_{ki}'s and the b_k's are jointly independent of each other. Hence, the random error for the fixed effects portion of Model (9.7) satisfies $e_{ki} = b_k + \epsilon_{ki}$. Although we could write the following discussion in terms of a general scale parameter (functional) and avoid the assumption of finite variances, for easier interpretation we simply use the variances. The variance components are:

$$\begin{aligned}
\sigma_b^2 &= \text{Var}(b_k) \\
\sigma_\epsilon^2 &= \text{Var}(\epsilon_{ki}) \\
\sigma_t^2 &= \sigma_b^2 + \sigma_\epsilon^2 \\
\rho &= \frac{\sigma_b^2}{\sigma_b^2 + \sigma_\epsilon^2}
\end{aligned} \tag{9.8}$$

The parameter ρ is often called the **intraclass correlation coefficient**, while the parameter σ_t^2 is often denoted as the **total variance**.

We discuss a general procedure for the estimation of the variance components based on the residuals from a fit of the fixed effects. Let $\hat{\theta}$ and $\hat{\eta}$ be, respectively, the given location and scale estimators. Denote the residuals of the fixed effects fit by $\hat{e}_{ki} = Y_{ki} - \hat{\alpha} - \boldsymbol{x}^T\hat{\boldsymbol{\beta}}$. Then for each cluster $k = 1, \ldots, m$, consider the pseudo-model

$$\hat{e}_{ki} = b_k + \epsilon_{ki}, \quad i = 1, \ldots, n_k. \tag{9.9}$$

Predict b_k by

$$\hat{b}_k = \hat{\theta}(\hat{e}_{k1}, \ldots, \hat{e}_{kn_k}). \tag{9.10}$$

This estimate is the prediction of the random effect. Then estimate the variance of b_k by the variation of the random effects, i.e.,

$$\hat{\sigma}_b^2 = \hat{\eta}^2(\hat{b}_1, \ldots, \hat{b}_m). \tag{9.11}$$

For estimation of the variance of ϵ_{ki}, consider Model (9.9), but now move the prediction of the random effect to the left-side; that is, consider the model

$$\hat{e}_{ki} - \hat{b}_k = \hat{\epsilon}_{ki}, \quad i = 1, \ldots, n_k. \tag{9.12}$$

Then, our estimate of the variance of ϵ_{ki} is given by

$$\hat{\sigma}_\epsilon^2 = \hat{\eta}^2(\hat{e}_{11} - \hat{b}_1, \ldots, \hat{e}_{mn_m} - \hat{b}_m). \tag{9.13}$$

Expressions (9.11) and (9.13) lead to our estimates of total variance given by $\hat{\sigma}_t^2 = \hat{\sigma}_b^2 + \hat{\sigma}_\epsilon^2$ and, hence, our estimate of intraclass correlation coefficient as $\hat{\rho} = \hat{\sigma}_b^2/\hat{\sigma}_t^2$.

Groggel (1983) and Groggel et al. (1988) proposed these estimates of the variance components except that the mean is used as the location functional and the sample variance as the scale functional. Assuming that the estimators are consistent estimators for the variance components in the case of iid errors, Groggel showed that they are also consistent for the same parameters when they are based on residuals under certain conditions. Dubnicka (2004) and Kloke et al. (2009) obtained estimates of the variance components using the median and the median absolute deviation from the median (MAD)[1] as the respective estimators of location and scale. Since these estimators are consistent for iid errors, they are consistent for our model based on the JR fit. The median and MAD comprise our first procedure for variance component estimation, and we label it as the MM procedure. As discussed below, we have written R functions which compute these estimates.

Although robust, simulation studies have shown, not surprisingly, that the median and MAD have low efficiency; see Bilgic (2012). A pair of estimators which have shown high efficiency in these studies are the Hodges–Lehmann location estimator and the rank-based dispersion estimator based on Wilcoxon scores. Recall from Chapter 1 that the Hodges–Lehmann location estimator of a sample $X_1, \ldots, X_n$ is the median of the pairwise averages

$$\hat{\theta}_{HL} = \text{med}_{i \le j} \left\{ \frac{X_i + X_j}{2} \right\}. \tag{9.14}$$

This is the estimator associated with the signed-rank Wilcoxon scores. It is a consistent robust estimator of its functional for asymmetric as well as symmetric error distributions. The associated scale estimator is the dispersion statistic given by

$$\hat{D}(\boldsymbol{X}) = \sqrt{\frac{\pi}{3}} \frac{1}{n} \sum_{i=1}^{n} \varphi \left[\frac{R(X_i)}{n+1} \right] X_i, \tag{9.15}$$

where $\varphi(u) = \sqrt{12}[u - (1/2)]$ is the Wilcoxon score function. Note that $D(\boldsymbol{X})$ is just a standardization of the norm, (7.4), of $\boldsymbol{X}$. It is a consistent estimator

[1]See Section 1.2 for defintion of the MAD.

of its functional for iid random errors as well as residuals; see Chapter 3 of Hettmansperger and McKean (2011). With the multiplicative factor $\sqrt{\pi/3}$ in expression (9.14), $\hat{D}(\boldsymbol{X})$ is a consistent estimator of σ provided that X_i is normally distributed with standard deviation σ. Although more efficient than MAD at the normal model, the statistic $\hat{D}(\boldsymbol{X})$ has an unbounded influence function in the Y-space; see Chapter 3 of Hettmansperger and McKean (2011) for discussion. Hence, it is not robust. We label the procedure based on the Hodges–Lehmann estimator of location and the dispersion function estimator of scale the DHL method.

The R function `vee` computes these variance component estimators for median-MAD (mm) and the HL-disp (dhl) procedures. The input to each consists of the residuals and the vector which identifies the center or cluster. The function returns the vector of variance component estimates and the estimates (predictions) of the random effects. We illustrate their use in the following example.

Example 9.4.1 (Variance Component Estimates). For the example, we generated a dataset for a mixed model with a treatment effect (2 levels) and a covariate. The data are over 10 clusters each with a cluster size of 10; i.e., $m = 10$, $n_i \equiv 10$, and $n = 100$. The errors ϵ_{ki} are iid $N(0,1)$, while the random effects are iid $N(0,3)$. Hence, the intraclass correlation coefficient is $\rho = 3/(1+3) = 0.75$. All fixed effects were set at 0. The following code segment computes the JR fit of the mixed model and the median-MAD and HL-dispersion variance component estimators. For both variance component estimators, we show the estimates $\hat{\sigma}_\epsilon^2$ and $\hat{\sigma}_b^2$.

```
> m<-10  # number of blocks
> n<-10  # number number
> k<-2   # number of treatments
> N<-m*n # total sample size
> x<-rnorm(N)                          # covariate
> w<-sample(c(0,1),N, replace=TRUE) # treatment indicator
> block<-rep(1:m,n)                    # m blocks of size n
> X<-cbind(x,w)
> Z<-model.matrix(~as.factor(block)-1)
> b<-rnorm(m,sd=3)
> e<-rnorm(N)
> y<-Z%*%b+e
> fit<-jrfit(X,y,block)
> summary(fit)

Coefficients:
   Estimate Std. Error t-value p.value
  -1.959030   2.835237 -0.6910  0.5053
x -0.098897   0.370353 -0.2670  0.7949
w  0.172955   0.766447  0.2257  0.8260
```

```
> vee(fit$resid,fit$block,method='mm')

$sigb2
[1] 22.85784

$sige2
[1] 0.6683255

> vee(fit$resid,fit$block)

$sigb2
         [,1]
[1,] 16.42025

$sige2
          [,1]
[1,] 0.9971398
```

Exercises 9.7.8–9.7.10 discuss the results of this example and two simulation investigations of the methods median-MAD and HL-dispersion for variance component estimation. ■

Of the two variance component methods of estimation, due to its robustness, we recommend the median-MAD procedure.

9.5 Multiple Rankings Estimator

A rank-based alternative to using the JR estimator of Section 9.3 is to use the MR estimator developed by Rashid et al. (2012). MR stands for multiple rankings, as it utilizes a separate ranking for each cluster; while the JR estimator uses the rankings of the entire dataset or the joint rankings.

The model is the same as in expression (9.3) that we repeat here for reference

$$Y_{ki} = \alpha + \boldsymbol{x}_{ki}^T\boldsymbol{\beta} + e_{ki} \text{ for } k = 1, \ldots, m, i = 1, \ldots, n_k. \tag{9.16}$$

The objective function is the sum of m separate dispersion functions, each having a separate ranking given by

$$D(\boldsymbol{\beta}) = \sum_{k=1}^{m} D_k(\boldsymbol{\beta}) \tag{9.17}$$

where $D_k(\boldsymbol{\beta}) = \sum_{i=1}^{n_k} a(R_k(Y_{ki} - \boldsymbol{x}_{ki}^T\boldsymbol{\beta}))(Y_{ki} - \boldsymbol{x}_{ki}^T\boldsymbol{\beta})$ and $R_k(Y_{ki} - \boldsymbol{x}_{ki}^T\boldsymbol{\beta})$ is the ranking of $Y_{ki} - \boldsymbol{x}_{ki}^T\boldsymbol{\beta}$ among $Y_{k1} - \boldsymbol{x}_{k1}^T\boldsymbol{\beta}, \ldots, Y_{kn_k} - \boldsymbol{x}_{kn_k}^T\boldsymbol{\beta}$.

For asymptotic theory of the rank-based fit, we need only assume that the distribution of the random errors have finite Fisher information and that the density is absolutely continuous; see Section 3.4 of Hettmansperger and McKean (2011) for discussion. In particular, as with the JR fit, the errors may have an asymmetric distribution or a symmetric distribution. Unlike the JR fit, though, for the MR fit, each cluster can have its own score function.

The MR fit obtains the estimates of the fixed effects of the model, while it is invariant to the random effects. The invariance of the MR estimate to the random effects is easy to see. Because the rankings are invariant to a constant shift, we have for center k that

$$R_j(Y_{ki} - \alpha - b_k - \boldsymbol{x}_{ki}^T\boldsymbol{\beta}) = R_j(Y_{ki} - \alpha - \boldsymbol{x}_{ki}^T\boldsymbol{\beta}).$$

Because the scores sum to 0, for each center, it follows that the objective function $D_M R$, and thus the MR estimator, are invariant to the random effects.

Rashid et al. (2012) show that the MR estimate is asymptotic normal with mean $\boldsymbol{\beta}$ and variance

$$\tau^2 \boldsymbol{V}_{MR} = \tau^2 \left(\sum_{k=1}^{m} \boldsymbol{X}_k^T \boldsymbol{X}_k \right)^{-1} \tag{9.18}$$

where the scale parameter τ is given by expression (3.19). If Wilcoxon scores are used, this is the usual parameter

$$\tau = \left[\sqrt{12} \int f(x)^2 \, dx \right]^{-1}, \tag{9.19}$$

where $f(x)$ is the pdf of random errors e_{ijl}. In expression (9.19), we are assuming that the same score function is used for each cluster. If this is not the case, letting τ_k denote the scale parameter for cluster k, the asymptotic covariance matrix is $(\sum_{k=1}^{m} \boldsymbol{X}_k^T \boldsymbol{X}_k / \tau_k)^{-1}$.

Model (9.16) assumes that there is no interaction between the center and the fixed effects. Rashid et al. (2012), though, developed a robust test for this interaction based on rank-based estimates which can be used in conjunction with the MR or JR analyses.

Estimation of Scale

A consistent estimator of the scale parameter τ can be obtained as follows. For the kth center, form the vector of residuals $\boldsymbol{r}_{MR}$ with components

$$r_{MR,ki} = y_{ki} - \boldsymbol{x}_{ki}^T \widehat{\boldsymbol{\beta}}_{MR}. \tag{9.20}$$

Denote by $\widehat{\tau}_k$ the estimator of τ proposed by Koul et al. (1987) for each of the m clusters. Note these estimates are invariant to the random effects. Furthermore, it is a consistent estimator of τ. As our estimator of τ, we take the average of these estimators, i.e.,

$$\widehat{\tau}_{MR} = \frac{1}{m}\sum_{k=1}^{m} \widehat{\tau}_j, \tag{9.21}$$

which is consistent for τ. Here we assume that the same score function is used for each cluster. If this is not the case, then, as noted above, each $1/\hat{\tau}_k$ appears within the sum in expression (9.18).

Inference

Inference based on the MR estimate can be done in the same way as with other linear models discussed in this book. For example, Wald type tests and confidence intervals based on the MR estimates can be formulated in the same way as those based on the JR fit discussed in Section 9.3. Another test statistic, not readily available for the JR procedures, is based on the reduction of dispersion in passing from the reduced to the full model. Denote the reduction in dispersion by

$$RD_{MR} = D_{MR}(\widehat{\boldsymbol{\beta}}_{MR,R}) - D_{MR}(\widehat{\boldsymbol{\beta}}_{MR,F}). \tag{9.22}$$

Large values of RD_{MR} are indicative of a lack of agreement between the collected data and the null hypothesis. As shown in Rashid et al. (2012), under H_0

$$D^*_{MR} = \tfrac{RD_{MR}}{\widehat{\tau}_{MR/2}} \text{ has an asymptotic } \chi^2(q) \text{ distribution.} \tag{9.23}$$

A nominal α decision rule is to reject H_0 in favor of H_A, if $D^*_{MR} > \chi^2_\alpha(q)$ where q is the number of constraints.

The drop-in-dispersion test was discussed in Section 7.1.3 and is analogous to the likelihood test statistic $-2\log\Lambda$, having similar interpretation. The use of a measure of dispersion to assess the effectiveness of a model fit to a set of data is common in regression analysis.

The code to compute the MR estimate is in the R package[2] `mrfit`. In the following example, the code segment demonstrates the analysis based on this R function.

Example 9.5.1 (Triglyceride Levels). The dataset **gly4gen** is a simulated dataset similar to an actual trial. Lipid levels for the patients were measured at specified times. The response variable of interest is the change in triglyceride level between the baseline and the week 4 visit. Five treatment groups were considered. The study was conducted at two centers. Centers form the random block effect. Group 1 is referenced.

```
> data(gly4gen)
```

[2]See **https://github.com/kloke/book** for more information.

```
> X<-with(gly4gen,model.matrix(~as.factor(group)-1))
> X<-X[,2:5]
> y<-gly4gen$diffgly4
> block<-gly4gen$center
> fit<-mrfit(X,y,block,rfit(y~X)$coef[2:5])
> summary(fit)

Coefficients:
                    Estimate Std. Error t-ratio
Xas.factor(group)2   0.28523    0.29624   0.96283
Xas.factor(group)3  -2.41176    0.29624  -8.14118
Xas.factor(group)4  33.03831    0.29238 112.99851
Xas.factor(group)5  26.11310    0.29624  88.14797
```

Notice for this simulated data, the triglyceride levels of Groups 3 through 5 differ significantly from Group 1. ■

9.6 GEE-Type Estimator

As in the previous sections of this chapter, we consider cluster-correlated data. Using the same notation, let Y_{ki} denote the ith response in the kth cluster, $i = 1, \ldots, n_k$ and $k = 1, \ldots, m$, and let $\boldsymbol{x}_{ki}$ denote the corresponding $p \times 1$ vector of covariates. For the kth cluster, stack the responses and covariates into the respective $n_k \times 1$ vector $\boldsymbol{Y}_k = (Y_{k1}, \ldots, Y_{kn_k})^T$ and $n_k \times p$ matrix $\boldsymbol{X}_k = (\boldsymbol{x}_{k1}^T, \ldots, \boldsymbol{x}_{kn_k}^T)^T$.

In the earlier sections of this chapter, we considered mixed (linear and random) models for $\boldsymbol{Y}_k$. For formal inference, these procedures require that the marginal distributions of the random error vectors for the clusters have the same distribution. In this section, we consider generalized linear models (glm) for cluster-correlated data. For our rank-based procedures, this assumption on the marginal distribution of the random error vector is not required.

Assume that the distribution of Y_{ki} is in the exponential family; that is, the pdf of Y_{ki} is of the form

$$f(y_{ki}) = \exp\{[y_{ki}\theta_{ki} - a(\theta_{ki}) + b(y_{ki})]\phi\}. \tag{9.24}$$

It easily follows that $E[Y_{ki}] = a'(\theta_{ki})$ and $\text{Var}[Y_{ki}] = a''(\theta_{ki})/\phi$. The covariates are included in the model using a specified function h in the following manner

$$\theta_{ki} = h(\boldsymbol{x}_{ki}^T \boldsymbol{\beta}).$$

The function h is called the **link function**. Often the canonical link is used where h is taken to be the identity function; i.e., the covariates are linked to the model via $\theta_{ki} = \boldsymbol{x}_{ki}^T\boldsymbol{\beta}$. The **Hessian** plays an important role in the fitting of the model. For the kth cluster, it is the $n_k \times p$ matrix defined by

$$\boldsymbol{D}_k = \frac{\partial a'(\boldsymbol{\theta}_k)}{\partial \boldsymbol{\beta}} = \left[\frac{\partial a'(\theta_{ki})}{\partial \beta_j}\right], \tag{9.25}$$

where $i = 1, \ldots, n_k$, $j = 1, \ldots, p$, and $\boldsymbol{\theta}_k = (\theta_{k1}, \ldots, \theta_{kn_k})^T$.

If the responses within a cluster are independent, then the above model is a generalized linear model. We are interested, though, in the cases where there is dependence within a cluster, which often occurs in practice. For the GEE estimates, we do not require the specific covariance of the responses, but, instead, we specify a dependence structure as follows. For cluster k, define the $n_k \times n_k$ matrix $\boldsymbol{V}_k$ by

$$\boldsymbol{V}_k = \boldsymbol{A}_k^{1/2}\boldsymbol{R}_k(\boldsymbol{\alpha})\boldsymbol{A}_k^{1/2}/\phi, \tag{9.26}$$

where $\boldsymbol{A}_k$ is a diagonal matrix with positive elements on the diagonal, $\boldsymbol{R}_k(\boldsymbol{\alpha})$ is a correlation matrix, and $\boldsymbol{\alpha}$ is a vector of parameters. The matrix $\boldsymbol{V}_k$ is called the **working covariance matrix** of $\boldsymbol{Y}_k$, but it need not be the covariance matrix of $\boldsymbol{Y}_k$. For example, in practice, $\boldsymbol{A}_k$ and $\boldsymbol{R}$ are not infrequently taken to be the identity matrices. In this case, we say the covariance structure is **working independence**.

Liang and Zeger (1986) develop an elegant fit of this model based on a set of generalized estimating equations (GEE) which lead to an iterated reweighted least squares (IRLS) solution. As shown by Abebe et al. (2016), each step of their solution minimizes the Euclidean norm for a nonlinear problem. Abebe et al. (2016) developed an analogous rank-based solution that leads to an IRLS robust solution.

Next, we briefly describe Abebe et al.'s solution. Assume that we have selected a score function $\varphi(u)$ which is odd about 1/2; i.e.,

$$\varphi(1-u) = -\varphi(u). \tag{9.27}$$

The Wilcoxon score function satisfies this property as do all score functions which are appropriate for symmetric error distributions. As discussed in Remark 9.6.1 this can be easily modified for score functions which do not satisfy (9.27). Suppose further that we have specified the working covariance matrix $\boldsymbol{V}$ and that we also have a consistent estimate $\hat{\boldsymbol{V}}$ of it. Suppose for cluster k that $\hat{\boldsymbol{V}}_k$ is the current estimate of the matrix $\boldsymbol{V}_k$. Let $\boldsymbol{Y}_k^* = \hat{\boldsymbol{V}}_k^{-1/2}\boldsymbol{Y}_k$ and let $g_{ki}(\boldsymbol{\beta}) = \boldsymbol{c}_i^T\mathbf{a}'(\boldsymbol{\theta}_\mathbf{k})$, where $\boldsymbol{c}_i^T$ is the ith row of $\hat{\boldsymbol{V}}_k^{-1/2}$. Then the rank-based estimate for the next step minimizes the norm

$$D(\boldsymbol{\beta}) = \sum_{k=1}^{m}\sum_{i=1}^{n_k}\varphi[R(Y_{ki}^* - g_{ki}(\boldsymbol{\beta}))/(n+1)][Y_{ki}^* - g_{ki}(\boldsymbol{\beta})]. \tag{9.28}$$

We next write the rank-based estimator as a weighted LS estimator. Let $e_{ki}(\boldsymbol{\beta}) = Y_{ki}^* - g_{ki}(\boldsymbol{\beta})$ denote the $(k,i)th$ residual and let $m_r(\boldsymbol{\beta}) = \text{med}_{(k,i)}\{e_{ki}(\boldsymbol{\beta})\}$ denote the median of all the residuals. Then, because the scores sum to 0 we have the identity,

$$\begin{aligned} D_R(\boldsymbol{\beta}) &= \sum_{k=1}^{m}\sum_{i=1}^{n_k} \varphi[R(e_{ki}(\boldsymbol{\beta}))/(n+1)][e_{ki}(\boldsymbol{\beta}) - m_r(\boldsymbol{\beta})] \\ &= \sum_{rki=1}^{m}\sum_{i=1}^{n_k} \frac{\varphi[R(e_{ki}(\boldsymbol{\beta}))/(n+1)]}{e_{ki}(\boldsymbol{\beta}) - m_r(\boldsymbol{\beta})}[e_{ki}(\boldsymbol{\beta}) - m_r(\boldsymbol{\beta})]^2 \\ &= \sum_{k=1}^{m}\sum_{i=1}^{n_k} w_{ki}(\boldsymbol{\beta})[e_{ki}(\boldsymbol{\beta}) - m_r(\boldsymbol{\beta})]^2 , \end{aligned} \tag{9.29}$$

where $w_{ki}(\boldsymbol{\beta}) = \varphi[R(e_{ki}(\boldsymbol{\beta}))/(n+1)]/[e_{ki}(\boldsymbol{\beta}) - m_r(\boldsymbol{\beta})]$ is a weight function. We set $w_{ki}(\boldsymbol{\beta})$ to be the maximum of the weights if $e_{ki}(\boldsymbol{\beta}) - m_r(\boldsymbol{\beta}) = 0$. Note that by using the median of the residuals in conjunction with property (9.27) ensures that the weights are positive.

Remark 9.6.1. To accommodate other score functions besides those that satisfy (9.27), quantiles other than the median can be used. For example, all rank-based scores are nondecreasing and sum to 0. Hence there are both negative and positive scores. So, for a given situation with sample size n, replace the median m_r with the i'th quantile where $a(i') \leq 0$ and $a(j) > 0$ for $j \geq i'$. Then the ensuing weights will be nonnegative. ■

Expression (9.29) establishes a sequence of IRLS estimates $\left\{\hat{\boldsymbol{\beta}}^{(j)}\right\}$, $j = 1, 2, \ldots$, which satisfy the generalized estimating equations (GEE) given by

$$\sum_{k=1}^{m} \boldsymbol{D}_k^T \hat{\boldsymbol{V}}_k^{-1/2} \hat{\boldsymbol{W}}_k \hat{\boldsymbol{V}}_k^{-1/2} \left[\boldsymbol{Y}_k - \boldsymbol{a}_k'(\boldsymbol{\theta}) - m_r^*\left(\hat{\boldsymbol{\beta}}^{(j)}\right)\right] = \boldsymbol{0}. \tag{9.30}$$

See Abebe et al. (2016) for details. We refer to these estimates as GEE rank-based estimates, (GEERB).

Also, a Gauss–Newton type algorithm can be developed based on the estimating equations (9.30). Since $\partial \boldsymbol{a}_k'(\boldsymbol{\theta})/\partial \boldsymbol{\beta} = \boldsymbol{D}_k$, a first-order expansion of $\boldsymbol{a}_k'(\boldsymbol{\theta})$ about the jth step estimate $\hat{\boldsymbol{\beta}}^{(j)}$ is

$$\boldsymbol{a}_k'(\boldsymbol{\theta}) = \boldsymbol{a}_k'(\hat{\boldsymbol{\theta}}^{(j)}) + \boldsymbol{D}_K\left(\boldsymbol{\beta} - \hat{\boldsymbol{\beta}}^{(j)}\right).$$

Substituting the right side of this expression for $\boldsymbol{a}_k'(\boldsymbol{\theta})$ in expression (9.30) and solving for $\hat{\boldsymbol{\beta}}^{(j+1)}$ yields

$$\begin{aligned} \hat{\boldsymbol{\beta}}^{(j+1)} &= \hat{\boldsymbol{\beta}}^{(j)} + \left[\sum_{k=1}^{m} \boldsymbol{D}_k^T \hat{\boldsymbol{V}}_k^{-1/2} \hat{\boldsymbol{W}}_k \hat{\boldsymbol{V}}_k^{-1/2} \boldsymbol{D}_k\right]^{-1} \\ &\quad \times \sum_{k=1}^{m} \boldsymbol{D}_{k'}^T \hat{\boldsymbol{V}}_k^{-1/2} \hat{\boldsymbol{W}}_k \hat{\boldsymbol{V}}_k^{-1/2} \left[\boldsymbol{Y}_k - \boldsymbol{a}_k'(\hat{\boldsymbol{\theta}}^{(j)}) - m_r^*\left(\hat{\boldsymbol{\beta}}^{(j)}\right)\right]. \end{aligned}$$

Abebe et al. (2016) developed the asymptotic theory for these rank-based GEERB estimates under the assumption of continuous responses. They showed that under regularity conditions, the estimates are asymptotically normal with mean $\boldsymbol{\beta}$ and with the variance-covariance matrix given by

$$\left\{\sum_{k=1}^{m}\mathbf{D}_k^T\mathbf{V}_k^{-1/2}\mathbf{W}_k\mathbf{V}_k^{-1/2}\mathbf{D}_k\right\}^{-1}\left\{\sum_{k=1}^{m}\mathbf{D}_k^T\mathbf{V}_k^{-1/2}\,Var(\boldsymbol{\varphi}_k^{\dagger})\mathbf{V}_k^{-1/2}\mathbf{D}_i\right\}$$
$$\times\left\{\sum_{k=1}^{m}\mathbf{D}_k^T\mathbf{V}_k^{-1/2}\mathbf{W}_k\mathbf{V}_k^{-1/2}\mathbf{D}_k\right\}^{-1}, \tag{9.31}$$

where $\boldsymbol{\varphi}_k^{\dagger}$ denotes the $n_k\times 1$ vector $(\varphi[R(e_{k1}^{\dagger})/(n+1)],\ldots,\varphi[R(e_{kn_k}^{\dagger})/(n+1)])^T$ and $e_{kn_k}^{\dagger}$ is defined by the following expressions:

$$\begin{aligned}\mathbf{Y}_k^{\dagger} &= \mathbf{V}_k^{-1/2}\mathbf{Y}_k = (Y_{k1}^{\dagger},\ldots,Y_{kn_k}^{\dagger})^T\\ \mathbf{G}_k^{\dagger}(\boldsymbol{\beta}) &= \mathbf{V}_k^{-1/2}\mathbf{a}_i'(\boldsymbol{\theta}) = [g_{ki}^{\dagger}]\\ e_{ki}^{\dagger} &= Y_{ki}^{\dagger} - g_{ki}^{\dagger}(\boldsymbol{\beta}).\end{aligned} \tag{9.32}$$

A practical implementation of inference based on the GEERB estimates is discussed in Section 9.6.4. As discussed in Abebe et al. (2016), the GEERB estimates are robust in the $\boldsymbol{Y}$ space (provided the score function is bounded) but not robust in the $\boldsymbol{X}$ space. If there are outliers in the $\boldsymbol{X}$ space, then we recommend using the option of the HBR GEE fit which provides a high breakdown fit.

The simulation studies presented in Abebe et al. (2016) showed that the rank-based GEE fits, including the HBR GEE fits, provided valid analyses over the situations simulated. These fits had high empirical efficiency relative to analogous LS procedures for normally distributed random errors and were much more efficient for contaminated normal errors.

The GEE estimates are quite flexible. The exponential family is a large family of distributions that is often used in practice. The choices of the link functions and working variance-covariance structures allow a large variety of models from which to choose. As shown in expression (9.31), the asymptotic covariance matrix of the estimate takes into account each of these choices; that is, the link function determines the Hessian matrices $\boldsymbol{D}_k$, (9.25); the working covariance structure determines the matrices $\boldsymbol{V}_k$, (9.26); and the pdf is reflected in the factor in the middle set of braces. We provided R code to compute GEERB estimates based on the Gauss–Newton step described above. The main driver is the R function `geerfit` which is included in the package[3] `rbgee`. It can be easily modified to compute different options. In the next several subsections, we discuss the weights, link functions, and the working covariance structure, and our R functions associated with these items. We

[3] See `https://github.com/kloke/book` for more information.

discuss some of the details of R code in the next three subsections which are followed by an illustrative example.

9.6.1 Weights

The R function `wtmat(Dmat,eitb,med=TRUE,scores=wscores)` computes the weights, where `eitb` is the vector of current residuals and `Dmat` is the current Hessian matrix. If the option `med` is set to `TRUE`, then it is assumed that the score function is odd about $\frac{1}{2}$ and the median of the current residuals is used in the calculation of the weights as given in expression (9.29). If `med` is `FALSE`, then the percentile discussed in Remark 9.6.1 is used. These are calculated at the current residuals, making use of the Hessian matrix $\boldsymbol{D}$, (9.25).

9.6.2 Link Function

The link function connects the covariance space to the distribution of the responses. Note that it affects the fitting algorithm in its interaction with the vectors $\boldsymbol{a}_k(\boldsymbol{\theta}_k)$, (9.24), and the Hessian matrices $\boldsymbol{D}_k$, (9.25). We have set the default link to a linear model. Thus, in the routine `getAp`, for cluster k and with $\boldsymbol{\beta}^{(j)}$ as the current estimate of $\boldsymbol{\beta}$, the vector $\boldsymbol{a}'_k(\boldsymbol{\theta}_k)$ is set to $\boldsymbol{X}_k\boldsymbol{\beta}^{(j)}$ and, in the routine `getD`, the matrix $\boldsymbol{D}_k$ is set to $\boldsymbol{X}_k$. For other link functions, these routines have to be changed.

9.6.3 Working Covariance Matrix

The working covariance matrix $\boldsymbol{V}$, (9.26), is computed in the function `veemat`. Currently, there are three options available: working independence, "WI"; compound symmetry (exchangeable), "CS"; and autoregressive order 1, "AR". Default is set at compound symmetry. For the compound symmetry case, this function also sets the method for computing the variance components. Currently, there are two options: the MAD-median option, "MM", and the dispersion and Hodges–Lehmann estimator, "DHL." The default option is the MAD-median option. Recall that the MAD-median option results in robust estimates of the variance components.

9.6.4 Standard Errors

Abebe et al. (2016) performed several Monte Carlo studies of procedures to standardize the GEERB in terms of validity. One procedure involved estimation of the asymptotic variance-covariance matrix, (9.31), of GEERB estimators, using the final estimates of the matrices $\boldsymbol{V}$, $\boldsymbol{W}$, and $\boldsymbol{D}$. For cluster k, a simple moment estimator of $Var(\varphi^{\dagger}_k)$ based on residuals is discussed in Abebe et al. (2016). The resulting estimator of the asymptotic variance-covariance matrix in the Monte Carlo studies, though, appeared to lead to a liberal inference. In the studies, a first-order approximation to the asymptotic variance-covariance matrix appeared to lead to a valid inference. The first-order approximation involves replacing the weight matrix $\hat{\boldsymbol{W}}$ by $\hat{\tau}^{-1}\boldsymbol{I}$, where $\hat{\tau}$ is the estimator of τ, and the matrix $Var(\boldsymbol{\varphi}_k^{\dagger})$ by $\boldsymbol{I}_k$. In our R function, the indicator of the variance-covariance procedure is the variable `varcovst`. The default setting is `varcovst=="var2"` which results is this approximation, while the setting `"var1"` results in the estimation of the asymptotic form. The third setting, `"var3"`, is a hybrid of the two where just $\boldsymbol{W}$ is approximated. This is similar to a sandwich type estimator.

9.6.5 Examples

The driver of our GEERB fit R function at defaults settings is given below.

```
geerfit(y,xmat,center,
        scores=wscores,geemod="LM",
        structure="CS",substructure="MM",
        med=TRUE,varcovst="var2",
        maxstp=50,eps=0.00001,hbrs=FALSE,delta=0.8,hparm=2)
```

The function `geerfit` assumes that `y` and `xmat` are sorted by center. For the HBR fit, set `hbrs=TRUE`.

The routine returns the estimates of the regression coefficients and their standard errors and t-ratios, along with the variance-covariance estimator and the history (in terms of estimates) of the Newton steps. We illustrate the routine with two examples.

Example 9.6.1. For this example, we simulated data with 5 clusters each of size 10. The covariance structure is compound symmetrical with the variances set at $\sigma_\varepsilon^2 = 1$ and $\sigma_b^2 = 3$, so that the intraclass coefficient is $\rho = 0.75$. The random errors and random effects are normally distributed, $N(0,1)$, with the true $\boldsymbol{\beta}$ set at $(0.5, 0.35, 0.0)^T$. There are 3 covariates which were generated from a standard normal distribution. The data can be found in the dataset `eg1gee`. The following R segment loads the data and computes the Wilcoxon GEE fit. We used the default settings for the GEERB fit. In particular, compound symmetry was the assumed covariance structure and the variance component estimates are returned in the code.

```
> xmat<- with(eg1gee,cbind(x1,x2,x3))
> gwfit <- geerfit(y,xmat,block)
> gwfit$tab

           Est        SE     t-ratio
x1  0.54067723 0.1118348  4.83460610
x2 -0.01492029 0.1576557 -0.09463847
x3 -0.20872498 0.1300584 -1.60485548

> vc <- gwfit$vc
> vc
```

```
[1] 2.2674708 1.0457868 1.2216840 0.4612129

> rho <- vc[2]/vc[1]
> rho

[1] 0.4612129
```

The GEERB estimates of the three components of $\boldsymbol{\beta}$ are 0.541, −0.015, and −0.209. Based on the standard error of the estimates, the true value of each component of $\hat{\boldsymbol{\beta}}$ is trapped within the respective 95% confidence interval. The estimates of the variance components are $\hat{\sigma}_t^2 = 2.267$, $\hat{\sigma}_b^2 = 1.046$, $\hat{\sigma}_\epsilon^2 = 1.222$, and $\hat{\rho} = 0.461$.

We next change the data so that $y_{11} = 53$ instead of 1.53 and rerun the fits:

```
> y[1] <-  53
> gwfit <- geerfit(y,xmat,block)
> gwfit$tab

             Est        SE      t-ratio
x1  0.6069074818 0.1491861  4.068123565
x2 -0.0006604922 0.2119288 -0.003116575
x3 -0.3447768897 0.1454904 -2.369756997

> gwfit$vc

[1] 2.1359250 0.6595324 1.4763927 0.3087807
```

There is little change in the estimate of $\boldsymbol{\beta}$, verifying the robustness of the GEERB estimator. ∎

Example 9.6.2 (Rounding Firstbase, Continued). Recall that in Example 9.2.1 three methods (round out, narrow angle, and wide angle) for rounding first base were investigated. Twenty-two baseball players served as blocks. The responses are their average times for two replications of each method. Thus, the design is a randomized block design. In Example 9.2.1, the data are analyzed using Friedman's test which is significant for treatment effect. Friedman's analysis, though, consists of only a test. In contrast, we next discuss the rank-based analysis based on the GEERB fit. In addition to a test for overall treatment effect, it offers estimates (with standard errors) of size effects, estimates of the variance components, and a residual analysis for checking quality of fit and in determining outliers. We use a design matrix which references the first method. The following code provides the Wald test, based on the fit, which tests for differences among the three methods:

```
> fit <- geerfit(y,xm,center)
> beta <- fit$tab[,1]
```

```
> tst <- t(beta)%*%solve(fit$varcov)%*%beta
> pv <- 1-pchisq(tst,2)
> c(tst,pv)

[1] 12.725285980  0.001724802
```

The Wald test is significant at the 5% level. With Friedman's method, this would be the end of the analysis. Let μ_i denote the mean time of method i. The effects of interest are the differences between these means. Because method 1 is referenced, the summary of the fit (**fit$tab**) provides the inference for $\mu_3 - \mu_1$ and $\mu_2 - \mu_1$. However, the next few lines of code yield the complete inference for comparison of the methods. The summary is displayed in Table 9.1.

```
> h <- matrix(c(-1,1),ncol=2); e32 <- h%*%beta
> se32 <- sqrt(h%*%fit$varcov%*%t(h))
> t32 <- e32/se32; c(e32,se32,t32)

[1] -0.07826494  0.02271888 -3.44492906
```

TABLE 9.1
Summary Table of Effects for the Firstbase Data.

	Effect Est.	SE	t-ratio
mu2 minus mu1	-0.02	0.03	-0.69
mu3 minus mu1	-0.10	0.03	-2.92
mu3 minus mu2	-0.08	0.02	-3.44

Based on Table 9.1, method 3 significantly differs from the other two methods, while methods 1 and 2 do not differ significantly. Hence, overall, method 3 (rounding first base using a wide angle) results in the quickest times. The top panel of Figure 9.3 displays comparison boxplots of the three methods. Outliers are prevalent. The bottom panel shows the $q-q$ plot of the residuals based on the GEERB fit. Notice that three outliers clearly stand out. A simple inspection of the residuals shows that these outliers correspond to the times of baseball player #22. He appears to be the slowest runner. The rank-based estimates of the variance components are:

```
> fit$vc

[1] 0.013402740 0.012364328 0.001038412 0.922522403
```

The estimate of the intraclass correlation coefficient is 0.92, indicating a strong correlation of running times within players over the three methods. ■

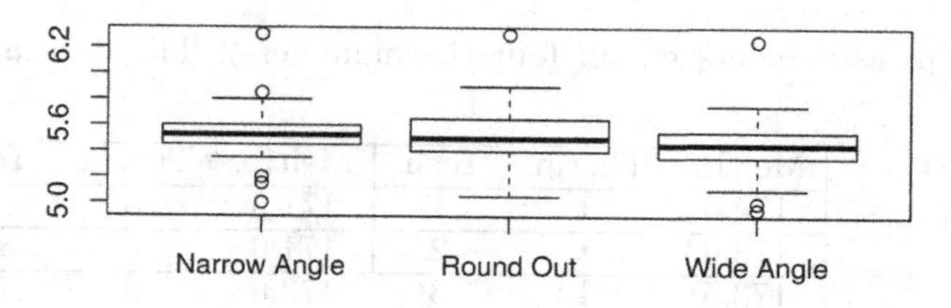

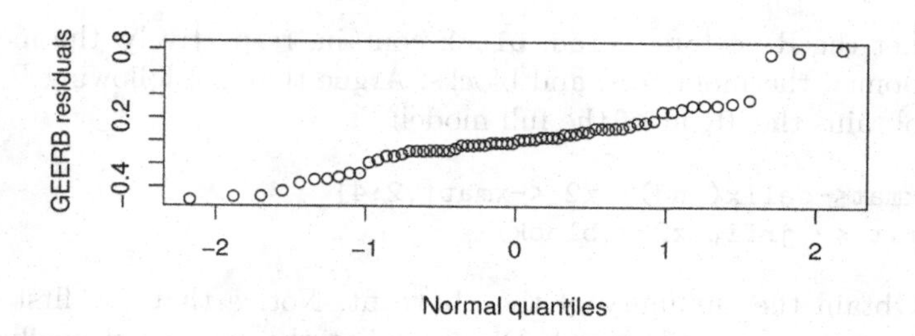

FIGURE 9.3
Comparison boxplots of the three methods and the $q-q$ plot of the residuals based on the GEERB fit.

9.7 Exercises

9.7.1. Transform the `firstbase` data frame into a vector and create categorical variables for treatment (rounding method) and subject. Obtain the results from `friedman.test` using these data objects.

9.7.2. Referring to Exercise 9.7.1, obtain estimates of the cell medians for each of the rounding methods. Also obtain estimates of the standard errors of these estimates of location using both a compound symmetry estimate as well as the sandwich estimate. Discuss.

9.7.3. It is straightforward to generate data from the simple mixed model. Write a function which generates a sample of size m of blocks each of size n which variance components σ_b and σ_ϵ. Assume normal errors.

9.7.4. Extend 9.7.3 to include errors of a t distribution.

9.7.5. Extend 9.7.3 to allow for different block sizes.

9.7.6. On page 418, Rasmussen (1992) discusses a randomized block design concerned with the readings of four thermometers for melting point of hydroquinone. Three technicians were used as a blocking factor (each technician obtained measurements for all four thermometers). The data are presented next.

MeltPt	Therm.	Tech.	MeltPt	Therm.	Tech.
174.0	1	1	171.5	3	1
173.0	1	2	171.0	3	2
173.5	1	3	173.0	3	3
173.0	2	1	173.5	4	1
172.0	2	2	171.0	4	2
173.0	2	3	172.5	4	3

(a) Let the R vectors `y`, `ind`, `block` contain, respectively, the melting points, thermometers, and blocks. Argue that the following R code obtains the JR fit of the full model:

```
xmat<-cellx(ind); x2 <-xmat[,2:4]
fit <- jrfit(x2,y,block)
```

(b) Obtain the summary of the above fit. Notice that the first thermometer was referenced. Discuss what the regression coefficients are estimating. Do there seem to be any significant differences in the thermometers?

(c) Obtain residual and normal $q-q$ plots of the Studentized residuals. Are there any outliers? Discuss the quality of fit.

(d) Argue that the following code obtains the Wald's test of no differences among the parameters. Is this hypothesis rejected at the 5% level?

```
beta<-fit$coef; b<-beta[2:4]; vc <- fit$varhat[2:4,2:4]
tst <- t(b)%*%solve(vc)%*%b/3; pv<-1-pf(tst,3,4)
```

9.7.7. Rasmussen (1992), (page 442), discusses a study on emotions. Each of eight volunteers were requested to express the emotions of fear, happiness, depression, and calmness. At the time of expression, their skin potentials in millivolts were measured; hence, there are 32 measurements in all. The data can be found in the dataset `emotion`.

(a) Discuss an appropriate model for the data and obtain a rank-based fit of it.

(b) Using Studentized residuals check quality of fit.

(c) Test to see if there is a significant difference in the skin potential of the four emotions. Conclude at the 5% level.

(d) Obtain a 95% interval for the shift between the emotions of fear and calmness.

(e) Obtain a 95% interval for the shift between the emotions of depression and calmness.

9.7.8. Consider the results of the `jrfit` in Example 9.4.1. Obtain 95% confidence intervals for the fixed effects coefficients. Did they trap the true values?

9.7.9. Run a simulation of size 10,000 on the model simulated in Example 9.4.1. In the simulations, collect the estimates of the variance components σ_b^2 and σ_ϵ^2 for both the median-MAD and the HL-dispersion methods. Obtain the empirical mean square errors. Which method, if any, performed better than the other?

9.7.10. Repeat Exercise 9.7.9, but for this simulation, use the t-distribution with 2 degrees of freedom for both the random errors ϵ_{ki} and the random effects b_k.

9.7.11. For Example 9.6.1, obtain the analyses using the HBR GEE fits. Discuss similarities and differences of the analyses.

9.7.12. For Example 9.6.1, substitute the outlier 10 for `x1[1]`. Compare the results of the HBR GEE fit with the Wilcoxon GEE fit with the changes in fits on the original data. Do residual plots of both fits show the outlier?

(a) Repeat using the dataset with both the $\boldsymbol{X}$-space outlier and the Y-space outliers (as in the example).

10

Multivariate Analysis

10.1 Introduction

We refer to multivariate analysis as the case when there are several outcome or response variables of interest, so that each of the measurements is collected on each of the experimental units in the study; i.e., the response for each subject is a vector.

Traditional least squares analyses are covered in the textbooks by Seber (1984) and Johnson and Wichern (2007). Nonparametric approaches are covered in Chapter 6 of Hettmansperger and McKean (2011) and Oja (2010).

As an example, consider the `nutrient` dataset in the R package `profileR` (Bulut and Desjardins 2018). The data were collected in 1985 for a study of women's nutrition; a sample of $n = 737$ females aged 25-50 was taken. There are five outcome measurements of interest in this case: calcium, iron, protein, vitamin A, and vitamin C. So the response for each participant in the study is a 5×1 vector.

In the following code segment, we view the first few records.

```
> library(profileR)
> head(nutrient)

  calcium   iron protein      a       c
1  522.29 10.188  42.561 349.13  54.141
2  343.32  4.113  67.793 266.99  24.839
3  858.26 13.741  59.933 667.90 155.455
4  575.98 13.245  42.215 792.23 224.688
5 1927.50 18.919 111.316 740.27  80.961
6  607.58  6.800  45.785 165.68  13.050
```

A `pairs` plot of the data is given in Figure 10.1. Notice, there is a substantial outlier: a value for vitamin A of almost 35000 — with the rest of the sample falling below half that level.

Often a useful summary in multivariate analysis is a correlation matrix, which simply provides the pairwise correlations of the variables in the data matrix. As an example, a correlation matrix of the `nutrient` data, using Spearman's method, is obtained in the next code segment.

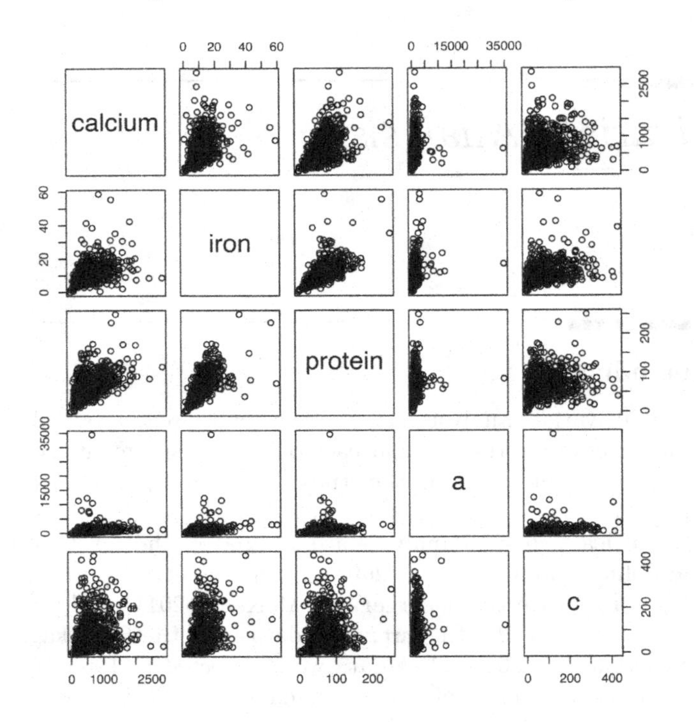

FIGURE 10.1
Pairs plot of the `nutrient` data.

```
> cor(nutrient,method='spearman')

         calcium    iron protein       a       c
calcium 1.00000 0.45570 0.51596 0.55123 0.31192
iron    0.45570 1.00000 0.71322 0.46673 0.36158
protein 0.51596 0.71322 1.00000 0.39709 0.28239
a       0.55123 0.46673 0.39709 1.00000 0.41393
c       0.31192 0.36158 0.28239 0.41393 1.00000
```

Note: the default in R is to use Pearson's correlation (or obtained using `method='pearson'`); Kendall's measure of association is also available (using `method='kendall'` as one might anticipate). The pairwise correlations range from 0.282 to 0.713. At the time of the study, the recommended intake was as given in Table 10.1.[1]

One research question might be how the median values compare with the recommended amounts. In this chapter we introduce multivariate approaches to address such a question.

[1] https://online.stat.psu.edu/stat505/lesson/7/7.1/7.1.1

TABLE 10.1
Recommended intake for health study.

Variable	Recommended
Calcium	1000
Iron	15
Protein	60
Vitamin A	800
Vitamin C	75

10.2 Brief Review of Multivariate Distributions

Before discussing methodology, we briefly discuss multivariate distributions as a review or introduction. For a more in depth-study, readers are referred to the texts by Seber (1984) as well as Johnson and Wichern (2007). Readers who know this material well may skip this section without loss of information.

10.2.1 Continuous Multivariate Distributions

The asymptotic distribution of the estimators discussed in this chapter, due to a multivariate central limit theorem, is multivariate normal. Suppose a study is conducted to measure k continuous responses for each subject or experimental unit. We denote the responses by the random variables $X_1, X_2, \ldots, X_k$ that form the $k \times 1$ vector $\boldsymbol{X} = [X_1, X_2, \ldots, X_k]^T$. We say that $\boldsymbol{X}$ is a **random vector**.

Let us first consider the case where $k = 2$ or a *bivariate distribution*; hence, the vector of responses is $\boldsymbol{X} = [X_1, X_2]^T$. A bivariate pdf extends the concept of a univariate pdf. The next two examples discuss examples of bivariate pdfs. These examples are followed by a discussion of general multivariate pdfs.

Example 10.2.1. Consider Figure 10.2 which displays a pdf for a bivariate random vector $\boldsymbol{X} = [X_1, X_2]^T$. The shape of the surface is ellipsoidal. Curves of constant height (same value of $z = f(x_1, x_2)$) on the surface are elliptical. From the elongated shape and direction of these ellipses, it seems that if X_1 is large, then it is likely that X_2 is, also. This indicates that the random variables are dependent. The surface peaks at the point $(x_1, x_2) = (0, 0)$ which is also the center of gravity of the surface, i.e., if we could cut this surface out the point of balance in the $x_1 \times x_2$ plane is $(0, 0)$. ■

Example 10.2.2. Figure 10.3 displays the surface of a pdf of a pair of nonnegative random variables X_1 and X_2. This distribution is discussed on Page 88 of Hogg et al. (2019). The joint pdf is

$$f(x_1, x_2) = 4x_1x_2e^{-(x_1^2+x_2^2)}, \quad x_1 > 0, x_2 > 0.$$

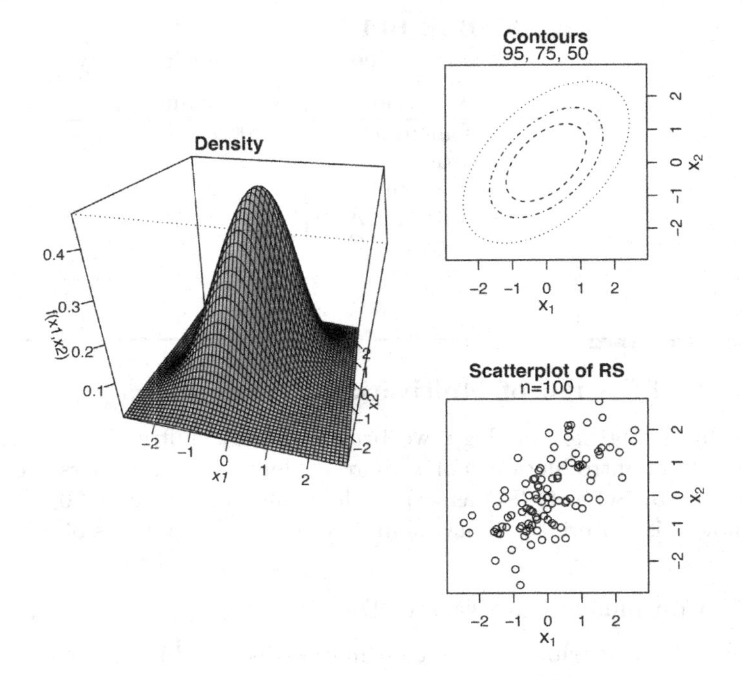

FIGURE 10.2
Graphics illustrating a bivariate normal distribution with $\boldsymbol{\mu} = [0,0]^T$, $\sigma_1^2 = \sigma_2^2 = 1$, and $\sigma_{12} = \sigma_{21} = \rho = 0.5$. (*Left*) Three-dimensional display of the surface of the pdf. (*Top-right*) Contours of 0.95, 0.75, 0.5. (*Bottom-right*) Scatterplot of a random sample of size $n = 100$.

As x_1 and x_2 increase from $(0,0)$, the pdf sharply rises and peaks at $(\sqrt{2}/2, \sqrt{2}/2)$ and then tapers decreasing close to 0 in all increasing directions. It appears to be right-skewed in both the x_1 and x_2 directions. For this distribution, the random variables X_1 and X_2 are independent, each having the pdf $f(x) = 2x\exp\{-x^2\}$, for $x > 0$, which is a right-skewed density function known as a Weibull pdf. The R function `rweibull` can be used to obtain a random sample. The bottom right-panel of Figure 10.3 displays the scatterplot of a random sample of 100 observations from this joint pdf. ■

Let $f(\boldsymbol{x}) = f(x_1, x_2, \ldots, x_k)$ be a non-negative function of k variables such that the total volume under the surface $z = f(\boldsymbol{x})$ is 1. Let A be a region in k-dimensional space. Then, we say $f(\boldsymbol{x})$ is the pdf of the random vector $\boldsymbol{X}$, if $P(\boldsymbol{X} \in A)$ is the volume under the surface $z = f(\boldsymbol{x})$ over the region A. We say that $f(\boldsymbol{x})$ is the **joint pdf** of $\boldsymbol{X}$.

From the joint distribution of $\boldsymbol{X}$, we can always obtain the univariate distributions of the random variables. For example, the cdf of X_1 is

$$F_1(x) = P(X_1 \le x) = P(X_1 \le x \text{ and } -\infty < X_i < \infty, \text{ for } i = 2, \ldots, k).$$

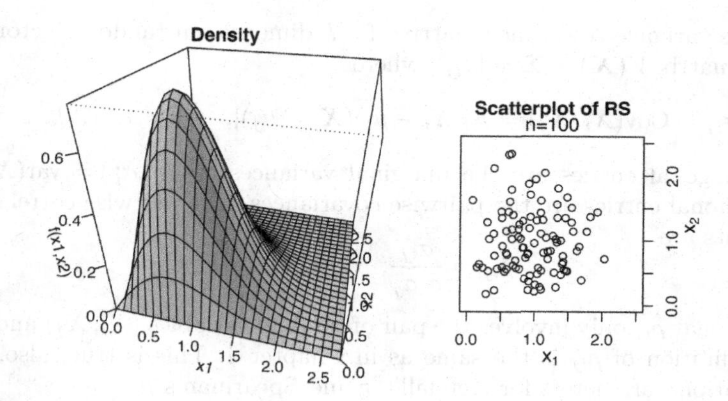

FIGURE 10.3
Graphics illustrating a bivariate Weibull distribution with independent components. (*Left*) Three-dimensional display of the surface of the pdf. (*Right*) Scatterplot of a random sample of size $n = 100$.

In this multivariate setting, we often call the distributions of the X_i's **marginal distributions**.

10.2.2 Expectation and Covariance

Let $\boldsymbol{X}$ be a k-dimensional random vector. Then we define the expected value of $\boldsymbol{X}$ to be

$$E(\boldsymbol{X}) = \begin{bmatrix} E(X_1) \\ E(X_2) \\ \vdots \\ E(X_k) \end{bmatrix}; \tag{10.1}$$

hence, expectations of random vectors are defined in terms of the marginal expectations. It follows that expectations are linear; i.e., $E(\boldsymbol{AX} + \boldsymbol{BY}) = \boldsymbol{A}E(\boldsymbol{X}) + \boldsymbol{B}E(\boldsymbol{Y})$. We denote the $E(\boldsymbol{X})$ by $\boldsymbol{\mu}$, so that $\boldsymbol{\mu} = [\mu_1, \ldots, \mu_k]^T$, where $\mu_i = E(X_i)$. The vector $\boldsymbol{\mu}$ is the centroid of the joint pdf.

For the random vector discussed in Example 10.2.1, it can be shown that $\boldsymbol{\mu} = \boldsymbol{0}$. So the origin is the point of balance of the joint pdf. For Example 10.2.2, it can be shown that $\boldsymbol{\mu} = [\sqrt{\pi}/2, \sqrt{\pi}/2]^T$.

The vector $\boldsymbol{\mu}$ serves as a location parameter for multivariate location models. We will use other location vectors, also. The vector of medians $\boldsymbol{\theta} = [\theta_1, \ldots, \theta_k]^T$, where θ_i is the median of X_i, will be often used. In Example 10.2.2, it can be shown that $\boldsymbol{\theta} = [\sqrt{\log 2}, \sqrt{\log 2}]^T$, while for Example 10.2.1, $\boldsymbol{\theta} = \boldsymbol{0}$.

The variance-covariance matrix of a k-dimensional random vector $\boldsymbol{X}$ is a $k \times k$ matrix $V(\boldsymbol{X}) = \boldsymbol{\Sigma} = [\sigma_{ij}]$, where

$$\sigma_{ij} = \mathrm{Cov}(X_i, X_j) = E[(X_i - \mu_i)(X_j - \mu_j)], \quad i, j = 1, \ldots, k. \tag{10.2}$$

The diagonal entries are the marginal variances $\sigma_{ii} = \sigma_i^2 = \mathrm{var}(X_i)$. The off-diagonal entries are the pairwise covariances. The pairwise correlation coefficients are

$$\rho_{ij} = \frac{\sigma_{ij}}{\sigma_i \sigma_j}, \quad i, j = 1, \ldots, k. \tag{10.3}$$

Notice that ρ_{ij} only involves the pair of random variables (X_i, X_j) and, hence, the definition of ρ_{ij} is the same as in Chapter 4. This is true, also, for the population parameters for Kendall's τ and Spearman's ρ.

Notice that $\sigma_{ij} = \sigma_{ji}$ so that the variance-covariance matrix is symmetric; i.e., $\boldsymbol{\Sigma}^T = \boldsymbol{\Sigma}$. Using linearity of expectations, various facts can be derived, including the following three:

$$V(\boldsymbol{X}) = E[\boldsymbol{X}\boldsymbol{X}^T] - \boldsymbol{\mu}\boldsymbol{\mu}^T \tag{10.4}$$

$$V(\boldsymbol{A}\boldsymbol{X}) = \boldsymbol{A}V(\boldsymbol{X})\boldsymbol{A}^T \tag{10.5}$$

$$E(\boldsymbol{X}^T\boldsymbol{C}\boldsymbol{X}) = \mathrm{tr}\boldsymbol{C}\boldsymbol{\Sigma} + \boldsymbol{\mu}^T\boldsymbol{C}\boldsymbol{\mu}, \text{ for a symmetric } \boldsymbol{C}. \tag{10.6}$$

In expression (10.6), the symbol, tr, refers to the **trace** operator; that is, if $\boldsymbol{A} = [a_{ij}]$ is an $n \times n$ matrix, then

$$\mathrm{tr}(\boldsymbol{A}) = \sum_{i=1}^{n} a_{ii}. \tag{10.7}$$

10.2.3 Multivariate Normal Distribution

Let $\boldsymbol{X}$ be a k-dimensional random vector, let $\boldsymbol{\mu}$ be a k-dimensional vector, and let $\boldsymbol{\Sigma}$ be a $k \times k$ positive definite symmetric matrix. Then, we say that $\boldsymbol{X}$ has a **multivariate normal distribution** with mean vector $\boldsymbol{\mu}$ and covariance matrix $\boldsymbol{\Sigma}$, if the pdf of $\boldsymbol{X}$ is given by

$$f(\boldsymbol{x}) = \frac{1}{(2\pi)^{k/2}|\boldsymbol{\Sigma}|^{1/2}} \exp\left\{-\frac{1}{2}(\boldsymbol{x} - \boldsymbol{\mu})^T\boldsymbol{\Sigma}^{-1}(\boldsymbol{x} - \boldsymbol{\mu})\right\}, \quad \boldsymbol{x} \in R^k. \tag{10.8}$$

We express this in notation as $\boldsymbol{X}$ is distributed $N_k(\boldsymbol{\mu}, \boldsymbol{\Sigma})$. Note that for c in the range of f, the contour $\{\boldsymbol{x} : f(\boldsymbol{x}) = c\}$ is an ellipsoid in k-space.

The pdf discussed in Example 10.2.1 and shown in Figure 10.2, is an example of a bivariate normal distribution. As we discussed, the plot shows the elliptical contours of a multivariate normal distribution. For this example, we set $\boldsymbol{\mu} = \boldsymbol{0}$, $\sigma_1 = \sigma_2 = 1$, and $\sigma_{12} = 0.5$. Hence, X_1 and X_2 are dependent random variables with correlation coefficient $\rho = 0.5$.

Suppose $\boldsymbol{X}$ has the multivariate normal distribution $N_k(\boldsymbol{\mu}, \boldsymbol{\Sigma})$. Then any linear function of $\boldsymbol{X}$ is normal. For example, any of the marginal distributions are univariate normal; i.e., $X_i \sim N_1(\mu_i, \sigma_i^2)$. More specifically, let $\boldsymbol{A}$ be an $m \times k$ matrix and $\boldsymbol{b}$ be a k-dimensional vector. Then $\boldsymbol{Y} = \boldsymbol{AX} + \boldsymbol{b}$ has a $N_m(\boldsymbol{A\mu} + \boldsymbol{b}, \boldsymbol{A\Sigma A}^T)$ distribution. Taking $\boldsymbol{A}$ to be the vector $\boldsymbol{\delta}_i^T = (0, \ldots, 0, 1, 0, \ldots, 0)$, where the position that the 1 occurs is the ith component of the vector. Then, as mentioned above, it follows that $X_i = \boldsymbol{\delta}_i^T \boldsymbol{X}$ is $N_1(\mu_i, \sigma_i^2)$.

Consider a random sample of size n on $\boldsymbol{X}$, $\boldsymbol{X}_1, \ldots, \boldsymbol{X}_n$. A representation of the sample is a **data matrix** which is the $n \times k$ matrix defined by

$$\boldsymbol{\mathcal{X}} = \begin{bmatrix} \boldsymbol{X}_1^T \\ \boldsymbol{X}_2^T \\ \vdots \\ \boldsymbol{X}_n^T \end{bmatrix}. \tag{10.9}$$

Notice that the jth column of the data matrix is the random sample on the jth component of the random vector $\boldsymbol{X}$. So, the data matrix consists of the entire multivariate sample.

The **multivariate Central Limit Theorem** (CLT) generalizes the univariate CLT. Assume $\boldsymbol{X}_1, \boldsymbol{X}_2, \ldots, \boldsymbol{X}_n, \ldots$ are independent k-dimensional random vectors with a common distribution that has mean $\boldsymbol{\mu}$ and a positive definite covariance matrix $\boldsymbol{\Sigma}$. Let

$$\overline{\boldsymbol{X}} = \frac{1}{n} \sum_{i=1}^{n} \boldsymbol{X}_i$$

denote the sample mean vector and

$$\boldsymbol{S} = \sum_{i=1}^{n} \boldsymbol{X}_i$$

denote the column sums of the data matrix $\boldsymbol{\mathcal{X}}$. Then

$$\begin{aligned} \boldsymbol{S} &\text{ is approximately } N_k(n\boldsymbol{\mu}, n\boldsymbol{\Sigma}) \\ \overline{\boldsymbol{X}} &\text{ is approximately } N_k(\boldsymbol{\mu}, n^{-1}\boldsymbol{\Sigma}). \end{aligned} \tag{10.10}$$

So, as long as the covariance matrix of the population exists (i.e., is finite), the vector of sample means has an approximate normal distribution. By the note above on marginal distributions, this implies that for each i, $1 \le i \le k$ that $\overline{X}_i$ has the approximate distribution $N(\mu_1, \sigma_i^2/n)$, which of course agrees with the univariate CLT.

As with the univariate normal, the multivariate normal has moderate tail length. Since most real data have outliers and/or are generated from distributions with heavier tails than normal distributions, the assumption of a multivariate normal population is often not attractive. By the CLT, though, an approximate LS-type analysis can still be constructed. Further, there are many types of Central Limit Theorems, so the robust estimators that we discuss in the chapter have approximate multivariate normal distributions.

Example 10.2.3. In the following code segment, we generate a three-dimensional array with 500 samples of size 100 from each of two bivariate pdfs: (1) the bivariate normal distribution discussed in Example 10.2.1 and (2) the bivariate Weibull discussed in Example 10.2.2.

```
> ### set overall constants
> nsim <- 500  #number of runs
> n <- 100       #number of multivariate samples (per run)
> ### bivariate normal w/ rho=0.5
> library(mvtnorm)
> rho <- 0.5
> sigma <- cbind(c(1,rho),c(rho,1))
> simDat <- array(rmvnorm(nsim*n,sigma=sigma),dim=c(nsim,n,2))
> xbarMat.normal <- apply(simDat,1,colMeans) #2 x nsim array of means
> ### Weibull w/ independent components
> shape <- 2
> simDat <- array(rweibull(nsim*n*2,shape=shape),dim=c(nsim,n,2))
> xbarMat.weibull <- apply(simDat,1,colMeans) #2 x nsim array of means
```

Figure 10.4 displays the 500 bivariate sample mean vectors for each of the distributions.

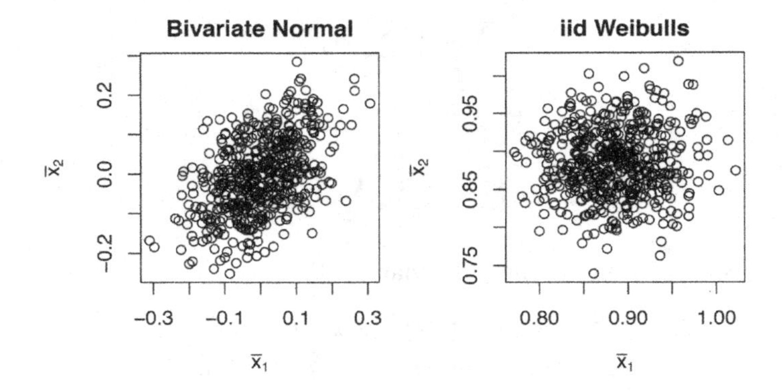

FIGURE 10.4
Scatterplots of sample means from 500 random samples of size 100 from the normal and Weibull distributions discussed in Example 10.2.3.

We close this section with another useful fact. Suppose that the k-dimensional random vector $\boldsymbol{X}$ has the multivariate normal distribution $N_k(\boldsymbol{\mu}, \boldsymbol{\Sigma})$. Then

$$Q = (\boldsymbol{X} - \boldsymbol{\mu})^T \boldsymbol{\Sigma}^{-1} (\boldsymbol{X} - \boldsymbol{\mu}) \text{ has a } \chi^2(k) \text{ distribution.} \tag{10.11}$$

This can be used for inference, also. For example, if $\overline{\boldsymbol{X}}$ is the sample average of n iid random vectors with common mean $\boldsymbol{\mu}$ and covariance matrix $\boldsymbol{\Sigma}$ then

$$Q = (\overline{\boldsymbol{X}} - \boldsymbol{\mu})^T n\boldsymbol{\Sigma}^{-1}(\overline{\boldsymbol{X}} - \boldsymbol{\mu}) \text{ has an approximate } \chi^2(k) \text{ distribution.} \tag{10.12}$$

10.3 Estimation in Multivariate Location Models

The median and Hodges–Lehmann estimators for the univariate location model are presented in Chapter 2, and in this section we generalize these to componentwise estimators including their associated inference procedures. The statistical inference for the procedures based on the sample mean follow from the central limit theorem. Central limit theorems also provide the basis for the inference for these robust procedures; for details, see, for example, Chapter 6 of the monograph by Hettmansperger and McKean (2011). The `mvrfit` function `cwmv` performs the multivariate estimation and confidence intervals of this section.

10.3.1 Multivariate Location Model

To begin, we define the multivariate location model. For inference on the distribution of the k-dimensional random vector $\boldsymbol{X}$, let $\boldsymbol{X}_1, \ldots, \boldsymbol{X}_n$ be a random sample on $\boldsymbol{X}$. For the **multivariate median location model** , assume that $\boldsymbol{X}_i$ follows the location model

$$\boldsymbol{X}_i = \boldsymbol{\theta} + \boldsymbol{e}_i, \quad i = 1, 2, \ldots, n, \tag{10.13}$$

where $\boldsymbol{e}_1, \boldsymbol{e}_2, \ldots, \boldsymbol{e}_n$ are iid with pdf f. We assume that the componentwise (population) medians of the random vector $\boldsymbol{e}_i$ are 0. This implies that the median of X_{ij} is θ_j.

For the **multivariate mean location model**, assume that $\boldsymbol{X}_i$ follows the model

$$\boldsymbol{X}_i = \boldsymbol{\mu} + \boldsymbol{e}_i, \quad i = 1, 2, \ldots, n, \tag{10.14}$$

where $\boldsymbol{e}_1, \boldsymbol{e}_2, \ldots, \boldsymbol{e}_n$ are iid with pdf $f(\boldsymbol{x})$ and $E(\boldsymbol{e}_i) = \boldsymbol{0}$. Denote the mean vector by $\boldsymbol{\mu} = E(\boldsymbol{X})$ and covariance matrix by $\boldsymbol{\Sigma} = \text{Cov}(\boldsymbol{X})$.[2] Hence, $f_{\boldsymbol{X}}(\boldsymbol{x}) = f(\boldsymbol{x} - \boldsymbol{\mu})$ and $E(\boldsymbol{X}_i) = \boldsymbol{\mu}$. The goal is inference on a location vector, either $\boldsymbol{\theta}$ or $\boldsymbol{\mu}$.

For discussion of different estimators, a convenient assumption on $f(\boldsymbol{x})$ is that it is **symmetric**; that is,

$$f(-\boldsymbol{x}) = f(\boldsymbol{x}), \quad \text{for all } \boldsymbol{x} \in R^k. \tag{10.15}$$

[2]For this section, we assume that the mean vector $\boldsymbol{\mu} = E(\boldsymbol{X})$ and covariance matrix $\boldsymbol{\Sigma} = \text{Cov}(\boldsymbol{X})$ exist.

This is equivalent to saying that $\boldsymbol{X}$ and $-\boldsymbol{X}$ have the same distribution. Note that if $f(\boldsymbol{x})$ is symmetric, then all the marginal pdfs are symmetric. Hence, simple but powerful graphical diagnostics for this assumption are to observe marginal boxplots and $q-q$ plots for asymmetric aspects. Plots that are strongly suggestive of asymmetry indicate that this assumption is not warranted. In Figure 10.5 these plots are displayed for the first two components of the `nutrient` data; i.e., calcium and iron. The distributions are clearly right-skewed and so the assumption of symmetry is not warranted.

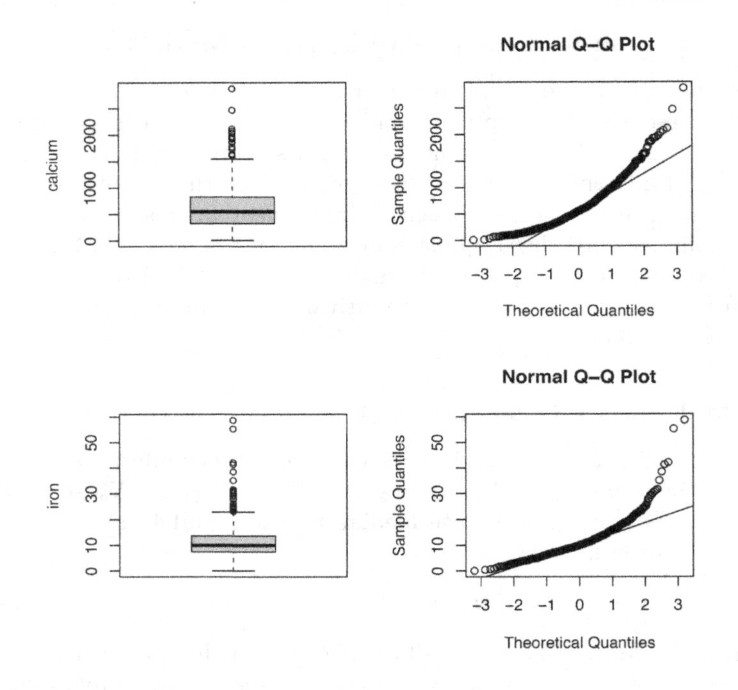

FIGURE 10.5
Univariate plots of the first two columns of the matrix `nutrient`.

10.3.2 Estimation and Confidence Intervals

A natural estimator of $\boldsymbol{\mu}$ is the vector of sample averages which can be written as

$$\overline{\boldsymbol{X}} = \frac{1}{n}\sum_{i=1}^{n}\boldsymbol{X}_i = \begin{bmatrix} \overline{X}_1 \\ \vdots \\ \overline{X}_k \end{bmatrix}, \tag{10.16}$$

where $\overline{X}_j = n^{-1}\sum_{i=1}^{n} X_{ij}$ is the sample mean of the jth component. The vector of sample means consists of the column means of the data matrix $\boldsymbol{\mathcal{X}}$.

Likewise, a natural estimator of the covariance matrix is the matrix of sample variances and covariances which we denote by $\widehat{\mathbf{\Sigma}} = [s_{jj'}]$, where

$$\begin{aligned} s_{jj} &= \frac{1}{n-1}\sum_{i=1}^{n}(X_{ij}-\overline{X}_j)^2,\ j=1,\ldots,k \\ s_{jj'} &= \frac{1}{n-1}\sum_{i=1}^{n}(X_{ij}-\overline{X}_j)(X_{ij'}-\overline{X}_{j'}),\ j\neq j', j,j'=1,\ldots,k. \end{aligned} \tag{10.17}$$

Because all the sample variances and covariances are consistent estimators of their parameter counterparts, $\widehat{\mathbf{\Sigma}} = [s_{jj'}]$ is a consistent estimator of $\mathbf{\Sigma}$.

Using the `nutrient` data, we calculate the sample mean vector and sample variance-covariance matrix in the following code segment.

```
> colMeans(nutrient)

calcium     iron protein       a       c
624.049   11.130  65.803 839.635  78.928

> var(nutrient)

             calcium     iron protein         a        c
calcium 157829.44  940.089 6075.82  102411.1  6701.62
iron       940.09   35.811  114.06    2383.2   137.67
protein   6075.82  114.058  934.88    7330.1   477.20
a       102411.13 2383.153 7330.05 2668452.4 22063.25
c         6701.62  137.672  477.20   22063.2  5416.26
```

By the multivariate Central Limit Theorem, $\overline{\boldsymbol{X}}$ has an approximate $N_k(\boldsymbol{\mu}, n^{-1}\mathbf{\Sigma})$ distribution. Hence, for component j, $j = 1,\ldots,k$, $\overline{X}_j$ has an approximate $N(\mu_j, n^{-1}\sigma_j^2)$ distribution. So marginal inference proceeds in the same way as in the univariate case. For example, an approximate $(1-\alpha)100\%$ confidence interval for μ_j using z-critical values is then given by

$$\overline{X}_j \pm z_{\alpha/2}\frac{s_j}{\sqrt{n}}, \quad j=1,\ldots,k. \tag{10.18}$$

Often, in practice, however, t-critical values with $n-1$ degrees of freedom are used.

In the following code segment, confidence intervals for each of the nutrients based on componentwise mean estimates are obtained.

```
> library(mvrfit)
> cwmv(nutrient,method='LS',conf.int=TRUE)

Componentwise Multivariate Procedure
```

```
Estimation method: Least Squares

Estimated Location Vector:
calcium    iron protein       a       c
624.049  11.130  65.803 839.635  78.928

Wald Confidence Intervals
        ci.lower ci.upper
calcium 595.320  652.778
iron     10.697   11.563
protein  63.592   68.015
a       721.506  957.765
c        73.606   84.250
CI adjust method: none
```

More generally, for a specified k-dimensional vector $\boldsymbol{h}$, an approximate $(1-\alpha)*100\%$ confidence interval for $\boldsymbol{h}^T\boldsymbol{\mu}$ is

$$\boldsymbol{h}^T\overline{\boldsymbol{X}} \pm t_{\alpha/2,n-1}\frac{\sqrt{\boldsymbol{h}^T\widehat{\boldsymbol{\Sigma}}\boldsymbol{h}}}{\sqrt{n}}, \qquad (10.19)$$

where we have used the conservative t-critical values. Intervals of the type in Equation (10.19) may be obtained by providing the contrast vector $\boldsymbol{h}$ as the `H` argument to `cwmv`.

10.3.3 Median Rank-Based Estimation of $\boldsymbol{\theta}$ and Confidence Intervals

Denote by $\widehat{\boldsymbol{\theta}}_m$ the vector of componentwise medians; i.e.,

$$\widehat{\boldsymbol{\theta}}_m = \begin{bmatrix} \text{med}\{X_{11},\ldots X_{n1}\} \\ \text{med}\{X_{12},\ldots X_{n2}\} \\ \vdots \\ \text{med}\{X_{1k},\ldots X_{nk}\} \end{bmatrix}. \qquad (10.20)$$

Based on theory, as found, for instance in Chapter 6 of Hettmansperger and McKean (2011), $\widehat{\boldsymbol{\theta}}_m$ has an asymptotic $N_k(\boldsymbol{\theta}, n^{-1}\boldsymbol{\Sigma}_m)$ distribution where $\boldsymbol{\Sigma}_m = [\sigma_{mjj'}]$, $j, j' = 1, \ldots, k$, is defined by

$$\sigma_{mjj'} = \begin{cases} \tau_{Sj}^2 & j = j' \\ \tau_{Sj}\tau_{Sj'}E[\text{sgn}(X_j-\theta_j)\text{sgn}(X_{j'}-\theta_{j'})] & j \neq j', \end{cases} \qquad (10.21)$$

For inference, an estimator of $\boldsymbol{\Sigma}_m$ is required. The diagonal parameters τ_{Sj} are estimated by $\widehat{\tau}_{Sj}$. An estimator of the expectation in (10.21) is the moment estimator:

$$\text{Estimator}\{E[\text{sgn}(X_j-\theta_j)\text{sgn}(X_{j'}-\theta_{j'})]\} = \frac{1}{n}\sum_{i=1}^{n}\text{sgn}(X_{ij}-\widehat{\theta}_j)\text{sgn}(X_{ij'}-\widehat{\theta}_{j'}). \tag{10.22}$$

Putting these estimators together, we obtain the estimator $\widehat{\boldsymbol{\Sigma}}_m = [\widehat{\sigma}_{mjj'}]$ of $\boldsymbol{\Sigma}_m$. Thus an approximate $(1-\alpha)100\%$ confidence interval θ_j is is

$$\hat{\theta}_{mj} \pm t_{n-1,\alpha/2}\frac{\sqrt{\widehat{\sigma}_{mjj}}}{\sqrt{n}}, \quad j = 1,\ldots,k, \tag{10.23}$$

where we have used the conservative t-critical value rather than the z-critical value. Note the close similarity of this confidence interval with that based on the sample mean given in expression (10.18).

In the following code segment, confidence intervals for each of the nutrients based on componentwise median estimates are obtained.

```
> cwmv(nutrient,method='Median',conf.int=TRUE)

Componentwise Multivariate Procedure

Estimation method: Median

Estimated Location Vector:
calcium     iron protein       a       c
548.290   10.033  61.072 524.030  53.594

Wald Confidence Intervals
        ci.lower ci.upper
calcium  517.774  578.806
iron       9.613   10.453
protein   58.509   63.635
a        483.726  564.334
c         47.086   60.102
CI adjust method: none
```

Note, as was apparent in the graphical exploration of the data, the marginal distributions are skewed, so the results of this method are expected to differ from those for the mean $\boldsymbol{\mu}$. The sample mean is affected by the outlier for the vitamin A variable, and it has affected the length of its confidence interval, also, which exceeds 230 units. On the other hand, the length of the confidence interval for the median is less than 80 units.

Confidence intervals based on the sample medians are easily constructed for linear functions of the form $\boldsymbol{h}^T\boldsymbol{\theta}$. That is,

$$\boldsymbol{h}^T\hat{\boldsymbol{\theta}}_m \pm t_{\alpha/2,n-1}\frac{\sqrt{\boldsymbol{h}^T\widehat{\boldsymbol{\Sigma}}_m\boldsymbol{h}}}{\sqrt{n}}, \tag{10.24}$$

Intervals of the type in Equation (10.24) may be obtained through specification of the `H` argument to `cwmv`.

10.3.4 Wilcoxon Rank-Based Estimation of θ and Confidence Intervals

The Hodges–Lehmann estimator of location for the univariate location model discussed in Chapter 2 generalizes to a componentwise for the multivariate location model. Let $\boldsymbol{X}_1, \ldots, \boldsymbol{X}_n$ be a random sample of k-dimensional vectors from Model (10.13). Assume additionally that the pdf of the random errors is symmetric; i.e., Condition (10.15) holds. Then the Hodges–Lehmann estimator of θ_j is

$$\widehat{\theta}_{wj} = \text{med}_{i \leq i'} \left\{ \frac{X_{ij} + X_{i'j}}{2} \right\}, \quad j = 1, 2 \ldots, k, \tag{10.25}$$

and, hence, the Hodges–Lehmann estimator of the vector $\boldsymbol{\theta}$ is $\widehat{\boldsymbol{\theta}}_w = (\widehat{\theta}_{w1}, \ldots, \widehat{\theta}_{wk})^T$. This is the rank-based estimator based on the Wilcoxon score function, so we have used the subscript w. The theory cited in this section can be found in Hettmansperger and McKean (2011). It follows that $\widehat{\boldsymbol{\theta}}_w$ is asymptotically normal with mean $\boldsymbol{\theta}$ and covariance matrix $n^{-1}\boldsymbol{\Sigma}_w$ where $\boldsymbol{\Sigma}_w = [\sigma_{wjj'}]$ is given by

$$\sigma_{wjj'} = \begin{cases} \tau_j^2 & j = j' \\ \tau_j \tau_{j'} 3\delta_{jj'} & j \neq j', \end{cases} \quad jj' = 1, \ldots, k, \tag{10.26}$$

where $\delta_{jj'} = \{4E[F_j(X_{1j} - \theta_j)F_{j'}(X_{1j'} - \theta_{j'})] - 1\}$ and τ_j is a scale parameter given in expression(2.21).

The estimation of the scale parameter τ_j is discussed below expression (3.20). The estimator of $\delta_{jj'}$ is the moment estimator,

$$\widehat{\delta}_{jj'} = \frac{4}{n} \sum_{i=1}^{n} \frac{R(X_{ij} - \widehat{\theta}_j) R(X_{ij'} - \widehat{\theta}_{j'})}{(n+1)^2} - 1 \tag{10.27}$$

where R denotes rank among the column of residuals.

Based on the approximate normal distribution of the sample median, an approximate $(1 - \alpha) * 100\%$ confidence interval θ_j is

$$\widehat{\theta}_{wj} \pm t_{n-1,\alpha/2} \frac{\sqrt{\widehat{\sigma}_{wjj}}}{\sqrt{n}}, \quad j = 1, \ldots, k. \tag{10.28}$$

As in the case of the median analysis, confidence intervals based on $\widehat{\boldsymbol{\theta}}_w$ may be constructed for linear functions of the form $\boldsymbol{h}^T\boldsymbol{\theta}$.

In the following code segment, confidence intervals for each of the nutrients based on componentwise Hodges–Lehmann estimates are obtained.

```
> cwmv(nutrient,method='HL',conf.int=TRUE)
```

```
Componentwise Multivariate Procedure

Estimation method: Hodges-Lehmann

Estimated Location Vector:
calcium     iron protein       a       c
581.590   10.489  63.455 610.590  66.764

Wald Confidence Intervals
         ci.lower ci.upper
calcium  557.036  606.144
iron      10.149   10.828
protein   61.488   65.422
a        580.947  640.233
c         63.501   70.028
CI adjust method: none
```

Similar to the analysis using the median, the Wilcoxon analysis is affected far less by the outlier than the sample mean. Note: a **Bonferroni Adjustment** is available in `cwmv` by specifying the argument `conf.adjust='bonferroni'` in which case the critical value is replaced with $t_{n-1,\alpha/(2k)}$.

10.4 Tests of Hypothesis for the Multivariate Location Model

At times, one may be interested in a test of hypothesis of several response variables at once. Such an example was mentioned in the introduction where it was suggested one might test the hypothesis that the median (or mean) intake was equal to the recommended (see Table 10.1). Such a hypothesis could be formed as

$$H_0 : \boldsymbol{\theta} = [1000, 15, 60, 800, 75]^T \text{ versus } H_A : \boldsymbol{\theta} \neq [1000, 15, 60, 800, 75]^T$$

where the alternative hypothesis means one or more of the individual parameters differ.

The `npsm` function `cwmv` performs the multivariate test of hypothesis discussed in this section.

10.4.1 Traditional Procedures

First, consider the hypotheses

$$H_0 : \boldsymbol{\mu} = \boldsymbol{\mu}_0 \text{ versus } H_A : \boldsymbol{\mu} \neq \boldsymbol{\mu}_0, \tag{10.29}$$

where $\boldsymbol{\mu}_0$ is a specified vector. In the univariate case, the LS test statistic is Student's t-statistic. The square of the t-test statistic for testing the null hypothesis $\mu = \mu^*$ can be expressed as $t^2 = n(\overline{X} - \mu^*)(s^2)^{-1}(\overline{X} - \mu^*)$. Hotelling's T^2 test statistic for the hypotheses (10.29) is

$$T^2 = n(\overline{\boldsymbol{X}} - \boldsymbol{\mu}_0)^T \widehat{\boldsymbol{\Sigma}}^{-1} (\overline{\boldsymbol{X}} - \boldsymbol{\mu}_0). \tag{10.30}$$

Assuming a normal population, the univariate test statistic t^2 has an F-distribution with 1 and $n - 1$ degrees of freedom under H_0; while the multivariate test statistic $[(n-k)/((n-1)k)]T^2$ has an F-distribution with k and $n - k$ degrees of freedom. Hence, the univariate test rejects at level α, if $t^2 \geq F_{\alpha,1,n-1}$; while the multivariate test rejects, if $[(n-k)/((n-1)k)]T^2 \geq F_{\alpha,k,n-k}$. If the population is not normal, Hotelling's T^2-test is an approximate level α-test. Actually, in this case, using the Central Limit Theorem, kT^2 has an asymptotically χ^2 distribution with k degrees of freedom under H_0.

In the following code segment, we illustrate the test using `cwmv`[3] to test the hypothesis

$$H_0 : \boldsymbol{\mu} = [1000, 15, 60, 800, 75]^T \text{ versus } H_A : \boldsymbol{\mu} \neq [1000, 15, 60, 800, 75]^T$$

```
> cwmv(nutrient,method='LS',test=TRUE,
+       g0=c(1000,15,60,800,75))

Componentwise Multivariate Procedure

Estimation method: Least Squares

Estimated Location Vector:
calcium     iron protein        a       c
624.049   11.130  65.803 839.635  78.928


 Wald  Test
F-Statistic      p-value
 3.4980e+02 2.9887e-191
Null Location Vector:
[1] 1000    15    60   800    75
```

10.4.2 Contrasts

It is often the case that a **contrast** of the individual means are of interest. For example,

$$H_0 : \boldsymbol{C\mu} = \boldsymbol{\gamma}_0 \text{ versus } H_A : \boldsymbol{C\mu} \neq \boldsymbol{\gamma}_0, \tag{10.31}$$

[3] Note the test is also available via the function `HotellingT2` in the package `ICSNP` Nordhausen et al. (2018).

for some prespecified full row rank contrast matrix $\boldsymbol{C}$ (row sums are 0) and vector γ_0. An interesting null hypothesis is that all the component means are the same. For example, consider a repeated measures design where the response is measured on an individual over k time points. Then, the hypothesis of equal components states that the mean response has not changed over these time periods. The analysis is naturally a two-stage analysis, because if this hypothesis is rejected, then interest settles on which of the pairwise contrasts $\mu_i - \mu_j$ $i \neq j$ are significantly different. The hypotheses are

$$H_0 : \mu_1 = \cdots = \mu_k \text{ versus } H_A : \mu_i \neq \mu_j \text{ for some } i \neq j. \tag{10.32}$$

The hypothesis H_0 is true if and only if $\mu_1 - \mu_j = 0$ for all $j = 2, \ldots, k$. We rewrite (10.32) in the form (10.31). Let $\boldsymbol{C}$ be the $(k-1) \times k$ matrix with the *ith* row $(1, 0, \ldots, 0, -1, 0, \ldots, 0)$, where the -1 is the $(i+1)th$ component for $i = 1, \ldots, (k-1)$. Then the hypotheses (10.32) can be expressed as

$$H_0 : \boldsymbol{C\mu} = \boldsymbol{0} \text{ versus } H_A : \boldsymbol{C\mu} \neq \boldsymbol{0}. \tag{10.33}$$

Further, because the rows of $\boldsymbol{C}$ are linearly independent, the covariance matrix is invertible. Hotelling's T^2 test statistic, (10.30), is

$$T^2 = n(\boldsymbol{C}\overline{\boldsymbol{X}})^T \left[(\boldsymbol{C}\widehat{\boldsymbol{\Sigma}}\boldsymbol{C}^T)\right]^{-1} \boldsymbol{C}\overline{\boldsymbol{X}}, \tag{10.34}$$

The test statistic rejects at approximately level α, if $[(n-k+1)/((n-1)(k-1))]T^2 \geq F_{\alpha,k-1,n-k+1}$. This test is exact, if the population is multivariate normal.

10.4.3 Robust Procedures

For the sign and Wilcoxon signed-rank procedures, there are two types of test statistics available: Wald and gradient types. Wald type test statistics are based on quadratic forms in the rank-based estimators similar to the Hotelling T^2 test statistics. Gradient type tests are generalizations of the sign and signed-rank Wilcoxon univariate test statistics discussed in Chapter 2.

Wald Type

Assume that the random sample $\boldsymbol{X}_1, \ldots, \boldsymbol{X}_n$ follows Model (10.13). For the signed-rank Wilcoxon tests, we also need to assume the symmetry condition (10.15). Consider the hypotheses

$$H_0 : \boldsymbol{\theta} = \boldsymbol{\theta}_0 \text{ versus } H_A : \boldsymbol{\theta} \neq \boldsymbol{\theta}_0, \tag{10.35}$$

where $\boldsymbol{\theta}_0$ is a specified vector.

Both the multivariate sign estimator and the Wilcoxon (Hodges–Lehmann) estimator have asymptotic normal distributions. Hence, the result (10.11) is used to form the following χ^2 test statistics:

$$T_S^2 = n(\widehat{\boldsymbol{\theta}}_S - \boldsymbol{\theta}_0)^T \widehat{\boldsymbol{\Sigma}}_S^{-1} (\widehat{\boldsymbol{\theta}}_S - \boldsymbol{\theta}_0) \quad (10.36)$$

$$T_W^2 = n(\widehat{\boldsymbol{\theta}}_W - \boldsymbol{\theta}_0)^T \widehat{\boldsymbol{\Sigma}}_W^{-1} (\widehat{\boldsymbol{\theta}}_W - \boldsymbol{\theta}_0). \quad (10.37)$$

Under H_0, T_S^2 and T_W^2 have asymptotic χ^2 distributions with k degrees of freedom. We recommend, though, similar to Hotelling's test based on the sample mean, that $[(n-k)/((n-1)k)]T_S^2$ and $[(n-k)/((n-1)k)]T_W^2$ be compared with $F_{\alpha,k,n-k}$ critical values. Recall a similar degrees of freedom correction is used for rank-based tests of linear hypotheses in Chapters 4 and 5. The result (10.11) can also be used to formulate confidence regions for $\boldsymbol{\theta}$ similar to the confidence region for $\boldsymbol{\mu}$.

Recall in Section 10.4.2, that we discussed the Hotelling test statistic for the contrast hypotheses (10.33). Analogously, sign and Wilcoxon tests of the contrast hypotheses $H_0 : \boldsymbol{C\theta} = \mathbf{0}$ versus $H_A : \boldsymbol{C\theta} \neq \mathbf{0}$, where $\boldsymbol{C}$ is a full row rank $(k-1) \times k$ contrast matrix, are easily formulated. The respective test statistics are:

$$T_S^2 = n\boldsymbol{C}\widehat{\boldsymbol{\theta}}_S^T \left[(\boldsymbol{C}\widehat{\boldsymbol{\Sigma}}_S \boldsymbol{C}^T\right]^{-1} \boldsymbol{C}\widehat{\boldsymbol{\theta}}_S \quad (10.38)$$

$$T_W^2 = n\boldsymbol{C}\widehat{\boldsymbol{\theta}}_W^T \left[(\boldsymbol{C}\widehat{\boldsymbol{\Sigma}}_W \boldsymbol{C}^T\right]^{-1} \boldsymbol{C}\widehat{\boldsymbol{\theta}}_W. \quad (10.39)$$

The test statistic T_W^2 rejects at approximately level α, if $[(n-k+1)/((n-1)(k-1))]T_W^2 \geq F_{\alpha,k-1,n-k+1}$. A similar statement can be made for the test statistic T_S^2. As with the test statistic based on $\overline{X}$, the test statistics T_S^2 and T_W^2 are invariant to any contrast matrix of full row rank. It is often thought of as a first-stage procedure because, upon rejection, confidence intervals for individual contrasts are of interest. This discussion holds for contrasts based on the sign and Wilcoxon estimators multivariate estimators of location, also.

Next, we compute these robust tests for the nutrient data.

```
> cwmv(nutrient,method='Median',test=TRUE,
+       g0=c(1000,15,60,800,75))

Componentwise Multivariate Procedure

Estimation method: Median

Estimated Location Vector:
calcium     iron protein       a       c
548.290  10.033  61.072 524.030  53.594

 Wald  Test
F-Statistic     p-value
 3.1992e+02 2.1326e-181
```

```
Null Location Vector:
[1] 1000    15    60   800    75

> cwmv(nutrient,method='HL',test=TRUE,
+       g0=c(1000,15,60,800,75))

Componentwise Multivariate Procedure

Estimation method: Hodges-Lehmann

Estimated Location Vector:
calcium     iron protein       a       c
581.590  10.489  63.455 610.590  66.764

 Wald  Test
F-Statistic       p-value
 6.2033e+02 2.4853e-260
Null Location Vector:
[1] 1000    15    60   800    75
```

Gradient Type

There are signed and signed-rank Wilcoxon tests for the hypotheses (10.35) that are the component-wise extensions of the sign and signed-rank Wilcoxon tests for univariate responses as discussed in Chapter 2. These are usually called gradient tests. As the discussion in Remark 10.4.1 points out, these gradient tests can be used for certain types of response data other than continuous data.

Assume that the random sample of k-dimensional random vectors $\boldsymbol{X}_1, \ldots, \boldsymbol{X}_n$ follows Model (10.13). For the signed-rank Wilcoxon test, assume further that the symmetry condition (10.15) holds. Recall that under Model (10.13), the median of X_{ij} is θ_j; hence, the null median of X_{ij} is θ_{0j}. We present the sign test statistic first.

Sign Test Statistic

The sign test statistic of the hypotheses (10.35) depends on the gradient vector of signs $\boldsymbol{S}_m$ given by

$$\boldsymbol{S}_m = \boldsymbol{S}_m(\boldsymbol{\theta}_0) = \begin{bmatrix} \sum_{i=1}^n \operatorname{sgn}(X_{i1} - \theta_{01}) \\ \vdots \\ \sum_{i=1}^n \operatorname{sgn}(X_{ik} - \theta_{0k}) \end{bmatrix} \tag{10.40}$$

The jth component of $\boldsymbol{S}_m$ can be expressed as $S_{mj} = 2S_{mj}^+ - n$, where $S_{mj}^+ = \#_i\{X_{ij} > \theta_{0j}\}$. Because under H_0 the median of X_{ij} is θ_{0j}, S_{mj}^+ has a binomial distribution with n trials and the probability of success $1/2$. It follows that the null mean and variance of $S_{mj}/\sqrt{n}$ are given by

$$\begin{aligned} E_{H_0}[S_{mj}/\sqrt{n}] &= (2(n/2) - n)/\sqrt{n} = 0 \\ V_{H_0}[S_{mj}/\sqrt{n}] &= (4(n/4))/n = 1. \end{aligned}$$

Hence, $E_{H_0}(\boldsymbol{S}_m/\sqrt{n}) = \boldsymbol{0}_k$ and the diagonal entries of the null covariance matrix of $\boldsymbol{S}_m$ have the value 1. The test, though, is not distribution free. The dependence appears in the covariances between the components of $\boldsymbol{S}_m$. Let $\boldsymbol{B}_m = [b_{mjj'}]$ denote the covariance matrix of $\boldsymbol{S}_m$. Its entries of the matrix are

$$b_{mjj'} = \begin{cases} 1 & j = j' \\ E[\text{sgn}(X_{ij} - \theta_{0j})\text{sgn}(X_{ij'} - \theta_{0j'})] & j \neq j', \end{cases} \tag{10.41}$$

Since the rows of the data matrix are identically distributed, the expectation is the same for all $i = 1, \ldots, n$. A natural estimator of $b_{mjj'}$ is the moment estimator given in expression (10.22), except that $\theta_{0j'}$ replaces $\hat{\theta}_j$. Denote the corresponding estimator by $\widehat{\boldsymbol{B}}_m$. Further, under H_0, $\boldsymbol{S}_m$ has an asymptotically normal distribution with mean $\boldsymbol{0}$ and covariance matrix $\boldsymbol{B}_m$. Based on the result (10.11), the χ^2 test statistic is

$$C_m = \frac{1}{n}\boldsymbol{S}_m^T \boldsymbol{B}_m^{-1} \boldsymbol{S}_m. \tag{10.42}$$

The level α asymptotic decision rule is to reject H_0 is $C_m \geq \chi^2_\alpha(k)$. As with the Wald-type rank-based tests, users may want to use the F-type test statistic which is $F_m = ((n-k)/(k*(n-1)))C_m$. In this case, the test statistic is compared with F-critical values based on k and $n-k$ degrees of freedom. This is similar to a Wald-type standardization.

```
> cwmv(nutrient,method='Median',test=TRUE,test.method="Gradient",
+        g0=c(1000,15,60,800,75))

Componentwise Multivariate Procedure

Estimation method: Median

Estimated Location Vector:
calcium     iron protein        a       c
548.290   10.033   61.072 524.030  53.594

 Gradient  Test
F-Statistic      p-value
 8.8580e+01  7.6136e-73
Null Location Vector:
[1] 1000    15    60   800    75
```

Recall in Chapter 2, that the median inverts the sign test; i.e., the median is the "most acceptable" hypothesis. The same is true in the multivariate case because the vector of medians $\widehat{\boldsymbol{\theta}}_m$ solves the estimating equations (EE):

$$\boldsymbol{S}_m(\widehat{\boldsymbol{\theta}}_m) = \mathbf{0}_k. \tag{10.43}$$

Signed-Rank Wilcoxon Test Statistic

The signed-rank Wilcoxon test of the hypotheses (10.35) depends on the gradient vector

$$\boldsymbol{S}_w = \boldsymbol{S}_w(\boldsymbol{\theta}_0) = \frac{1}{n+1}\begin{bmatrix} \sum_{i=1}^n \mathrm{sgn}(X_{i1} - \theta_{01})R(|X_{i1} - \theta_{01}|) \\ \vdots \\ \sum_{i=1}^n \mathrm{sgn}(X_{ik} - \theta_{0k})R(|X_{ik} - \theta_{0k}|) \end{bmatrix}. \tag{10.44}$$

Denote the jth component of $\boldsymbol{S}_w$ by S_{wj}. This, of course, is the univariate signed-rank Wilcoxon test statistic to test $\theta_j = \theta_{0j}$ that was discussed in Chapter 2. So, the null mean and variance of S_{wj} are 0 and $(2n+1)/[6(n+1)]$, respectively. Furthermore assuming H_0 is true, it can be shown that $\boldsymbol{S}'_w/\sqrt{n}$ is asymptotically normal with mean $\mathbf{0}_k$ and covariance matrix $\boldsymbol{B}_w = [b_{wjj'}]$ where

$$b_{wjj'} = \begin{cases} 1/3 & j = j' \\ \delta_{jj'} & j = j', \end{cases} \tag{10.45}$$

where $\delta_{jj'}$ is defined after expression (10.26), except that $\boldsymbol{\theta}_0$ replaces $\boldsymbol{\theta}$. Likewise, an estimator of δ is given in expression (10.27) with $\boldsymbol{\theta}_0$ replacing $\widehat{\boldsymbol{\theta}}$. A Wald-type standardization uses $\widehat{\delta}$ given by expression (10.27).

The χ^2 test statistic is

$$C_w = \frac{1}{n}\boldsymbol{S}_w^T \boldsymbol{B}_w^{-1} \boldsymbol{S}_w. \tag{10.46}$$

The level α asymptotic decision rule is to reject H_0, (10.35), if $C_w \geq \chi^2_\alpha(k)$. As with the Wald type tests based on estimators, for these tests we compute the F-test version; i.e., the F-test statistic given by $F_w = ((n-k)/(k*(n-1)))C_w$, with degrees of freedom k and $n-k$.

```
> cwmv(nutrient,method='HL',test=TRUE,test.method="Gradient",
+      g0=c(1000,15,60,800,75))

Componentwise Multivariate Procedure

Estimation method: Hodges-Lehmann

Estimated Location Vector:
calcium     iron protein        a       c
581.590   10.489  63.455 610.590  66.764
```

```
 Gradient  Test
F-Statistic      p-value
 2.4679e+02 2.4256e-154
Null Location Vector:
[1] 1000   15   60  800   75
```

Note, similar to the median, discussed above, the Hodges–Lehmann estimator solves the estimating equations:

$$\boldsymbol{S}_w(\widehat{\boldsymbol{\theta}}_w) = \mathbf{0}_k. \tag{10.47}$$

Remark 10.4.1 (Types of Data for Gradient Tests)**.** Unlike their Wald type counterparts, response data for the gradient tests need not be continuous. The sign-type gradient test requires few assumptions. For example, suppose we are testing the null hypothesis $H_0 : \boldsymbol{\theta} = \mathbf{0}$. Then all we need to know for a response is whether or not it is greater than 0 or less than 0, in order to construct the sign-test statistic C_m, (10.42). Similar to the case in Chapter 2, observed response vectors with at least one zero observation would be omitted. The type of response data for the signed-rank test statistic C_w (10.46) is not as general as that for the sign test. To compute this test statistic, the rank of the absolute value of an observation must be known; i.e., not only its sign but where an observation is absolutely located relative to the absolute location of the other observations in that column of the response matrix. If it is known, then C_w can be computed for the signed-rank case. ■

10.5 Multivariate Linear Models

In this section, we discuss rank-based procedures for the multivariate linear model, extending the univariate linear model of Chapter 4 to the multivariate setting. As in Chapter 4, these methods are valid for skewed as well as symmetric random error distributions. Recall that the univariate two-sample problem and the ANOVA-type designs are also regression models. Our discussion will extend these design models to the multivariate setting, also.

Assume that we observe k responses on each of n subjects. Denote the vector of measurements for the ith subject as $\boldsymbol{Y}_i^T = [Y_{i1}, \ldots, Y_{ik}]$. Stack $\boldsymbol{Y}_i^T$ row-by-row in the $n \times k$ data matrix $\boldsymbol{Y}$ — the response *matrix*. For the regression part of the model, suppose that we have p explanatory variables which are measured on each subject. Let $\boldsymbol{x}_i^T = [x_{i1}, \ldots, x_{ip}]$ be the vector of measurements for the ith subject. As in the case of iid regression (e.g., Chapter 4), stack $\boldsymbol{x}_i^T$ row-by-row in the $n \times p$ data matrix $\boldsymbol{X}$. For the vector of intercepts, let $\mathbf{1}_n$ denote an $n \times 1$ vector of ones. Then our **multivariate regression model** is given by

$$\boldsymbol{Y} = \mathbf{1}_n \boldsymbol{\alpha}^T + \boldsymbol{X}\boldsymbol{\beta} + \boldsymbol{e}, \tag{10.48}$$

where $\boldsymbol{\beta}$ is a $p \times k$ matrix of regression coefficients, $\boldsymbol{\alpha}$ is a $k \times 1$ vector of intercept parameters, and $\boldsymbol{e}$ is a $n \times k$ matrix of random errors. The rows of $\boldsymbol{e}$ are assumed to be independent but within a row there is generally dependence. For theory, we assume that rows of $\boldsymbol{e}$ are identically distributed with pdf and cdf $f(t_1, \ldots, t_k)$ and $F(t_1, \ldots, t_k)$, respectively. Because we have an intercept vector, without loss of generality, we assume that the true median of e_{ij} is 0.

This is a componentwise multiple regression model. To see this, set j, for $j = 1, \ldots k$, and define the indicator vector $\boldsymbol{\delta}_j^T = (0, \ldots, 0, 1, 0, \ldots, 0)$, where the 1 is the value of the jth component. Multiplying both sides of equation (10.48) on the right by $\boldsymbol{\delta}_j$, we get

$$\boldsymbol{Y}_j^{(c)} = \alpha_j + \boldsymbol{X}\boldsymbol{\beta}_j^{(c)} + \boldsymbol{e}_j^{(c)}, \tag{10.49}$$

where $\boldsymbol{Y}_j^{(c)}$, $\boldsymbol{\beta}_j^{(c)}$ and $\boldsymbol{e}_j^{(c)}$ are the jth columns of the matrices $\boldsymbol{Y}$, $\boldsymbol{\beta}$, and $\boldsymbol{e}$, respectively, for $j = 1, \ldots, k$. Because the true median of e_{ij} is 0, note that α_j is the adjusted median of Y_{ij}, i.e., $\alpha_j = \text{med}\{Y_{ij} - \boldsymbol{x}_{ij}\boldsymbol{\beta}_j^{(c)}\}$. For the jth variable, Model (10.49) is a univariate multiple regression model as discussed in Chapter 4. These componentwise models suggest estimating the matrix $\boldsymbol{\beta}$ column-wise by computing the univariate componentwise estimates. That is, for $j = 1, \ldots k$, let $\widehat{\boldsymbol{\beta}}_j$ be the rank-based estimator for Model (10.49) and let $\hat{\alpha}_j$ be the median of the residuals based on this fit. Then the rank-based estimators of $\boldsymbol{\beta}$ and $\boldsymbol{\alpha}$ for the multivariate Model (10.48) are

$$\widehat{\boldsymbol{\beta}} = [\hat{\boldsymbol{\beta}}_1 \cdots \hat{\boldsymbol{\beta}}_j \cdots \hat{\boldsymbol{\beta}}_k] \tag{10.50}$$

$$\widehat{\boldsymbol{\alpha}}^T = [\hat{\alpha}_1, \ldots, \hat{\alpha}_j, \ldots, \hat{\alpha}_k]. \tag{10.51}$$

The matrices of fitted values and residuals are given by

$$\widehat{\boldsymbol{Y}} = \boldsymbol{1}_n\widehat{\boldsymbol{\alpha}}^T + \boldsymbol{X}\widehat{\boldsymbol{\beta}} \text{ and } \widehat{\boldsymbol{e}} = \boldsymbol{Y} - \widehat{\boldsymbol{Y}}. \tag{10.52}$$

Hence, a loop using the function `rfit` may be used to compute the rank-based fit of Model (10.49). We illustrate such a calculation in the next example.

Example 10.5.1 (Illustrative Dataset for Multivariate Regression). Consider the simple dataset presented by Johnson and Wichern (2007), consisting of a 7×2 data matrix $\boldsymbol{Y}$ and 2 predictors that for convenience we have displayed in Table 10.2.

TABLE 10.2
Illustrative dataset for multivariate regression.

id	$\boldsymbol{Y}$		$\boldsymbol{X}$	
1	141.50	301.80	123.50	2.11
2	168.90	396.10	146.10	9.21
3	154.80	328.20	133.90	1.91
4	146.50	307.40	128.50	0.81
5	172.80	362.40	151.50	1.06
6	160.10	369.50	136.20	8.60
7	108.50	229.10	92.00	1.12

In the next code segment, we define the matrices `ymat` and `xmat`; notice we `rbind` the individual subject's data (response vector and explanatory vector).

```
> ymat <- rbind(c(141.50, 301.80),
+               c(168.90, 396.10),
+               c(154.80, 328.20),
+               c(146.50, 307.40),
+               c(172.80, 362.40),
+               c(160.10, 369.50),
+               c(108.50, 229.10)
+              )
> xmat <- rbind(c(123.50, 2.11),
+               c(146.10, 9.21),
+               c(133.90, 1.91),
+               c(128.50, 0.81),
+               c(151.50, 1.06),
+               c(136.20, 8.60),
+               c(92.00, 1.12)
+              )
> colnames(ymat) <- c('Y1','Y2')
> colnames(xmat) <- c('x1','x2')
```

Now we illustrate how the multivariate model may be fit by fitting each of the columns of `ymat` versus the design matrix `xmat` using `rfit`. We first convert the columns of `ymat` to a list, then use the R command `lapply` to obtain the fits.

```
> library(Rfit)
> ylist <- as.list(as.data.frame(ymat))
> rfit1 <- function(y,x=x) rfit(y~x)
> fit <- lapply(ylist,rfit1,x=xmat)
> fit

$Y1
Call:
rfit.default(formula = y ~ x)

Coefficients:
(Intercept)         xx1         xx2
    8.57130     1.08109     0.41793


$Y2
Call:
```

```
rfit.default(formula = y ~ x)

Coefficients:
(Intercept)         xx1         xx2
    16.1695      2.2460      5.6231
```

The result is a list of lists stored in `fit` where each list contains a separate fit based on `rfit`. When `fit` is printed, the default summary for each fit is printed. Note: the core R function `lm` may be used directly to obtain the multivariate fit.

```
> lm(ymat~xmat)

Call:
lm(formula = ymat ~ xmat)

Coefficients:
             Y1     Y2
(Intercept)  8.43  14.20
xmatx1       1.08   2.25
xmatx2       0.42   5.67
```

We have also written an R package `mvrfit` which performs the multivariate rank-based regression fit described above and also has some addition functionatlity for inference. In the following R segment, the rank-based fit is computed by the R function `mv_rfit`.

```
> library(mvrfit)
> mv_rfit(xmat,ymat)

Call:
mv_rfit(x = xmat, y = ymat)

Coefficients:
                 Y1      Y2
(Intercept) 8.57130 16.1695
x1          1.08109  2.2460
x2          0.41793  5.6231
```

Notice, as with the `lm` fit, a matrix of the regression coefficiencts is printed by default. ■

10.5.1 Asymptotic Distribution of $\widehat{\beta}$

Davis and McKean (1993) obtained the asymptotic distribution of the rank-based fit of Model (10.48) for general score functions. For a specified score function $\varphi(u)$, let $\widehat{\boldsymbol{\beta}}_\varphi$ be the rank-based estimator of $\boldsymbol{\beta}$. The estimator $\widehat{\boldsymbol{\beta}}_\varphi$ is a $p \times k$ matrix. Its asymptotic distribution can be described by rolling $\widehat{\boldsymbol{\beta}}_\varphi$ out (join row-by-row) into a vector, but, for our purposes, we prefer the following matrix description:

$$\widehat{\boldsymbol{\beta}}_\varphi \text{ is approximately distributed } N_{p,k}(\boldsymbol{\beta}, (\boldsymbol{X}_c^T\boldsymbol{X}_c)^{-1}, \boldsymbol{T}\boldsymbol{\Sigma}_\varphi\boldsymbol{T}). \tag{10.53}$$

The notation $N_{p,k}$ is read as matrix normal (see Arnold (1981)). There are three parts. The first part $\boldsymbol{\beta}$ is of course the asymptotic mean of $\widehat{\boldsymbol{\beta}}_\varphi$. In the second part, $\boldsymbol{X}_c$ is the centered design matrix. For the third part, the matrix $\boldsymbol{T}$ is the diagonal matrix $\{\tau_{\varphi,1}, \ldots, \tau_{\varphi,k}\}$, where $\tau_{\varphi,j}$ is the scale parameter (3.19) for the jth variable. To define $\boldsymbol{\Sigma}_\varphi$, let $F_j(t)$ be the marginal cdf for the random errors $e_{1j}, \ldots, e_{nj}$, i.e., the jth column of the random error matrix $\boldsymbol{e}$ of Model (10.48). Then, the elements of the covariance matrix $\boldsymbol{\Sigma}_\varphi$ are:

$$\sigma_{\varphi,j,j'} = \begin{cases} \text{cov}\{\varphi[F_j(e_{ij})], \varphi[F_{j'}(e_{ij'})]\} & j, j' = 1, \ldots, k; j \neq j' \\ 1 & j = j', j = 1, \ldots, k. \end{cases} \tag{10.54}$$

The estimator of the intercept, (10.51), is also asymptotically normal and its joint asymptotic distribution with $\widehat{\boldsymbol{\beta}}_\varphi$ is given in Davis and McKean (1993). In particular, it is a $\sqrt{n}$ consistent estimator of $\boldsymbol{\alpha}$, the vector of adjusted medians in expression (10.48).

For inference, estimators of $\boldsymbol{T}$ and $\boldsymbol{\Sigma}_\varphi$ are required. For $j = 1, \ldots, k$, let $\hat{\tau}_{\varphi,j}$ be the estimator of $\tau_{\varphi,j}$, (10.49), discussed in Section 3.2.3. Then use the diagonal matrix $\widehat{\boldsymbol{T}}$ with the diagonal $\{\hat{\tau}_{\varphi,1}, \ldots, \hat{\tau}_{\varphi,k}\}$ as the estimator of $\boldsymbol{T}$. To estimate the covariances $\sigma_{\varphi,j,j'}$, $j; j \neq j'$, we shall use a simple moment (nonparametric) estimator based on the residuals. In the definition for $\sigma_{\varphi,j,j'}$, replace e_{ij} with the residual $\hat{e}_{ij}$ and replace F_j with the empirical cdf $\hat{F}_j$ of the residuals $\hat{e}_{1j}, \ldots, \hat{e}_{nj}$. Recall that $n\hat{F}_j(\hat{e}_{1j})$ is the rank of $\hat{e}_{1j}$ among the residuals $\hat{e}_{1j}, \ldots, \hat{e}_{nj}$ which we denote by $R_j(\hat{e}_{nj})$. This leads to the simple moment estimator

$$\hat{\sigma}_{\varphi,j,j'} = \frac{1}{n+1}\sum_{i=1}^{n} \varphi\left[\frac{R_j(\hat{e}_{ij})}{n+1}\right]\varphi\left[\frac{R_{j'}(\hat{e}_{ij'})}{n+1}\right], \quad j, j' = 1, \ldots, k; j \neq j'. \tag{10.55}$$

Using these estimators, denote the resulting estimator of $\boldsymbol{\Sigma}_\varphi$ by $\widehat{\boldsymbol{\Sigma}}_\varphi$. These estimators are returned by the R function `mv_rfit`.

We will find the following general fact useful. Suppose $\boldsymbol{M}$ and $\boldsymbol{K}$ are specified $q \times p$ and $k \times s$ matrices. Assume that $\boldsymbol{M}$ is full row rank and $\boldsymbol{K}$ has full column rank. Then using linear multivariate normal properties, it follows from the asymptotic distribution of $\widehat{\boldsymbol{\beta}}$, (10.53), that $\boldsymbol{M}\widehat{\boldsymbol{\beta}}_\varphi\boldsymbol{K}$ is approximately distributed

$$N_{q,s}(\boldsymbol{M}\boldsymbol{\beta}\boldsymbol{K}, \boldsymbol{M}(\boldsymbol{X}_c^T\boldsymbol{X}_c)^{-1}\boldsymbol{M}^T, \boldsymbol{K}^T\boldsymbol{T}\boldsymbol{\Sigma}_\varphi\boldsymbol{T}\boldsymbol{K}). \tag{10.56}$$

We will make use of (10.56) in the next section on hypotheses testing, but, for now, we use it to find a confidence interval for the linear function $\boldsymbol{h}^T\boldsymbol{\beta}\boldsymbol{l}$

where $\boldsymbol{h}$ and $\boldsymbol{l}$ are vectors of order p and k, respectively. In this case $q = s = 1$, so the transformed matrix normal is a univariate normal with mean $\boldsymbol{h}'\boldsymbol{\beta}\boldsymbol{l}$ and variance $[\boldsymbol{h}^T(\boldsymbol{X}_c^T\boldsymbol{X}_c)^{-1}\boldsymbol{h}][\boldsymbol{l}^T\boldsymbol{T}\boldsymbol{\Sigma}_\varphi\boldsymbol{T}\boldsymbol{l}]$. Substituting in the estimates discussed above leads to the approximate confidence interval for $\boldsymbol{h}^T\boldsymbol{\beta}\boldsymbol{l}$ given by

$$\boldsymbol{h}^T\widehat{\boldsymbol{\beta}}\boldsymbol{l} \pm t_{\alpha/2,n-p-1}\left\{\boldsymbol{h}^T(\boldsymbol{X}_c^T\boldsymbol{X}_c)^{-1}\boldsymbol{h}\boldsymbol{l}^T\widehat{\boldsymbol{T}}\widehat{\boldsymbol{\Sigma}}_\varphi\widehat{\boldsymbol{T}}\boldsymbol{l}\right\}^{1/2}. \tag{10.57}$$

We next consider two special cases of this interval.

Suppose first that we are interested in the standard error of β_{ij}. The estimate is $\widehat{\beta}_{ij}$ obtained from the fit of the componentwise model, (10.49). Hence from Chapter 4, the standard error of the estimate is the square-root of $\hat{\tau}_{\varphi,j}^2(\boldsymbol{X}_c^T\boldsymbol{X}_c)_{ii}^{-1}$. To see that this agrees with the standard error in (10.57), write $\beta_{ij} = \boldsymbol{\delta}_{p,i}^T\boldsymbol{\beta}\boldsymbol{\delta}_{k,j}$, where $\boldsymbol{\delta}_{p,i}$ is a vector of order p with all components 0 except that the ith component is 1; it follows that $\boldsymbol{\delta}_{k,j}$ is a vector of order k with all components 0 except that the jth component is 1. In this case, the term in braces in expression (10.57) simplifies to $\hat{\tau}_{\varphi,j}^2(\boldsymbol{X}_c^T\boldsymbol{X}_c)_{ii}^{-1}$, thus agreeing with the standard error obtained from fitting the componentwise model. For β_{ij} only variable j is involved.

Now consider the effect $\beta_{11} - \beta_{12}$. Since $\beta_{11} - \beta_{12} = \boldsymbol{\delta}_{p,1}^T\boldsymbol{\beta}\boldsymbol{v}$, where $\boldsymbol{v} = [1, -1, 0, \ldots, 0]^T$, the quantity in braces in expression (10.57) simplifies to $(\boldsymbol{X}_c^T\boldsymbol{X}_c)_{11}^{-1}(\hat{\tau}_{\varphi,1}^2 + \hat{\tau}_{\varphi,2}^2 - 2\hat{\sigma}_{\varphi,12})$, which depends on both response variables 1 and 2. The R function `lincombci.mv_rfit` computes the confidence interval (10.57).

10.5.2 General Linear Hypotheses

For this section, we assume that the multivariate linear model given in expression (10.48) holds. The general linear hypotheses in the multivariate regression setting is expressed as follows. Suppose $\boldsymbol{M}$ and $\boldsymbol{K}$ are, respectively, $q \times p$ and $k \times s$ specified matrices, where $\boldsymbol{M}$ has full row rank and $\boldsymbol{K}$ has full column rank. Then the general linear hypotheses are

$$H_0: \boldsymbol{M}\boldsymbol{\beta}\boldsymbol{K} = \boldsymbol{0} \text{ versus } H_A: \boldsymbol{M}\boldsymbol{\beta}\boldsymbol{K} \neq \boldsymbol{0}. \tag{10.58}$$

As an example, suppose for a multivariate linear model that we are interested in testing whether or not the regression coefficients of the first two response variables are the same. If we take $\boldsymbol{M} = \boldsymbol{I}_p$ and $\boldsymbol{K}$ to be the one column matrix with entries $[1, -1, 0, \ldots, 0]^T$, then under H_0 the regression coefficients of the first two response variables are the same.

As in the LS analysis, there are several rank-based tests for these hypotheses. We consider the following rank-based test statistic for hypotheses (10.58). It is similar to the traditional Lawley-Hotelling (e.g., Seber (1984)) test. For a specified score function $\varphi(u)$, let $\widehat{\boldsymbol{\beta}}$ be the rank-based estimator of $\boldsymbol{\beta}$ in Model

(10.48). The rank-based test statistic is given by:

$$Q = \text{tr}\left\{ \left[\boldsymbol{M}\widehat{\boldsymbol{\beta}}\boldsymbol{K}\right]^T \left[\boldsymbol{M}\left(\boldsymbol{X}^T\boldsymbol{X}\right)^{-1}\boldsymbol{M}^T\right]^{-1} \boldsymbol{M}\widehat{\boldsymbol{\beta}}\boldsymbol{K}\left[\boldsymbol{K}^T\widehat{\boldsymbol{T}}\widehat{\boldsymbol{\Sigma}}\widehat{\boldsymbol{T}}\boldsymbol{K}\right]^{-1}\right\}. \tag{10.59}$$

An asymptotic level α test is given by

$$\text{Reject } H_0 \text{ in favor of } H_A, \text{ if } Q \geq \chi^2(\alpha, qs). \tag{10.60}$$

This test and its asymptotic theory are developed in Davis and McKean (1993); see also, Chapter 6 of Hettmansperger and McKean (2011). The R function `quad.test` in `mvrfit` computes this test.

```
> library(mvrfit)
> quad.test

function (fit, M, K, ...)
{
    if ((!is.matrix(M)) | (!is.matrix(K)))
        stop("invalid data type on input")
    q <- nrow(M)
    s <- ncol(K)
    b <- coef(fit)[-1, ]
    mbk <- M %*% b %*% K
    v1 <- solve(M %*% fit$xpxi %*% t(M))
    v2 <- solve(t(K) %*% fit$TST %*% K)
    test.statistic <- sum(diag((t(mbk) %*% v1 %*% mbk %*% v2)))
    p.value <- pchisq(test.statistic, q * s, lower.tail = FALSE)
    result <- list(test.statistic = test.statistic,
        p.value = p.value, M = M, K = K)
    class(result) <- "mv_rfit.quad.test"
    result
}
<bytecode: 0x55630efa1aa0>
<environment: namespace:mvrfit>
```

Required input is (1) the multivariate rank-based fit as a result of a call to `mv_rfit`, (2) the hypothesis matrix $\boldsymbol{M}$, and (3) the hypothesis matrix $\boldsymbol{K}$. It returns the value of the test statistic Q and the corresponding p-value.

As an example, consider the following dataset drawn from a pharmaceutical study.

Example 10.5.2 (Study of the Drug Amitriptyline). Johnson and Wichern (2007) present the results of a study on 17 patients who were admitted to the hospital after taking an overdose of the antidepressant amitriptyline. The data are in the dataset `amidata`. The two response variables recorded are total TCAD plasma level (`tot`) and the amount of amitriptyline in TCAD (`amt`).

Several predictors were also recorded. In this example, we consider the two predictors gender (**gen**) and the PR (**pr**) electrocardiogram measurement. In the next R segment, we obtain the rank-based (Wilcoxon scores) fit of linear model, displaying the table of coefficients. The first explanatory variable is gender and the second is PR. The first response variable is total TCAD plasma level, while the second is the amount of amitriptyline in TCAD.

```
> ymat <- amidata[,1:2]
> xmat <- amidata[,c('gen','pr')]
> fit <- mv_rfit(xmat,ymat)
> summary(fit)

$tot
             Estimate Std. Error t.value  p.value
(Intercept) -19.4400     9.8253 -1.9786 0.067884
gen           3.9865     2.5491  1.5639 0.140168
pr           15.2750     5.1570  2.9620 0.010298

$ami
             Estimate Std. Error t.value   p.value
(Intercept) -24.9388    10.4988 -2.3754 0.0323558
gen           5.5267     2.7409  2.0164 0.0633660
pr           17.0523     5.5449  3.0753 0.0082262
```

The results for the predictor PR are significant for both variables. Based on the estimates, for every unit increase of PR, we estimate that total TCAD increases by 15 units and the amount of amitriptyline in TCAD increases by 17 units. The dummy variable for gender is 1 if the patient is a female, 0 otherwise. Note that gender is not significant at level 0.05 for either response variable. As an illustration of the test (10.60), the next R segment tests that gender can be dropped from the model.

```
> mmat <- matrix(c(1,0),ncol=2)
> kmat <- diag(rep(1,2))
> quad.test(fit,mmat,kmat)

Multivariate Rank-Based Regression Quadratic Test
Test_Statistic          p-value
      4.763021         0.092411
```

The test is not significant at the 5% level. ∎

The R function `rstudent.mv_rfit`, in library `mvrfit`, computes the Studentized residuals. The call is `rstudent.mv_rfit(fit)`, where `fit` is the rank-based fit of the multivariate linear model. In Exercise 10.8.9, the reader is asked to perform a residual analysis on the rank-based fit in Example 10.5.2.

10.6 One-Way Rank-Based MANOVA

In this section we consider the rank-based fit and hypotheses tests for a one-way layout for multivariate responses on k variables. The data consist of p subsamples or groups. We can also refer to the design as having a factor at p levels or that there are p treatments of interest. Let n_i denote the sample size of group i and take $n = \sum_{i=1}^{p} n_i$ as the total sample size. Let $\boldsymbol{Y}_{ij}$ denote the $k \times 1$ vector of responses on the jth subject in the ith group and assume that $\boldsymbol{Y}_{ij}$ follows the location model:

$$\boldsymbol{Y}_{ij} = \boldsymbol{\mu}_i + \boldsymbol{e}_{ij}, \quad i = 1, \ldots, p; j = 1, \ldots, n_i, \tag{10.61}$$

where the random errors $\boldsymbol{e}_{ij}$ are iid. Without loss of generality, assume that the marginal pdfs of $\boldsymbol{e}_{ij}$ have true median 0, so that the components of $\boldsymbol{\mu}_i$ are adjusted medians. Stack the $\boldsymbol{Y}_{ij}^T$s group-by-group into the data matrix $\boldsymbol{Y}$ and let $\boldsymbol{\mu}$ be the $p \times k$ vector with the ith row $\boldsymbol{\mu}_i^T$. Let $\boldsymbol{W}$ be the $n \times p$ incidence matrix of group membership. Then we can write Model (10.61) as

$$\boldsymbol{Y} = \boldsymbol{W}\boldsymbol{\mu} + \boldsymbol{e}. \tag{10.62}$$

For the fitting of Model (10.62), we prefer the rank-based fit discussed in Section 10.5 that uses a model with an intercept. Its theory allows for skewed as well as symmetric error distributions. Let $\boldsymbol{E}$ be the matrix that is formed by taking the identity $\boldsymbol{I}_p$ and replacing its first column by a column of 1's. Then $\boldsymbol{W}\boldsymbol{E} = [\boldsymbol{1}_n \ \boldsymbol{X}]$ where the columns of $\boldsymbol{X}$ are columns 2 through p of $\boldsymbol{W}$. The inverse of $\boldsymbol{E}$ is

$$\boldsymbol{E}^{-1} = \begin{bmatrix} \boldsymbol{\delta}_{p1}^T \\ \boldsymbol{C} \end{bmatrix} \text{ where } \boldsymbol{C} = \begin{bmatrix} -1 & 1 & 0 & \cdots & 0 \\ -1 & 0 & 1 & \cdots & 0 \\ \vdots & \vdots & \vdots & \vdots & \vdots \\ -1 & 0 & 0 & \cdots & 1 \end{bmatrix}, \tag{10.63}$$

and $\boldsymbol{\delta}_{p1}^T = [1, 0, 0, \ldots, 0]$. The rows of the $(p-1) \times p$ matrix $\boldsymbol{C}$ sum to 0, so $\boldsymbol{C}$ is called a **contrast matrix**. Further, the rows of $\boldsymbol{C}$ are linearly independent. It follows that

$$\boldsymbol{W}\boldsymbol{\mu} = \boldsymbol{W}\boldsymbol{E}\boldsymbol{E}^{-1}\boldsymbol{\mu} = [\boldsymbol{1}_n \ \boldsymbol{X}] \begin{bmatrix} \boldsymbol{\delta}_{p1}^T\boldsymbol{\mu} \\ \boldsymbol{C}\boldsymbol{\mu} \end{bmatrix} = [\boldsymbol{1}_n \ \boldsymbol{X}] \begin{bmatrix} \boldsymbol{\mu}_1^T \\ \boldsymbol{\Delta} \end{bmatrix} = \boldsymbol{1}_n\boldsymbol{\mu}_1^T + \boldsymbol{X}\boldsymbol{\Delta}, \tag{10.64}$$

where the $(p-1) \times k$ matrix $\boldsymbol{\Delta}$ has as its ith row $(\boldsymbol{\mu}_{i+1} - \boldsymbol{\mu}_1)^T$, $i = 1, 2, \ldots, p-1$. Hence, Model (10.62) is equivalent to the model

$$\boldsymbol{Y} = \boldsymbol{1}_n\boldsymbol{\mu}_1^T + \boldsymbol{X}\boldsymbol{\Delta} + \boldsymbol{e}. \tag{10.65}$$

We use the R function `mvrfit` discussed in Section 10.5 for the rank-based fit of this model.

Usually for a one-way model, we are interested in testing that $\boldsymbol{\mu}_1 = \cdots = \boldsymbol{\mu}_p$. If this equality is true, then by Model (10.61) the distributions that generated the p subsamples are the same. The equality is true, if and only if $\boldsymbol{\Delta} = \mathbf{0}$. Then, because $\boldsymbol{C}$ has full row rank, the equality is true, if and only if $\boldsymbol{C}\boldsymbol{\mu} = \mathbf{0}$. Hence, the hypotheses of interest are

$$H_0 : \boldsymbol{C}\boldsymbol{\mu}\boldsymbol{I}_k = \mathbf{0} \text{ versus } H_A : \boldsymbol{C}\boldsymbol{\mu}\boldsymbol{I}_k \neq \mathbf{0}. \tag{10.66}$$

The rank-based procedure to test these hypotheses is to first obtain the rank-based fit of the one-way model, (10.65). Then, because $\boldsymbol{C}\boldsymbol{\mu}\boldsymbol{I}_{k1} = \boldsymbol{I}_{p-1}\boldsymbol{\Delta}\boldsymbol{I}_k$, the rank-based test statistic is Q, (10.59). In this situation, an asymptotic level-α test rejects H_0 if $Q \geq \chi^2(\alpha, k(p-1))$.

Example 10.6.1 (Anteaters). We consider a real dataset presented on page 436 of Seber (1984), concerning the skull size of anteaters from three different areas of South America. The original source of the data is Reeve (1941). Three measurements (basal length (Y_1), occipitionasal length (Y_2), and greatest length of nasal (Y_3) of skull size were recorded. The logs of the data are presented in the next code segment

```
   basal occipitionasal nasal               location
1  2.068          2.070 1.580  Minas Graes, Brazil
2  2.068          2.074 1.602  Minas Graes, Brazil
3  2.090          2.090 1.613  Minas Graes, Brazil
4  2.097          2.093 1.613  Minas Graes, Brazil
5  2.117          2.125 1.663  Minas Graes, Brazil
6  2.140          2.146 1.681  Minas Graes, Brazil
7  2.045          2.054 1.580 Matto Grosso, Brazil
8  2.076          2.088 1.602 Matto Grosso, Brazil
9  2.090          2.093 1.643 Matto Grosso, Brazil
10 2.111          2.114 1.643 Matto Grosso, Brazil
11 2.093          2.098 1.653  Santa Cruz, Bolivia
12 2.100          2.106 1.623  Santa Cruz, Bolivia
13 2.104          2.101 1.653  Santa Cruz, Bolivia
```

The areas sampled were: Minas Graes, Brazil; Matto Grosso, Brazil; and Santa Cruz, Bolivia. Note that the sample sizes, given in the next R code segment, from each area are small, so caution should be used in the interpretation of the inference.

```
> table(anteaters[,'location'])

 Minas Graes, Brazil Matto Grosso, Brazil
                   6                    4
 Santa Cruz, Bolivia
                   3
```

Comparison boxplots for each measurement are presented in Figure 10.6.

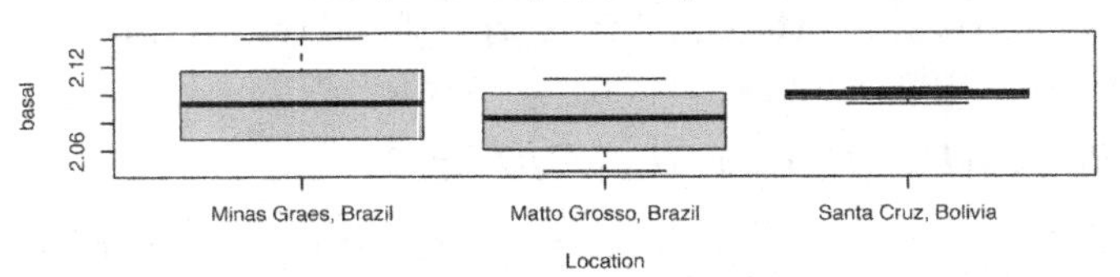

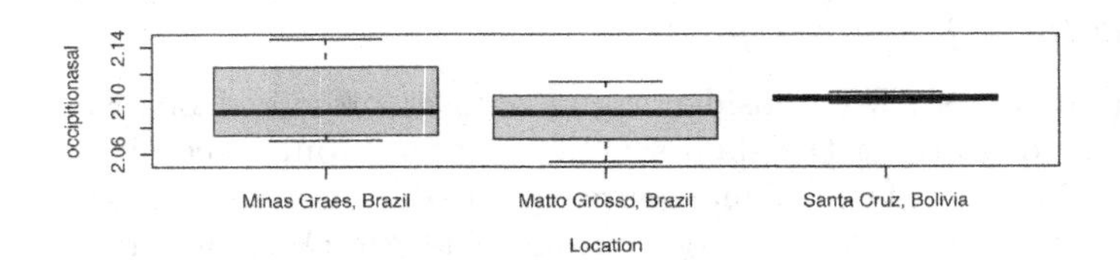

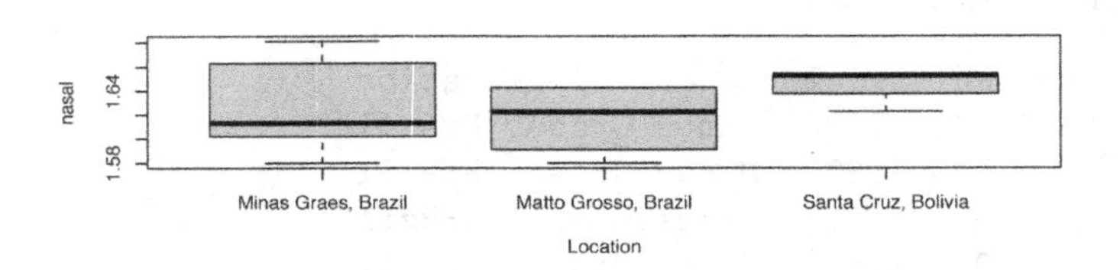

FIGURE 10.6
For each skull measurement in the anteater dataset comparison boxplots versus location.

In the next R segment, based on the Wilcoxon rank-based fit, we compute the test statistic, (10.59), of the one-way hypotheses (10.66).

```
> ymat <- anteaters[,-4]
> xmat <- model.matrix(~anteaters[,4])[,-1]
> colnames(xmat) <- levels(anteaters[,4])[-1]
> fit <- mv_rfit(xmat,ymat)
> fit

Call:
mv_rfit(x = xmat, y = ymat)

Coefficients:
                           basal occipitionasal      nasal
(Intercept)                2.097          2.093  1.6130000
`Matto Grosso, Brazil` -0.014         -0.005 -0.0083333
`Santa Cruz, Bolivia`   0.003          0.008  0.0176667
```

```
> p <- ncol(xmat)
> Ip <- diag(rep(1,p))
> Ik <- diag(rep(1,ncol(ymat)))
> quad.test(fit,Ip,Ik)

Multivariate Rank-Based Regression Quadratic Test
Test_Statistic          p-value
       6.74746          0.34483
```

The test is not significant at level α=0.05. ■

10.7 Two-Way Rank-Based MANOVA

In Section 10.6 we discussed rank-based procedures for the one-way MANOVA model. We now extend this discussion to the two-way MANOVA model. This is also a generalization of the univariate two-way model discussed in Chapter 5, so our discussion is brief. The design remains the same, i.e., the experiment contains two factors, say, A and B, with A having a levels and B having b levels. In the multivariate case, vectors of k responses are collected at each of the ab factor combinations. Let n_{ij} denote the number of observations at ith and jth levels of factors A and B, respectively, and let $n = \sum_i \sum_j n_{ij}$ denote the total sample size. Let $\boldsymbol{Y}_{ijl}$ denote the $k \times 1$ vector of responses for the lth sampled item at the ith and jth levels of A and B, respectively. Assume that $\boldsymbol{Y}_{ijl}$ follows the linear model

$$\boldsymbol{Y}_{ijl} = \boldsymbol{\mu}_{ij} + \boldsymbol{e}_{ijl}, \quad l = 1, \ldots, n_{ij}, i = 1, \ldots, a; j = 1, \ldots, b, \tag{10.67}$$

where the random errors $\boldsymbol{e}_{ijl}$ are iid. Without loss of generality assume that the marginal pdfs of $\boldsymbol{e}_{ijl}$ have true median 0, so that the components of $\boldsymbol{\mu}_{ij}$ are adjusted medians. Assume that $f(\boldsymbol{t})$ and $F(\boldsymbol{t})$ are, respectively, the pdf and cdf of $\boldsymbol{e}_{ijlk}$. Further, assume that at least one $n_{ij} > 1$.

Model (10.67) is a one-way model with ab levels, and it is the full model for our discussion. In order to bring in the two-way structure, we proceed as in Section 5.3. First define

$$\begin{aligned} \overline{\boldsymbol{\mu}}_{i\cdot} &= \frac{1}{b}\sum_{j=1}^{b} \boldsymbol{\mu}_{ij}, \quad i = 1, \ldots, a, & (10.68)\\ \overline{\boldsymbol{\mu}}_{\cdot j} &= \frac{1}{a}\sum_{i=1}^{a} \boldsymbol{\mu}_{ij}, \quad j = 1, \ldots, b. & (10.69)\end{aligned}$$

For the additive model, $\boldsymbol{\mu}_{ij}$ is modeled as

$$\boldsymbol{\mu}_{ij} = \overline{\boldsymbol{\mu}}_{\cdot\cdot} + (\overline{\boldsymbol{\mu}}_{i\cdot} - \overline{\boldsymbol{\mu}}_{\cdot\cdot}) + (\overline{\boldsymbol{\mu}}_{\cdot j} - \overline{\boldsymbol{\mu}}_{\cdot\cdot})\,. \tag{10.70}$$

The additive model is highly interpretable. The effect of level i of factor A is the same for all levels of factor B and, likewise, the effect of level j of factor B is the same for all levels of factor A. In order to test for additivity, we define the interaction parameters to be the difference between $\boldsymbol{\mu}_{ij}$ of the full model, (10.67), and the additive model; i.e.,

$$\begin{aligned} \gamma_{ij} &= \boldsymbol{\mu}_{ij} - [\overline{\boldsymbol{\mu}}_{\cdot\cdot} + (\overline{\boldsymbol{\mu}}_{i\cdot} - \overline{\boldsymbol{\mu}}_{\cdot\cdot}) + (\overline{\boldsymbol{\mu}}_{\cdot j} - \overline{\boldsymbol{\mu}}_{\cdot\cdot})] \\ &= \boldsymbol{\mu}_{ij} - \overline{\boldsymbol{\mu}}_{i\cdot} - \overline{\boldsymbol{\mu}}_{\cdot j} + \overline{\boldsymbol{\mu}}_{\cdot\cdot}, \quad i = 1, \ldots, a, \; j = 1, \ldots, b. \qquad (10.71) \end{aligned}$$

Note that $\boldsymbol{\mu}_{ij}$ of the full model, (10.67), can then be written as

$$\boldsymbol{\mu}_{ij} = \overline{\boldsymbol{\mu}}_{\cdot\cdot} + (\overline{\boldsymbol{\mu}}_{i\cdot} - \overline{\boldsymbol{\mu}}_{\cdot\cdot}) + (\overline{\boldsymbol{\mu}}_{\cdot j} - \overline{\boldsymbol{\mu}}_{\cdot\cdot}) + \gamma_{ij}. \qquad (10.72)$$

The hypotheses of interest are:

$$\begin{aligned} H_{0A}: \overline{\boldsymbol{\mu}}_{1\cdot} = \cdots = \overline{\boldsymbol{\mu}}_{a\cdot} &\text{ vs. } H_{1A}: \overline{\boldsymbol{\mu}}_{i\cdot} \neq \overline{\boldsymbol{\mu}}_{i'\cdot}, \text{for some } i \neq i' \; (10.73) \\ H_{0B}: \overline{\boldsymbol{\mu}}_{\cdot 1} = \cdots = \overline{\boldsymbol{\mu}}_{\cdot b} &\text{ vs. } H_{1B}: \overline{\boldsymbol{\mu}}_{\cdot j} \neq \overline{\boldsymbol{\mu}}_{\cdot j'}, \text{for some } j \neq j' (10.74) \\ H_{0AB}: \gamma_{ij} = \mathbf{0}, \text{ for all } i, j &\text{ vs. } H_{1AB}: \gamma_{ij} \neq \mathbf{0}, \text{ for some } i, j. \quad (10.75) \end{aligned}$$

There are $a-1$ constraints for the hypothesis H_{0A}, $b-1$ for H_{0B}, and $(a-1)(b-1)$ for H_{0AB}. Generally, the interaction hypothesis, H_{0AB}, is tested first. If H_{0AB} is not rejected, then there is evidence that the additive model holds and the interpretation of the main effect hypotheses is clear. For example, if H_{0A} is also not rejected, then the effect of Factor A is considered not statistically significant for all levels of Factor A (i.e., $\boldsymbol{\mu}_{ij}$ does not vary with i). On the other hand, if H_{0AB} is rejected then, for example, the interpretation of H_{0A} is that the averages of the row effects over the levels of Factor B are the same, which is not easy to interpret. Hypotheses (10.73)–(10.75) are the multivariate versions of Type III hypotheses discussed in Section 7.4.

All of the null hypotheses can be expressed in the form $\boldsymbol{H}\boldsymbol{\mu}\boldsymbol{I}_k = \mathbf{0}$ for appropriate contrast matrices $\boldsymbol{H}$ such as those discussed in Section 7.4. These are generalized linear hypotheses. We consider, as the rank-based test statistics, the trace statistics of the Lawley-Hotelling type as given in expression (10.59) along with the associated decision rule (10.60). The degrees of freedom for the tests of the main effect hypotheses H_{0A} and H_{0B} are $k(a-1)$ and $k(b-1)$, respectively, while the degrees of freedom for the interaction hypothesis H_{0AB} are $k(a-1)(b-1)$.

The R function `twoway.mvrfit` computes the rank-based tests of hypotheses (10.73)–(10.75). Input consists of the $n \times k$ matrix of responses and $n \times 2$ matrix of factor levels. Use of `twoway.mvrfit` is illustrated in the following code segment.

Example 10.7.1 (Two-Way Multivariate)**.** We generated data for a 2×3 design; i.e., $a = 2$ and $b = 3$. The responses consist of vectors of three measurements. The design is balanced with 5 observations in each of the $ab = 6$ cells. Some of the true interaction contrasts are not zero in the model that generated the responses. The random errors have a multivariate t-distribution with 5 degrees of freedom. The simulated dataset is in `sim_two_way`. Some of the data are listed in the R segment below.

```
> head(sim_two_way)

     y1    y2    y3 f1 f2
1  8.51 11.34 10.15  1  1
2  0.09  4.94  6.39  1  1
3 14.97 15.82 16.32  1  1
4 -0.88 -2.04  3.73  1  1
5 14.31 27.11 26.39  1  1
6 10.91 14.60 16.11  1  2

> with(sim_two_way,table(f1,f2))

   f2
f1  1 2 3
  1 5 5 5
  2 5 5 5
```

Computation of the rank-based tests of interaction and main effects are illustrated in the following R segment.

```
> ymat <- sim_two_way[,1:3]
> gmat <- sim_two_way[,4:5]
> twoway.mvrfit(ymat,gmat)

Multivariate Rank-Based Regression Quadradic Tests
Main Effects and Interaction
      Test_Statistic DF         p-value
f1      9.6964543     3.0000000  0.0213307
f2      7.9072023     6.0000000  0.2449811
f1xf2 18.2629003      6.0000000  0.0056077
```

The rank-based test for interaction is significant at most levels of α commonly used in practice. Note: the attribute `muhat` contains the rank-based estimates of the full model $\boldsymbol{\mu}_{ij}$'s as shown below.

```
> twoway.mvrfit(ymat,gmat)$muhat

                y1      y2     y3
f1_1.f2_1  8.5100 11.1802 11.018
f1_1.f2_2  9.0349 12.0108 14.502
f1_1.f2_3  4.5349  6.3902 10.498
f1_2.f2_1 12.2100 11.5308 14.588
f1_2.f2_2 16.0049 15.7098 15.788
f1_2.f2_3 13.9749 21.8098 21.742
```

The full model fit from `mvrfit` is also returned in the attribute `fit` (not displayed). ∎

10.8 Exercises

10.8.1. Consider the following 10 samples of a $k = 4$ random vector:

v_1	v_2	v_3	v_4
7	17	4	8
9	6	11	11
11	10	10	15
14	12	8	8
9	8	9	10
8	9	12	9
63	33	29	44
8	7	9	9
9	3	14	12
10	9	9	13

(a) Obtain a `pairs` plot of the data and the six correlation coefficients, using both Pearson's and Spearman's. Discuss the results.

(b) Obtain the point estimates and 95% confidence intervals for the location parameters using the signed-rank Wilcoxon analysis.

(c) Repeat Part (b) using the sign analysis. Compare with Part (b).

10.8.2. For the dataset in Exercise 10.8.1, consider the hypotheses:

$$H_0 : \mu_1 = \cdots = \mu_4 \text{ versus } H_A : \mu_i \neq \mu_j \text{ for some } i \neq j.$$

(a) Test at nominal $\alpha = 0.05$ using the signed-rank Wilcoxon procedure. Discuss the results.

(b) Repeat Part (b) using the sign procedure.

10.8.3. For the dataset in Exercise 10.8.1, consider the hypotheses

$$H_0 : \mu_1 = \cdots = \mu_4 = 10 \text{ versus } H_A : \mu_i \neq 10 \text{ for some } i = 1, \ldots, 4.$$

(a) Test at nominal $\alpha = 0.05$ using the signed-rank Wilcoxon procedure. Discuss the results.

(b) Repeat Part (b) using the sign procedure.

10.8.4. The dataset `anteaters` is discussed in Example 10.6.1. Consider the variable `nasal` length. Using the two-sample Wilcoxon analyses of Chapter 3, obtain confidence intervals for the pairwise differences for this variable over the three locations.

10.8.5. Continuing with Exercise 10.8.4, consider the `Mineas Graes` location and the variables `basal` length and `occipitionasal`. Test at level 0.05 whether or not these variables differ using the Wilcoxon signed-rank test.

.

10.8.6. Exercise 10.8.5 concerns a univariate analysis. To obtain a multivariate test over all the locations on the difference between `basal` length and `occipitionasal nasal` length, fit the model as in Example 10.6.1, then use the hypotheses (10.58) with

$$\boldsymbol{M} = \begin{bmatrix} 1 & 0 & 0 \\ 0 & 1 & 0 \end{bmatrix} \text{ and } \boldsymbol{K} = \begin{bmatrix} 1 \\ -1 \end{bmatrix}.$$

1. State explicitly what the null and alternative hypotheses are in terms of the μ_{ij} and in words.
2. Carry out the test concluding in terms of the Wilcoxon signed-rank test statistic and its p-value using nominal $\alpha = 0.05$.

10.8.7. Similar to Exercise 10.8.6, but consider all three differences between all three variables.

10.8.8. Consider the rank-based fit computed in Example 10.5.1.

(a) Use the R function `lincomb.mv_rfit` to find 95% confidence intervals for all non intercept regression coefficients.

(b) Next, get 99% confidence intervals. Compare these with the confidence intervals found in Part (a).

10.8.9. For the rank-based fit discussed in Example 10.5.2, use the function `rstudent.mv_rfit` to compute the Studentized residuals. Using the Studentized residuals, obtain residual plots and $q-q$ for the fits of both variables. Discuss in terms of the problem.

10.8.10. In Example 10.5.2, considered the model with gender and PR as the predictors. Add the interaction column (`gen*pr`) to the model as the third predictor.

(a) Test to see if there is interaction between gender and PR.

(b) Test to see if there is interaction between gender and PR for the variable `tot`.

(c) Test to see if there is interaction between gender and PR for the variable `ami`.

10.8.11. In Example 10.5.2, we considered two predictors, gender and PR. There are three other predictors in the dataset. These can be found in the dataset `amidata` with the names: `amt`, amount of antidepressants taken at the time of overdose; `diap`, diastolic blood pressure; and `qrs`, QRS wave measurements.

(a) Add these three predictors to the model with predictors gender and PR that was fitted in Example 10.5.2 and compute the rank-based fit using Wilcoxon scores. Is the new model a better fit? Answer based on the Studentized residual plots and normal $q-q$ plots.

(b) Answer the question in the last part by testing the null hypothesis that each of the regression coefficients of the new predictors is 0. Use nominal $\alpha = 0.05$ for your conclusion.

10.8.12. Johnson and Wichern (2007) present a dataset concerning crude oil. Three zones (groups) of sandstone were sampled: Wilhelm (1), Sub-Mulinia (2), and Upper (3). For each sampled item, five measurements were obtained: trace amount of vanadium, trace amount of iron, trace amount of beryllium, saturated hydrocarbons (in percent area), and aromatic hydrocarbons (in percent area). The data are in the dataset `crudeoil`.

(a) Obtain the rank-based test of the hypotheses (10.66). Conclude in terms of the problem using $\alpha = 0.05$.

(b) Use an appropriate MCP for all 15 pairwise differences between medians for this dataset. State why you think it is appropriate. Summarize the results of the MCP.

10.8.13. The dataset `sim_two_way` contains the results of a two-way experimental design over two variables.

(a) Obtain the interaction plots for each variable; see Exercise 10.8.16 for code. Comment on the plots in terms of interaction.

(b) Obtain the rank-based tests of the main effects and interaction. Conclude on the results of the tests using $\alpha = 0.05$.

(c) If we assume no interaction, what are the interpretations of the main effect results.

10.8.14. For Exercise 10.8.13, for each variable obtain the rank-based estimate and confidence interval for the interaction contrast $\mu_{11}+\mu_{22}-\mu_{12}-\mu_{21}$. Discuss the result.

10.8.15. The dataset `paspalum` contains the results of a 2×4 experimental design performed on plants of *paspalum* grass that was discussed by Seber (1984). Forty-eight plants were selected for the experiment and each of the eight treatment combinations had six of these plants. Factor A is the treatment indicator: 1 if control and 2 if the plant was inoculated with a fungal infection. Factor B is the temperature at which the inoculation took place (14, 18, 22, and 26 degrees C). After a specified amount of time, three measurements were taken: the fresh weight of the roots of the plant (gm); the maximum rootlength of the plant (mm); and the fresh weight of the tops of the plant (gm). The responses and the level indicators are in the file `paspalum.rda`.

(a) Use the function `twoway.mvrfit` to obtain the rank-based fit of this dataset. Then test interactions and main effects of the factors.

(b) Obtain a diagnostic analysis of this fit by obtaining a 3×2 page of plots consisting of normal q–q plots of the residuals and residual plots.

(c) Based on the residual plots, is the linear model adequate? Comment on homoscedasticity and possible distributions of the random errors.

(d) Repeat Parts (a) and (b) using the logs of the responses. Is the fit an improvement? Why?

10.8.16. For the data on *paspalum* grass of Exercise 10.8.15, obtain the rank-based analysis based on Wilcoxon scores. Use the log of the responses; i.e., `ymat <- log(resp)`.

(a) For each variable, obtain the profile (of Factor A over Factor B) interaction plot. Are the profiles approximately parallel? Comment. Here is code for the first variable:
First, obtain the rank-based fit of the two-way model and assign it to `fit`.

```
> tst <- twoway.mvrfit(ymat,levsind)
> muhat <- tst$muhat
> xs <- 1:4
> all <- muhat[,1]
> x2<-c(xs,xs)
> plot(all~x2,pch=" ",ylab="muhat",xlab="Factor B")
> title(main="Profile Plot for Var 1")
> all1 <-all[1:4]
> lines(all1~xs,lty=1)
> all2 <- all[5:8]
> lines(all2~xs,lty=2)
> legend(2.50,1.6,c("A1","A2"),lty=c(1,2))
```

(b) Obtain the two-way MANOVA. Discuss the analysis using the 0.05 level.

(c) The contrast for the average main effects of Factor A is $\boldsymbol{h} = (1,1,1,1,-1,-1,-1,-1)^T$. Use this along with the R function `lincomb.mv_rfit` to obtain estimates of this contrast for each of the variables. In the absence of interaction, these are estimates of the shifts between control and treatment for this problem. Interpret the analysis with this in mind.

10.8.17. For the last exercise:

(a) Investigate the significance, $\alpha = 0.05$, of the simple interaction contrast $\mu_{11} - \mu_{12} - \mu_{21} + \mu_{22}$ for each variable.

(b) Investigate the significance, $\alpha = 0.05$, of the three differences between variables for the contrast $\mu_{11} - \mu_{12} - \mu_{21} + \mu_{22}$.

10.8.18. For Example 10.7.1, investigate the significance, $\alpha = 0.05$, of the simple interaction contrast $\mu_{11} - \mu_{12} - \mu_{21} + \mu_{22}$ for each variable.

11

Big Data

In this chapter we introduce some basic concepts of big data and revisit some analyses discussed in earlier chapters using big data specific software. In particular, big data implementation for our usual linear regression model, introduced in Chapter 4, is considered. We begin with a discussion of a rank-based estimation algorithm intended to be used for big data. Following that, we go over some general R packages that are suited for big data. Next, we cover some of the computational aspects for implementing rank-regression model fitting in a big data setting and provide example usage of our R package `bigRfit`. We close by discussing what is one of the main software tools for big data analytics, Spark, as well as an R interface to Spark, `sparklyr` (Luraschi et al. 2020).

Over the years, there has been some debate as to what constitutes *big data*. Currently, there seems to be some consensus around the definition: too big to fit in memory (RAM) on a single machine (i.e., computer). A solution to this problem is to distribute the data over a set of machines; i.e., use a cluster of compute nodes. While, in a production setting, one will likely make use of a cluster, it is also possible to run Spark on a single local machine.[1]

11.1 Approximate Scores

In this section, for our usual linear model, a rank-based estimation algorithm is outlined which does not require a full ranking, but instead depends on partitioning the residuals into bins. The fit and associated analyses lose little in terms of efficiency compared to that based on the full ranking of residuals even with a small number of bins (e.g., 20). Details of the method are presented in the manuscript by Kapenga, Kloke, and McKean (2024).

Consider, again, the estimation of the regression coefficients of a linear model. Let $\boldsymbol{Y}$ be an $N \times 1$ vector of observations and assume it follows our usual the linear model

$$\boldsymbol{Y} = \alpha\mathbf{1} + \boldsymbol{X}\boldsymbol{\beta} + \boldsymbol{e}, \tag{11.1}$$

[1] In the development of this book, this was the approach taken.

where $\boldsymbol{X}$ is an $N \times p$ design matrix; $\mathbf{1}$ is an $N \times 1$ vector of ones; $\boldsymbol{e}$ is an $N \times 1$ vector of random errors with probability density function (pdf) $f(x)$ and cumulative distribution function (cdf) $F(x)$; α is the intercept parameter; and $\boldsymbol{\beta}$ is $p \times 1$ vector of regression coefficients. For big data, N may be in the millions or more. Ultimately, our goal is similar to that of Chapter 4 in that we are interested in estimation and inference (esp. confidence intervals) on the parameter $\boldsymbol{\beta}$ as well as diagnostics to assess the fit. The interpretation is the same as in previous chapters — the algorithm is that which differs. We go over, in moderate detail, the key components of the algorithms for estimation of $\boldsymbol{\beta}$ — diagnostic procedures in a big data setting may be considered in future work.

For estimation we use Jaeckel's dispersion function, i.e.,

$$\widehat{\boldsymbol{\beta}}_\varphi = \text{Argmin}\|\boldsymbol{Y} - \boldsymbol{X}\boldsymbol{\beta}\|_\varphi. \tag{11.2}$$

The *pseudo-norm* $\|\cdot\|$ is given by

$$\|\boldsymbol{v}\|_\varphi = \sum_{i=1}^{N} a[R(v_i)]v_i, \quad \boldsymbol{v} \in R^N, \tag{11.3}$$

where $R(v_i)$ denotes the rank of v_i among $v_1, v_2, \ldots, v_N$; the scores are $a_\varphi(i) = \varphi[i/(N+1)]$ with φ the chosen score function. As before, the dispersion function is given by

$$D(\boldsymbol{\beta}) = \sum_{i=1}^{N} a[R(y_i - \boldsymbol{x}_i^T\boldsymbol{\beta})](y_i - \boldsymbol{x}_i^T\boldsymbol{\beta}). \tag{11.4}$$

A full ranking may be avoided by approximating $\varphi(u)$ with a score function selected from one of two classes: (1) *Class I*, linear approximation, and (2) *Class II*, step score approximation, which we discuss in this section.

Both types assume a partitioning based on the range of the residuals which we now define. Denote the initial estimate of $\boldsymbol{\beta}$ by $\widehat{\boldsymbol{\beta}}^{(0)}$ and let $\widehat{\boldsymbol{e}}^{(0)}$ denote the corresponding vector of residuals. Denote the residuals at the kth step of the algorithm by $\widehat{\boldsymbol{e}}^{(k)}$.

For a given step, let $\widehat{e}_{(1)}$ and $\widehat{e}_{(N)}$ denote the minimum and maximum of residuals, respectively. Partition the data into B subintervals (or bins) based on the range of all N residuals. Let $d_0 < d_1 < \cdots < d_B$ denote the points defining the subintervals. Denote the subintervals **bins**; by B_j, defined by

$$B_j = (d_{j-1}, d_j], \quad j = 1, \ldots, B. \tag{11.5}$$

Let n_j denote the number of residuals in bin B_j, $j = 1, \ldots, B$. In practice the d's are the approximate quantiles so that there will be approximately an equal number of residuals in each of the bins.

For a given bin, say, j ($j = 1, \ldots, B$), the lowest rank is $R_j^L = \sum_{k<j}(n_k+1)$ and the highest rank is $R_j^H = \sum_{k \le j} n_k$. The **step scores** are given by

$$s_j = \left[a\left(R_j^H\right) + a\left(R_j^L\right)\right]/2 = \left[a\left(\sum_{l=1}^{j-1}(n_l+1)\right) + a\left(\sum_{l=1}^{j} n_l\right)\right]/2. \quad (11.6)$$

Computation of step scores is illustrated using `data.table` (Dowle and Srinivasan 2022) in Section 11.2.3.

Another option is to use a linear approximation to the score function. The line segment joining the points $\left(d_{j-1}, a\left(\sum_{l=1}^{j-1} n_{\cdot l} + 1\right)\right)$ and $\left(d_j, a\left(\sum_{l=1}^{j} n_{\cdot l}\right)\right)$ is given by

$$g_j(x) = \frac{a\left(\sum_{l=1}^{j} n_{\cdot l}\right) - a\left(\sum_{l=1}^{j-1} n_{\cdot l} + 1\right)}{d_j - d_{j-1}}(x - d_{j-1}) + a\left(\sum_{l=1}^{j-1} n_{\cdot l} + 1\right). \quad (11.7)$$

On the kth step of the algorithm, the score for a residual $\widehat{e}_l^{(k)} \in \boldsymbol{B}_j$ is $g_j(\widehat{e}_l^{(k)})$.

Note that there are different ways of choosing the bins; however, we recommend bins based on percentiles. For example, if $B = 100$, then the percentiles $0.00, 0.01, \ldots, 0.99, 1.00$ are used. Thus, the bins are preset and, in particular, the score function is not data driven.

When using either class of approximating score function, denote the minimizer of the objective function (11.2) as $\widehat{\boldsymbol{\beta}}_{\varphi_B}$ where B is the number of bins used.

11.1.1 Asymptotic Distribution

The estimator $\widehat{\boldsymbol{\beta}}_{\varphi_B}$ has the same asymptotic properties as discussed in Chapter 4, which we summarize here for convenience. Note, we use the median to estimate the intercept. Let $\boldsymbol{X}_c$ denote the center design matrix. Under regularity conditions (see Kapenga et al. (2024)), the asymptotic distribution of the estimators is given by

$$\begin{bmatrix} \widehat{\alpha}_S^* \\ \widehat{\boldsymbol{\beta}}_{\varphi_B} \end{bmatrix} \sim N_{p+1}\left(\begin{pmatrix} \alpha_0 \\ \boldsymbol{\beta}_0 \end{pmatrix}, \begin{bmatrix} \kappa_n & -\tau_\varphi^2 \overline{\mathbf{x}}^T (\mathbf{X}_c^T \mathbf{X}_c)^{-1} \\ -\tau_\varphi^2 (\mathbf{X}_c^T \mathbf{X}_c)^{-1} \overline{\mathbf{x}} & \tau_\varphi^2 (\mathbf{X}_c^T \mathbf{X}_c)^{-1} \end{bmatrix}\right), \quad (11.8)$$

where $\kappa_n = n^{-1}\tau_S^2 + \tau_\varphi^2 \overline{\mathbf{x}}^T (\mathbf{X}_c^T \mathbf{X_c})^{-1} \overline{\mathbf{x}}$ and the scale parameters τ_φ and τ_S are defined by

$$\tau_\varphi = \left[\int \varphi(u)\varphi_f(u)\,du\right]^{-1} \text{ and } \tau_S = [2f(0)]^{-1}, \quad (11.9)$$

where

$$\varphi_f(u) = -\frac{f'[F^{-1}(u)]}{f[F^{-1}(u)]}. \quad (11.10)$$

Estimators of the scale parameters τ_φ for big data are discussed in Section 11.1.2.

From here, statistical inference ensues. For example, a $(1-\alpha)*100\%$ confidence interval for β_j is

$$\widehat{\beta}_{\varphi_B,j} \pm t_{\alpha/2,N-p-1}\widehat{\tau}_{\varphi_B}\sqrt{(\boldsymbol{X}^T\boldsymbol{X})^{-1}_{jj}}, \tag{11.11}$$

where $(\boldsymbol{X}^T\boldsymbol{X})^{-1}_{jj}$ is the jth diagonal entry of the matrix $(\boldsymbol{X}^T\boldsymbol{X})^{-1}$ and the estimator $\widehat{\tau}_{\varphi_B}$ is discussed below (see expression (11.12)).

11.1.2 Estimation of Scale Parameters for Big Data

Consider a kernel type density estimator of τ_{φ_B} developed by Aubuchon and Hettmansperger (1984) based on the residuals from a rank-based fit of a linear model (c.f., Section 5.2 of Hettmansperger (1984)). We discuss estimation of τ_φ via its reciprocal $\gamma_\varphi = 1/\tau_\varphi$ — the **efficacy**.

Using integration by parts, the efficacy can be written as

$$\gamma_{\varphi_B} = \int_{-\infty}^{\infty} \varphi'[F(x)]f(x)dF(x).$$

Let F_N denote the empirical distribution function of the residuals $\widehat{e}_1,\ldots,\widehat{e}_N$; i.e., $F_N(t) = \frac{1}{N}\sum_{i=1}^{N} I(e_i \le t)$ where $I(\cdot)$ is an indicator function. Then our estimator is the Stieltjes integral

$$\widehat{\gamma}_{\varphi_B} = \int_{-\infty}^{\infty} \varphi'[F_N(x)]f_N(x)dF_N(x),$$

where

$$f_N(x) = \frac{1}{Nh_N}\sum_{i=1}^{N} w\left(\frac{x-\widehat{e}_i}{h_N}\right)$$

and the kernel w is a symmetric density function about the origin and h_N is the bandwidth. For the step scores, using the fact that F_N is the empirical distribution function of the residuals, it follows that

$$\begin{aligned}\widehat{\gamma}_{\varphi_B} &= \frac{1}{N^2h_N}\sum_{i=1}^{N}\sum_{j=1}^{N}\varphi'[F_N(\widehat{e}_j)]w\left(\frac{\widehat{e}_j-\widehat{e}_i}{h_N}\right)\\ &= \frac{1}{N^2h_N}\sum_{l=1}^{B} s_l'\sum_{i_l=1}^{n_l}\sum_{j=1}^{N} w\left(\frac{\widehat{e}_j-\widehat{e}_{i_l}}{h_N}\right),\end{aligned} \tag{11.12}$$

where for the last equality the sum is through the bins, and s_l' denotes the constant derivative of φ_B within bin l. An estimator of τ_{φ_B} is $\widehat{\gamma}^{-1}_{\varphi_B}$. The same argument holds for the linear approximate scores. Hence, for our big data algorithm and for either the linear approximation scores or the step scores, a full ranking of the residuals is not needed. For implementation, we use a variant of the data-driven bandwidth function `bw.nrd` in base R and the standard normal density function for the kernel. We have plans for implementations using `data.table` and `Rcpp` (Eddelbuettel and François 2011); information on developmental versions may be found at `github.com/kloke/bigRfit`.

For very large N, the estimator (11.12) can be slow to compute. As an approximation, consider

$$\widehat{\gamma}_{\varphi_B}^{B} \doteq \frac{1}{N^2 h_N} \sum_{i=1}^{B} s_i n_i \sum_{j=1}^{B} n_j w \left(\frac{m_j - m_i}{h_N} \right) \tag{11.13}$$

where m_i is the mid-point and n_i is the count for bin i. In `bigRfit`, the function `tauhatDT` may be used to calculate the approximate estimate (11.13).

Based on a small simulation study, for large B (e.g., $B = 1000$)[2] the estimator (11.13) provides a good approximation and is much faster to compute than (11.12) for big data (e.g., $N \geq 10000$). At the time of writing, we are not aware of a proof of consistency of this approximation estimator and consider it an open problem. However, intuitively, it suggests that an estimate of the density based on $B = 1000$ bins is close to the one based on the entire dataset.

Another option to decrease computation time is to notice that bins that are far apart will have values of w in expression (11.13) of approximately zero. Since the bins offer a partial sort to the data, and we may loop through them in order, we need only consider observations in bins which are close (e.g., where $w > \epsilon_h$). This method could be applied to either (11.12) or (11.13).

Estimation of the scale parameter τ_S is given by

$$\widehat{\tau}_S = \frac{\sqrt{N}}{\sqrt{N-p-1}} \frac{\sqrt{N}[\widehat{e}_{(N-c)} - \widehat{e}_{(c+1)}]}{2 z_{\alpha/2}}, \tag{11.14}$$

where c is the greatest integer less than or equal to $(N/2) - z_{\alpha/2}\sqrt{N/4} - 0.5$, $z_{\alpha/2} = \Phi^{-1}(1 - \alpha/2)$, and Φ is the cdf of a standard normal distribution function. For our big data algorithm, we know the bins containing $\widehat{e}_{(N-c)}$ and $\widehat{e}_{(c+1)}$.

We have discussed a number of ways in which computational requirements can be reduced using binned residuals rather than a full ranking of the residuals.

11.1.3 Efficiency Based on the Number of Bins

In this section, we address the number of bins in terms of asymptotic efficiency. This study indicates that 100 bins seem to suffice for moderate sample sizes to millions of observations. We set the default to `B=1000` in `bigRfit` based

[2] In the next subsection, we discuss how there is little loss in efficiency, relative to the full ranking, when using bins in the upper 100s or 1000, suggesting 1000 is a large number of bins for most practical purposes.

on empirical evaluation of simulated datasets using ranges of values of B — considering computational time and relative efficiency.[3]

Let $\varphi(u)$ be the selected score function for estimation and associated inference. Intuitively, the step and linear score functions which approximate $\varphi(u)$ converge to $\varphi(u)$ as the number of bins goes to ∞ (see Kapenga et al. (2024) for more details).

In the next two subsections, we obtain the efficacy of the step scores which approximate the Wilcoxon scores for normal and logistic distributed errors, respectively.

Efficiencies of the Step Scores at the Normal Distribution

We determine the efficiencies of the step scores approximating the Wilcoxon score function ($\varphi(u) = \sqrt{12}[u-(1/2)]$) at the normal distribution. As discussed previously in the text, at the normal distribution, the ARE of the Wilcoxon estimator relative to the LS estimator is 0.9549297.

For illustration, we use a sample size of $N = 10000$. Table 11.1 displays the calculated efficiencies of the step scores for various values of B. It also tables their ARE's relative to the Wilcoxon and to LS. Of course, LS dominates since the error distribution is normal. Note that the step scores for $B = 2$ are the sign scores; with ARE relative to LS of 0.637. At $B = 100$ there is little difference between the step scores and the Wilcoxon scores in terms of efficiency.

TABLE 11.1
Table of efficiencies of step scores for normal errors. ARE is the asymptotic relative efficiency; B is the number of bins; SS denotes the step scores; W denotes the Wilcoxon scores (based on a full ranking); and LS denotes least squares.

	Efficiency		ARE	
B	SS	W	SS to W	SS to LS
2	0.797885	0.977205	0.666667	0.636620
4	0.925281	0.977205	0.896553	0.856145
5	0.942297	0.977205	0.929831	0.887923
10	0.967152	0.977205	0.979531	0.935383
15	0.972493	0.977205	0.990380	0.945743
20	0.974368	0.977205	0.994203	0.949394
50	0.976687	0.977205	0.998939	0.953917
100	0.977064	0.977205	0.999711	0.954654
200	0.977167	0.977205	0.999922	0.954855
400	0.977195	0.977205	0.999979	0.954910

[3]Computational time was similar when using 100 and 1000 bins, for example, on a mid-range laptop.

Wilcoxon Scores at the Logistic Distribution

In the last subsection, we assumed a normal population while in this section we consider a logistic population. Recall that the Wilcoxon scores are optimal in this case. The efficacy for the Wilcoxon score function $\varphi_W(u) = \sqrt{12}[u-(1/2)]$ at the logistic probability model is 0.5773503.

Table 11.2 displays the efficiencies of step scores for various values of B relative to the efficacy of the Wilcoxon at the logistic distribution. At $B = 2$ the step scores are the sign scores and, of course, they are inefficient relative to the Wilcoxon scores in the case of logistic errors. Note, though, at $B = 5$ this ARE has climbed to 0.96. At $B = 100$ the efficiency of step score analyses is essentially the same as the Wilcoxon analyses. Also at $B = 4$, the step scores analyses are more efficient than LS analyses.

TABLE 11.2
Table of efficiencies of step scores for logistic errors. ARE denotes the asymptotic relative efficiency; B is number of bins; SS denotes the step scores; W denotes the Wilcoxon scores (based on a full ranking); and LS stands for least squares.

	Efficiency		ARE	
B	SS	W	SS to W	SS to LS
2	0.500000	0.577350	0.750000	0.822467
4	0.559017	0.577350	0.937500	1.028084
5	0.565685	0.577350	0.960000	1.052758
10	0.574456	0.577350	0.990000	1.085656
15	0.576122	0.577350	0.995751	1.091963
20	0.576628	0.577350	0.997500	1.093881
50	0.577235	0.577350	0.999600	1.096184
100	0.577321	0.577350	0.999900	1.096513
200	0.577343	0.577350	0.999975	1.096595
400	0.577348	0.577350	0.999994	1.096616

11.1.4 Big Rfit Algorithm

The algorithm for computing the rank-based estimates is an iterative Newton-type algorithm (see Kapenga, Kloke, and McKean (2024); Hettmansperger and McKean (2011)).

Implementation of developmental (at the time of writing) versions are discussed further at: `github.com/kloke/bigRfit`. Currently, we consider the primary interface to be `bigRreg`. The package `data.table` (Dowle and Srinivasan 2022) is used for binning the residuals once the `cuts` have been made as well as scores calculations. This algorithm is illustrated in Subsection 11.2.3 Main concepts of the `bigRreg` interface are outlined in Section 11.3. A one-step estimate using `sparklyr` (Luraschi, Kuo, Ushey, Allaire, and The Apache Software Foundation 2020) is outlined in Section 11.5.

The software, `bigRreg`, employs the step score function corresponding to the score function selected. In Section 11.1.3, we discussed the number of bins in terms of efficiency which indicated about 100 bins is sufficient; however, with massive data sets in mind, we have set the default number of bins at 1000. For the associated analysis (e.g., confidence intervals), the software uses the estimates of the scale parameters provided in expressions (11.13) and (11.14).

11.1.5 Examples

Taxicab Data

Sheather (2016) explored a large dataset consisting of taxicab fares for one day in New York City, Tuesday, January 15, 2013. He selected this dataset with very specific characteristics including: the standard city rate was used to determine the fare; the payment type was either credit card or cash; the rounded trip distance was less than 3 miles, where the rounding was down to the nearest 1/5 mile; and the average trip speed was greater than or equal to 25 miles per hour (mph). The initial charge for a fare is \$2.50, and then the standard rate is \$2.50 per mile. For slow traffic (less than 12 mph), the rate also includes a rate per time; but, the last condition eliminates such observations in the selected set. The are 49,800 observations in the dataset. The standard rate and the conditions imposed indicate that the model should be

$$\text{median}(\text{Fareamount}) = 2.50 + 2.50(\text{RoundedTripDistance}). \qquad (11.15)$$

Due to the predictors being rounded, there are only 15 possible values for them, (0 to 2.8 in steps of 0.2). Figure 11.1 displays the comparison boxplots where the boxes correspond to the responses at each value of the predictor. Notice how narrow the boxes are around the medians. Furthermore, it is easy to show that the 15 medians versus the 15 values of the predictors follow the deterministic linear model specified in Equation (11.15). There are some outliers, but the bulk of the data follows Model (11.15). Evidently, a robust fit should also follow this model.

The first three rows of Table 11.3 display the fitted regression coefficients of the Wilcoxon fit, the simple scores $SS_W(100)$ fit approximating the Wilcoxon

TABLE 11.3
Table of estimates and 95% confidence intervals for fits of the taxicab data. $SS_W(100)$ means step scores for Wilcoxon; W means Wilcoxon scores; LS means least squares; $SS_{ns}(100)$ means step scores for normal scores.

	$\widehat{\alpha}$	95% CI	$\widehat{\beta}$	95% CI
$SS_W(100)$	2.500	(2.500, 2.500)	2.500	(2.500, 2.500)
W	2.500	(2.499, 2.500)	2.500	(2.499, 2.500)
LS	2.276	(2.258, 2.294)	2.603	(2.592, 2.613)
$SS_{ns}(100)$	2.500	(2.500, 2.500)	2.500	(2.500, 2.500)

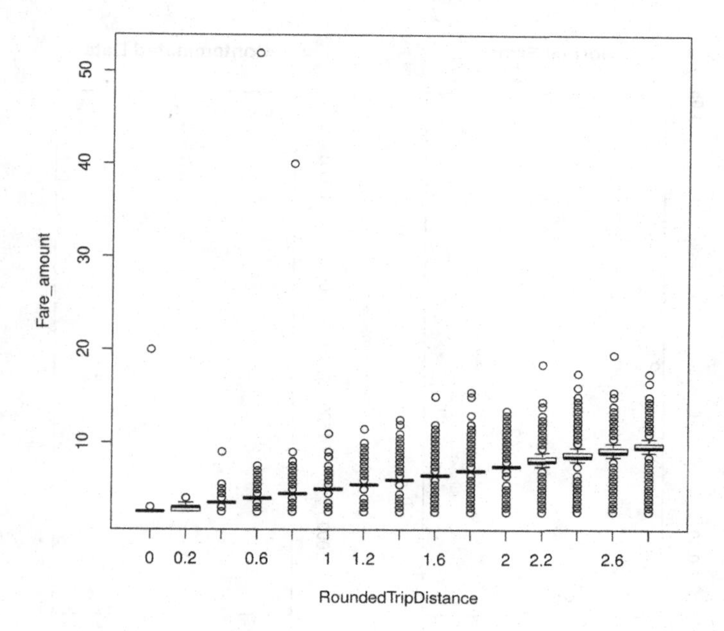

FIGURE 11.1
Comparison boxplots of fare amounts at each value of the rounded trip distances for the taxicab example.

using 100 bins, and the LS fit, respectively. The intercept and slope estimates based on the Wilcoxon and $SS_W(100)$ fits are both 2.5 with confidence intervals having length less than 1 penny ($0.01). On the other hand, the LS estimates each differ from 2.5 with confidence intervals not capturing the true value. Hence, the rank-based procedure correctly estimates the parameters, while the LS procedure does not. Clearly, the outliers affected the LS fit. The fourth row in Table 11.3 shows the step scores estimates of the regression coefficients when the rank-based fit is based on step scores approximation of the normal scores. The fit agrees with those of the other rank-based fits.

Simulation Study

Here we carry out a small simulation study using bigger datasets ($N =$ 20,000,000). We generate data from the following linear model

$$Y_i = 10 + \boldsymbol{x}_i^T[1, 0.25, 0.25, 0.25, 0.25, 0.05, 0.05, 0, 0]^T + e_i \qquad (11.16)$$

where the covariates were generated with $x_{i1} \sim U(-5, 5)$ and $x_{i2}, \ldots, x_{i9}$ from a multivariate normal distribution with mean vector $\boldsymbol{\mu} = \boldsymbol{0}$ and a compound

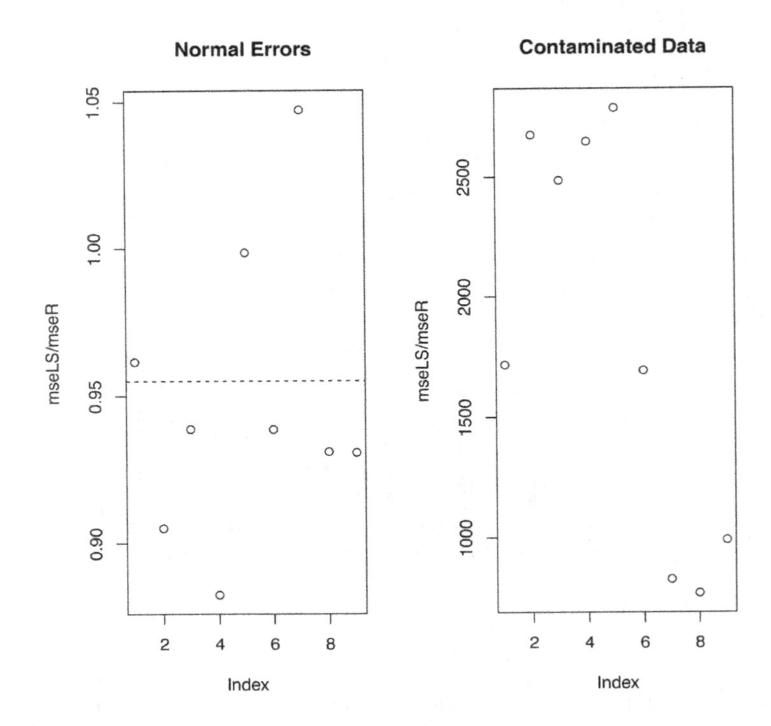

FIGURE 11.2
Results of a small simulation study to compare least squares and one-step Wilcoxon estimates with $B = 1000$ bins.

symmetric variance-covariance matrix with $\sigma = 1$ and $\rho = 0.2$. The error terms were generated from a standard normal distribution; i.e., $e_i \sim N(0, 1)$.

Two sets of simulations were carried out. The first set was as described above and the second with 0.1% of the data contaminated by multiplying the outcome variable by 100. For each set, the simulation size was 40. Estimates for least squares and a one-step Wilcoxon estimate (using $B = 1000$ bins) were carried out using `sparklyr` (Luraschi et al. 2020) as discussed in Section 11.5. The results are plotted in Figure 11.2. The vertical axis is the ratio of the MSEs of the two estimates across the 40 studies and the horizontal axis is the index of the regression coefficient. The scatterplot on the left is for the noncontaminated set and the scatterplot on the right is for the contaminated set. The horizontal line indicates the true ARE of Wilcoxon to least squares for normal errors $(3/\pi)$. Based on these results we can see that, as is expected, with normal errors the least squares estimates are more efficient than the Wilcoxon estimates; although, the relative efficiency is close to 95%. With contaminated samples, however, the Wilcoxon estimates are much more efficient than the least squares estimates.

11.2 Big Data Packages

In this section, we discuss some of the R packages we have encountered that we use in our big data application. There are undoubtedly many more useful packages available.[4]

Packages that we discuss in the coming sections include:

- `data.table` (Dowle and Srinivasan 2022) - used for data processing; fast reads and writes.
- `biglm` (Lumley 2020) - big data algorithm for least squares.
- `dbplot` (Ruiz 2020) - big data graphics; seemless calls to Spark.
- `dplyr` (Wickham et al. 2023) - data processing; seemless calls to Spark.

In this chapter, we will use the package `data.table` (Dowle and Srinivasan 2022) to illustrate the computation of the step scores discussed in Section 11.1. While we have delayed this discussion to the present Chapter on big data, we want to note that this package offers a convenient syntax for data processing (for datasets of all sizes); e.g., merge, subsetting, processing by a group variable. By default, the package uses multiple cores for parallel processes.[5] Note: some R packages have options to connect to an external database or Spark.[6]

11.2.1 `data.table::fread, fwrite`

In Subsection 11.2.3 we use the package `data.table` (Dowle and Srinivasan 2022) to compute the step scores discussed in Section 11.1. In this subsection we illustrate the package's fast read and fast write functions, `fread` and `fwrite`, respectively. The function `fread` has similar arguments to the `read.csv` function in base R. We will illustrate the amount of time used by the two methods to read in a dataset with dimensions given in the following code segment.

```
> dim(flights14)

[1] 253316      11
```

[4]See for example: `cran.r-project.org/web/views/HighPerformanceComputing.html` for Large memory and out-of-memory data in the High Performance Computing TaskView.

[5]See `help(getDTthreads)`.

[6]We do much of the analysis in this section in memory. Advanced R users familiar with out-of-memory analyses should have little difficulty translating into an out-of-memory solution for the methods discussed.

We have downloaded the file `flights14.csv`[7] into a local directory `data`. The following code segment demonstrates the `system.time` to read the file using each of `read.csv` in core R and `fread` in `data.table`.

```
> system.time(read.csv('data/flights14.csv'))

   user  system elapsed
  0.251   0.000   0.251

> system.time(fread('data/flights14.csv'))

   user  system elapsed
  0.112   0.000   0.018
```

Even with about a quarter of a million records, `fread` appears to be much faster than `read.csv`. With larger datasets, the difference may be even more dramatic.

Example 11.2.1 (xy7). In the next code chunk, we generate a dataset which will be used as an example throughout this Chapter. The example briefly illustrates the use of `fwrite`.

```
> library(mvtnorm)
> set.seed(20221022)
> n <- 10^7
> beta <- c(3,1.5,0,0,2,0,0,0)
> p <- length(beta)
> Sigma_x <- 0.5^abs(outer(1:p,1:p,'-'))
> x <- rmvnorm(n,sigma=Sigma_x)
> n1 <- 0.5*n
> n2 <- n-n1
> e <- sample(c(rcn(n1,0.1,10),rcnx100(n2)))
> y <- drop(round(x%*%beta + e,3))
> x <- round(x,2)
> xy <- data.frame(x0=1,x,y)
> fwrite(xy,file='data/xy7.csv')
```

Note: the `x`'s are measured to 2 decimal places and `y` to 3. ■

11.2.2 `biglm`

In this subsection, we illustrate the use of `biglm`, (Lumley 2020), which implements a big data algorithm so that the amount of data in memory is a function of p rather than of N. For comparison, we also include a call to the core R function `lm`. Later in the chapter, we use `biglm` in the computation of the step scores estimate of $\boldsymbol{\beta}$ discussed in Section 11.1.

[7] raw.githubusercontent.com/Rdatatable/data.table/master/vignettes/flights14.csv

```
> library(biglm)
> f_ <- formula(y~X1+X2+X3+X4+X5+X6+X7+X8)
> fito <- lm(f_,data=xy)
> fitb <- biglm(f_,data=xy)
> print(summary(fitb),digits=4)

Large data regression model: biglm(f_, data = xy)
Sample size =  10000000
               Coef    (95%     CI)     SE      p
(Intercept) -0.0007 -0.0028 0.0014 0.0010 0.4988
X1           3.0001  2.9977 3.0025 0.0012 0.0000
X2           1.4988  1.4961 1.5015 0.0013 0.0000
X3          -0.0012 -0.0039 0.0015 0.0013 0.3750
X4           0.0002 -0.0025 0.0029 0.0013 0.8656
X5           2.0005  1.9978 2.0032 0.0013 0.0000
X6           0.0004 -0.0023 0.0031 0.0013 0.7703
X7          -0.0017 -0.0044 0.0010 0.0013 0.2109
X8          -0.0004 -0.0028 0.0020 0.0012 0.7264
```

11.2.3 Computation of Step Scores via data.table

The package data.table (Dowle and Srinivasan 2022) is used to compute the step scores discussed in Section 11.1; in this section we illustrate this computation. Briefly, a data table is a generalization of a data frame in base R, but also implements a convenient syntax for fast data processing.[8]

As an initial fit for considering computational speed, we use an LS fit. For illustration, consider the example of the previous section. In the following code segment, we use the biglm fit from the previous section as our initial fit and calculate the residuals. A data table is created with a single column for the residuals.

```
> x1 <- cbind(1,x)
> ehat <- y-drop(as.matrix(x1)%*%coef(fitb))
> ehatDT <- data.table(ehat)
```

In the next segment, we determine bin endpoints (breaks). The quantile function is used so that we have the same number of observations in each bin in our case due to N being a multiple of B. In general, there should be approximately the same number in each bin. Notice we use a *fudge factor* to ensure the min and max are contained in a bin — this is so that cut may be used without error.

```
> B <- 1001 # number of breaks (number of bins + 1)
> breaks <- quantile(ehat,seq(0,1,length=B))
```

[8]See vignettes available at https://cran.r-project.org/web/packages/data.table/.

```
> # ensure min an max are contained in bin
> ind <- c(1,length(breaks))
> breaks[ind] <- breaks[ind] + 0.001*c(-1,1)
> breaks <- unique(breaks)
```

Next, we bin the data.

```
> ehatDT[,bin := cut(ehat,breaks,label=FALSE)]

               ehat bin
       1:  0.018268 507
       2:  1.362755 896
       3:  0.345310 629
       4: -1.606670  74
       5:  0.159627 561
      ---
 9999996: -0.990519 176
 9999997: -0.372551 362
 9999998: 12.160608 995
 9999999:  1.143349 857
10000000: -0.900414 198
```

Note: all bins contain the same number of observations in this case.

We now illustrate the calculation of the scores using **data.table**. Note: using repeated uses of the bracket (`[]`) is referred to as *chaining*.[9] We have two main steps in our calculation: (1) **Low and High Rank** and (2) `Score Calculation`; as specified by comments in the code chunk.

We use the function **getScoresNS** to indicate that we do not wish our scores to be scaled; instead, we center manually as the last steps (center and scale the scores) in the Score Calculation step.

```
> library(bigRfit)
> scores <- wscores  # use wilcoxon scores
> ##### Low and High Rank #####
> # calculate bin counts and order by bin id
> scoreMat <- ehatDT[,.(count=.N),by=bin][order(bin)]
> # high and low rank for the bin
> scoreMat[,rnkH := cumsum(count)][,rnkL := rnkH-count+1]

      bin count    rnkH    rnkL
   1:   1 10000   10000       1
   2:   2 10000   20000   10001
   3:   3 10000   30000   20001
   4:   4 10000   40000   30001
```

[9]See cran.r-project.org/web/packages/data.table/vignettes/datatable-intro.html.

```
   5:    5 10000    50000  40001
  ---
 996:  996 10000  9960000 9950001
 997:  997 10000  9970000 9960001
 998:  998 10000  9980000 9970001
 999:  999 10000  9990000 9980001
1000: 1000 10000 10000000 9990001

> ##### Score Calculation #####
> # high and low percentiles for the bin
> scoreMat[, `:=`(rnkH=rnkH/(n+1),rnkL=rnkL/(n+1))]

       bin count  rnkH      rnkL
   1:    1 10000 0.001 0.0000001
   2:    2 10000 0.002 0.0010001
   3:    3 10000 0.003 0.0020001
   4:    4 10000 0.004 0.0030001
   5:    5 10000 0.005 0.0040001
  ---
 996:  996 10000 0.996 0.9950000
 997:  997 10000 0.997 0.9960000
 998:  998 10000 0.998 0.9970000
 999:  999 10000 0.999 0.9980000
1000: 1000 10000 1.000 0.9990000

> # raw scores
> scoreMat[,scrs := 0.5*(getScoresNS(scores,rnkH)+
+                        getScoresNS(scores,rnkL))]
> # center the scores
> m <- scoreMat[,sum(count*scrs)/n]
> scoreMat[,scrs := scrs - m]
> # scale the scores
> sd <- sqrt(scoreMat[,sum(scrs*scrs*count/(n+1)) ])
> scoreMat[,scrs := scrs / sd]

> scoreMat

       bin count  rnkH      rnkL    scrs
   1:    1 10000 0.001 0.0000001 -1.7303
   2:    2 10000 0.002 0.0010001 -1.7269
   3:    3 10000 0.003 0.0020001 -1.7234
   4:    4 10000 0.004 0.0030001 -1.7199
   5:    5 10000 0.005 0.0040001 -1.7165
  ---
 996:  996 10000 0.996 0.9950000  1.7165
 997:  997 10000 0.997 0.9960000  1.7199
```

```
 998:  998 10000 0.998 0.9970000  1.7234
 999:  999 10000 0.999 0.9980000  1.7269
1000: 1000 10000 1.000 0.9990000  1.7303
```

As we are using step scores, all of the observations within a bin use the same score. So, for example, the dispersion function (11.4) may be calculated by merging **ehatDT** and **scoreMat** as illustrated next. Since, when using step scores, the score is constant within a bin, we may first sum the residuals within in a bin then merge.

```
> # residual totals by bin
> ehatDT_sb <- ehatDT[,.(D=sum(ehat)),bin]
> # merge scores with totals by bin
> dsDT <- merge(scoreMat,ehatDT_sb,by='bin',sort=FALSE)
> # dispersion based on (initial) LS fit
> dsDT[,sum(D*scrs)]

[1] 16332459
```

A Newton-type algorithm can be used to minimize the dispersion function with the scores updated at each step.

11.3 bigRreg

As mentioned previously, we are developing several different implementatoins of the rank-based methods discussed in Section 11.1. In this section we discuss an implementation, in development, that uses **biglm** (Lumley 2020).

At the time of writing, this implementation only calls **biglm** to compute the projection once per iteration; i.e., it does not process the projection in steps.[10]

As an empirical illustration, we performed a small simulation to compare two implementations: **rfit** and **bigRreg**. While **rfit** will fit a general linear model, **bigRreg** requires an intercept (column of ones) be in the model (design). Our goal is primarily to compare the time to run each of the implementations. But, first, we do a quick comparison on the relative efficiencies. The model is our usual linear model with parameters as in Model(11.1). The simulation size was 20 runs. The selected sample sizes used are:

```
5000 25000 100000
```

[10]An implementation which processes the projection in chunks seems possible but would require a number of modifications to the current code. Advanced R users with experience implementing out-of-memory regression using **biglm** may be able to modify the code (under GPL) for an out of memory solution for the computation of rank-based regression.

The results for empirical relative efficiency of the full ranking method (via `rfit`) and the step-scores using a partial ranking (via `bigRreg`) are provided in Table 11.4. In terms of empirical relative efficiency, there is not much difference between the two methods of estimation.

TABLE 11.4
Table of relative efficiencies for selected sample sizes.

	beta1	beta2	beta3	beta4	beta5	beta6	beta7	beta8
5000	1.0006	1.0001	0.9999	1.0002	1.0000	0.9991	0.9999	1.0003
25000	0.9995	0.9997	1.0002	0.9997	0.9992	1.0000	1.0005	0.9999
100000	1.0003	1.0003	1.0001	1.0005	0.9998	1.0005	1.0006	1.0004

Both `Rfit` and `bigRreg` were run with the option `TAU='N'` specifying that τ was not to be estimated in the case of `Rfit` and a final estimate of tau not be estimated in the case of `bigRreg` (an initial estimate is required for the Newton algorithm; however, a final estimate based on the final residuals is not estimated). In Figure 11.3 times for each of the runs are plotted for two of the selected sample sizes.

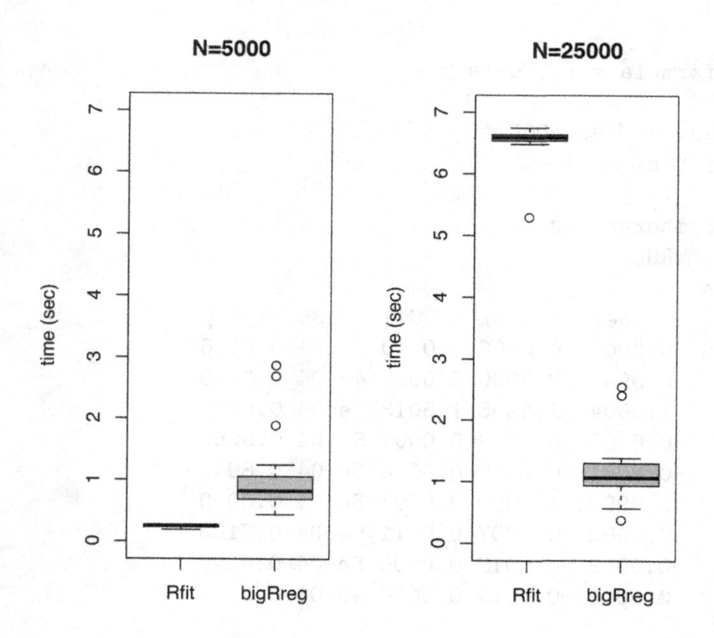

FIGURE 11.3
Comparison boxplots for timing simulation for selected sample sizes. Sample sizes given in the titles.

The average time for the 20 runs for `Rfit` is given in the following code chunk.

```
   5000       25000      100000
  0.24395   6.52305 105.04115
```

The average time for the 20 runs for **bigRreg** is given in the following code chunk.

```
5000   25000  100000
1.0522 1.1684 1.2987
```

As one would expect, the computation time for each method increases with sample size. The increase for Rfit is much larger than for **bigRreg**. In the case of **bigRreg**, for this range of sample sizes, the difference is quite small (and perhaps would be difficult to observe without using a *stopwatch*; a.k.a. system.time).

Example 11.3.1 (xy7 (continued)). In the following code chunk we demonstrate the usage of **bigRreg** on the xy dataset generated in Example 11.2.1.

```
> fitR <- bigRreg(f_,data=xy)
> print(summary(fitR),digits=4)

Call:
bigRreg(formula = f_, data = xy)

Sample size =  10000000
Number of bins =  1000

Number of iterations: 2
Converge: TRUE
Estimates:
             Coef     (95%     CI)     SE      p
Intercept  0.0002 -0.0006 0.0010 4e-04 0.6126
X1         2.9998  2.9990 3.0006 4e-04 0.0000
X2         1.5004  1.4995 1.5013 5e-04 0.0000
X3         0.0000 -0.0009 0.0009 5e-04 0.9262
X4        -0.0001 -0.0010 0.0008 5e-04 0.8942
X5         1.9999  1.9990 2.0007 5e-04 0.0000
X6         0.0002 -0.0007 0.0011 5e-04 0.7193
X7        -0.0003 -0.0012 0.0006 5e-04 0.4620
X8        -0.0004 -0.0012 0.0004 4e-04 0.2970
```

Note that both methods get the *right* answer in that both sets of CIs capture the true parameters; however, the rank-based method is much more efficient.

11.3.1 Approximation of τ

As mentioned in Subsection 11.1.2, an approximation of the scale parameter τ may be based on Equation (11.13) provided N is sufficiently large. In `bigRfit` an implementation of this approximation is provided. At the time of writing, the function is defined as given in the following code chunk.

```
> approxtauAH

function (scoreMat, mids, n, h = bw_bigRfit(mids, n, scoreMat[,
    count]))
{
    pc <- scoreMat[, count]/n
    D <- (dnorm(outer(mids, mids, "-")/h)/h) %*% pc
    tauhat0 <- 1/drop(crossprod(pc * unlist(scoreMat[, "scrsD"]),
        D))
    return(tauhat0)
}
<bytecode: 0x556311cb98a0>
<environment: namespace:bigRfit>
```

Required arguments are a `data.table` of the score matrix (`scoreMat`) — calculation of which is illustrated in Subsection 11.2.3 — a vector of the bin mid-points (`mids`) and the total sample size (`n`). Optionally, the bandwidth may be provided; otherwise, `bw_bigRfit` provides the data-based bandwidth using `bw.nrd` in core R. Another option available is `gettauSSF0` which uses the KSM (Koul et al. 1987) estimate in `Rfit` based on the step scores. In most big data settings, `approxtauAH` should be faster and is the default in `bigRreg`.

Note: it is not recommended to use normal scores (i.e., `nscores`) for big data at this time, as an adjustment may be needed (under investigation; see Kapenga et al. (2024)).

11.4 Spark Basics

Spark was designed for large-scale data processing and can be run on a single computer or on a large cluster of compute nodes. When run with multiple cores (or on multiple nodes), data are able to be processed in parallel. There are several interfaces to Spark, including PySpark, SparkR, and sparklyr. We make use of the `sparklyr` (Luraschi et al. 2020) R interface to Spark. General references to Spark include *Spark the Definitive Guide* (Chambers and Zaharia 2018) and *Learning Spark* (Karau et al. 2015). As the software is constantly being developed, it is recommended to refer to the official Spark documentation `spark.apache.org`. *Mastering Spark with R* (Luraschi et al. 2019) offers an introduction to using the `sparklyr` (Luraschi et al. 2020) package. See also `spark.rstudio.com`.

In the production big data setting — where the size of the input data is larger than a single computer can handle — a cluster of machines or nodes are connected. Those without access to a cluster may still become familiar with these tools and try out the methods in this section using a single machine. Spark may be downloaded from the Apache Spark website or installed through `sparklyr` functions; see Chapter 2 of Luraschi et al. (2019) for more information.

Under the hood, Apache Spark[11] uses *resilient distributed datasets (RDDs)* where, in a cluster setting, the data are distributed across the nodes and may be processed in parallel.[12] In addition, the use of RDDs allows for one node to fail without crashing the entire system; i.e., resiliency. We will primarily use Spark DataFrames which are built on top of RDDs and are conceptually similar to R data frames.

In the next code chunk, we read our data file created in Section 11.2.1 `xy7.csv` into a Spark dataset. First, a connection to Spark is requested using the command `spark_connect`. The connection may be configured in terms of the amount of cluster resources (e.g., cpus, memory). In our example, we use the default options, but specify that our local machine is used via the `master` argument. The option, `inferSchema=TRUE` specifies that the data types should be inferred. The data types are printed following the column names; i.e., `<int>` for integer and `<dbl>` for double precision (floating point). So, in our case, the data types were correctly inferred.

```
> library(sparklyr)
> sc <- spark_connect(master='local')
> d <- spark_read_csv(sc,'data/xy7.csv',header=TRUE,
+                          inferSchema=TRUE)
> d

# Source: spark<xy7_936fa31d_fe5d_4eb5_a6a4_ecdf2facf793>
#   [?? x 10]
      x0    X1    X2    X3    X4    X5    X6    X7    X8
   <int> <dbl> <dbl> <dbl> <dbl> <dbl> <dbl> <dbl> <dbl>
 1     1 -0.82  0.83  1.21  0.79  0.23 -0.55  0.05  0.01
 2     1  4.4   3.14  2.31  0.65  0.58  0.24 -0.12 -0.67
 3     1 -0.01  0     0.13 -0.06 -0.3  -0.76 -0.53 -0.3
 4     1  1.3   1    -0.86  0.27  0.21 -0.6   1.25  0.97
 5     1  0.33 -0.34 -1.19 -0.85  0.03  0.26 -0.46 -1.19
 6     1 -0.2   0.09 -1.07 -0.99 -0.21 -0.13  0.54  0.34
 7     1  0.49 -0.07  0.49  0.07  0.22 -0.44 -0.79  1.11
 8     1  0.87  0.22 -0.9   0.13 -0.24 -1.78 -0.13 -0.59
```

[11] spark.apache.org/docs/latest/sparkr.html
[12] spark.apache.org/docs/latest/rdd-programming-guide.html

```
9         1 -0.98  0.26  0.58  0.01  0.45  0.44  0.92 -0.48
10        1 -0.5   1.16  0.81  1.55  1.16  1.85 -0.12  0.65
```

The data displayed are in the form of a tibble with an unknown number of rows. This is because the dataset is a Spark DataFrame and not stored in R. When we request the object `d` to be printed, a few rows of the dataset are returned to R and then printed in the form of a tibble. Notice in the print of `d`, `??` is displayed as the number of rows in `d`. The majority of the data remain in a Spark DataFrame; so that, in particular, the number or records in the dataset is unknown to R at this point. It is recommended that data read into R from Spark DataFrames be limited, if possible.

Note: `sc` may be printed out the way many R objects can be. The reader is encouraged to print `sc` out and look at the contents. Among other cluster configuration details, listed are the number of cores allocated to this instance of the Spark cluster, the memory (ram), and the number of partitions.

In the following section, we return summaries (e.g., bin counts) from Spark into R and use the score calculations via `data.table` discussed in Section 11.2.3. Note: the `dplyr` data wrangling tools work with a Spark DataFrame, sending the request to Spark which performs the operation (on the distributed dataset) and creates a new Spark DataFrame, so that computations occur outside of R and the datasets remain outside of R unless a specific request is made to read them into R.

Next, we provide an overview of `dbplot`.

11.4.1 `dbplot`

The package `dbplot` (Ruiz 2020) may be used for graphics of big datasets (e.g., in a SQL database or a Spark DataFrame). When used on a Spark DataFrame via `sparklyr`, the calculations of the summary statistics for the plot are performed in Spark with the required summaries sent back to R for graphing. For big data, this method is preferred as a relatively small amount of data (summary statistics) are then read into R for graphing instead of the entire dataset. Under the hood, the `dplyr` (Wickham et al. 2023) interface is used to create the summary statistics in Spark and `ggplot2` (Wickham 2009) is used to create the graphic in R.

As an example, consider Figure 11.4 where `y` versus `x1` from our DataFrame `d` are plotted. By default, a 100×100 grid is created and the number of records in each rectangle are counted (in Spark) and then returned to R. A graphic is then produced in R using `ggplot2`. The number of observations in each rectangle is indicated using a grayscale with darker areas indicating a larger number of data points. No matter how large the dataset is in Spark, only 100*100 observations would be returned (other resolutions are possible, see `help(dbplot_raster)`).

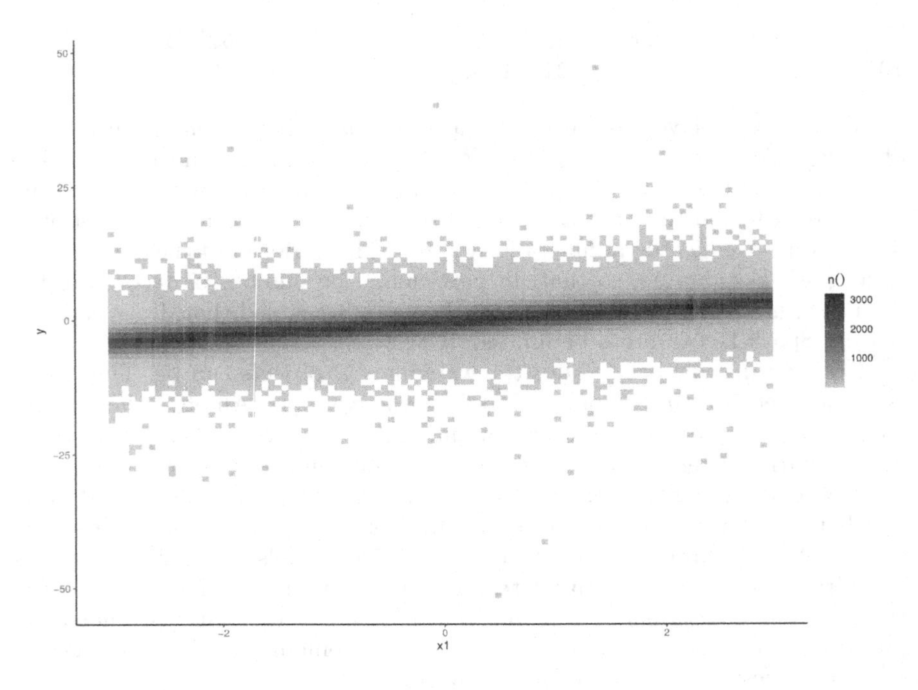

FIGURE 11.4
Example raster plot.

11.4.2 Machine Learning Tools

Spark, along with the `sparklyr` interface, offer a wide range of machine learning tools for distributed data. Modeling functions in `sparklyr` use a familiar syntax allowing the user to specify a `model` which is convenient for data analysts coming to Spark from R. For example, the function `ml_random_forest` may be used to fit random forest models, previously discussed, with the data and computations in Spark. There is functionality for data processing/wrangling. The `sparklyr` interface to Spark includes functions which begin with `ft_` for **feature transformer**. Also, `dplyr` (Wickham et al. 2023) may be used to interface with Spark using the well-known data wrangling tool in `tidyverse`.[13] There is functionality to create training and validation splits of a dataset as well as model selection tools implementing commonly used metrics. We illustrate some of the basics in the next section. Our goal is to provide a brief introduction to `sparklyr` and illustrate the basics of a big data implementation of rank-regression. Least squares linear regression is fit using the function `ml_linear_regression` which is used as an initial fit in the Wilcoxon one-step estimate calculated in the next section.

[13] www.tidyverse.org

11.4.3 spark_disconnect

To disconnect from Spark, use `spark_disconnect` as illustrated in the following.

```
> spark_disconnect(sc)
```

The Spark process is closed with temporary Spark datasets removed. In the next section, we will start a new session and will reread the data into a Spark DataFrame.

11.5 One-step Wilcoxon Estimate Using `sparklyr`

In this section we illustrate a one-step Wilcoxon rank-based estimation for a linear regression problem using `sparklyr`.

We use our simulated dataset `data/xy7.csv` that is read into a Spark data frame in the next code segment.

```
> library(sparklyr)
> library(dplyr)
> sc <- spark_connect(master = "local")
> d <- spark_read_csv(sc,path='./data/xy7.csv')
```

In the following code segment, we create a Spark data frame which includes a column for the centered design matrix combined into lists and a column for our response. Recall that the names of the predictor variables are `X1`, ..., `X8`. The function `ft_vector_assembler` is used to combine the values of the predictor variable into a vector (list) for each row in the data. The function `ft_standard_scalar` is used to center the design matrix similar to how `scale` in core R is used to center the design matrix in `rfit`.

```
> xnames <- paste0('X',1:8)
> d <- ft_vector_assembler(d,input_cols=xnames,output_col='xv')
> df1 <- ft_standard_scaler(d,input_col='xv',output_col='xvc',
+                              with_mean=TRUE,with_std=FALSE)
> df1 <- select(df1,'xvc','y')
> df1

# Source: spark<?> [?? x 2]
  xvc            y
  <list>     <dbl>
1 <dbl [8]> -0.74
2 <dbl [8]> 20.4
3 <dbl [8]> -0.285
```

```
 4 <dbl [8]>  4.21
 5 <dbl [8]>  0.702
 6 <dbl [8]> -1.34
 7 <dbl [8]>  1.26
 8 <dbl [8]>  3.82
 9 <dbl [8]> -1.60
10 <dbl [8]>  2.21
```

In the next code chunk, we obtain a least squares fit which will be used as our initial fit.

```
> fitLS <- ml_linear_regression(df1,
+                                features_col='xvc',
+                                label_col='y',
+                                standardization=FALSE,
+                                prediction_col='yhatLS')
> betahatLS <- coef(fitLS)
> betahatLS

[1]  3.00011728  1.49878030 -0.00119602  0.00022823
[5]  2.00051701  0.00039383 -0.00168768 -0.00042203
```

Note, alternatively, one could use the `model` argument as illustrated in the following. In this case, we use the `magrittr` (Bache and Wickham 2022) operator, %>%, which passes the output of the function on the left as the first argument to the function on the right.

```
> ml_linear_regression(df1,y~xvc,fit_intercept=FALSE,
+                      standardization=FALSE) %>% coef()

      xvc_0       xvc_1       xvc_2       xvc_3       xvc_4
 3.00011728  1.49878030 -0.00119602  0.00022823  2.00051701
      xvc_5       xvc_6       xvc_7
 0.00039383 -0.00168768 -0.00042203
```

In the next code chunk, the `ml_summary` function is used to obtain the standard errors of the estimated regression coefficients and the estimate of residual standard deviation.

```
> seLS <- ml_summary(fitLS)$coefficient_standard_errors()
> sigmahat <- ml_summary(fitLS)$root_mean_squared_error
```

Next, the least squares fitted values and residuals are calculated to be used as initial fits for the rank-based fit.

```
> pred0 <- ml_predict(fitLS,df1)  # initial fitted values
> pred0 <- mutate(pred0,residLS=y-yhatLS)  # initial residuals
> pred0
```

```
# Source: spark<?> [?? x 4]
   xvc               y yhatLS residLS
   <list>        <dbl>  <dbl>   <dbl>
 1 <dbl [8]> -0.74   -0.758  0.0183
 2 <dbl [8]> 20.4    19.1    1.36
 3 <dbl [8]> -0.285 -0.630  0.345
 4 <dbl [8]>  4.21    5.82  -1.61
 5 <dbl [8]>  0.702  0.542  0.160
 6 <dbl [8]> -1.34  -0.886 -0.451
 7 <dbl [8]>  1.26    1.80  -0.549
 8 <dbl [8]>  3.82    2.46   1.36
 9 <dbl [8]> -1.60   -1.65   0.0508
10 <dbl [8]>  2.21    2.56  -0.351
```

Next, we prepare to calculate the step scores by binning the residuals.

```
> B <- 1000  # number of bins
> pred <- ft_quantile_discretizer(pred0,'residLS','bin',
+                                  num_buckets=B)
> pred
```

```
# Source: spark<?> [?? x 5]
   xvc               y yhatLS residLS   bin
   <list>        <dbl>  <dbl>   <dbl> <dbl>
 1 <dbl [8]> -0.74   -0.758  0.0183    506
 2 <dbl [8]> 20.4    19.1    1.36      895
 3 <dbl [8]> -0.285 -0.630  0.345     628
 4 <dbl [8]>  4.21    5.82  -1.61       73
 5 <dbl [8]>  0.702  0.542  0.160     560
 6 <dbl [8]> -1.34  -0.886 -0.451     334
 7 <dbl [8]>  1.26    1.80  -0.549     301
 8 <dbl [8]>  3.82    2.46   1.36      895
 9 <dbl [8]> -1.60   -1.65   0.0508    519
10 <dbl [8]>  2.21    2.56  -0.351     368
```

In the next code chunk, we calculate the bin counts and approximate the midpoints. The function dplyr is used and, since pred is a Spark data frame, the computation is executed in Spark. The result is stored in bc, a Spark data frame.

```
> bc <- pred %>% group_by(bin) %>%
+        summarize(count=n(),m=0.5*(max(residLS)+min(residLS)))
```

In the next segment, the data are read from Spark into R using the command sdf_collect. We convert the returned data to a data.table and then use steps similar to those in Section 11.2.3. Note, the size of the dataset read from Spark to R is a fraction of the entire dataset (i.e., scoreMat is a data table with (approximately) B rows [14]).

```
> scoreMat <- data.table(sdf_collect(bc))
> dim(scoreMat)

[1] 999   3
```

The next segment is similar to that in Section 11.2.3.

```
> scores <- wscores #use Wilcoxon scores
> scoreMat <- scoreMat[order(bin)]
> scoreMat[,rnkH := cumsum(count)]
> n <- scoreMat[.N,rnkH]
```

The sample size will now be available in the variable n (as it is the value of the highest rank).

The remaining steps in the calculation of the step scores are the same as in Section 11.2.3.

```
> scoreMat[,scrs := 0.5*(getScoresNS(scores,rnkH/(n+1))+
+                     getScoresNS(scores,(rnkH-count+1)/(n+1)))]
> # center the scores
> sbar <- scoreMat[,sum(count*scrs)/n]
> scoreMat[,scrs := scrs - sbar]
> # scale the scores
> sd <- sqrt(scoreMat[,sum(scrs*scrs*count/(n+1)) ])
> scoreMat[,scrs := scrs / sd]
```

In the following segment, we calculate a Newton step. We use the nested function calls rather than the magrittr %>%.

```
> pred <- select(
+           left_join(pred,scoreMat[,c('bin','scrs')],
+                     by='bin',copy=TRUE),
+           xvc,'y','residLS','scrs')
> pred

# Source: spark<?> [?? x 4]
   xvc              y residLS    scrs
   <list>       <dbl>   <dbl>   <dbl>
 1 <dbl [8]> -1.34   -0.451   -0.575
 2 <dbl [8]>  4.21   -1.61    -1.48
 3 <dbl [8]> -1.60    0.0508   0.0668
 4 <dbl [8]>  3.91    0.833    0.984
```

[14] Note: the bin ranges, in Spark, are determined by an approximation algorithm (see help(ft_quantile_discretizer)).

```
 5 <dbl [8]>  2.21   -0.351  -0.457
 6 <dbl [8]> 20.4     1.36    1.37
 7 <dbl [8]>  1.26   -0.549  -0.689
 8 <dbl [8]>  3.82    1.36    1.37
 9 <dbl [8]> -0.285  0.345   0.444
10 <dbl [8]>  0.702  0.160   0.209
```

```
> fit1 <- ml_linear_regression(pred,
+                              features_col='xvc',
+                              label_col='scrs',
+                              standardization=FALSE,
+                              prediction_col='ahat')
> predR1 <- ml_predict(fit1,pred)
> tauhat0 <- approxtauDT(scoreMat,scoreMat[['m']],n)
> predR1 <- mutate(predR1,d=-1*tauhat0*ahat)    # direction
> predR <- mutate(predR1,residR = residLS + d) # Newton step
```

At this point, **predR** is a Spark data frame which contains the estimated residuals. In the next segment, we calculate the fitted values and regression coefficients. We also present a simple way to calculate the standard errors of the regression coefficients. Note: we have not estimated an intercept for the Wilcoxon analysis.

```
> predR <- mutate(predR,yhatR = y-residR)
> fitR <- ml_linear_regression(predR,
+                              features_col='xvc',
+                              label_col='yhatR',
+                              standardization=FALSE)
> betahatR <- coef(fitR)
> tauhat0/sigmahat

[1] 0.3382

> seR <- seLS[-1]/sigmahat*tauhat0
> coef <- cbind(betahatR,seR)
> round(coef,4)

     betahatR   seR
[1,]   2.9998 4e-04
[2,]   1.5004 5e-04
[3,]   0.0000 5e-04
[4,]  -0.0001 5e-04
[5,]   1.9999 5e-04
[6,]   0.0002 5e-04
[7,]  -0.0003 5e-04
[8,]  -0.0004 4e-04
```

11.6 Exercises

11.6.1. Argue, for Wilcoxon (linear) scores, that the step scores defined in expression (11.6) may be computed as

$$s_j = a\left[\left(\sum_{l=1}^{j-1}(n_l + 1) + \sum_{l=1}^{j} n_l\right)/2\right].$$

11.6.2. Formulate, using concise notation, the linear model from which the sample in Example 11.2.1 was drawn.

11.6.3. Redo Example 11.2.1 using a sample size of $n = 10333333$, if possible.

11.6.4. Using the data from the last exercise, calculate `scoreMat` using `data.table` as illustrated in Section 11.2.3. Calculate a frequency distribution of the bin counts.

11.6.5. Redo Example 11.2.1 with `n <- 10^8`, if possible. *Hint:* Use a loop to generate the data and write to disk in chunks.

11.6.6. Using the data from the last exercise and `sparklyr`, if available, calculate the least squares regression coefficients and associated standard errors along with the Wilcoxon one-step counter parts as illustrated in Section 11.5.

11.6.7. Consider the model fit in Section 11.2.2 as the initial fit to an iterative algorithm for a rank-based fit using Wilcoxon step scores. An initial estimate of the scale parameter is needed — calculate an estimate using the Aubuchon and Hettmansperger (1984) approximation of τ discussed in Section 11.1.2 (Equation (11.13)).

Appendix - R Version Information

The majority of the material in this book was developed using Sweave (Leisch 2002). We did not rerun all of the R code chunks from the first edition for the second edition.

The following code chunk lists, in alphabetical order, all packages loaded at the end of running the current edition, i.e., these packages were used in developing new material or fixing one or more bugs from the first edition.

```
> sort(.packages())

 [1] "base"          "biglm"         "bigRfit"
 [4] "car"           "carData"       "class"
 [7] "data.table"    "datasets"      "DBI"
[10] "dbplot"        "dplyr"         "ellipse"
[13] "ggplot2"       "graphics"      "grDevices"
[16] "hbrfit"        "ISLR"          "jrfit"
[19] "lattice"       "lavaan"        "MASS"
[22] "methods"       "mvrfit"        "mvtnorm"
[25] "npsm"          "npsmReg2"      "plyr"
[28] "profileR"      "quantreg"      "randomForest"
[31] "RColorBrewer"  "reshape"       "Rfit"
[34] "robustbase"    "sparklyr"      "SparseM"
[37] "stats"         "survival"      "tree"
[40] "utils"         "xtable"
```

Most of the new material for the second edition of this book was developed using the following version of R.

```
> version

                 _
platform         x86_64-pc-linux-gnu
arch             x86_64
os               linux-gnu
system           x86_64, linux-gnu
status
major            4
minor            3.0
year             2023
month            04
day              21
svn rev          84292
language         R
version.string   R version 4.3.0 (2023-04-21)
nickname         Already Tomorrow
```

Bibliography

Abebe, A., Crimin, K., McKean, J. W., Vidmar, T. J., and Haas, J. V. (2001), "Rank-Based procedures for linear models: Applications to pharmaceutical science data," *Drug Information Journal*, 35, 947–971.

Abebe, A. and McKean, J. W. (2007), "Highly efficient nonlinear regression based on the Wilcoxon norm," in *Festschrift in Honor of Mir Masoom Ali on the Occasion of his Retirement*, ed. Umbach, D., Ball State University, pp. 340–357.

— (2013), "Weighted Wilcoxon estimators in nonlinear regression," *Australian & New Zealand Journal of Statistics*, 55, 401-420.

Abebe, A., McKean, J. W., Kloke, J. D., and Bilgic, Y. K. (2016), "Iterated reweighted rank-based estimates for GEE Models," in *Robust Rank-Based and Nonparametric Methods*, eds. Liu, R. and McKean, J. W., Switzerland: Springer, pp. 61–79.

Adichie, J. N. (1978), "Rank tests of sub-hypotheses in the general linear regression," *The Annals of Statistics*, 6, 1012–1026.

Agresti, A. (1996), *An Introduction to Categorical Analysis*, New York: John Wiley & Sons, Inc.

— (2002), *Categorical Data Analysis*, New York: John Wiley & Sons, Inc.

Al-Shomrani, A. (2003), "A Comparison of Different Schemes for Selecting and Estimating Score Functions Based on Residuals," PhD thesis, Western Michigan University, Department of Statistics.

Arnold, S. (1981), *The Theory of Linear Models and Multivariate Analysis*, New York: John Wiley & Sons, Inc.

Aubuchon, J. and Hettmansperger, T. (1984), "A note on the estimation of the integarl of $f(x)^2$," *Journal of Statistical Planning and Inference*, 9, 321–331.

Azzalini, A. (1985), "A class of distributions which includes the normal ones," *Scandinavian Journal of Statistics*, 12, 171–178.

— (2014), *The R* `sn` *package : The skew-normal and skew-t distributions (version 1.0-0)*, Università di Padova, Italia.

Bache, S. M. and Wickham, H. (2022), *magrittr: A Forward-Pipe Operator for R*, R package version 2.0.3.

Belsley, D. A., Kuh, K., and Welsch, R. E. (1980), *Regression Diagnostics*, New York: John Wiley & Sons, Inc.

Bilgic, Y. (2012), "Rank-based estimation and predicition for mixed effects models in nested designs," PhD thesis, Western Michigan University, Department of Statistics.

Bliss, C. (1952), *The Statistics of Bioassay*, Academic Press.

Bowerman, B. L., O'Connell, R. T., and Koehler, A. B. (2005), *Forecasting, Time Series, and Regression: An Applied Approach*, Australia: Thomson.

Bowman, A. and Azzalini, A. (1997), *Applied smoothing techniques for data analysis: The kernel approach with S-Plus illustrations*, Oxford: Oxford University Press.

Bowman, A. W. and Azzalini, A. (2014), *R package sm: Nonparametric smoothing methods (version 2.2-5.4)*, University of Glasgow, UK and Università di Padova, Italia.

Box, G. E. P., Jenkins, G. M., and Reinsel, G. M. (2008), *Time Series Analysis: Forecasting and Control 4th Edition*, Hoboken, NJ: John Wiley & Sons, Inc.

Breslow, N. E., Day, N. E., et al. (1980), *Statistical methods in cancer research. Vol. 1. The analysis of case-control studies.*, Distributed for IARC by WHO, Geneva, Switzerland.

Bulut, O. and Desjardins, C. D. (2018), *profileR: Profile Analysis of Multivariate Data in R*, R package version 0.3-5.

Canty, A. and Ripley, B. (2013), *boot: Bootstrap R (S-Plus) Functions*, R package version 1.3-9.

Carroll, R. A. and Ruppert, D. (1982), "Robust estimation in heteroscedastic linear models," *The Annals of Statistics*, 10, 424–441.

Chambers, B. and Zaharia, M. (2018), *Spark: The Definitive Guide*, Sebastopol, CA: O'Reilly Media, Inc.

Chambers, J. M. (2008), *Software for Data Analysis: Programming with R*, New York: Springer Verlag.

Chang, W. H., McKean, J. W., Naranjo, J. D., and Sheather, S. J. (1999), "High-breakdown rank regression," *Journal of the American Statistical Association*, 205–219.

Chiang, C. Y. and Puri, M. L. (1984), "Rank procedures for testing subhypotheses in linear regression," *Annals of the Institute of Statistical Mathematics*, 36, 35–50.

Cleveland, W., Grosse, E., and Shyu, W. (1992), "Local regression models," in *Statistical Models in S*, eds. Chambers, J. and Hastie, T., Wadsworth & Brooks/Cole, pp. 309–376.

Cobb, G. W. (1998), *Introduction to Design and Analysis of Experiments*, New York: Springer Verlag.

Collett, D. (2003), *Modeling Survival Data in Medical Research*, vol. 57, Boca Raton, FL: Chapman & Hall.

Conover, W. (1980), *Practical Nonparametric Statistics, 2nd Edition*, New York: John Wiley & Sons, Inc.

Conover, W. J., Johnson, M. E., and Johnson, M. M. (1983), "A comparative study of tests for homogeneity of variances, with applications to the outer continental shelf bidding data," *Technometrics*, 23, 351–361.

Cook, T. D. and DeMets, D. L. (2008), *Introduction to Statistical Methods for Clinical Trials*, Boca Raton, FL: Chapman & Hall.

Cortez, P., Cerdeira, A., Almeida, F., Matos, T., and Reis., J. (2009), "Modeling wine preferences by data mining from physicochemical properties," *Decision Support Systems*, 47, 547–553.

Cox, D. R. (1972), "Regression models and life-tables," *Journal of the Royal Statistical Society. Series B (Methodological)*, 187–220.

Crimin, K., McKean, J. W., and Vidmar, T. J. (2012), "Rank-based estimate of 4-parameter logistic model," *Pharmaceutical Statistics*, 214–221.

Cui, Y. and Huang, B. (2023), *WINS: The R WINS Package*, R package version 1.3.3.

Daniel, W. W. (1978), *Applied Nonparametric Statistics*, Boston: Houghton Mifflin Co.

Davis, J. and McKean, J. (1993), "Rank-based methods for multivariate linear models," *Journal of the American Statistical Association*, 88, 245–251.

Davison, A. C. and Hinkley, D. (1997), *Bootstrap Methods and Their Application*, vol. 1, Cambridge: Cambridge University Press.

Devore, J. (2012), *Probability and statistics for engineering and the sciences*, Boston: Brooks/Cole, 8th ed.

Dixon, S. L. and McKean, J. W. (1996), "Rank based analysis of the heteroscedastic linear model," *Journal of the American Statistical Association*, 91, 699–712.

Dobson, A. J. (1983), *An Introduction to Statistical Modeling*, London: Chapman & Hall.

Doksum, K. (2013), "Asymptotic optimality of Hodges–Lehmann inverse rank likelihood estimators," *Statistical Modelling*, 13, 397–407.

Dowle, M. and Srinivasan, A. (2022), *data.table: Extension of 'data.frame'*, R package version 1.14.4.

Draper, N. L. and Smith, H. (1966), *Applied Regression Analysis*, New York: John Wiley & Sons, Inc.

Dua, D. and Graff, C. (2017), "UCI Machine Learning Repository," http://archive.ics.uci.edu/ml.

Dubnicka, S. R. (2004), "A rank-based estimation procedure for linear models with clustered data," *Journal of Modern Applied Statistics*, 3, 39–48.

Dupont, W. D. (2002), *Statistical Modeling for Biomedical Researchers: A Simple Introduction to the Analysis of Complex Data*, Cambridge: Cambridge University Press.

Eddelbuettel, D. and François, R. (2011), "Rcpp: Seamless R and C++ Integration," *Journal of Statistical Software*, 40, 1–18.

Efron, B. and Tibshirani, R. (1993), *An Introduction to the Bootstrap*, vol. 57, Boca Raton, FL: Chapman & Hall.

Everitt, B. S. and Hothorn, T. (2014), *HSAUR2: A Handbook of Statistical Analyses Using R 2nd Edition*, R package version 1.1-9.

Faraway, J. (2006), *Extending the Linear Model with R*, Boca Raton, FL: Chapman & Hall.

Finkelstein, D. M. and Schoenfeld, D. A. (1999), "Combining mortality and longitudinal measures in clinical trials," *Statistics in Medicine*, 18, 1341–1354.

Fligner, M. A. and Killeen, T. J. (1976), "Distribution-free two-sample test for scale," *Journal of the American Statistical Association*, 71, 210–213.

Fligner, M. A. and Policello, G. E. (1981), "Robust rank procedures for the Behrens–Fisher problem," *Journal of the American Statistical Association*, 76, 162–168.

Fox, J. and Weisberg, S. (2011), *An R and S-PLUS Companion to Applied Regression, 2nd Edition*, Thousand Oaks, CA: SAGE Publications, chap. Bootstrapping Regression Models.

Friedman, M. (1937), "The use of ranks to avoid the assumption of normality implicit in the analysis of variance," *Journal of the American Statistical Association*, 32, 675–701.

Graybill, F. A. (1976), *Theory and Application of the Linear Model*, North Scituate, MA: Duxbury Press.

Graybill, F. A. and Iyer, H. K. (1994), *Regression Analysis: Concepts and Applications*, Belmont, CA: Duxbury Press.

Greenwood, M. (1926), "The errors of sampling of the survivorship tables," *Reports on Public Health and Statistical Subjects*, 33, 26.

Groggel, D. J. (1983), "Asymptotic nonparametric confidence intervals for the ratio of scale parameters in balanced one-way random effects models," PhD thesis, University of Florida, Department of Statistics.

Groggel, D. J., Eackerly, D., and Rao, P. (1988), "Nonparametric estimation in one-way random effects models," *Communications in Statistics: Simulation and Computation*, 17, 887–903.

Guimarães, P. O., Quirk, D., Furtado, R. H., Maia, L. N., Saraiva, J. F., Antunes, M. O., Kalil Filho, R., Junior, V. M., Soeiro, A. M., Tognon, A. P., Veiga, V. C., Martins, P. A., Moia, D. D., Sampaio, B. S., Assis, S. R., Soares, R. V., Piano, L. P., Castilho, K., Momesso, R. G., Monfardini, F., Guimarães, H. P., Ponce de Leon, D., Dulcine, M., Pinheiro, M. R., Gunay, L. M., Deuring, J. J., Rizzo, L. V., Koncz, T., and Berwanger, O. (2021), "Tofacitinib in Patients Hospitalized with Covid-19 Pneumonia," *New England Journal of Medicine*, 385, 406–415, pMID: 34133856.

Hájek, J. and Šidák, Z. (1967), *Theory of Rank Tests*, New York: Academic Press.

Hamilton, L. C. (1992), *Regression with Graphics*, Pacific Grove, CA: Brooks/Cole.

Harvey, A. (1989), "Forecasting, structual time series models, and the Kalman filter," *Cambridge University Press*, 519–523.

Hastie, T., Tibshiani, R., and Friedman, J. (2017), *The Elements of Statistical Learning: Data Mining, Inference, and Prediction, Second Edition*, New York: Springer.

Hawkins, D. M., Bradu, D., and Kass, G. V. (1984), "Location of several outliers in multiple regression data using elemental sets," *Technometrics*, 26, 197–208.

Hettmansperger, T. (1984), *Statistical Inference Based on Ranks*, New York: John Wiley & Sons.

Hettmansperger, T. P. and McKean, J. W. (1973), "On testing for significant change in CxC tables," *Communications in Statistics*, 2, 551–560.

— (2011), *Robust Nonparametric Statistical Methods, 2nd Edition*, Boca Raton, FL: Chapman & Hall.

Higgins, J. J. (2003), *Introduction to Modern Nonparametric Statistics*, Duxbury Press.

Hocking, R. R. (1985), *The Analysis of Linear Models*, Brooks/Cole Publishing Company: Monterey, CA.

Hodges, J. L. and Lehmann, E. L. (1962), "Rank methods for combination of independent experiments in analysis of variance," *The Annals of Mathematical Statistics*, 33, 482–497.

— (1963), "Estimation of location based on rank tests," *The Annals of Mathematical Statistics*, 34, 598–611.

Hogg, R. V., McKean, J. W., and Craig, A. T. (2019), *Introduction to Mathematical Statistics, 8th Edition*, Boston: Pearson.

Hollander, M. and Wolfe, D. A. (1999), *Nonparametric Statistical Methods, 2nd Edition*, New York: John Wiley & Sons, Inc.

Høyland, A. (1965), "Robustness of the Hodges-Lehmann estimates for shift," *The Annals of Mathematical Statistics*, 36, 174–197.

Huber, P. (1981), *Robust Statistics*, New York: John Wiley & Sons, Inc.

Huitema, B. E. (2011), *The Analysis of Covariance and Alternatives, 2nd Edition*, New York: John Wiley & Sons, Inc.

Jaeckel, L. A. (1972), "Estimating regression coefficients by minimizing the dispersion of residuals," *The Annals of Mathematical Statistics*, 43, 1449–1458.

James, G., Witten, D., Hastie, T., and Tibshirani, R. (2013), *An Introduction to Statistical Learning with Applications in R*, New York: Springer.

Jin, Z., Lin, D., Wei, L., and Ying, Z. (2003), "Rank-based inference for the accelerated failure time model," *Biometrika*, 90, 341–353.

Johnson, B. A. and Peng, L. (2008), "Rank-based variable selection," *Journal of Nonparametric Statistics*, 20, 241–252.

Johnson, R. and Wichern, D. (2007), *Applied Multivariate Statistical analysis, 6th Edition*, Upper Saddle River, NJ: Pearson.

Jurečková, J. (1971), "Nonparametric estimate of regression coefficients," *The Annals of Mathematical Statistics*, 42, 1328–1338.

Kalbfleisch, J. D. and Prentice, R. L. (2002), *The Statistical Analysis of Failure Time Data 2nd Edition*, New York: John Wiley & Sons, Inc.

Kapenga, J., Kloke, J., and McKean, J. (2024), "Big Data Algorithms for Rank-Based Fitting and Analyses," In preparation.

Kapenga, J. A. and McKean, J. W. (1989), "Spline estimation of the optimal score function for linear models," in *American Statistical Association: Proceedings of the Statistical Computing Section*, pp. 227–232.

Kaplan, E. L. and Meier, P. (1958), "Nonparametric estimation from incomplete observations," *Journal of the American Statistical Association*, 53, 457–481.

Karau, H., Konwinski, A., Wendell, P., and Zaharia, M. (2015), *Learning Spark*, Sebastopol, CA: O'Reilly Media, Inc.

Kendall, M. and Stuart, A. (1979), *The Advanced Theory of Statistics*, New York: Macmillan.

Kloke, J. and Cook, T. (2014), "Nonparametric, covariate-adjusted hypothesis tests using R estimation for clinical trials," In preparation.

Kloke, J., McKean, J., Kimes, P., and Parker, H. (2010), "Adaptive nonparametric statistics with applications to gene expression data," Conference Presentation at useR! 2010.

Kloke, J. D. and McKean, J. W. (2011), "Rank-based estimation for Arnold transformed data," in *Nonparametric Statistics and Mixture Models: A Festschrift in Honor of Thomas P. Hettmansperger*, eds. Hunter, D. R., Richards, D. P., and Rosenberger, J. L., Word Press, pp. 183–203.

— (2012), "Rfit: Rank-based estimation for linear models," *The R Journal*, 4, 57–64.

— (2013), "Small sample properties of JR estimators," in *JSM Proceedings*, Alexandria, VA: American Statistical Association.

Kloke, J. D., McKean, J. W., and Rashid, M. (2009), "Rank-based estimation and associated inferences for linear models with cluster correlated errors," *Journal of the American Statistical Association*, 104, 384–390.

Koenker, R. (2013), *quantreg: Quantile Regression*, R package version 5.05.

Koul, H. L. and Saleh, A. K. M. E. (1993), "R-estimation of the parameters of autoregressive $[AR(p)]$ models," *The Annals of Statistics*, 21, 534–551.

Koul, H. L., Sievers, G. L., and McKean, J. W. (1987), "An estimator of the scale parameter for the rank analysis of linear models under general score functions," *Scandinavian Journal of Statistics*, 14, 131–141.

Kruskal, W. and Wallis, W. (1952), "Use of ranks in one-criterion variance analysis," *Journal of the American Statistical Association*, 47, 583–621.

Kuhn, M. and Johnson, K. (2013), *Applied Predictive Modeling*, New York: Springer.

Lee, E. (1992), *Statistical Methods for Survival Data Analysis 2nd Edition*, New York: John Wiley & Sons, Inc.

Lehmann, E. (1999), *Elements of Large-Sample Theory*, New York: Springer.

Leisch, F. (2002), "Sweave: Dynamic generation of statistical reports using literate data analysis," in *Compstat 2002 — Proceedings in Computational Statistics*, eds. Härdle, W. and Rönz, B., Physica Verlag, Heidelberg, pp. 575–580.

Liang, K. Y. and Zeger, S. L. (1986), "Longitudinal data analysis using generalized linear models," *Biometrika*, 73, 13–22.

Liaw, A. and Wiener, M. (2002), "Classification and Regression by randomForest," *R News*, 2, 18–22.

Lumley, T. (2020), *biglm: Bounded Memory Linear and Generalized Linear Models*, R package version 0.9-2.1.

Luraschi, J., Kuo, K., and Ruiz, E. (2019), *Mastering Spark with R*, Sebastopol, CA: O'Reilly Media, Inc.

Luraschi, J., Kuo, K., Ushey, K., Allaire, J., and The Apache Software Foundation (2020), *sparklyr: R Interface to Apache Spark*, R package version 1.1.0.

Mao, L. and Kim, K. (2021), "Statistical models for composite endpoints of death and non-fatal events: a review," *Statistics in Biopharmaceutial Research*, 13, 260–269.

Mao, L. and Wang, T. (2023), *WR: Win Ratio Analysis of Composite Time-to-Event Outcomes*, R package version 1.0.

McKean, J. and Schrader, R. (1984), "A comparison of methods for Studentizing the sample median," *Communications in Statistics, Part B Simulation and Computation*, 6, 751–773.

McKean, J. and Sheather, S. (1991), "Small sample properties of robust analyses of linear models based on R-estimates: A survey," in *Directions in Robust Statistics and Diagnostics, Part II*, eds. Stahel, W. and Weisberg, S., New York: Springer Verlag, pp. 1–19.

McKean, J. W., Huitema, B. E., and Naranjo, J. D. (2001), "A robust method for the analysis of experiments with ordered treatment levels," *Psychological Reports*, 89, 267–273.

McKean, J. W. and Kloke, J. D. (2014), "Efficient and adaptive rank-based fits for linear models with skew-normal errors," *Journal of Statistical Distributions and Applications*, in press.

McKean, J. W., Naranjo, J. D., and Sheather, S. J. (1996), "Diagnostics to detect differences in robust fits of linear models," *Computational Statistics*, 11, 223–243.

McKean, J. W. and Schrader, R. (1980), "The geometry of robust procedures in linear models," *Journal of the Royal Statistical Society, Series B, Methodological*, 42, 366–371.

McKean, J. W. and Sheather, S. J. (2009), "Diagnostic procedures," *Wiley Interdisciplinary Reviews: Computational Statistics*, 1(2), 221–233.

McKean, J. W., Sheather, S. J., and Hettmansperger, T. P. (1994), "Robust and high breakdown fits of polynomial models," *Technometrics*, 36, 409–415.

McKean, J. W. and Sievers, G. L. (1989), "Rank scores suitable for analysis of linear models under asymmetric error distributions," *Technometrics*, 31, 207–218.

McKean, J. W., Vidmar, T. J., and Sievers, G. (1989), "A robust two-stage multiple comparison procedure with application to a random drug screen," *Biometrics*, 45, 1281–1297.

Miettinen, O. and Nurminen, M. (1985), "Comparative analysis of two rates," *Statistics in Medicine*, 4, 213–226.

Miliken, G. A. and Johnson, D. E. (1984), *Analysis of Messy Data, Volume 1*, New York: Van Nostrand Reinhold Company.

Mood, A. (1950), *Introduction to the Theory of Statistics*, New York: McGraw-Hill.

Morrison, D. F. (1983), *Applied Linear Statistical Models*, Englewood Cliffs, New Jersey: Prentice Hall.

Naranjo, J. D. and McKean, J. W. (1997), "Rank regression with estimated scores," *Statistics and Probability Letters*, 33, 209–216.

Nordhausen, K., Sirkia, S., Oja, H., and Tyler, D. E. (2018), *ICSNP: Tools for Multivariate Nonparametrics*, R package version 1.1-1.

Oja, H. (2010), *Multivariate Nonparametric Methods with R*, New York: Springer.

Okyere, G. (2011), "Robust adaptive schemes for linear mixed models," PhD thesis, Western Michigan University, Department of Statistics.

Pocock, S. J., Ariti, C. A., Collier, T. J., and Wang, D. (2011), "The win ratio: a new approach to the analysis of composite endpoints in clinical trials based on clinical priorities," *European Heart Journal*, 33, 176–182.

Puri, M. L. and Sen, P. K. (1985), *Nonparametric Methods in General Linear Models*, New York: John Wiley & Sons, Inc.

R Core Team (2023), *R: A Language and Environment for Statistical Computing*, R Foundation for Statistical Computing, Vienna, Austria.

Rashid, M. M., McKean, J. W., and Kloke, J. D. (2012), "R estimates and associated inferences for mixed models with covariates in a multicenter clincal trial," *Statistics in Biopharmaceutical Research*, 4, 37–49.

Rasmussen, S. (1992), *An Introduction to Statistics with Data Analysis*, Pacific Grove, CA: Brooks/Cole Publishing Co.

Reeve, E. (1941), "A statistical analysis of taxonomic differences within the genus *Tamandua* Gray (Xenarthra)," *Proc. Zool. Soc. London A*, 111, 279–302.

Ripley, B. (2023), *tree: Classification and Regression Trees*, R package version 1.0-43.

Rousseeuw, P. and Van Driessen, K. (1999), "A fast algorithm for the minimum covariance determinant estimator," *Technometrics*, 41, 212–223.

Rousseeuw, P. J., Leroy, A. M., and Wiley, J. (1987), *Robust Regression and Outlier Detection*, vol. 3, Wiley Online Library.

Rousseeuw, P. J. and van Zomeren, B. C. (1990), "Unmasking multivariate outliers and leverage points," *Journal of the American Statistical Association*, 85, 633–648.

Ruiz, E. (2020), *dbplot: Simplifies Plotting Data Inside Databases*, R package version 0.3.3.

Seber, G. A. F. (1984), *Multivariate Observations*, New York: John Wiley & Sons, Inc.

Seber, G. A. F. and Wild, C. J. (1989), *Nonlinear Regression*, New York: John Wiley & Sons, Inc.

Sheather, S. (2016), "Applications of robust regression to "Big" data problems," in *Robust rank-based and nonparametric methods*, eds. Liu, R. and McKean, J., Switzerland: Springer, pp. 101–120.

Sheather, S. and Jones, M. (1991), "A reliable data-based bandwidth selection method for kernel density estimation," *Journal of Royal Statistical Society, B*, 53, 683–690.

Sheather, S. J. (2009), *A Modern Approach to Regression with R*, New York: Springer.

Siegel, S. (1956), *Nonparametric Statistics*, New York: McGraw-Hill.

Silverman, B. (1986), *Density Estimation*, London: Chapman & Hall.

Stokes, M. E., Davis, M. E. S. C. S., Koch, G. G., and Davis, C. S. (1995), *Categorical Data Analysis Using the SAS System*, Cary, NC: SAS institute.

Terpstra, J., McKean, J. W., and Anderson, K. (2003), "Studentized autoregressive time series residuals," *Computational Statistics*, 18, 123–141.

Terpstra, J., McKean, J. W., and Naranjo, J. D. (2000), "Highly efficient weighted Wilcoxon estimates for autoregression," *Statistics*, 35, 45–80.

— (2001), "GR-Estimates for an autoregressive time series," *Statistics and Probability Letters*, 51, 165–172–80.

Terpstra, J. F. and McKean, J. W. (2005), "Rank-based analyses of linear models using R," *Journal of Statistical Software*, 14, 1–26.

Theil, H. (1950), "A rank-invariant method of linear and polynomial regression analysis, III," *Proc. Kon. Ned. Akad. v. Wetensch*, 53, 1397–1412.

Therneau, T. (2013), *A Package for Survival Analysis in S*, R package version 2.37-4.

Tryon, P. V. and Hettmansperger, T. P. (1973), "A class of non-parametric tests for homogeneity against ordered alternatives," *The Annals of Statistics*, 1, 1061–1070.

Venables, W. N. and Ripley, B. D. (2002), *Modern Applied Statistics with S*, New York: Springer, 4th ed., iSBN 0-387-95457-0.

Verzani, J. (2014), *Using R for Introductory Statistics 2nd Edition*, Boca Raton, FL: CRC Press.

Wahba, G. (1990), *Spline Models for Observational Data*, vol. 59, Philadelphia: Society for Industrial and Applied Mathematics.

Watcharotone, K., McKean, J., and Huitema, B. (2017), "A Monte Carlo study of traditional and rank-based picked-point analyses," *Communications in Statistics B*, 46, 4050–4066.

Wells, J. M. and Wells, M. A. (1967), "Note on project SCUD," *Proceedings 5th Berkeley Symposium*, V, 357–369.

Wickham, H. (2009), *ggplot2: Elegant Graphics for Data Analysis*, New York: Springer.

Wickham, H., Danenberg, P., Csárdi, G., and Eugster, M. (2020), *roxygen2: In-Line Documentation for R*, R package version 7.1.1.

Wickham, H., François, R., Henry, L., Müller, K., and Vaughan, D. (2023), *dplyr: A Grammar of Data Manipulation*, R package version 1.1.2.

Wood, S. (2006), *Generalized Additive Models: An Introduction with R*, Boca Raton, FL: Chapman & Hall.

Index

Printed in the United States
by Baker & Taylor Publisher Services

Printed in the United States
by Baker & Taylor Publisher Services